EQUILIBRIUM STATISTICAL PHYSICS

3rd Edition

EQUILIBRIUM STATISTICAL PHYSICS

3rd Edition

Michael Plischke
Simon Fraser University, Canada

Birger Bergersen
University of British Columbia, Canada

NEW JERSEY • LONDON • SINGAPORE • BEIJING • SHANGHAI • HONG KONG • TAIPEI • CHENNAI

Published by

World Scientific Publishing Co. Pte. Ltd.
5 Toh Tuck Link, Singapore 596224
USA office: 27 Warren Street, Suite 401-402, Hackensack, NJ 07601
UK office: 57 Shelton Street, Covent Garden, London WC2H 9HE

Library of Congress Cataloging-in-Publication Data
Plischke, Michael.
Equilibrium statistical physics / Michael Plischke, Birger Bergersen.--3rd ed.
p. cm.
ISBN-13 978-981-256-048-3 -- ISBN-10 981-256-048-3 (alk. paper)
ISBN-13 978-981-256-155-8 (pbk) -- ISBN-10 981-256-155-2 (pbk. : alk. paper)
1. Statistical physics--Textbooks. 2. Critical phenomena (Physics)--Textbooks. I. Bergersen, Birger. II. Title.

QC174.8 .P55 2005
530.13--dc22 2004056775

British Library Cataloguing-in-Publication Data
A catalogue record for this book is available from the British Library.

First published 2006
Reprinted 2007

Printed in Singapore by B & JO Enterprise

Contents

Preface to the First Edition xi

Preface to the Second Edition xiv

Preface to the Third Edition xv

1 Review of Thermodynamics 1

1.1 State Variables and Equations of State 1
1.2 Laws of Thermodynamics 3
1.2.1 First law 3
1.2.2 Second law 5
1.3 Thermodynamic Potentials 9
1.4 Gibbs–Duhem and Maxwell Relations 12
1.5 Response Functions 14
1.6 Conditions for Equilibrium and Stability 16
1.7 Magnetic Work 18
1.8 Thermodynamics of Phase Transitions 20
1.9 Problems 24

2 Statistical Ensembles 29

2.1 Isolated Systems: Microcanonical Ensemble 30
2.2 Systems at Fixed Temperature: Canonical Ensemble 35
2.3 Grand Canonical Ensemble 40
2.4 Quantum Statistics 43
2.4.1 Harmonic oscillator 44
2.4.2 Noninteracting fermions 44
2.4.3 Noninteracting bosons 45
2.4.4 Density matrix 46

2.5 Maximum Entropy Principle 48
2.6 Thermodynamic Variational Principles 53
2.6.1 Schottky defects in a crystal 53
2.7 Problems 54

3 Mean Field and Landau Theory **63**
3.1 Mean Field Theory of the Ising Model 64
3.2 Bragg–Williams Approximation 67
3.3 A Word of Warning 69
3.4 Bethe Approximation 71
3.5 Critical Behavior of Mean Field Theories 74
3.6 Ising Chain: Exact Solution 77
3.7 Landau Theory of Phase Transitions 83
3.8 Symmetry Considerations 86
3.8.1 Potts model 87
3.9 Landau Theory of Tricritical Points 90
3.10 Landau–Ginzburg Theory for Fluctuations 94
3.11 Multicomponent Order Parameters: n-Vector Model 98
3.12 Problems 100

4 Applications of Mean Field Theory **109**
4.1 Order–Disorder Transition 110
4.2 Maier–Saupe Model 114
4.3 Blume–Emery–Griffiths Model 120
4.4 Mean Field Theory of Fluids: van der Waals Approach 123
4.5 Spruce Budworm Model 129
4.6 A Non-Equilibrium System: Two Species Asymmetric Exclusion Model 132
4.7 Problems 137

5 Dense Gases and Liquids **143**
5.1 Virial Expansion 145
5.2 Distribution Functions 151
5.2.1 Pair correlation function 151
5.2.2 BBGKY hierarchy 157
5.2.3 Ornstein–Zernike equation 158
5.3 Perturbation Theory 161
5.4 Inhomogeneous Liquids 163
5.4.1 Liquid–vapor interface 164

5.4.2 Capillary waves 169

5.5 Density-Functional Theory 171

5.5.1 Functional differentiation 171

5.5.2 Free-energy functionals and correlation functions 174

5.5.3 Applications 179

5.6 Problems 181

6 Critical Phenomena I **183**

6.1 Ising Model in Two Dimensions 184

6.1.1 Transfer matrix 184

6.1.2 Transformation to an interacting fermion problem . . . 188

6.1.3 Calculation of eigenvalues 191

6.1.4 Thermodynamic functions 194

6.1.5 Concluding remarks 199

6.2 Series Expansions 199

6.2.1 High-temperature expansions 200

6.2.2 Low-temperature expansions 206

6.2.3 Analysis of series 206

6.3 Scaling 211

6.3.1 Thermodynamic considerations 211

6.3.2 Scaling hypothesis 212

6.3.3 Kadanoff block spins 215

6.4 Finite-Size Scaling 218

6.5 Universality 223

6.6 Kosterlitz–Thouless Transition 226

6.7 Problems 233

7 Critical Phenomena II: The Renormalization Group **237**

7.1 The Ising Chain Revisited 238

7.2 Fixed Points 242

7.3 An Exactly Solvable Model: Ising Spins on a Diamond Fractal 248

7.4 Position Space Renormalization: Cumulant Method 258

7.4.1 First-order approximation 262

7.4.2 Second-order approximation 264

7.5 Other Position Space Renormalization Group Methods 267

7.5.1 Finite lattice methods 267

7.5.2 Adsorbed monolayers: Ising antiferromagnet 268

7.5.3 Monte Carlo renormalization 272

7.6 Phenomenological Renormalization Group 275
7.7 The ϵ-Expansion 279
7.7.1 The Gaussian model 281
7.7.2 The S^4 model 284
7.7.3 Conclusion 290
Appendix: Second Order Cumulant Expansion 292
7.8 Problems 295

8 Stochastic Processes **303**
8.1 Markov Processes and the Master Equation 304
8.2 Birth and Death Processes 306
8.3 Branching Processes 309
8.4 Fokker–Planck Equation 313
8.5 Fokker–Planck Equation with Several Variables: SIR Model . . 316
8.6 Jump Moments for Continuous Variables 321
8.6.1 Brownian motion 323
8.6.2 Rayleigh and Kramers equations 326
8.7 Diffusion, First Passage and Escape 328
8.7.1 Natural boundaries: The Kimura–Weiss model for genetic drift 329
8.7.2 Artificial boundaries 331
8.7.3 First passage time and escape probability 332
8.7.4 Kramers escape rate 337
8.8 Transformations of the Fokker–Planck Equation 340
8.8.1 Heterogeneous diffusion 340
8.8.2 Transformation to the Schrödinger equation 343
8.9 Problems 345

9 Simulations **349**
9.1 Molecular Dynamics 350
9.1.1 Conservative molecular dynamics 351
9.1.2 Brownian dynamics 353
9.1.3 Data analysis 355
9.2 Monte Carlo Method 357
9.2.1 Discrete time Markov processes 358
9.2.2 Detailed balance and the Metropolis algorithm 359
9.2.3 Histogram methods 363
9.3 Data Analysis 365
9.3.1 Fluctuations 365

9.3.2 Error estimates 367
9.3.3 Extrapolation to the thermodynamic limit 368
9.4 The Hopfield Model of Neural Nets 371
9.5 Simulated Quenching and Annealing 376
9.6 Problems 379

10 Polymers and Membranes **383**
10.1 Linear Polymers 384
10.1.1 The freely jointed chain 386
10.1.2 The Gaussian chain 389
10.2 Excluded Volume Effects: Flory Theory 391
10.3 Polymers and the n-Vector Model 395
10.4 Dense Polymer Solutions 400
10.5 Membranes 405
10.5.1 Phantom membranes 406
10.5.2 Self-avoiding membranes 409
10.5.3 Liquid membranes 415
10.6 Problems 418

11 Quantum Fluids **421**
11.1 Bose Condensation 422
11.2 Superfluidity 430
11.2.1 Qualitative features of superfluidity 430
11.2.2 Bogoliubov theory of the ^{4}He excitation spectrum . . . 439
11.3 Superconductivity 442
11.3.1 Cooper problem 443
11.3.2 BCS ground state 445
11.3.3 Finite-temperature BCS theory 449
11.3.4 Landau–Ginzburg theory of superconductivity 453
11.4 Problems 456

12 Linear Response Theory **461**
12.1 Exact Results 462
12.1.1 Generalized susceptibility and the structure factor . . . 462
12.1.2 Thermodynamic properties 469
12.1.3 Sum rules and inequalities 470
12.2 Mean Field Response 472
12.2.1 Dielectric function of the electron gas 473
12.2.2 Weakly interacting Bose gas 475

12.2.3 Excitations of the Heisenberg ferromagnet 477
12.2.4 Screening and plasmons 480
12.2.5 Exchange and correlation energy 486
12.2.6 Phonons in metals 487
12.3 Entropy Production, the Kubo Formula, and the Onsager Relations for Transport Coefficients 490
12.3.1 Kubo formula 490
12.3.2 Entropy production and generalized currents and forces 492
12.3.3 Microscopic reversibility: Onsager relations 494
12.4 The Boltzmann Equation 498
12.4.1 Fields, drift and collisions 498
12.4.2 DC conductivity of a metal 500
12.4.3 Thermal conductivity and thermoelectric effects 503
12.5 Problems 507

13 Disordered Systems **513**
13.1 Single-Particle States in Disordered Systems 515
13.1.1 Electron states in one dimension 516
13.1.2 Transfer matrix 517
13.1.3 Localization in three dimensions 523
13.1.4 Density of states 525
13.2 Percolation 530
13.2.1 Scaling theory of percolation 533
13.2.2 Series expansions and renormalization group 536
13.2.3 Rigidity percolation 540
13.2.4 Conclusion 542
13.3 Phase Transitions in Disordered Materials 542
13.3.1 Statistical formalism and the replica trick 544
13.3.2 Nature of phase transitions 546
13.4 Strongly Disordered Systems 551
13.4.1 Molecular glasses 552
13.4.2 Spin glasses 554
13.4.3 Sherrington–Kirkpatrick model 558
13.5 Problems 565

A Occupation Number Representation **569**

Bibliography **583**

Index **603**

Preface to the First Edition

During the last decade each of the authors has regularly taught a graduate or senior undergraduate course in statistical mechanics. During this same period, the renormalization group approach to critical phenomena, pioneered by K. G. Wilson, greatly altered our approach to condensed matter physics. Since its introduction in the context of phase transitions, the method has found application in many other areas of physics, such as many-body theory, chaos, the conductivity of disordered materials, and fractal structures. So pervasive is its influence that we feel that it now essential that graduate students be introduced at an early stage in their career to the concepts of scaling, universality, fixed points, and renormalization transformations, which were developed in the context of critical phenomena, but are relevant in many other situations.

In this book we describe both the traditional methods of statistical mechanics and the newer techniques of the last two decades. Most graduate students are exposed to only one course in statistical physics. We believe that this course should provide a bridge from the typical under-graduate course (usually concerned primarily with noninteracting systems such as ideal gases and paramagnets) to the sophisticated concepts necessary to a researcher.

We begin with a short chapter on thermodynamics and continue, in Chapter 2, with a review of the basics of statistical mechanics. We assume that the student has been exposed previously to the material of these two chapters and thus our treatment is rather concise. We have, however, included a substantial number of exercises that complement the review.

In Chapter 3 we begin our discussion of strongly interacting systems with a lengthy exposition of mean field theory. A number of examples are worked out in detail. The more general Landau theory of phase transitions is developed and used to discuss critical points, tricritical points, and first-order phase transitions. The limitations of mean field and Landau theory are described and the role of fluctuations is explored in the framework of the Landau–Ginzburg model.

Chapter 4 is concerned with the theory of dense gases and liquids. Many of the techniques commonly used in the theory of liquids have a long history and are well described in other texts. Nevertheless, we feel that they are sufficiently important that we could not omit them. The traditional method of viral expansions is presented and we emphasize the important role played in both theory and experiment by the pair correlation function. We briefly describe some of the useful and still popular integral equation methods based on the Ornstein–Zernike equation used to calculate this function as well as the modern perturbation theories of liquids. Simulation methods (Monte Carlo and molecular dynamics) are introduced. In the final section of the chapter we present an interesting application of mean field theory, namely the van der Waals theory of the liquid-vapor interface and a simple model of roughening of this interface due to capillary waves.

Chapters 5 and 6 are devoted to continuous phase transitions and critical phenomena. In Chapter 5 we review the Onsager solution of the two-dimensional Ising model on the square lattice and continue with a description of the series expansion methods, which were historically very important in the theory of critical phenomena. We formulate the scaling theory of phase transitions following the ideas of Kadanoff, introduce the concept of universality of critical behavior, and conclude with a mainly qualitative discussion of the Kosterlitz–Thouless theory of phase transitions in two-dimensional systems with continuous symmetry.

Chapter 6 is entirely concerned with the renormalization group approach to phase transitions. The ideas are introduced by means of technically straightforward calculations for the one- and two-dimensional Ising models. We discuss the role of the fixed points of renormalization transformations and show how the theory leads to universal critical behavior. The original ϵ-expansion of Wilson and Fisher is also discussed. This section is rather detailed, as we have attempted to make it accessible to students without a background in field theory.

In Chapter 7 we turn to quantum fluids and discuss the ideal Bose gas, the weakly interacting Bose gas, the BCS theory of superconductivity, and the phenomenological Landau–Ginzburg theory of superconductivity. Our treatment of these topics (except for the ideal Bose gas) is very much in the spirit of mean field theory and provides more challenging applications of the formalism developed in Chapter 3.

Chapter 8 is devoted to linear response theory. The fluctuation-dissipation theorem, the Kubo formalism, and the Onsager relations for transport coefficients are discussed. This chapter is consistent with our emphasis on equilibrium phenomena — in the linear response approximation the central role is played by equilibrium correlation functions. A number of applications of the formalism, such as the dielectric response of an electron gas, the elementary excitations of a Heisenberg ferromagnet, and the excitation spectrum of an interacting Bose fluid, are discussed in detail. The complementary approach to transport via the linearized Boltzmann equation is also presented.

Chapter 9 provides an introduction to the physics of disordered materials. We discuss the effect of disorder on the quantum states of a system and introduce (as an example) the notion of localization of electronic states by an explicit calculation for a one-dimensional model. Percolation theory is introduced and its analogy to thermal phase transitions is elucidated. The nature of phase transitions in disordered materials is discussed and we conclude with a very brief and qualitative description of the glass and spin-glass transitions. These subjects are all very much at the forefront of current research and we do not claim to be at all comprehensive in our treatment. In compensation, we have provided a more extensive list of references to recent articles on these topics than elsewhere in the book.

We have found the material presented here suitable for an introductory graduate course, or with some selectivity, for a senior undergraduate course. A student with a previous course in statistical mechanics, some background in quantum mechanics, and preferably, some exposure to solid state physics should be adequately prepared. The notation of second quantization is used extensively in the latter part of the book and the formalism is developed in detail in the Appendix. The instructor should be forewarned that although some of the problems, particularly in the early chapters, are quite straightforward, those toward the end of the book can be rather challenging.

Much of this book deals with topics on which there is a great deal of recent research. For this reason we have found it necessary to give a large number of references to journal articles. Whenever possible, we have referred to recent review articles rather than to the original sources.

The writing of this book has been an ongoing (frequently interrupted) process for a number of years. We have benefited from discussion with, and critical comments from, a number of our colleagues. In particular, Ian Affleck, Leslie Ballentine, Robert Barrie, John Berlinsky, Peter Holdsworth, Zoltán Rácz, and Bill Unruh have been most helpful. Our students Dan Ciarniello, Victor Finberg and Barbara Frisken have also helped to decrease the number of errors, ambiguities, and obscurities. The responsibility for the remaining faults rests entirely with the authors.

MICHAEL PLISCHKE
BIRGER BERGERSEN

Preface to the Second Edition

During the five years that have passed since the first edition of this book was published, we have received numerous helpful suggestions from friends and colleagues both at our own institutions and at others. As well, the field of statistical mechanics had continued to evolve. In composing this second edition we have attempted to take all of this into account. The purpose of the book remains the same: to provide an introduction to state-of-the-art techniques in statistical physics for graduate students in physics, chemistry and materials science.

While the general structure of the second edition is very similar to that of the first edition, there are a number of important additions. The rather abbreviated treatment of computer simulations has been expanded considerably and now forms a separate Chapter 7. We have included an introduction to density-functional methods in the chapter on classical liquids. We have added an entirely new Chapter 8 on polymers and membranes. In the discussion of critical phenomena, we have corrected an important omission of the first edition and have added sections on finite-size scaling and phenomenological renormalization group. Finally, we have considerably expanded the discussion of spin-glasses and have also added a number of new problems. We have also compiled a solution manual which is available from the publisher.

It goes without saying that we have corrected those errors of the first edition that we are aware of. In this task we have been greatly helped by a number of individuals. In particular, we are grateful to Vinay Ambegaokar, Leslie Ballentine, David Boal, Bill Dalby, Zoltán Rácz, Byron Southern and Philip Stamp.

Michael Plischke
Birger Bergersen
Vancouver, Canada

Preface to the Third Edition

In the third edition we have added a significant amount of new material. There are also numerous corrections and clarifications throughout the text. We have also added several new problems.

In Chapter 1 we have added a section on magnetic work, while in Chapter 2 we have added to the discussion of the maximum entropy principle, emphasizing the importance of the assumption that the entropy is extensive in a normal thermodynamic system.

In Chapter 3 we have replaced the derivation of the Bragg Williams approximation from the density matrix, to a more intuitive one, stressing the mean field assumption of statistical independence of spins at different sites. We have also added a section on the Potts model. The sections on the Maier-Saupe model for liquid crystals, the Blume-Emery-Griffiths model for ^{3}He-^{4}He mixtures and van der Waals fluid have been moved to a new chapter called "Applications of Mean Field Theory" that includes a section on an insect infestation model in ecology and also includes a non-equilibrium system: the two species asymmetric exclusion model. This section illustrates the application of mean field theory outside the scope of equilibrium statistical mechanics.

The new Chapters 5 and 6 only contain relatively minor changes to the old Chapters 4 and 5. In Chapter 7, the section on the epsilon expansion in the old Chapter 6 has been rewritten, and we have added a section on the Ising model on the diamond fractal.

Because of the growing importance of the field we have added a new Chapter 8 on stochastic processes. We start with a description of discrete birth and death processes, and we return to the insect infestation model of Chapter 4. Most of the remainder of the chapter is concerned with the Fokker-Planck equation for both discrete and continuous processes. We apply the theory both to a genetics problems and diffusion of particles in fluids. Other applications involve the rate of excape from a metastable state and problems of heterogeneous diffusion. Finally we show how the Fokker-Planck equation can be transformed into a form similar to the Schrödinger equation, allowing the application of techniques familiar from quantum mechanics.

In Chapter 9 (old Chapter 7) we have rewritten the section on molecular dynamics and added a subsection on Brownian dynamics. There are only relatively minor changes to Chapters 10 and 11 (old Chapters 8 and 9) except that we have updated the references to the literature in view of important new developments in superconductivity and Bose condensation. A section on rigidity percolation has been added to Chapter 13 (old Chapter 11).

Helpful comments and suggestions from Ian Affleck, Marcel Franz, Michel Gingras, Margarita Ifti, Greg Lakatos, Zoltán Rácz, Fei Zhou and Martin Zuckermann are gratefully acknowledged.

Updated information of interest to readers will be displayed on our website http://www.physics.ubc.ca/~birger/equilibrium.htm.

Michael Plischke
Birger Bergersen
Vancouver, Canada

EQUILIBRIUM STATISTICAL PHYSICS

3rd Edition

Chapter 1

Review of Thermodynamics

This chapter presents a brief review of elementary thermodynamics. It complements Chapter 2, in which the connection between thermodynamics and statistical mechanical ensembles is established. The reader may wish to use this chapter as a short refresher course and may wish to consult one of the many books on thermodynamics, such as that of Callen [55] or Chapters 2 to 4 of the book by Reichl [254], for a more complete discussion of the material. The outline of the present chapter is as follows. In Section 1.1 we introduce the notion of state variables and equations of state. Section 1.2 contains a discussion of the laws of thermodynamics, definition of thermodynamic processes, and the introduction of entropy. In Section 1.3 we introduce the thermodynamic potentials that are most useful from a statistical point of view. The Gibbs–Duhem equation and a number of useful Maxwell relations are derived in Section 1.4. In Section 1.5 we turn to the response functions, such as the specific heat, susceptibility, and compressibility, which provide the common experimental probes of macroscopic systems. Section 1.6 contains a discussion of some general conditions of equilibrium and stability. We discuss the issue of magnetic work in Section 1.7 and we conclude, in 1.8, with a brief discussion of the thermodynamics of phase transitions and the Gibbs phase rule.

1.1 State Variables and Equations of State

A macroscopic system has many degrees of freedom, only a few of which are measurable. Thermodynamics thus concerns itself with the relation between a small number of variables which are sufficient to describe the bulk behavior of the system in question. In the case of a gas or liquid the appropriate variables are the pressure P, volume V, and temperature T. In the case of a magnetic solid the appropriate variables are the magnetic field $\mathbf{H}$, the magnetization $\mathbf{M}$, and the temperature T. In more complicated situations, such as when a liquid is in contact with its vapor, additional variables may be needed: such as the volume of both liquid and gas V_L, V_G, the interfacial area A, and surface tension σ. If the thermodynamic variables are independent of time, the system is said to be in a *steady state.* If, moreover, there are no macroscopic currents in the system, such as a flow of heat or particles through the material, the system is in *equilibrium.* Any quantity which, in equilibrium, depends only on the thermodynamic variables, rather than on the history of the sample, is called a *state function.* In subsequent sections we shall meet a number of such quantities. For a large system, the state variables can normally be taken to be either *extensive* (*i.e.*, proportional to the size of the system) or *intensive* (*i.e.*, independent of system size). Examples of extensive variables are the internal energy, the entropy, and the mass of the different constituents or their number, while the pressure, the temperature, and the chemical potentials are intensive. The postulate that quantities like the internal energy and entropy are extensive and independent of shape is equivalent to an assumption of additivity or, as we shall see in Section 2.1, of the existence of the *thermodynamic limit.* In the process of taking the thermodynamic limit, we let the size of the system become infinitely large, with the densities (of mass, energy, magnetic moment, polarization, *etc.*) remaining constant.

In equilibrium the state variables are not all independent and are connected by equations of state. The rôle of statistical mechanics is the derivation, from microscopic interactions, of such equations of state. Simple examples are the ideal gas law,

$$PV - Nk_BT = 0 \tag{1.1}$$

where N is the number of molecules in the system and k_B is Boltzmann's constant; the van der Waals equation,

$$\left(p + \frac{aN^2}{V^2}\right)(V - Nb) - Nk_BT = 0 \tag{1.2}$$

where a, b are constants; the virial equation of state

$$P - \frac{Nk_BT}{V}\left[1 + \frac{NB_2(T)}{V} + \frac{N^2B_3(T)}{V^2} + \dots\right] = 0 \tag{1.3}$$

where the functions $B_2(T)$, $B_3(T)$ are called virial coefficients; and in the case of a paramagnet, the Curie law,

$$M - \frac{CH}{T} = 0 \tag{1.4}$$

where C is a constant called the Curie constant. Equations (1.1), (1.2), and (1.4) are approximations, and we shall use them primarily to illustrate various principles. Equation (1.3) is, in principle, exact, but as we shall see in Chapter 4, calculation of more than a few of the virial coefficients is very difficult.

1.2 Laws of Thermodynamics

In this section we explore the consequences of the zeroth, first, and second laws of thermodynamics. The zeroth law can be thought of as the statement that matter in equilibrium can be assigned values for the temperature, pressure and chemical potentials, which in principle can be measured. Formally the law can be stated as:

> If system A is in equilibrium with systems B and C then B is in equilibrium with C.

The zeroth law allows us to introduce universal scales for temperature, pressure *etc.*

Another way of looking at the zeroth law is through an analogy with mechanics. In equilibrium the forces are balanced. This implies that the intensive variables are constant throughout the system. In particular:

$$\begin{aligned} T &= \text{const.} \;\rightarrow\; \text{Thermal equilibrium} \\ P &= \text{const.} \;\rightarrow\; \text{Mechanical equilibrium} \\ \mu &= \text{const.} \;\rightarrow\; \text{Chemical equilibrium} \end{aligned}$$

As we shall see in the next chapter, the zeroth law has a fairly straightforward statistical interpretation and this will allow us to make contact between the thermodynamic and statistical description.

1.2.1 First law

The first law of thermodynamics restates the law of conservation of energy. However, it also partitions the change in energy of a system into two pieces, heat and work:

$$dE = đQ - đW \ . \tag{1.5}$$

In (1.5) dE is the change in internal energy of the system, $đQ$ the amount of heat *added to* the system, and $đW$ the amount of work *done by* the system during an infinitesimal process[1]. Aside from the partitioning of the energy into two parts, the formula distinguishes between the infinitesimals dE and $đQ$, $đW$. The difference between the two measurable quantities $đQ$ and $đW$ is found to be the same for any process in which the system evolves between two given states, independently of the path. This indicates that dE is an exact differential or, equivalently, that the internal energy is a state function. The same is not true of the differentials $đQ$ and $đW$, hence the difference in notation.

Consider a system whose state can be specified by the values of a set of state variables x_j (*e.g.*, the volume, the number of moles of the different constituents, the magnetization, the electric polarization, *etc.*) and the temperature. As mentioned earlier, thermodynamics exploits an analogy with mechanics and we write, for the work done during an infinitesimal process,

$$đW = -\sum_j X_j dx_j \tag{1.6}$$

where the X_j's can be thought of as generalized forces and the x_j's as generalized displacements.

Before going on to discuss the second law, we pause to introduce some terminology. A thermodynamic transformation or process is any change in the state variables of the system. A *spontaneous* process is one that takes place without any change in the external constraints on the system, and is due simply to the internal dynamics. An *adiabatic* process is one in which no heat is exchanged between the system and its surroundings. A process is *isothermal* if the temperature is held fixed, *isobaric* if the pressure is fixed, *isochoric* if the density is constant, and *quasistatic* if the process is infinitely slow. A *reversible* process is by nature quasistatic and follows a path in thermodynamic space which can be exactly reversed. If this is not possible, the process is irreversible. An example of a reversible process is the slow adiabatic expansion of a gas against a piston on which a force is exerted externally. This force is infinitesimally less than PA, where P is the pressure of the gas and A the area of the piston. An example of an irreversible process is the free adiabatic expansion of a gas into a vacuum. In this case the initial state of the gas can be recovered if one compresses it and removes excess heat. This is, however, not the same thermodynamic path.

[1]The sign convention for the work term is not universally accepted, some authors consider work to be positive if done *on* the system.

1.2.2 Second law

The second law of thermodynamics introduces the *entropy* S as an extensive state variable and states that for an infinitesimal reversible process at temperature T, the heat given to the system is

$$đQ|_{rev} = TdS \tag{1.7}$$

while for an irreversible process

$$đQ|_{irrev} \leq TdS \ .$$

If we are only interested in thermodynamic equilibrium states we can use (1.7) and treat the entropy S as the generalized displacement which is coupled to the 'force' T. The above formulation of the second law is due to Gibbs[2].

We present next two equivalent statements of the second law of thermodynamics. The *Kelvin* version is:

> There exists no thermodynamic process whose sole effect is to extract a quantity of heat from a system and to convert it entirely to work.

The equivalent statement of *Clausius* is:

> No process exists in which the sole effect is that heat flows from a reservoir at a given temperature to a reservoir at a higher temperature.

A corollary of these statements is that the most efficient engine operating between two reservoirs at temperatures T_1 and T_2 is the Carnot engine. The Carnot engine is an idealized engine in which all the steps are reversible. We show the Carnot cycle for an ideal gas working substance in Figure 1.1. In step AB heat Q_1 is absorbed by the gas, which expands isothermally and does work in the process. The next step, BC, is adiabatic and further work is done. In step CD heat $(-Q_2)$ is given off to the low-temperature reservoir and work is done on the gas. Step DA returns the working substance adiabatically to its original state.

The efficiency, η, of the engine is defined to be the ratio of the total work done in one cycle to the heat absorbed from the high-temperature reservoir:

$$\eta = \frac{W}{Q_1} = \frac{Q_1 + Q_2}{Q_1} \tag{1.8}$$

[2]Readers who are interested in the history of thermodynamics will enjoy the article by M. J. Klein [157] on J. Willard Gibbs.

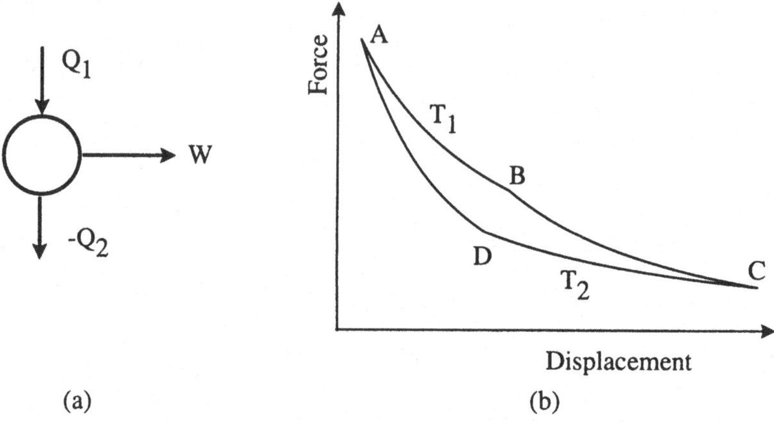

Figure 1.1: Carnot cycle for an ideal gas working substance.

In (1.8) we have followed the convention of the first law that heat transfer is positive if added to the working system. Suppose now that a second more efficient engine operates between the same two temperatures. We can use this engine to drive the Carnot engine backwards—since it is reversible, Q_1, Q_2, and W will simply change sign and η will remain the same.

In Figure 1.2(a) the Carnot engine is denoted by C, the other hypothetical super-engine, with efficiency $\eta_S > \eta_C$ is denoted by S. We use all the work done by engine S to drive engine C. Let the heat absorbed from the reservoirs be Q_{1C}, Q_{1S}, Q_{2C}, Q_{2S}. By assumption we have

$$\eta_S = \frac{W}{Q_{1S}} > \frac{-W}{Q_{1C}} = \eta_C \ . \tag{1.9}$$

The inequality implies that $|Q_{1C}| > Q_{1S}$ and the net effect of the entire process is to transfer heat from the low-temperature reservoir to the high-temperature reservoir. This violates the Clausius statement of the second law. Similarly, if we take only part of the work output of engine S, and adjust it so that there is no net heat transfer to the low-temperature reservoir, a contradiction of the Kelvin statement of the second law results. We conclude that no engine operating between two reservoirs at fixed temperatures is more efficient than a Carnot engine. Equivalently, all reversible engines operating between fixed temperatures have the same efficiency and are Carnot engines. The result that all Carnot engines operating between two temperatures have

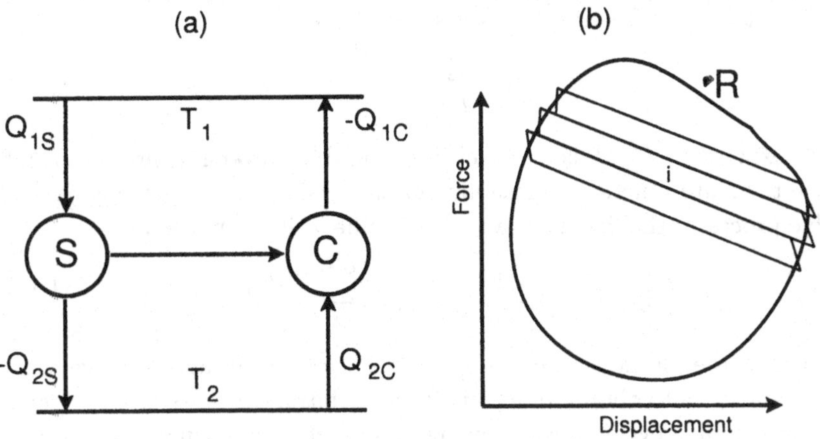

Figure 1.2: (a) Carnot engine (C) driven in reverse by an irreversible engine (S). (b) Arbitrary reversible process covered by infinitesimal Carnot cycles.

the same efficiency can be used to define a temperature scale. One possible definition is

$$\frac{T_2}{T_1} = 1 - \eta_C(T_1, T_2) \tag{1.10}$$

where $\eta_C(T_1, T_2)$ is the Carnot efficiency. Using an ideal gas as a working substance, one can easily show (Problem 1.1) that this temperature scale is identical with the ideal gas (or absolute) temperature scale. Substituting for η in equation (1.8), we have, for a Carnot cycle,

$$\frac{Q_1}{T_1} + \frac{Q_2}{T_2} = 0 \ . \tag{1.11}$$

With this equation we are in a position to define the entropy. Consider an arbitrary reversible cyclic process such as the one drawn in Figure 1.2(b). We can cover the region of the $P-V$ plane, enclosed by the reversible cycle R in Figure 1.2(b), with a set of Carnot cycles operating between temperatures arbitrarily close to each other. For each Carnot cycle we have, from (1.11),

$$\sum_i \frac{Q_i}{T_i} = 0 \ . \tag{1.12}$$

As the number of Carnot cycles goes to infinity, the integral of $đQ/T$ over the uncompensated segments of these cycles approaches

$$\int_R \frac{đQ}{T} = 0 \; . \tag{1.13}$$

Thus the expression $đQ/T$ is an exact differential for reversible processes and we define the state function, whose differential it is, to be the entropy S. For reversible processes the first law can therefore be written in the form

$$dE = TdS - đW = TdS + \sum_j X_j dx_j \; . \tag{1.14}$$

The fact that the Carnot cycle is the most efficient cycle between two temperatures allows us to obtain an inequality for arbitrary processes. Consider a possibly irreversible cycle between two reservoirs at temperatures T_1 and T_2.

$$\frac{Q_1 + Q_2}{Q_1} \leq \frac{Q_{1C} + Q_{2C}}{Q_{1C}} = \eta_C \; . \tag{1.15}$$

This implies that $Q_2/Q_1 \leq -T_2/T_1$ and

$$\frac{Q_1}{T_1} + \frac{Q_2}{T_2} \leq 0 \; . \tag{1.16}$$

Generalizing to an arbitrary process, we obtain

$$\oint \frac{đQ}{T} \leq 0 \tag{1.17}$$

where the equality holds for reversible processes. Since the entropy is a state function, $\oint dS = 0$ for any reversible closed cycle. We can imagine an arbitrary process combined with a reversible process to form a cycle and we therefore obtain for an arbitrary infinitesimal process $T\Delta S \geq \Delta Q$. Combining this with the first law we have, for arbitrary infinitesimal processes,

$$T\Delta S \geq \Delta E + \Delta W \tag{1.18}$$

where, once again, the equality holds for reversible processes.

A further consequence of the foregoing discussion is that the entropy of an isolated system cannot decrease in any spontaneous process. Imagine a spontaneous process in which the system evolves from point A to point B (Figure 1.3) in the thermodynamic space. (Note that the irreversible path cannot be represented as a curve in the $P-T$ plane. The dotted line represents a reversible path connecting the same endpoints.) Since the system is isolated $\Delta Q = 0$ and

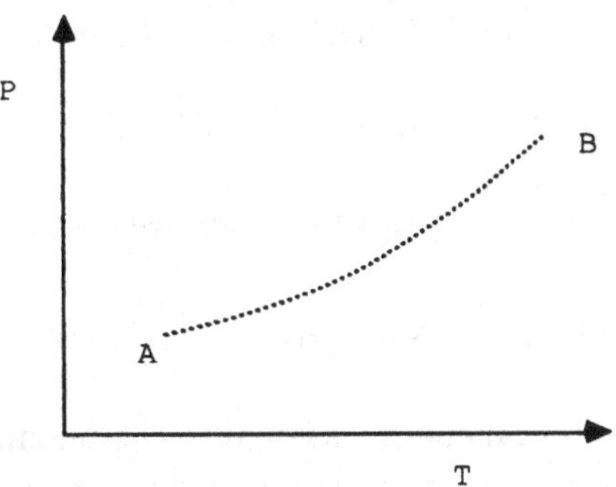

Figure 1.3: Thermodynamic path.

$$\int_A^B dS \geq \int_A^B \frac{đQ}{T} = 0 \tag{1.19}$$

or

$$S(B) - S(A) \geq 0\ . \tag{1.20}$$

Since spontaneous processes tend to drive a system toward equilibrium, we conclude that the equilibrium state of an isolated system is the state of maximum entropy.

1.3 Thermodynamic Potentials

The term thermodynamic potential derives from an analogy with mechanical potential energy. In certain circumstances the work obtainable from a macroscopic system is related to the change in the appropriately defined thermodynamic potential. The simplest example is the internal energy $E(S,V)$ for a PVT system. The second law for reversible processes reads

$$dE = TdS - PdV = đQ - đW\ . \tag{1.21}$$

In a reversible adiabatic transformation the decrease in internal energy is equal to the amount of work done by the expanding system. If the transformation is adiabatic but not reversible, $đQ = 0$ and the first law yields

$$\Delta E = -(\Delta W)_{irrev} \tag{1.22}$$

with the same change in E as in a reversible transformation connecting the same endpoints in the thermodynamic space. However, the change in entropy is not necessarily zero and must be calculated along a reversible path:

$$\Delta E = (\Delta Q)_{rev} - (\Delta W)_{rev} \ . \tag{1.23}$$

Subtracting and using $(đQ)_{rev} = TdS$, we find that

$$(\Delta W)_{rev} - (\Delta W)_{irrev} = \int TdS \geq 0 \ . \tag{1.24}$$

Thus the decrease in internal energy is equal to the maximum amount of work obtainable through an adiabatic process, and this maximum is achieved if the process is reversible.

We now generalize the formulation to allow other forms of work, as well as the exchange of particles between the system under consideration and its surroundings. This more general internal energy is a function of the entropy, the extensive generalized displacements, and the number of particles of each species: $E = E(S, \{x_i\}, \{N_j\})$ with a differential (for reversible processes)

$$dE = TdS + \sum_i X_i dx_i + \sum_j \mu_j dN_j \ . \tag{1.25}$$

Here N_j is the number of molecules of type j and the *chemical potential* μ_j is defined by (1.25). We are now in a position to introduce a number of other useful thermodynamic potentials. The *Helmholtz free energy* [3], A, is related to the internal energy through a *Legendre transformation*:

$$A = E - TS \ . \tag{1.26}$$

The quantity A is a state function with differential

$$\begin{aligned} dA &= dE - TdS - SdT \\ &= -SdT + \sum_i X_i dx_i + \sum_j \mu_j dN_j \ . \end{aligned} \tag{1.27}$$

As in the case of the internal energy, the change in Helmholtz free energy may be related to the amount of work obtainable from the system. In a general infinitesimal process

$$\begin{aligned} dA &= dE - d(TS) \\ &= đQ - TdS - SdT - đW \ . \end{aligned} \tag{1.28}$$

[3]Some authors use the symbol F to represent the Helmholtz free energy.

Thus

$$đW = (đQ - TdS) - SdT - dA \ . \tag{1.29}$$

In a reversible transformation $đQ = TdS$. If the process is isothermal as well as reversible we have $đW = -dA$ and the Helmholtz free energy plays the role of a potential energy for reversible isothermal processes. If the process in question is isothermal but not reversible, we have $đQ - TdS \leq 0$ and

$$(đW)_{irrev} = đQ - TdS - dA \leq -dA \tag{1.30}$$

which shows that $-dA$ is the maximum amount of work that can be extracted, at constant temperature, from the system. We also see, from (1.30), that if the temperature and generalized displacements are fixed ($đW = 0$), a spontaneous process can only decrease the Helmholtz free energy and conclude that the equilibrium state of a system at fixed $(T, \{x_i\}, \{N_j\})$ is the state of minimum Helmholtz free energy.

Another thermodynamic potential which is often useful is the *Gibbs free energy* G. For a PVT system we write

$$G = A + PV \ . \tag{1.31}$$

This function is again a state function with a differential

$$dG = dA + PdV + VdP = -SdT + VdP \ . \tag{1.32}$$

In a general process

$$\begin{aligned} dG &= dE - d(TS) + d(PV) & (1.33)\\ &= (đQ - TdS) - (đW - PdV) + VdP - SdT \ . & (1.34) \end{aligned}$$

We see that the relations

$$\begin{aligned} đW - PdV &= 0 \\ đQ - TdS &\leq 0 \end{aligned} \tag{1.35}$$

imply that the Gibbs potential can only decrease in a spontaneous process at fixed T and P.

In many applications one considers processes which take place at ambient pressure. In such a process there may be volume change (*e.g.* due to release of gases in a chemical reaction). The PdV work then represents work against the environment and is commonly not considered to be available work. We write

$$đW = \int PdV + W_{other}$$

where W_{other} could represent *e.g.* electric energy in a fuel cell. It is then easy to show that $-\Delta G$ is the maximum amount of other work that can be extracted at fixed T. The maximum occurs when the process is reversible.

One further potential that is very useful in statistical physics is the grand potential $\Omega_G(T, V, \{\mu\})$. This potential is obtained from the internal energy through the transformation

$$\Omega_G(T, V, \{\mu\}) = E - TS - \sum_i N_i\mu_i \tag{1.36}$$

and has the differential

$$d\Omega_G = -SdT - PdV - \sum_i N_i d\mu_i \ . \tag{1.37}$$

The grand potential is necessary for the description of *open systems* (systems that can exchange particles with their surroundings).

1.4 Gibbs–Duhem and Maxwell Relations

The internal energy E has as its natural independent variables the entropy S, the volume V, and other generalized displacements which are all extensive variables. If these quantities are rescaled by a factor λ, the internal energy must itself change by the same factor:

$$E(\lambda S, \{\lambda x_i\}, \{\lambda N_j\}) = \lambda E(S, \{x_i\}, \{N_j\}) \ . \tag{1.38}$$

Differentiating both sides with respect to λ using (1.25) on the right-hand side, we obtain the *Gibbs–Duhem equation*:

$$E(S, \{x_i\}, \{N_j\}) = TS + \sum_i X_i x_i + \sum_j \mu_j N_j \ . \tag{1.39}$$

For a single-component PVT system, (1.39) reduces to

$$E = TS - PV + \mu N \tag{1.40}$$

or

$$G(P, T, N) = \mu N \ . \tag{1.41}$$

Taking the differential of (1.39) and using (1.25), we find that

$$0 = SdT + \sum_i x_i dX_i + \sum_j N_j d\mu_j \tag{1.42}$$

which illustrates the fact that the intensive variables T, $\{X_i\}$, $\{\mu_j\}$ are not all independent. An r-component PVT system thus has $r+1$ independent intensive thermodynamic variables. Another consequence is that at least one extensive variable is needed to specify completely the state of the system.

It follows from the differential form (1.27) for a single-component PVT system that

$$\begin{aligned} \left(\frac{\partial A}{\partial T}\right)_{N,V} &= -S \\ \left(\frac{\partial A}{\partial V}\right)_{T,N} &= -P \\ \left(\frac{\partial A}{\partial N}\right)_{T,V} &= \mu\,. \end{aligned} \tag{1.43}$$

It is a well-known result from the theory of partial differentiation that higher order derivatives are independent of the order in which the differentiation is carried out; that is, if ϕ is a single-valued function of the independent variables $x_1, x_2, ..., x_n$, then

$$\frac{\partial}{\partial x_i}\left(\frac{\partial \phi}{\partial x_j}\right) = \frac{\partial}{\partial x_j}\left(\frac{\partial \phi}{\partial x_i}\right)\,. \tag{1.44}$$

By applying this result to (1.43) we immediately obtain the *Maxwell relations*:

$$\begin{aligned} \left(\frac{\partial S}{\partial V}\right)_{T,N} &= \left(\frac{\partial P}{\partial T}\right)_{V,N} \\ \left(\frac{\partial S}{\partial N}\right)_{V,T} &= -\left(\frac{\partial \mu}{\partial T}\right)_{V,N} \\ \left(\frac{\partial P}{\partial N}\right)_{V,T} &= -\left(\frac{\partial \mu}{\partial V}\right)_{T,N}\,. \end{aligned} \tag{1.45}$$

Similarly, in the case of the Gibbs potential we find from (1.32)

$$\begin{aligned} \left(\frac{\partial G}{\partial T}\right)_{N,P} &= -S \\ \left(\frac{\partial G}{\partial P}\right)_{T,N} &= V \\ \left(\frac{\partial G}{\partial N}\right)_{T,P} &= \mu \end{aligned} \tag{1.46}$$

from which we have the additional Maxwell relations:

$$\begin{aligned}
\left(\frac{\partial S}{\partial P}\right)_{T,N} &= -\left(\frac{\partial V}{\partial T}\right)_{P,N} \\
\left(\frac{\partial V}{\partial N}\right)_{P,T} &= \left(\frac{\partial \mu}{\partial P}\right)_{T,N} \\
\left(\frac{\partial S}{\partial N}\right)_{P,T} &= -\left(\frac{\partial \mu}{\partial T}\right)_{P,N} .
\end{aligned} \tag{1.47}$$

Further equations of this type can be found for magnetic systems, using (1.80), or, in the case of PVT systems, by using the internal energy or the grand potential. The usefulness of these relations is demonstrated in the next section, in which we derive relations between some of the most commonly measured response functions.

1.5 Response Functions

A great deal can be learned about a macroscopic system through its response to various changes in externally controlled parameters. Important response functions for a PVT system are the specific heats at constant volume and pressure,

$$C_V = \left(\frac{đQ}{\partial T}\right)_V = T\left(\frac{\partial S}{\partial T}\right)_V \tag{1.48}$$

$$C_P = \left(\frac{đQ}{\partial T}\right)_P = T\left(\frac{\partial S}{\partial T}\right)_P \tag{1.49}$$

the isothermal and adiabatic compressibilities,

$$K_T = -\frac{1}{V}\left(\frac{\partial V}{\partial P}\right)_T \tag{1.50}$$

$$K_S = -\frac{1}{V}\left(\frac{\partial V}{\partial P}\right)_S \tag{1.51}$$

and the coefficient of thermal expansion

$$\alpha = \frac{1}{V}\left(\frac{\partial V}{\partial T}\right)_{P,N} . \tag{1.52}$$

Intuitively, we expect the specific heats and compressibilities to be positive and $C_P > C_V, K_T > K_S$. In this section we derive relations between these response functions. The intuition that the response functions are positive will be justified in the following section in which we discuss thermodynamic stability. We begin with the assumption that the entropy has been expressed in terms of T and V and that the number of particles is kept fixed. Then

$$dS = \left(\frac{\partial S}{\partial T}\right)_V dT + \left(\frac{\partial S}{\partial V}\right)_T dV \tag{1.53}$$

and

$$T\left(\frac{\partial S}{\partial T}\right)_P = T\left(\frac{\partial S}{\partial T}\right)_V + T\left(\frac{\partial S}{\partial V}\right)_T \left(\frac{\partial V}{\partial T}\right)_P \tag{1.54}$$

or

$$C_P - C_V = T\left(\frac{\partial S}{\partial V}\right)_T \left(\frac{\partial V}{\partial T}\right)_P . \tag{1.55}$$

We now use the Maxwell relation (1.45) and the chain rule

$$\left(\frac{\partial z}{\partial x}\right)_y \left(\frac{\partial y}{\partial z}\right)_x \left(\frac{\partial x}{\partial y}\right)_z = -1 \tag{1.56}$$

which is valid for any three variables obeying an equation of state of the form $f(x,y,z)=0$ to obtain

$$\left(\frac{\partial S}{\partial V}\right)_T = \left(\frac{\partial P}{\partial T}\right)_V = -\left(\frac{\partial P}{\partial V}\right)_T \left(\frac{\partial V}{\partial T}\right)_P \tag{1.57}$$

and

$$C_P - C_V = -T\left(\frac{\partial P}{\partial V}\right)_T \left(\frac{\partial V}{\partial T}\right)_P^2 = \frac{TV}{K_T}\alpha^2 . \tag{1.58}$$

In a similar way we obtain a relation between the compressibilities K_T and K_S. Assume that the volume V has been obtained as function of S and P. Then

$$dV = \left(\frac{\partial V}{\partial P}\right)_S dP + \left(\frac{\partial V}{\partial S}\right)_P dS \tag{1.59}$$

and

$$-\frac{1}{V}\left(\frac{\partial V}{\partial P}\right)_T = -\frac{1}{V}\left(\frac{\partial V}{\partial P}\right)_S - \frac{1}{V}\left(\frac{\partial V}{\partial S}\right)_P \left(\frac{\partial S}{\partial P}\right)_T \tag{1.60}$$

or

$$K_T - K_S = -\frac{1}{V}\left(\frac{\partial V}{\partial S}\right)_P \left(\frac{\partial S}{\partial P}\right)_T . \tag{1.61}$$

The Maxwell relations (1.47) and the equation

$$\left(\frac{\partial V}{\partial S}\right)_P = \left(\frac{\partial V}{\partial T}\right)_P \left(\frac{\partial S}{\partial T}\right)_P^{-1} \tag{1.62}$$

yield

$$K_T - K_S = \frac{TV}{C_P}\alpha^2 . \tag{1.63}$$

Thus (1.58) and (1.63) together produce the interesting and useful exact results

$$C_P(K_T - K_S) = K_T(C_P - C_V) = TV\alpha^2 \tag{1.64}$$

and

$$\frac{C_P}{C_V} = \frac{K_T}{K_S} . \tag{1.65}$$

1.6 Conditions for Equilibrium and Stability

We consider two systems in contact with each other. It is intuitively clear that if heat can flow freely between the two systems and if the volumes of the two systems are not separately fixed, the parameters will evolve so as to equalize the pressure and temperature of the two systems. These conclusions can easily be obtained from the principle of maximum entropy. Suppose that the two systems have volumes V_1, and V_2, energies E_1 and E_2, and that the number of particles in each, as well as the combined energy and total volume, are fixed. In equilibrium, the total entropy

$$S = S_1(E_1, V_1) + S_2(E_2, V_2) \tag{1.66}$$

must be a maximum. Thus

$$dS = \left(\frac{\partial S_1}{\partial E_1}\right)_{V_1} dE_1 + \left(\frac{\partial S_2}{\partial E_2}\right)_{V_2} dE_2 + \left(\frac{\partial S_1}{\partial V_1}\right)_{E_1} dV_1 + \left(\frac{\partial S_2}{\partial V_2}\right)_{E2} dV_2$$

$$= \left[\left(\frac{\partial S_1}{\partial E_1}\right)_{V_1} - \left(\frac{\partial S_2}{\partial E_2}\right)_{V_2}\right] dE_1 + \left[\left(\frac{\partial S_1}{\partial V_1}\right)_{E_1} - \left(\frac{\partial S_2}{\partial V_2}\right)_{E_2}\right] dV_1 = 0 \tag{1.67}$$

where we have used the constraint $E_1 + E_1 = const.$, $V_1 + V_2 = const.$ We have

$$\left(\frac{\partial S_j}{\partial E_j}\right)_{V_j} = \left(\frac{\partial E_j}{\partial S_j}\right)_{V_j}^{-1} = \frac{1}{T_j} \tag{1.68}$$

and

$$\left(\frac{\partial S_j}{\partial V_j}\right)_{E_j} = \frac{P_j}{T_j} \tag{1.69}$$

which together with (1.67) yield

$$\frac{1}{T_1} = \frac{1}{T_2} \tag{1.70}$$

$$\frac{P_1}{T_1} = \frac{P_2}{T_2} \tag{1.71}$$

or $T_1 = T_2$, $P_1 = P_2$, which is the expected result. More generally, one finds that when the conjugate displacements are unconstrained, all generalized forces of two systems in equilibrium must be equal.

To this point we have required only that the equilibrium state correspond to a *stationary state* of the entropy. Requiring this stationary state to be a *maximum* will provide conditions on the second derivatives of the entropy. These conditions are local in nature. A stronger (global) condition is that the entropy be a *concave* function of the generalized displacements (see Problem 1.10).

Some of the most useful stability criteria are obtained from the Gibbs potential rather than from the entropy and we proceed to consider a small (but macroscopic) system in contact with a much larger reservoir. This reservoir is assumed to be so large that fluctuations in the small system do not change the temperature or pressure of the reservoir, which we denote by T_0 and P_0. The Gibbs potential, as we have seen in Section 1.3, is a minimum in equilibrium, and for the small system we have

$$G_1(P_0, T_0) = E_1 - T_0 S_1 + P_0 V_1 \ . \tag{1.72}$$

Suppose now that there is a fluctuation in the entropy and volume of this system. To second order in the fluctuating quantities,

$$\delta G_1 = \delta S_1 \left(\frac{\partial E_1}{\partial S_1} - T_0 \right) + \delta V_1 \left(\frac{\partial E_1}{\partial V_1} + P_0 \right)$$

$$+\frac{1}{2}\left[(\delta S_1)^2 \left(\frac{\partial^2 E_1}{\partial S_1^2} \right) + 2\delta S_1 \delta V_1 \left(\frac{\partial^2 E_1}{\partial S_1 \partial V_1} \right) + (\delta V_1)^2 \left(\frac{\partial^2 E_1}{\partial V_1^2} \right) \right] \tag{1.73}$$

which must be greater than zero if the state specified by P_0, T_0 is the state of minimum Gibbs potential. Since $\partial E_1/\partial S_1 = T_0$ and $\partial E_1/\partial V_1 = -P_0$ we obtain the condition

$$(\delta S)^2 \left(\frac{\partial^2 E}{\partial S^2} \right) + 2\delta S \delta V \left(\frac{\partial^2 E}{\partial S \partial V} \right) + (\delta V)^2 \left(\frac{\partial^2 E}{\partial V^2} \right) > 0 \tag{1.74}$$

where we have dropped the subscripts. The fluctuations in the entropy and volume are independent of each other, and we can guarantee that the expression

(1.74) is positive if we require that $E(S,V)$ satisfies the conditions

$$\begin{aligned} &\frac{\partial^2 E}{\partial S^2} > 0 \\ &\frac{\partial^2 E}{\partial V^2} > 0 \\ \frac{\partial^2 E}{\partial S^2}\frac{\partial^2 E}{\partial V^2} - &\left(\frac{\partial^2 E}{\partial S \partial V}\right)^2 > 0 \, . \end{aligned} \tag{1.75}$$

The first inequality reduces to

$$\left(\frac{\partial T}{\partial S}\right)_V = \frac{T}{C_V} > 0 \text{ or } C_V > 0 \tag{1.76}$$

while the second implies

$$-\left(\frac{\partial P}{\partial V}\right)_S = \frac{1}{VK_S} > 0 \text{ or } K_S > 0 \tag{1.77}$$

and the final inequality yields

$$\frac{T}{VK_SC_V} > \left(\frac{\partial T}{\partial V}\right)_S^2 \, . \tag{1.78}$$

These inequalities are special cases of Le Châtelier's principle, which states that if a system is in equilibrium, any spontaneous changes in its parameters will bring about processes that tend to restore the system to equilibrium. In our situation such spontaneous processes raise the Gibbs potential. Other stability criteria can be obtained by using one of the other thermodynamic potentials of Section 1.3.

1.7 Magnetic Work

In our microscopic statistical treatment of magnetic materials we will model these materials by systems of magnetic moments (or "spins") which can orient themselves in an applied magnetic field $\mathbf{H}$. The work done by the system, if the applied field is held constant, is then done by a magnet during a process in which its magnetization is changed

$$đW = -\mathbf{H} \cdot d\mathbf{M} \, . \tag{1.79}$$

Instead of defining yet another potential, we modify the definition of the Gibbs potential for a purely magnetic system to read

$$G(\mathbf{H},T) = E(S,\mathbf{M}) - TS - \mathbf{M}\cdot\mathbf{H} \tag{1.80}$$

$$dG = -SdT - \mathbf{M}\cdot d\mathbf{H}\ . \tag{1.81}$$

It should be noted that (1.80) is not a universally accepted convention. Some authors refer to (1.80) as the Helmholtz free energy.

The appropriate response functions are then the isothermal and adiabatic susceptibilities, $\chi_T = (\partial M/\partial H)_T$, $\chi_S = (\partial M/\partial H)_S$. A derivation analogous to what we did in Section 1.5, in the case of magnetic systems (see also Problem 1.4) produces the equations

$$C_H(\chi_T - \chi_S) = \chi_T(C_H - C_M) = T\left(\frac{\partial M}{\partial T}\right)_H^2 \tag{1.82}$$

where C_H and C_M are the specific heats at constant applied field and constant magnetization, respectively. The identification of the magnetization $\mathbf{M}$ to be analogous to a displacement x, and the generalized force to be the applied magnetic field $\mathbf{H}$, is strictly speaking not correct, since we neglect energy stored in the field[4]. To see why (1.80) leads to inconsistent results consider the question of thermodynamic stability. In order that the Gibbs free energy be a minimum in a constant magnetic field we must also have

$$\frac{\partial^2 E}{\partial M^2} = \frac{\partial H}{\partial M} = \chi^{-1} > 0$$

since thermodynamic stability requires that the susceptibility be positive. But, this cannot be true, since some materials such as superconductors are known to be diamagnetic, with negative susceptibility. To resolve this question we start by writing down Faraday's law

$$\nabla\times\mathbf{E} = -\frac{\partial\mathbf{B}}{\partial t}$$

and Ampère's law

$$\mathbf{j} = \nabla\times\mathbf{H}$$

where $\mathbf{j}$ is the current density, and the magnetic induction $\mathbf{B}$ is related to the field through $\mathbf{B} = \mu\mathbf{H} = \mu_0(\mathbf{H}+\mathbf{M})$, where μ and μ_0 are the permeability and vacuum permeability, respectively, (not to be confused with the chemical

[4]See also Appendix B in Callen [55].

potential). The relationship between the permeability and the susceptibility is $\mu = \mu_0(1 + \chi)$. The work done by the system in a short time interval δt is that of currents moving against the field

$$đW = -\delta t \int \mathbf{E} \cdot \mathbf{j} dV = -\delta t \int \mathbf{E} \cdot (\nabla \times \mathbf{H}) dV \ .$$

This expression can be rewritten

$$đW = \delta t \int \nabla \cdot (\mathbf{E} \times \mathbf{H}) dV - \delta t \int \mathbf{H} \cdot (\nabla \times \mathbf{E}) dV \, .$$

With suitable boundary conditions at infinity the first term on the left-hand side vanishes and we find

$$đW = \int \mathbf{H} \cdot \delta \mathbf{B} dV$$

i.e. the conjugate thermodynamic variables are **H** and **B**, not **H** and **M**. The correct thermodynamic stability requirement is thus

$$\frac{\partial B}{\partial H} = \mu > 0$$

which allows for negative susceptibilities as long as $\chi > -1$. We are not going to discuss magnetic interactions in detail here, and since we now know how to do magnetic work correctly if pressed to do so, we will revert to the cruder form (1.80).

1.8 Thermodynamics of Phase Transitions

A typical phase diagram for a one-component PVT system looks like Figure 1.4. The solid lines separate the $P - T$ plane into regions in which different phases are the stable thermodynamic states. As the system passes through one of these lines, called *coexistence curves*, a *phase transition* occurs, generally accompanied by the absorption or liberation of latent heat. In Figure 1.4 there are two special points, the *triple point* P_t, T_t, and the *critical point*, P_c, T_c. At the critical point the properties of the fluid and vapor phase become identical and much of our study of phase transitions in later chapters will focus on the region of the phase diagram around this point. We note that the properties of the system vary smoothly along any curve which does not cross a coexistence curve. Thus, it is possible to pass continuously from

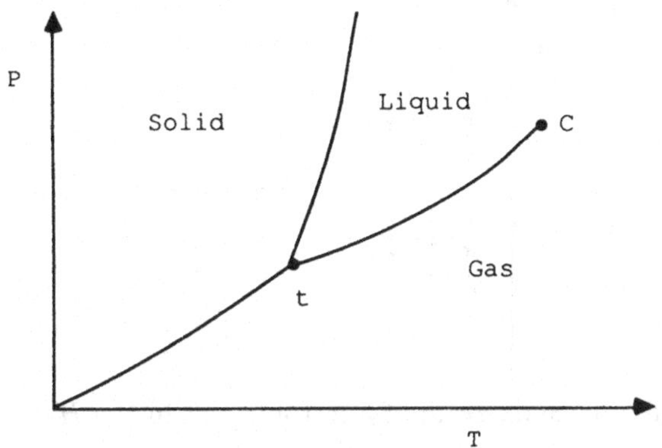

Figure 1.4: Schematic phase diagram of a simple one-component PVT system.

the vapor to the liquid phase by taking the system to high enough temperature, increasing the pressure, and then lowering the temperature again. It is not possible to avoid the liquid-solid coexistence curve—this curve extends to $P = \infty$, $T = \infty$ (as far as we know). The analogous phase diagram for a ferromagnetic substance is shown in Figure 1.5 in the $H - T$ plane with a critical point at $H_c = 0$, T_c.

The phase diagrams of Figures 1.4 and 1.5 are particularly simple, partly because of the choice of variables. The fields H and P as well as the temperature T are required (see Section 1.6) to be equal in the coexisting phases. Conversely, the densities conjugate to these fields (*i.e.*, the magnetization M, density ρ, specific volume v, or the entropy per particle s) can take on different values in the two phases. Thus, the phase diagram of the ferromagnet of Figure 1.5 takes the form shown in Figure 1.6 when drawn in the $T - M$ plane. The points A, B on the bell-shaped curve represent the stable states as the coexistence curve of Figure 1.5 is approached from one phase or the other. A state inside the bell-shaped region, such as the one marked $\times$ on the vertical line, is not a stable single-phase state—the system separates into two regions, one with the magnetization of point A, the other with the magnetization of point B.

Similarly, the phase diagram of the simple fluid of Figure 1.4, drawn in the $\rho - T$ plane, is schematically shown in Figure 1.7. The liquid-gas coexistence curve in Figure 1.7, while lacking the symmetry of the bell-shaped curve of Figure 1.6, has many of the same properties. Indeed, we will find, in subsequent chapters, that certain magnetic systems behave in a way similar to liquid-gas systems as the critical point is approached. We now discuss some of the

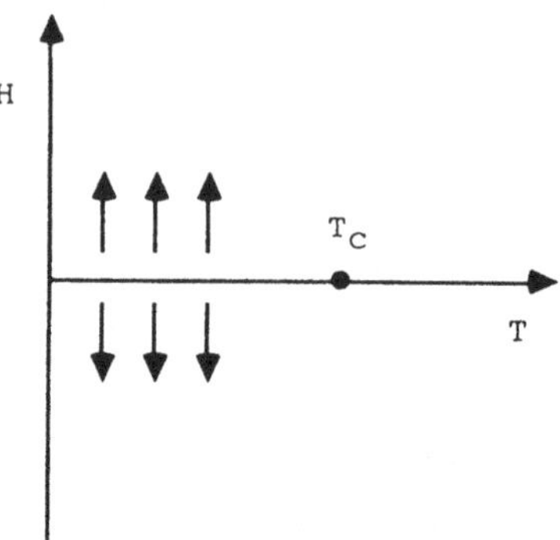

Figure 1.5: Phase diagram of a ferromagnet in the $H - T$ plane.

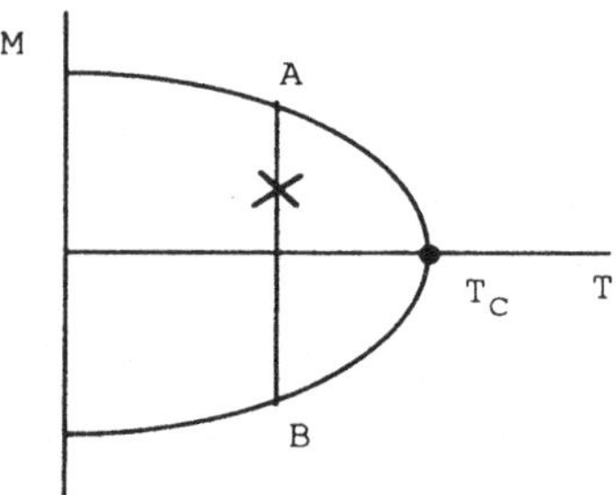

Figure 1.6: Phase diagram of the ferromagnet in the $M - T$ plane.

properties of coexistence curves.

We consider a single-component PVT system on either side of the liquid-gas or liquid-solid coexistence curve. The coexisting phases may be thought of as two equilibrium systems in contact with each other. We therefore have

$$\begin{aligned} T_1 &= T_2 \\ P_1 &= P_2 \\ \mu_1 &= \mu_2 \end{aligned} \tag{1.83}$$

where the subscripts 1 and 2 refer to the two phases. From the Gibbs–Duhem equation (1.41) we obtain

$$g_1(T,P) = g_2(T,P) \tag{1.84}$$

where g_1, and g_2 are the Gibbs potential per particle in phases 1 and 2, respectively. The equality (1.84) must hold along the entire coexistence curve

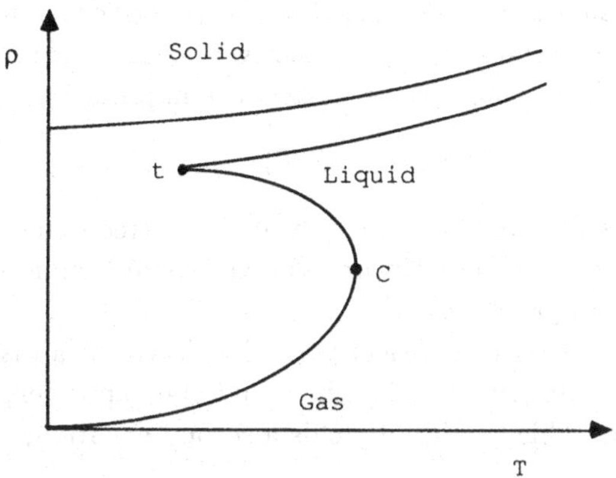

Figure 1.7: Phase diagram of the $P-V-T$ system in the $\rho-T$ plane.

and hence

$$dg_1 = -s_1 dT + v_1 dP = dg_2 = -s_2 dT + v_2 dP \tag{1.85}$$

for differentials (dT, dP) along the coexistence curve. Thus

$$\left(\frac{dP}{dT}\right)_{coex} = \frac{s_1 - s_2}{v_1 - v_2} = \frac{L_{12}}{T(v_1 - v_2)} \tag{1.86}$$

where L_{12} is the latent heat per particle needed to transform the system from phase 2 to phase 1. Equation (1.86) is known as the Clausius–Clapeyron equation. As a simple example, consider a transition from liquid to vapor with $v_1 \gg v_2$ and with $v_1 = k_B T/P$. Then

$$\frac{dP}{dT} \approx \frac{PL_{12}}{k_B T^2} \tag{1.87}$$

and if L_{12} is roughly constant along the coexistence curve, we have

$$P(T) \approx P_0 \exp\left\{-L_{12}\left(\frac{1}{T} - \frac{1}{T_0}\right)\right\} \tag{1.88}$$

where P_0, T_0 is a reference point on the coexistence curve. Using approximate equations for the solid and liquid phases, one can derive an equation similar to (1.88) for the solid-liquid coexistence curve.

As a final topic we now briefly discuss the Gibbs phase rule. This rule allows one to limit the topology of a phase diagram on the basis of some very general considerations. Consider first a single-component PVT system with a phase diagram as shown in Figure 1.4. For two-phase coexistence the chemical potential $\mu(P,T)$ must be the same in the two phases, yielding a curve in the $P-T$ plane. Similarly, three-phase coexistence implies that

$$\mu_1(P,T) = \mu_2(P,T) = \mu_3(P,T) \tag{1.89}$$

which, in general, will have a solution only at an isolated point, the triple point. Four-phase coexistence is ruled out unless there are hidden fields separate from the temperature and pressure.

One can also see that the critical point P_c, T_c will be an isolated point for a PVT system. At the critical point the liquid and vapor densities, or specific volumes, are equal. This condition yields a second equation,

$$v_1(P_c,T_c) = \left.\frac{\partial g_1}{\partial P}\right|_{P_c,T_c} = v_2(P_c,T_c) = \left.\frac{\partial g_2}{\partial P}\right|_{P_c,T_c} \tag{1.90}$$

which together with $\mu_1(P_c,T_c) = \mu_2(P_c,T_c)$ determines a unique point in the $P-T$ plane.

In a multicomponent system the situation is more complicated. We take as thermodynamic variables P, T, and c_{ij}, $i = 1, 2, ..., r$, where c_{ij} is the mole fraction of constituent i in phase j of an r-component system. Suppose that there are s coexisting phases. Since

$$\sum_{i=1}^{r} c_{ij} = 1 \tag{1.91}$$

there are $s(r-1)+2$ remaining independent variables. Equating the chemical potentials for the r components gives $r(s-1)$ equations for these variables. If a solution is to exist, we must have at least as many variables as equations, that is,

$$s(r-1)+2 \geq r(s-1) \tag{1.92}$$

or

$$s \leq r+2\ . \tag{1.93}$$

Therefore, at most, $r+2$ phases can coexist in a mixture of r constituents.

1.9 Problems

1.1. *Equivalence of Carnot and Ideal Gas Temperature Scale.*

Consider a Carnot engine working between reservoirs at temperatures T_1 and T_2. The working substance is an ideal gas obeying the equation of state (1.1), which may be taken to be a definition of a temperature scale. Show explicitly that the efficiency of the cycle is given by

$$\eta = 1 - \frac{T_2}{T_1}$$

where $T_1 > T_2$.

1.2. *Adiabatic Processes in Paramagnets.*

(a) Show that the energy $E(M,T)$ is a function of T only for a Curie paramagnet, ($M = CH/T$). Assuming the specific heat C_M to be constant, show that during a reversible adiabatic process

$$\frac{1}{T} \exp\left(\frac{M^2}{2CC_M}\right) = \text{const.}$$

(b) Verify that $\eta = \frac{Q_1+Q_2}{Q_1} = 1 - T_2/T_1$ for a Carnot cycle with the magnet as the working substance.

1.3. *Stability Analysis for an Open System.*

Analyze the stability of a system that is kept at a fixed volume but is free to exchange energy and particles with a reservoir. The temperature and chemical potential of the reservoir are not affected by the fluctuations. In particular, show that

$$\begin{aligned} C_{V,N} &\geq 0 \\ \left(\frac{\partial N}{\partial \mu}\right)_{V,S} &\geq 0 \\ \frac{C_{V,N}}{T}\left(\frac{\partial N}{\partial \mu}\right)_{V,S} &\leq \left(\frac{\partial S}{\partial \mu}\right)^2_{V,N} . \end{aligned}$$

1.4. *Specific Heats of Magnets.*

Derive the relations

$$\begin{aligned} C_H - C_M &= \frac{T}{\chi_T}\left(\frac{\partial M}{\partial T}\right)^2_H \\ \chi_T - \chi_S &= \frac{T}{C_H}\left(\frac{\partial M}{\partial T}\right)^2_H \\ \frac{\chi_T}{\chi_S} &= \frac{C_H}{C_M} \end{aligned}$$

for a magnetic system.

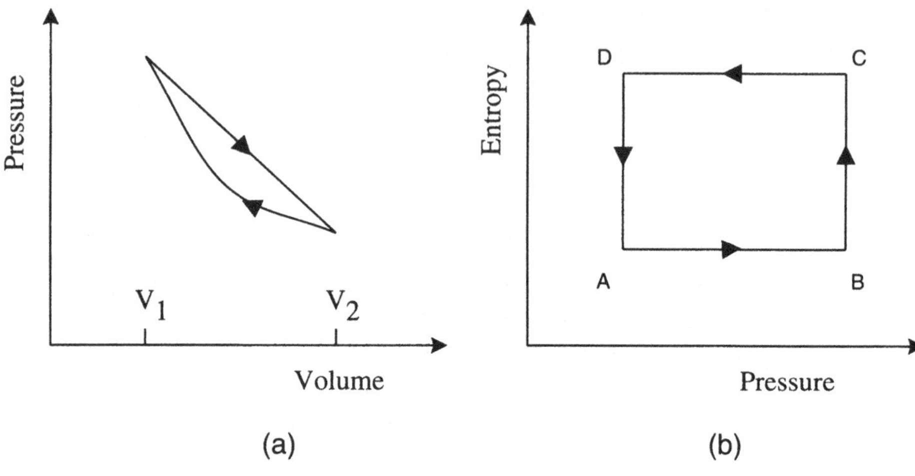

Figure 1.8: (a) Cycle of Problem 1.5. (b) Cycle of Problem 1.6.

1.5. *Efficiency of a Thermodynamic Cycle.*

Consider a heat engine with a two-step cycle in which an adiabatic compression is followed by an expansion along a straight line in a $P - V$ diagram (Figure 1.8(a)).

(a) When integrated over the straight line segment, the heat supplied must, since energy is conserved over a cycle, be equal to the network during the cycle. Why does this not give rise to an efficiency $\eta = W/Q = 1$ in violation of the second law of thermodynamics?

(b) The working substance is a monatomic ideal gas ($\gamma = \frac{5}{3}$) and the ratio of the initial and final volume is 2. Find the efficiency of the process.

1.6. *Brayton Cycle.*

The Joule or Brayton cycle is shown in the $P - S$ plane in Figure 1.8(b). Assuming that the working substance is an ideal gas, show that the efficiency of the cycle is

$$\eta = 1 - \left(\frac{P_A}{P_B}\right)^{\frac{C_P - C_V}{C_P}}$$

where C_P and C_V are the heat capacities at constant pressure and volume, respectively. You may assume that C_P and C_V are constant.

1.7. *Ideal Gas Response Functions.*

Find the thermal expansion coefficient and the isothermal compressibility for an ideal gas and show that in this case (1.58) reduces to $C_P - C_V = Nk_B$ for the molar specific heats.

1.8. *Effect of an Inert Gas on Vapor Pressure.*

A liquid is in equilibrium with its vapor at temperature T and pressure P. Suppose that an inert ideal gas, which is insoluble in the liquid, is introduced into the container and has partial pressure P_i. The temperature is kept fixed. Show that if the volume per particle, v_L, in the liquid is much smaller than the specific volume, v_G, in the gas, the vapor pressure will increase by an amount δP given by

$$\frac{\delta P}{P} = \frac{P_i v_L}{k_B T}$$

if P_i is small enough.

1.9. *Entropy of Mixing.*

Calculate the entropy of mixing of two volumes V_A and V_B of an ideal gas of species A and B, respectively, both initially at the same temperature T and pressure P and, with a final volume $V = V_A + V_B$.

1.10. *Concavity and Local Stability.*

The equilibrium state of a system is the state of maximum entropy. It is easily seen that this statement implies the relation

$$S(E + \Delta E, V, N) + S(E - \Delta E, V, N) - 2S(E, V, N) \leq 0$$

which can be derived by considering an energy fluctuation between two subsystems of the same size in thermal contact with each other. More generally,

$$S(E + \Delta E, V + \Delta V, N) + S(E - \Delta E, V - \Delta V, N) - 2S(E, V, N) \leq 0$$

which is the mathematical statement that S is a *concave* function of its extensive variables. The local stability requirements are special cases of the concavity condition.

$$\left(\frac{\partial^2 S}{\partial E^2}\right)_{V,N} \leq 0$$

$$\left(\frac{\partial^2 S}{\partial V^2}\right)_{E,N} \leq 0$$

$$\left(\frac{\partial^2 S}{\partial E^2}\right)\left(\frac{\partial^2 S}{\partial V^2}\right) - \left(\frac{\partial^2 S}{\partial E \partial V}\right)^2 \leq 0 \, .$$

(a) Show that at fixed S and V the equilibrium state of a system is the state of minimum internal energy and that this implies that E is a *convex* function of S and V.

(b) Show that the Gibbs potential is a concave function of P and T.

1.11. *Derivation of Equation of State.*

In a certain system the internal energy E is related to the entropy S, particle number N, and volume V through

$$E = \text{const} N \left(\frac{N}{V}\right)^d \exp\left[\frac{dS}{Nk_B}\right] \, .$$

(a) Show that the system satisfies the ideal gas law independently of the value of the constant d.

(b) Find the coefficient γ in the adiabatic equation of state $PV^\gamma =$ *const.* and the molar specific heats C_P and C_V of the system.

Chapter 2

Statistical Ensembles

In this chapter we develop the foundations of equilibrium statistical mechanics. No attempt will be made to derive the equilibrium theory from the detailed microscopic dynamics. Instead, we will content ourselves with postulating the statistical laws. In Section 2.1 we consider a closed classical system and develop the concept of the microcanonical ensemble. We define the entropy in the thermodynamic limit, and make contact with thermodynamics. In Section 2.2 we extend the formalism to systems in thermal contact with the surroundings and introduce the canonical ensemble. We also discuss fluctuations of thermodynamic variables. The grand canonical ensemble is discussed in Section 2.3 for systems that are free to exchange particles with the outside, and we demonstrate the general relation between fluctuations in particle number and the compressibility. In Section 2.4 we modify the formalism to include quantum systems and introduce the density matrix. An alternative information-theoretic approach is discussed in Section 2.5 and a number of useful thermodynamic variational principles are introduced in Section 2.6.

We assume that the reader has had some previous exposure to the material under discussion. Our treatment is therefore fairly condensed. Some general references that we have found particularly useful are [136], [161], [254], [304] and [318].

2.1 Isolated Systems: Microcanonical Ensemble

We consider first a system of $3N$ degrees of freedom described by the canonical variables $q_1, \ldots, q_j, \ldots, q_{3N}$, $p_1, \ldots, p_j, \ldots, p_{3N}$. These variables are assumed to obey classical Hamiltonian dynamics:

$$\dot{p}_j = -\frac{\partial H}{\partial q_j} \tag{2.1}$$

$$\dot{q}_j = \frac{\partial H}{\partial p_j} \tag{2.2}$$

for $j = 1, 2, \ldots, 3N$. A possible example of such a system is that of N particles confined to a volume V by an idealized surface which does not permit the flow of heat or particles.

In general, a Hamiltonian system will have a number of conserved quantities, the most obvious and universal being the total energy E. If the system is fully integrable, the motion can be completely expressed in terms of $6N$ constants such as the action variables[1] and the initial values of the $3N$ angles. Most systems are not completely integrable, but there may still be a few first integrals expressible as conservation laws, such as those of total linear and angular momentum. If we restrict our attention to a particular frame of reference (*e.g.*, the one in which there is no net rotational or translational motion), the energy will typically be the only remaining easily identifiable conserved quantity.

We define a $6N$-dimensional phase space and represent the state of the system, at a particular instant, by a $6N$-dimensional vector $\mathbf{x}$ in this space, with components given by the generalized coordinates and momenta q_j, p_j. If we specify the energy E of the system, the motion of the N-particle system will be confined to a $(6N-1)$-dimensional surface given by

$$H(\mathbf{x}) = E \ . \tag{2.3}$$

We denote this surface by $\Gamma(E)$. If a system is in equilibrium its macroscopic properties should be time independent. One way to measure a macroscopic property of the system (*e.g.*, the pressure associated with N particles in a box) is to carry the experiment out over a period of time $t_0 < t < t_0 + \tau$ during which the phase point covers some part of $\Gamma(E)$. If the instantaneous value of

[1] For a discussion of action-angle variables see *e.g.* [113]

the property in question is $\phi(x(t))$, the measured value will be

$$\langle \phi \rangle = \frac{1}{\tau} \int_{t_0}^{t_0+\tau} \phi(\mathbf{x}(t)) \; . \tag{2.4}$$

As we will discuss later in Section 9.1 this type of measurement can be simulated using the technique of molecular dynamics. Most often we will be unable to work out the detailed dynamics $\mathbf{x}(t)$, but we may be able to study averages over the surface $\Gamma(E)$. The simplest assumption, which allows us to proceed is the *ergodic hypothesis:*

> During any significant time interval τ the phase point $\mathbf{x}(t)$ will spend equal time intervals in all regions of $\Gamma(E)$.

There is nothing obvious about this hypothesis. In Problem 2.1 the reader will find one example of a system that satisfies this hypothesis and one that does not. Both examples are trivial in the sense that the equations of motion of the dynamical variables are integrable, whereas in real systems, the equations of motion are invariably nonintegrable and relatively little is known in general about the behavior of large ($\sim 10^{23}$ particles) systems of equations. Suffice to say that a proof of the ergodic hypothesis under fairly general conditions is lacking. In any case, there is no clear connection between the measuring time and the time it takes for an ergodic system to visit all parts of the surface $\Gamma(E)$. Indeed, it may not even be necessary for the measurement to take very long, since the instantaneous measurement of a macroscopic property such as the pressure of particles in a box involves averaging over contributions from all the degrees of freedom of the system.

A hypothesis that leads to statistical mechanics, as we know it, is the *mixing hypothesis.* Briefly stated, mixing refers to the notion that an initial (compact) distribution of points on $\Gamma(E)$ very quickly distorts into a convoluted object which permeates the entire surface while still occupying the same volume (as required by Liouville's theorem). Mixing can be proven to occur, for example, for systems of hard spheres [274] [275]. When one talks about a distribution of points on $\Gamma(E)$, one is no longer discussing a single system, and for a truly isolated system, mixing is of course irrelevant. However, the sensitivity of $\mathbf{x}(t)$ to minute changes in initial conditions, will allow small amounts of noise in an imperfectly insulated system to distort the trajectory sufficiently that in effect one obtains mixing.

The foregoing discussion is quite incomplete and was included only to highlight some of the difficulties in the foundation of statistical mechanics. Further discussion of these topics can be found in [252] and in the Appendix of the book by Balescu [23].

We shall henceforth simply assume that we can replace the time average (2.4) by an average over the surface $\Gamma(E)$. This average is a special case of an *ensemble average.* An ensemble can be visualized as a collection of snapshots of the system at different times. A number of measurement techniques (such as neutron and x-ray scattering) emulate an ensemble average more closely than they do the time averaging process (2.4). Similarly, in computer experiments the Monte Carlo methods of Section 9.2 simulate ensemble averages rather than time averages. We specify an ensemble through the probability density $\rho(\mathbf{x})$ for a selected member of the ensemble to occupy the phase space point $\mathbf{x}$. The ensemble average of an observable $\phi(\mathbf{x}(t))$ is then

$$\langle\phi\rangle = \int d^{6N}x\rho(\mathbf{x})\phi(\mathbf{x}) \tag{2.5}$$

where the integral is carried out over the entire $6N$-dimensional phase space. In the special case that we are considering: a closed classical system of energy E,

$$\rho(\mathbf{x}) = C\delta(H(x) - E) \tag{2.6}$$

where C is a normalizing factor. The ensemble corresponding to (2.6) is called the *microcanonical ensemble.*

As we shall see, the entropy S plays a particularly fundamental role when the microcanonical ensemble is used. To arrive at an acceptable definition of the entropy, which can easily be extended to quantum systems, and which can be used in semiclassical arguments, we make a slight generalization of the microcanonical ensemble.

Instead of working with the constant energy surface $\Gamma(E)$ we assume that all $\mathbf{x}$ such that

$$E < H(\mathbf{x}) < E + \delta E \tag{2.7}$$

occur with equal probability in the ensemble. We discuss the tolerance δE in greater detail later in this section. Consider first a system of *distinguishable* particles, for example, a perfect crystal or a polymer chain where the particles can be identified by their positions in the lattice or in the chain. We define

$$\Omega(E) = \frac{1}{h^{3N}} \int_{E \leq H(\mathbf{x}) \leq E+\delta E} d^{6N}x\ . \tag{2.8}$$

Ω is proportional to the volume of phase space which satisfies (2.7). The entropy $S(E)$ is then defined as

$$S(E) = k_B \ln \Omega(E) \tag{2.9}$$

where k_B is Boltzmann's constant. The factor in front of the integral in (2.8) requires some comment. To make $\Omega(E)$ dimensionless a constant of dimension $[action]^{-3N}$ is required. The particular value chosen is arbitrary in a purely classical context. We have chosen the value h^{-3N} to make contact with the Bohr-Sommerfeld semiclassical theory. According to this theory the quantization rule for a single degree of freedom is $\oint pdq = rh$, where r is an integer. A volume of phase space $W = \int_W dpdq$ which is large enough that the discreteness of the levels is unimportant will then on the average contain W/h quantum states. It is intuitively straightforward[2] to make the generalization that a volume

$$\int_W dp_1 dp_2 ... dp_{3N} dq_1 dq_2 ... dq_{3N}$$

will contain on the average W/h^{3N} states.

When writing (2.8) we assumed that the system was composed of distinguishable particles. If we are dealing with a system of identical particles, we do not distinguish between states which differ only by the labeling of the particles occupying the various regions of phase space. In this case our measure of the number of available states in the ensemble becomes[3]

$$\Omega(E) = \frac{1}{h^{3N} N!} \int_{E \leq H(\mathbf{x}) \leq E + \delta E} d^{6N}x \ . \tag{2.10}$$

The factor of $N!$ corrects for the non-distinguishability of identical particles in the *classical* limit that the number of thermally possible states is much larger than the number of particles in the system. We then don't need to worry about the possibility of there being more than one particle in the same state (see problem 2.13). The entropy, S, is, as before, given by (2.9).

In a purely classical treatment the tolerance δE is, strictly speaking, not necessary. The energy levels of a large but finite quantum system, on the other hand, are discrete but closely spaced. To obtain a continuous function

---

[2]The result is obvious when the action-angle variables of an integrable system are used. Since phase space volumes are invariant under canonical transformations, the result will hold for any choice of generalized coordinates of an integrable system. In the case of nonintegrable systems (which are the ones we are mainly interested in since we are assuming the mixing hypothesis), the program of semiclassical quantization is remarkably difficult to carry out [34].

[3]Historically, the $N!$ in (2.10) was introduced by Gibbs to remove a spurious entropy of mixing of two volumes of the same ideal gas (see Problem 2.2). If we are dealing with a mixture consisting of N_1 particles of type 1, N_2 particles of type 2, and so on, we must replace $N!$ by $N_1!N_2!\ldots$.

$\Omega(E)$ in the quantum case, we have introduced this tolerance δE. However, we must require that δE plays no role in the *thermodynamic limit* $N \to \infty$. This requires that $E/N < \delta E \ll E$. We show this explicitly for a monatomic ideal gas of N particles in a volume V for which

$$H(\mathbf{x}) = \sum_{j=1}^{N} \frac{p_j^2}{2m} = E \,. \tag{2.11}$$

The equation $\sum_j p_j^2 = 2mE$ defines the surface of a $3N$-dimensional sphere of radius $\sqrt{2mE}$. If $\delta E \ll E$, we have

$$\Omega(E,V,N) = \frac{V^N}{h^{3N}N!} A_{3N}\left[\sqrt{2mE}\right] \frac{m\delta E}{\sqrt{2mE}} \tag{2.12}$$

where

$$A_n(r) = \frac{2\pi^{n/2} r^{n-1}}{\Gamma(n/2)} \tag{2.13}$$

is the surface area of an n-dimensional sphere of radius r and, for integer n, $\Gamma(n) = (n-1)!$. We obtain

$$\Omega(E,V,N) = \frac{V^N}{h^{3N}N!} \frac{(2m\pi E)^{3N/2}}{(\frac{3N}{2}-1)!} \frac{\delta E}{E} \,. \tag{2.14}$$

If $|\ln(\delta E/E)| \ll N$, we find, using Stirling's formula, $\ln N! \approx N \ln N - N$,

$$S(E,V,N) = Nk_B \ln \frac{V}{N} + \frac{3N}{2} k_B \ln \frac{4\pi m E}{3Nh^2} + \frac{5}{2} N k_B \,. \tag{2.15}$$

This expression is independent of δE.

We conclude that if δE is chosen as specified, the entropy is an extensive variable; that is, $S \propto N$ in the limit $N \to \infty$ if the number of particles per unit volume is kept constant and the energy is proportional to N. The expression (2.15) for the entropy of an ideal gas is referred to as the Sackur–Tetrode formula.

We are now in a position to make contact with thermodynamics. The differential of the entropy for a system of N particles in a volume V is (1.25)

$$dS(E,N,V) = \frac{dE}{T} + \frac{PdV}{T} - \frac{\mu dN}{T} \,. \tag{2.16}$$

Once we have calculated the microcanonical entropy, (2.16) provides the statistical definition of the temperature, pressure, and chemical potential:

$$T = \left(\frac{\partial S}{\partial E}\right)_{N,V}^{-1} \qquad \mu = -T\left(\frac{\partial S}{\partial N}\right)_{E,V} \qquad P = T\left(\frac{\partial S}{\partial V}\right)_{N,E} \,. \tag{2.17}$$

Returning to our example of the classical ideal gas, we find the expected results:

$$\frac{1}{T} = \frac{3Nk_B}{2E} \quad \text{or} \quad E = \frac{3}{2}Nk_BT \tag{2.18}$$

$$P = \frac{Nk_BT}{V} \quad \text{or} \quad PV = Nk_BT \tag{2.19}$$

$$\mu = k_BT \ln\left[\frac{N\lambda^3}{V}\right] \tag{2.20}$$

where

$$\lambda = \sqrt{\frac{h^2}{2\pi mk_BT}} \tag{2.21}$$

has dimension length and is called the *thermal wavelength.*

The variables T, P, and μ are independent of the size of the system in the limit $N \to \infty$ and are thus *intensive* (Section 1.1). The ideal gas is an example of a *normal* system. We will consider a system to be normal if in the thermodynamic limit, $N \to \infty$ with N/V = constant, the energy and entropy are extensive (*i.e.*, proportional to N), and T, P, and μ are intensive. Not all systems are normal in this sense. Examples of systems which are not are (i) a system that is self-bound by gravitational forces (see Problem 2.12), and (ii) a system with a net macroscopic charge. In both cases there are contributions to the energy which increase more rapidly than linearly with the number of particles (charges). Neutral systems are, however, normal [168] [172]. An important property of normal thermodynamic systems is the Gibbs–Duhem relation (see Section 1.4).

2.2 Systems at Fixed Temperature: Canonical Ensemble

We now consider two systems which are free to exchange energy but which are isolated from the rest of the universe by an ideal insulating surface. The particle numbers N_1, N_2 and volumes V_1, V_2 are fixed for each subsystem. The total energy will be constant under our assumptions and we assume further that the two subsystems are sufficiently weakly interacting that we can write

$$E_T = E_1 + E_2 \tag{2.22}$$

where E_1 and E_2 are the energies of the subsystems. We again assume a tolerance δE chosen suitably so that the statistical weights Ω, Ω_1, Ω_2 are proportional to δE. We then have

$$\Omega(E) = \int_{E<E_T<E+\delta E} \frac{dE_T}{\delta E} \int_{-\infty}^{E_T} \frac{dE_1}{\delta E} \Omega_2(E_T - E_1)\Omega_1(E_1) \ . \tag{2.23}$$

If the subsystems are sufficiently large, the product $\Omega_2(E_T - E_1)\Omega_1(E_1)$ will be a sharply peaked function of E_1. The reason for this is that Ω_1 and Ω_2 are rapidly increasing functions of E_1 and $E_T - E_1$ respectively.[4] From the definition (2.9) we note that the entropy is a monotonically increasing function of Ω and that the product $\Omega_1\Omega_2$ will be at a maximum when the total entropy

$$S(E, E_1) = S_1(E_1) + S_2(E - E_1) \tag{2.24}$$

is at a maximum. We can now make an argument, similar to Section 1.6, that the most likely value $\langle E_1 \rangle$ of E_1 is the one for which

$$\frac{\partial S_1}{\partial E_1} + \frac{\partial S_2}{\partial E_2}\frac{\partial E_2}{\partial E_1} = 0 \ . \tag{2.25}$$

Since $\partial E_2/\partial E_1 = -1$ we find, using (2.17), that

$$\frac{1}{T_1} - \frac{1}{T_2} = 0 \tag{2.26}$$

or $T_1 = T_2 = T$. The most probable partition of energy between the two systems is the one for which the two temperatures are the same. This is the basis of the zeroth law of thermodynamics, which we see follows naturally from the ensemble concept.

In the preceding section we gave a statistical mechanical definition of the temperature, pressure, and chemical potential and we showed that these definitions produced the correct thermodynamic results in the special case of a classical ideal gas. We can now appeal to the zeroth law of thermodynamics and imagine that an arbitrary system is in thermal contact with an ideal gas reservoir. If the system is sufficiently large, it is overwhelmingly probable that the partition of energy between the system and the reservoir will be such as to leave the temperatures effectively the same, and we conclude that our definition of temperature is in agreement with thermodynamics. In Section 2.3 we make a similar argument with regard to the chemical potential, and it is also

[4]The reader is encouraged to verify this statement explicitly for the case of two ideal gases in contact using (2.14).

easy to show that the pressure must be the same in two subsystems that are free to adjust their volumes.

To establish the *canonical ensemble*, we again consider two systems in thermal contact in such a way that the volume and particle number in each subsystem are held fixed. We now assume that subsystem 2 is very much larger than subsystem 1. The probability $p(E_1)dE_1$ that subsystem 1 has energy between E_1 and $E_1 + dE_1$ is[5]

$$p_C(E_1)dE_1 = \frac{\Omega_1(E_1)\Omega_2(E - E_1)dE_1}{\int dE_1 \Omega_1(E_1)\Omega_2(E - E_1)} . \tag{2.27}$$

We have

$$\Omega_2(E - E_1) = \exp\left\{\frac{S_2(E - E_1)}{k_B}\right\} . \tag{2.28}$$

Since $E_1 \ll E$ we may expand S_2 in a Taylor series:

$$S_2(E - E_1) = S_2(E) - E_1\frac{\partial S_2}{\partial E} + \frac{1}{2}E_1^2\frac{\partial^2 S_2}{\partial E^2} + \cdots . \tag{2.29}$$

The temperature of the large system is T and we have

$$\frac{\partial S_2}{\partial E} = \frac{1}{T}$$

$$\frac{\partial^2 S_2}{\partial E^2} = \frac{\partial(1/T)}{\partial E} = -\frac{1}{T^2}\left(\frac{\partial T}{\partial E}\right)_{V_2,N_2} = -\frac{1}{T^2 C_2} \tag{2.30}$$

where C_2 is the heat capacity of system 2 at constant V and N. Since the second system is very much larger than the first, we have $E_1 \ll C_2 T$ and

$$\Omega_2(E - E_1) = \text{const.} \exp\left\{\frac{-E_1}{k_B T}\right\} . \tag{2.31}$$

With the notation $\beta = 1/(k_B T)$ we thus find that (2.27) can be rewritten

$$p_C(E_1) = \frac{1}{\delta E Z_C}\Omega_1(E_1)\exp\{-\beta E_1\} . \tag{2.32}$$

The probability density $p_C(E_1)$ is called the *canonical distribution*, and the normalizing term

$$Z_C = \int \frac{dE_1}{\delta E}\Omega_1(E_1)\exp\{-\beta E_1\} \tag{2.33}$$

---

[5] It is now understood that dE_1 is much larger than the tolerance δE.

is the *canonical partition function.* The canonical distribution is often a more convenient tool than the microcanonical distribution. In the microcanonical ensemble we were dealing with states with a specified energy $E(S,V,N)$ and T, P, and μ were derived quantities. In the canonical ensemble the system is kept at fixed temperature. We have already seen in Section 1.3 that the change of independent variable from S to T is achieved by replacing the energy E as dependent variable by the Helmholtz free energy A:

$$A = E - TS\ . \tag{2.34}$$

Using (2.16), we find

$$\begin{aligned} dA &= dE - TdS - SdT \\ &= \mu dN - PdV - SdT\ . \end{aligned} \tag{2.35}$$

The statistical mechanical definition of the free energy is

$$A = -k_B T \ln Z_C\ . \tag{2.36}$$

We will now show that the two definitions (2.34) and (2.36) agree in the sense that for a large system the free energy defined by (2.36) is given by

$$A = \langle E \rangle - T\langle S \rangle \tag{2.37}$$

where $\langle E \rangle$ and $\langle S \rangle$ are the most probable values of E and S in the canonical ensemble. To see this, let us rewrite the partition function

$$\begin{aligned} Z_C &= \int \frac{dE}{\delta E}\Omega(E,V,N)\exp\left\{\frac{-E}{k_B T}\right\} \\ &= \int \frac{dE}{\delta E}\exp\left\{-\frac{1}{k_B}\left[\frac{E}{T} - S(E,V,N)\right]\right\}\ . \end{aligned} \tag{2.38}$$

If the system is large, it is overwhelmingly probable that the energy will be close to the most likely value $\langle E \rangle$ given by

$$-\frac{1}{k_B}\left[\frac{\langle E \rangle}{T} - S(\langle E \rangle, V, N)\right] = \text{maximum}\ . \tag{2.39}$$

We expand the exponent around its maximum value at $\langle E \rangle$:

$$-\frac{1}{k_B}\left(\frac{E}{T} - S\right) = -\frac{1}{k_B}\left(\frac{\langle E \rangle}{T} - \langle S \rangle\right) + \frac{1}{2k_B}\left(E - \langle E \rangle\right)^2 \left.\frac{\partial^2 S}{\partial E^2}\right|_{E=\langle E \rangle} + \cdots\ . \tag{2.40}$$

Using (2.33), we find that

$$\begin{aligned} Z_C &\approx \exp\{-\beta(\langle E\rangle - T\langle S\rangle)\}\int \frac{dE}{\delta E}\exp\left\{-\frac{(E-\langle E\rangle)^2}{2Ck_BT^2}\right\} \\ &\approx \frac{\sqrt{2\pi k_BT^2C}}{\delta E}\exp\{-\beta(\langle E\rangle - T\langle S\rangle)\} \end{aligned}$$

or

$$-k_BT\ln Z_C = \langle E\rangle - T\langle S\rangle - k_BT\ln\frac{\sqrt{2\pi k_BT^2C}}{\delta E}\ . \tag{2.41}$$

In the limit that the system is very large, we can make sure the logarithmic term is small compared to the other terms, by choosing $\delta E \propto \sqrt{\langle E\rangle}$. This choice would make the logarithmic term of order unity while both $\langle E\rangle$ and $\langle S\rangle$ are extensive. We thus have the desired result.

We have argued that in the canonical ensemble the energy and entropy fluctuate about their mean values. Instead of using the condition (2.39), we can calculate the mean energy by taking an ensemble average over all possible values of the energy:

$$\begin{aligned} \langle E\rangle &= \frac{\int dEE\,\Omega(E)\exp\{-\beta E\}}{\int dE\,\Omega(E)\exp\{-\beta E\}} \\ &= -\frac{\partial\ln Z_C}{\partial\beta} = \frac{\partial(\beta A)}{\partial\beta}\ . \end{aligned}\tag{2.42}$$

The mean-square fluctuation in the energy is given by

$$\begin{aligned} \langle(E-\langle E\rangle)^2\rangle &= \langle E^2\rangle - \langle E\rangle^2 \\ &= -\frac{\partial\langle E\rangle}{\partial\beta} = k_BT^2\frac{\partial\langle E\rangle}{\partial T} = k_BT^2C_{V,N} \end{aligned}\tag{2.43}$$

where $C_{V,N}$ is the heat capacity at constant N and V. The heat capacity is an extensive variable, as is the energy, and the root-mean-square (rms) fluctuation in the energy will thus be proportional to $\sqrt{\langle E\rangle}$. The mean fluctuation is therefore large for a large system, but is a vanishingly small fraction of the total energy in the thermodynamic limit:

$$\frac{\sqrt{\langle(E-\langle E\rangle)^2\rangle}}{\langle E\rangle} \sim \frac{1}{\sqrt{N}}\ .$$

The relationship (2.43) between the response function $C_{V,N}$ and the mean-square fluctuation of the energy is a special case of a very general result known as the *fluctuation-dissipation theorem* (see Chapter 12). We shall encounter a number of such relations.

2.3 Grand Canonical Ensemble

In a number of instances it is not convenient or possible to fix the number of particles in a system. An example is the case of chemical equilibrium between a number of different species. As an external parameter, such as the temperature, is varied, the concentration of the various constituents will change and we must formulate the statistical treatment in terms of a partition function which allows concentrations to adjust. Moreover, we shall see in Section 2.4 that in quantum statistical mechanics a description in terms of a variable number of particles is usually more practical.

We proceed, as in Sections 2.2 and 1.6, by considering two systems in contact. They are free to exchange energy and particles, but the total energy and particle number are held fixed, as are the volumes of the two subsystems. We first derive an equilibrium condition and then let one system be extremely large compared to the other to obtain the probability density in phase space for the smaller system.

Let systems 1 and 2 have energies E_1, E_2 and particle numbers N_1, N_2 with

$$\begin{aligned} E_1 + E_2 &= E_T \\ N_1 + N_2 &= N_T \ . \end{aligned} \tag{2.44}$$

The microcanonical partition function for the composite system is given by

$$\Omega(E, N_T) = \sum_{N_1=0}^{N_T} \int_E^{E+\delta E} \frac{dE_T}{\delta E} \int_{-\infty}^{E_T} \frac{dE_1}{\delta E} \Omega_1(E_1, N_1)\Omega_2(E_T - E_1, N_T - N_1) \ . \tag{2.45}$$

The product $\Omega_1(E_1, N_1)\Omega_2(E_T - E_1, N_T - N_1)$ will be sharply peaked near the values of E_1 and N_1 that maximize it. This occurs for values of E_1 and N_1 near the ones that maximize the total entropy

$$S_1(E_1, N_1) + S_2(E - E_1, N_T - N_1) \ . \tag{2.46}$$

Differentiation using (2.17) yields

$$\begin{aligned} \frac{1}{T_1} &= \frac{1}{T_2} \\ \frac{\mu_1}{T_1} &= \frac{\mu_2}{T_2} \ . \end{aligned} \tag{2.47}$$

In addition to the previous condition, that at equilibrium two systems in thermal contact have the same temperature, we now have a second condition, namely that the chemical potentials μ_1, μ_2 of the two systems must be equal.

We can construct the *grand canonical ensemble* and partition function by using arguments similar to those of the preceding section. We let subsystem 2 be very much larger than subsystem 1. This allows us to expand

$$
\begin{aligned}
\Omega_2(E-E_1,N_T-N_1) &= \exp\left\{\frac{1}{k_B}S(E-E_1,N_T-N_1)\right\} \\
&= \Omega_2(E,N_T)\exp\left\{-\frac{1}{k_B}\left(E_1\frac{\partial S}{\partial E}+N_1\frac{\partial S}{\partial N}\right)\right\} \\
&= \text{const.}\exp\{-\beta(E_1-\mu N_1)\}\ . \qquad (2.48)
\end{aligned}
$$

We can thus write for the grand canonical probability density,

$$p_G(E_1,N_1)=\frac{\Omega_1(E_1,N)}{\delta E Z_G}\exp\{-\beta(E_1-\mu N_1)\} \qquad (2.49)$$

where the normalizing factor, the *grand partition function*, is

$$Z_G(\mu,T,V)=\sum_{N_1=0}^{\infty}\exp\{\beta\mu N_1\}Z_C(T,N_1)\ . \qquad (2.50)$$

In (2.50) we have, for convenience, taken the upper limit of the summation to be ∞ rather than N_T because the summand becomes negligibly small for values of N_1 comparable to or larger than N_T.

In the grand canonical ensemble the system is kept at constant volume, temperature, and chemical potential in distinction from the canonical ensemble, where V, T, N are kept fixed. In thermodynamics the change from N to μ as an independent variable is accomplished by a Legendre transformation in which the dependent variable A, the Helmholtz free energy, is replaced by the grand potential, $\Omega_G=A-\mu N$. For a normal thermodynamic system we have, from the Gibbs–Duhem relation (1.39),

$$\Omega_G=-PV\ . \qquad (2.51)$$

Using (2.35), we obtain

$$d\Omega_G=-SdT-PdV-Nd\mu \qquad (2.52)$$

and

$$N=-\left(\frac{\partial\Omega_G}{\partial\mu}\right)_{V,T}\qquad S=-\left(\frac{\partial\Omega_G}{\partial T}\right)_{V,\mu}\qquad P=-\left(\frac{\partial\Omega_G}{\partial V}\right)_{\mu,T}\ . \qquad (2.53)$$

In analogy with the treatment of the preceding section, we postulate the statistical mechanical definition of the grand potential:

$$\Omega_G = -k_B T \ln Z_G \ . \tag{2.54}$$

It is a straightforward matter to show that the definitions (2.51) and (2.54) are equivalent in the sense that for a large system,

$$k_B T \ln Z_G = \mu \langle N \rangle - \langle A \rangle \ . \tag{2.55}$$

The proof is left as an exercise for the reader (Problem 2.5). In the grand canonical ensemble the number of particles fluctuates about the mean particle number

$$\langle N \rangle = \frac{1}{Z_G} \sum_N N \exp\{\beta \mu N\} Z_C(N) = k_B T \frac{\partial}{\partial \mu} \ln Z_G \ . \tag{2.56}$$

A convenient measure of the expected magnitude of the fluctuations is

$$\begin{aligned} (\Delta N)^2 &= \langle (N - \langle N \rangle)^2 \rangle = \langle N^2 \rangle - \langle N \rangle^2 \\ &= k_B T \frac{\partial \langle N \rangle}{\partial \mu} \ . \end{aligned} \tag{2.57}$$

It is clear from this that since $\partial \langle N \rangle / \partial \mu \sim \langle N \rangle$, we must have

$$\frac{\Delta N}{\langle N \rangle} \approx \frac{1}{\sqrt{\langle N \rangle}} \ . \tag{2.58}$$

A useful equation is obtained if we rewrite (2.58) in terms of the isothermal compressibility

$$K_T = -\frac{1}{V} \left(\frac{\partial V}{\partial P} \right)_{N,T} \ . \tag{2.59}$$

To do this we note that from the Gibbs–Duhem equation,

$$d\Omega_G = -SdT - PdV - Nd\mu = -PdV - VdP \tag{2.60}$$

and have

$$d\mu = \frac{V}{N} dP - \frac{S}{N} dT \ . \tag{2.61}$$

Let $v = V/N$ be the specific volume. Since μ is intensive we can express it as $\mu(v, T)$ and obtain

$$\left(\frac{\partial \mu}{\partial v} \right)_T = v \left(\frac{\partial P}{\partial v} \right)_T \ . \tag{2.62}$$

We can change v by changing either V or N:

$$\begin{aligned}\left(\frac{\partial}{\partial v}\right)_{V,T} &= \left(\frac{\partial N}{\partial v}\right)_{V,T}\left(\frac{\partial}{\partial N}\right)_{V,T} = -\frac{N^2}{V}\left(\frac{\partial}{\partial N}\right)_{V,T} \\ \left(\frac{\partial}{\partial v}\right)_{N,T} &= \left(\frac{\partial V}{\partial v}\right)_{N,T}\left(\frac{\partial}{\partial V}\right)_{N,T} = N\left(\frac{\partial}{\partial V}\right)_{N,T} .\end{aligned} \tag{2.63}$$

However, the way in which v is changed cannot affect (2.62). Therefore,

$$-\frac{N^2}{V}\left(\frac{\partial \mu}{\partial N}\right)_{V,T} = V\left(\frac{\partial P}{\partial V}\right)_{N,T} . \tag{2.64}$$

Substitution of this result into (2.57) and (2.59) finally yields

$$\frac{(\Delta N)^2}{\langle N\rangle} = \frac{k_B T K_T}{v} . \tag{2.65}$$

Equation (2.65) illustrates that the mean-square fluctuation in a thermodynamic variable is proportional to a response function. In the case of the energy fluctuations (2.43) the relevant response function is the heat capacity. In the present case, where the thermodynamic variable is the number of particles in a fixed subvolume of an open system, the appropriate response function is the compressibility.

2.4 Quantum Statistics

The simplest consequence of a quantum-mechanical treatment is that the energy levels are discrete. A second consequence originates in the symmetry requirements on the many particle wave functions imposed by the Pauli principle for fermions and by the corresponding condition for bosons. As we shall see, this leads to combinatorical factors which are different from the simple $N!$ with which we avoided the Gibbs paradox in classical statistics.

The modifications of our statistical formulation due to the discreteness of the energy spectrum are rather simple when the canonical and grand canonical ensembles are used. If E_γ is the energy of the γ'th quantum state and its degeneracy is g_γ we write the canonical partition function as

$$Z_c = \sum_\gamma g_\gamma \exp\{-\beta E_\gamma\} \tag{2.66}$$

where the sum extends over all energy levels of the system. The correct combinatorical factors for particles obeying Bose–Einstein or Fermi–Dirac statistics are most easily obtained if we consider our system to be quantum states in contact with a heat bath. For a noninteracting many-fermion system the solutions of the Schrödinger equation, consistent with the requirements of quantum statistics, are Slater determinants of single-particle wave functions. In the case of bosons they are symmetrized linear combinations of product states (see the Appendix). A unique many-particle state can thus be completely specified through the occupation numbers of the single-particle basis states. In the canonical ensemble the sum of these occupation numbers must be equal to the number of particles, N, whereas in the grand canonical ensemble we do not have this restriction and we may freely sum over all allowed values of the occupation number for each single-particle state. We will illustrate these principles by a number of examples (see also Problems 2.7 and 2.8).

2.4.1 Harmonic oscillator

The energy levels of a single oscillator are $(n+1/2)h\nu$, where $n = 0, 1, 2, \ldots$ and the states are nondegenerate. The partition function is

$$Z = \sum_{n=0}^{\infty} e^{-\beta(n+1/2)h\nu} = \frac{e^{-\beta h\nu/2}}{1 - e^{-\beta h\nu}} . \tag{2.67}$$

The ensemble for this partition function can be thought of as grand canonical in the sense that the number of quanta is variable, or canonical in the sense that there is only one oscillator. The average energy is

$$\begin{aligned} \langle E \rangle &= \frac{1}{Z}\sum_{n=0}^{\infty}\left(n+\frac{1}{2}\right) h\nu e^{-\beta h\nu(n+1/2)} \\ &= -\frac{\partial}{\partial\beta}\ln Z = \frac{1}{2}h\nu + \frac{h\nu}{e^{\beta h\nu}-1} . \end{aligned} \tag{2.68}$$

2.4.2 Noninteracting fermions

We suppose that the single-particle states are labeled by a wave vector $\mathbf{k}$ and a spin index σ. The Pauli principle requires the occupation number of each such state to be 0 or 1 and the contribution of this state to the grand partition function is

$$\sum_{n=0}^{1} \exp\{-n\beta(E_{\mathbf{k},\sigma} - \mu)\} = 1 + \exp\{-\beta(E_{\mathbf{k},\sigma} - \mu)\} \tag{2.69}$$

where the energy $E_{\mathbf{k},\sigma} = \hbar^2k^2/2m$ in the case of free particles. The grand partition function is the product of such terms,

$$Z_G = \prod_{\mathbf{k},\sigma}[1 + \exp\{-\beta(E_{\mathbf{k},\sigma} - \mu)\}] \tag{2.70}$$

and the mean number of particles $\langle N\rangle$ is (2.56)

$$\langle N\rangle = \sum_{\mathbf{k},\sigma} \frac{1}{\exp\{\beta(E_{\mathbf{k},\sigma} - \mu)\} + 1} \,. \tag{2.71}$$

Thus the term

$$\langle n_{\mathbf{k},\sigma}\rangle \equiv \frac{1}{\exp\{\beta(E_{\mathbf{k},\sigma} - \mu)\} + 1} \tag{2.72}$$

is the mean occupation number of the state $(\mathbf{k}, \sigma)$ or, equivalently, the probability that it is occupied, while the probability that it is unoccupied is

$$1 - \langle n_{\mathbf{k},\sigma}\rangle = \frac{1}{1 + \exp\{-\beta(E_{\mathbf{k},\sigma} - \mu)\}} \,. \tag{2.73}$$

We note that as $T \to 0$ at fixed density $\langle N\rangle/V$ the chemical potential μ must approach a finite positive limit, the *Fermi energy* ϵ_F. In this limit the occupation number (2.72) is 1 for energies less than ϵ_F and 0 otherwise. Conversely, from (2.73) we see that if $e^{-\beta\mu} \gg 1$, the probability of occupation of any state is small (note that $E_{\mathbf{k},\sigma}$ is positive). In this case the system of fermions is nondegenerate and approximately satisfies Boltzmann statistics:

$$\langle n_{\mathbf{k},\sigma}\rangle \approx \exp\{-\beta(E_{\mathbf{k},\sigma} - \mu)\} \,. \tag{2.74}$$

From (2.20) we see that equation (2.74) will be valid for all $\mathbf{k}$ if the thermal wavelength (2.21) λ satisfies $\lambda^3 \ll V/N$. Conversely, if $\lambda^3 \gg V/N$ the system will be degenerate.

2.4.3 Noninteracting bosons

In the case of bosons (integer spin, which we take to be zero) a single particle state can have an arbitrary occupation number. The contribution of the state labeled by $\mathbf{k}$ (energy $\hbar^2k^2/2m$) to the grand partition function is

$$\sum_{n=0}^{\infty} \exp\{-n\beta(E_{\mathbf{k}} - \mu)\} = \frac{1}{1 - \exp\{-\beta(E_{\mathbf{k}} - \mu)\}} \tag{2.75}$$

and, as in the case of fermions, the grand partition function is the product of such terms:

$$Z_G = \prod_{\mathbf{k}} [1 - \exp\{-\beta(E_{\mathbf{k}} - \mu)\}]^{-1} \ . \tag{2.76}$$

The average number of particles occupying the state **k** is

$$\langle n_{\mathbf{k}} \rangle = -k_B T \frac{\partial}{\partial \mu} \ln[1 - \exp\{-\beta(E_{\mathbf{k}} - \mu)\}] = \frac{1}{\exp\{\beta(E_{\mathbf{k}} - \mu)\} - 1} \ . \tag{2.77}$$

A Bose system, like a Fermi system, obeys Boltzmann statistics to a good approximation if the thermal wavelength (2.21) is small compared to typical interparticle separation $(V/N)^{1/3}$. The statistical mechanics of a noninteracting Bose system is discussed in more detail in Section 11.1.

2.4.4 Density matrix

We conclude this section by introducing the density matrix. Let us first assume that we know all the quantum states $\mid n\rangle$ of a given system. These states are eigenstates of the Hamiltonian

$$H \mid n\rangle = E_n \mid n\rangle \tag{2.78}$$

and we assume that they are *orthonormal*:

$$\langle n \mid n'\rangle = \delta_{n,n'} \ . \tag{2.79}$$

We also assume that these states are *complete*, that is, all possible states are expressible as a linear combination of this basis set. If $\mid a\rangle$ is an arbitrary state, we can thus write

$$\mid a\rangle = \sum_n a_n \mid n\rangle \tag{2.80}$$

where, from (2.79), we have

$$a_n = \langle n \mid a\rangle \tag{2.81}$$

and

$$\mid a\rangle = \sum_n \mid n\rangle\langle n \mid a\rangle \ . \tag{2.82}$$

Formally, this allows us to express the completeness requirement as

$$\tilde{1} = \sum_n \mid n\rangle\langle n \mid \tag{2.83}$$

where $\tilde{1}$ is the unit operator. In the canonical ensemble the probability that the system is in state $| n\rangle$ is given by

$$p_n = \frac{\exp\{-\beta E_n\}}{\sum_n \exp\{-\beta E_n\}} \tag{2.84}$$

and the thermal average of an arbitrary operator $\tilde{A}$ is

$$\langle \tilde{A} \rangle = \sum_n \langle n \mid \tilde{A} \mid n \rangle \frac{\exp\{-\beta E_n\}}{\sum_n \exp\{-\beta E_n\}} = \frac{1}{Z_C} \sum_n \langle n \mid \tilde{A} \mid n \rangle \exp\{-\beta E_n\} \, . \tag{2.85}$$

The operator

$$\tilde{\rho} = \frac{1}{Z_C} \sum_n | n \rangle \langle n | \exp\{-\beta E_n\} \tag{2.86}$$

is commonly referred to as the *density operator* and its matrix representation as the *density matrix*. This formalism can easily be generalized to the grand canonical ensemble. We define the grand Hamiltonian ($\tilde{N}$ is the number operator):

$$K = H - \mu \tilde{N} \, . \tag{2.87}$$

A complete set of states then contains states with any number of particles. If $\{| n\rangle\}$ is a complete set of eigenstates of K with

$$K \mid n \rangle = K_n \mid n \rangle \tag{2.88}$$

we may define the grand canonical density operator to be

$$\tilde{\rho} = \frac{1}{Z_G} \sum_n | n \rangle \langle n | \exp\{-\beta K_n\} \, . \tag{2.89}$$

Frequently, the complete set of eigenstates of the Hamiltonian is not known, but it may be possible to find a set of states $| v\rangle$ which are not eigenstates of the Hamiltonian, but are complete. The density matrix is in general not diagonal in this representation:

$$\rho_{v,v'} = \langle v \mid \tilde{\rho} \mid v' \rangle = \frac{1}{Z_G} \langle v \mid e^{-\beta K} \mid v' \rangle \, . \tag{2.90}$$

We see that the density matrix can be written, in *any* representation, as

$$\tilde{\rho} = \frac{1}{Z_G} e^{-\beta K} \, . \tag{2.91}$$

It is easy to see that the thermal average of an operator $\tilde{A}$ can be expressed, in an arbitrary basis, as

$$\begin{aligned} \langle A \rangle &= \frac{1}{Z_G} \sum_n \langle n \mid \tilde{A} \mid n \rangle \exp\{-\beta K_n\} \\ &= \frac{1}{Z_G} \sum_{v,v'} \langle v \mid \tilde{A} \mid v' \rangle \langle v' \mid e^{-\beta K} \mid v \rangle \\ &= \frac{1}{Z_G} \mathrm{Tr} \tilde{A} e^{-\beta K} = \mathrm{Tr}\, \tilde{A} \tilde{\rho} \, . \end{aligned} \tag{2.92}$$

In this notation, the partition function is given by

$$Z_G = \mathrm{Tr}\, e^{-\beta K} \tag{2.93}$$

and the normalization condition on the density matrix is

$$\mathrm{Tr}\, \tilde{\rho} = 1 \, . \tag{2.94}$$

2.5 Maximum Entropy Principle

The link provided by the ergodic hypothesis between the time evolution of a dynamical system and the ensemble concept is a weak one. It is therefore of interest to show that it is possible to formulate the foundation of equilibrium statistical mechanics in terms of information theory [142], in which the thermodynamic variables are inferred from a least-possible-bias estimate on the basis of available information. In this way the link with dynamics is severed altogether and the approach leads to a maximum entropy principle that yields results equivalent to the ones found earlier in this chapter.

Consider first the situation in which a state variable x is capable of attaining any one of Ω discrete values $x_i \, (i = 1, 2, \ldots, \Omega)$. Let p_i be the associated probability distribution. This distribution must be normalized:

$$\sum_{i=1}^{\Omega} p_i = 1 \tag{2.95}$$

and there may be a number of other constraints of the form

$$f_{avg} = \langle f(x) \rangle = \sum_{i=1}^{\Omega} p_i f(x_i) \, . \tag{2.96}$$

For example, the energy, pressure, or concentrations of different constituents may have known mean values established by the appropriate contact with the surroundings. In general, nothing else is known about the p_j's. In the case of the microcanonical ensemble the only constraints were that the total energy should lie within a tolerance δE of a prescribed value and that the particle number N and volume V be fixed. We then have no reason to prefer one allowed state over another and have no other choice but to assume that all states are equally likely or $p_i = 1/\Omega$. Indeed, it is intuitively obvious that any other choice would represent a bias, or additional information about the system.

From the above it is clear that we require an unambiguous measure of "the amount of uncertainty" associated with the probability distribution for the accessible states. Imagine that we prepare the system N times and measure its state each time. Since this is theory, we can assume that N is enormously large. By the law of large numbers the i-th state will come up $N_i \approx Np_i$ times. A possible measure of the uncertainty would then be the number of different sequences of outcomes that are compatible with the probability distribution

$$Z_N = \frac{N!}{\prod_{i=1}^{\Omega}(Np_i!)} \quad . \tag{2.97}$$

This is not the only possible measure of uncertainty; any monotonic function $S(Z_N)$ may serve. Consider next a system made up of two independent subsystems. We imagine that we measure the state of subsystem 1 $N(1)$ times and subsystem 2 $N(2)$ times. The number of possible sequences is now $Z = Z(1)Z(2)$. We next limit the choice of possible functions S by requiring our uncertainty to be the sum of the uncertainties associated with independent subsystems, *i.e.* we require that S is an *extensive variable*:

$$S(Z(1)Z(2)) = S(Z(1)) + S(Z(2)) \quad . \tag{2.98}$$

The choice of the function S is now essentially unique. To see this let us differentiate (2.98) two different ways

$$Z(1)\frac{dS(Z(1)Z(2))}{d(Z(1))} = Z(1)\frac{dS(Z(1))}{dZ(1)} = Z(1)Z(2)\frac{dS(Z(1)Z(2))}{d(Z(1)Z(2))}$$

and, if differentiating instead with respect to $Z(2)$ we obtain

$$Z(1)\frac{dS(Z(1))}{dZ(1)} = Z(2)\frac{dS(Z(2))}{dZ(2)} \quad . \tag{2.99}$$

Since the left side of (2.99) is independent of $Z(2)$, and the right hand side is independent of $Z(1)$ they must both be equal to a constant. We call this constant K. We now drop the subscripts and find

$$Z\frac{dS(Z)}{dZ} = K \ .$$

We integrate this equation to obtain

$$S(Z) = K \ln Z + c_1$$

where c_1 is another constant. If there is no uncertainty, *i.e.*, if we know the outcome then $Z = 1$ and we require that $S(1) = 0$. This gives $c_1 = 0$ and we are left with

$$S(Z) = K \ln Z \ .$$

The choice of the constant K is arbitrary at this stage, but, since we will identify S with the entropy, we choose K to be equal to the Boltzmann constant k_B.

The entropy that we have defined is the one associated with N repeated experiments. Since S is extensive we now define the "uncertainty" associated with the probability distribution itself to be $1/N$ times the uncertainty associated with N repetitions of the experiment. This gives for the "uncertainty" of the probability distribution

$$S = k_B \lim_{N\to\infty} \frac{1}{N} \ln \frac{N!}{\prod_{i=1}^{\Omega}(NP_i)!} \ .$$

Since we only need to evaluate the logarithm of the factorials to accuracy N we can use the simple version of Stirling's formula

$$\ln N! \approx N \ln N - N$$

and we find after some algebra

$$S = -k_B \sum_{i=1}^{\Omega} p_i \ln p_i \ . \tag{2.100}$$

Equation (2.100) was derived by Shannon [271] shortly after the second world war and is generally considered to be one of the fundamental results of information theory. Note the key rôle of the requirement of extensivity in establishing the uniqueness of the Boltzmann-Shannon entropy. As pointed out at the end of section 2.1 not all system are "normal" with an extensive energy. It is therefore natural that attempts have been made to replace the Boltzmann-Shannon entropy with a nonextensive measure. The most significant such attempt is that of Costantino Tsallis [305] and there is now a very large literature on the subject [6].

It is easy to generalize the Boltzmann-Shannon definition to a continuous probability distribution $\rho(x)$:

$$\int_{E<H(\mathbf{x})<E+\delta E} d\mathbf{x}\rho(\mathbf{x}) = 1 \,. \tag{2.101}$$

Consider the case of N identical particles in a 6N-dimensional phase space. We have seen that according to the semiclassical rules there are

$$\frac{d^{6N}x}{h^{3N}N!} \tag{2.102}$$

states in the phase space volume $d^{6N}x$. Therefore, we find for the entropy

$$S = -k_B \int_{E<H(\mathbf{x})<E+\delta E} d^{6N}x\rho(\mathbf{x}) \ln[h^{3N}N!\rho(\mathbf{x})] \,. \tag{2.103}$$

It is easy to see that if we let

$$\rho(\mathbf{x}) = \frac{1}{h^{3N}N!\Omega(E)} \tag{2.104}$$

where $\Omega(E)$ is given by (2.10), we get back the expression (2.9) for the entropy of the microcanonical ensemble. To see that the probability density (2.104) maximizes the expression (2.103) subject to the constraint (2.101), we use the method of Lagrange multipliers. Requiring that the functional derivative

$$\frac{\delta}{\delta\rho(x)}\left\{-k_B \int d^{6N}x[\rho\ln(h^{3N}N!\rho) - \lambda\rho]\right\}$$

should vanish yields

$$\rho = \frac{e^{\lambda-1}}{h^{3N}N!} \,.$$

The Lagrange multiplier λ is determined by the normalization condition (2.101) from which (2.104) follows. The extension to the case in which there are one or

[6] A fairly complete list of references can be found at http://tsallis.cat.cbpf.br/biblio.htm

more constraints of the form (2.96) is straightforward. If we wish to constrain the expectation value of the function $f(x)$, we maximize the expression

$$-k_B \sum_i [p_i \ln p_i - \lambda p_i + \beta p_i f(x_i)]$$

where β and λ are Lagrange multipliers. This procedure yields

$$p_i = \exp\{\lambda - 1 - \beta f(x_i)\} \ . \tag{2.105}$$

With the definition

$$Z(\beta) = \sum_i \exp\{-\beta f(x_i)\}$$

we find from the normalizing constraint (2.95),

$$Z(\beta) = e^{1-\lambda} \ .$$

The remaining Lagrange multiplier β is determined from the condition

$$\langle f(x) \rangle = \sum_i p_i f(x_i) = -\frac{\partial}{\partial \beta} \ln Z(\beta) \ .$$

If we let $f(x_i)$ be the energy of the state x_i, we see that we recover the canonical distribution with the identification

$$\beta = \frac{1}{k_B T} \quad \text{and} \quad \lambda = \beta A + 1$$

where A is the Helmholtz free energy. It is easy to see that the second variation of S is negative, indicating that we have found a true maximum of the entropy subject to the constraint. The extension of the method to the grand canonical ensemble is left as an exercise (Problem 2.9).

It should be noted that while the information-theoretic approach and the approach taken in Sections 2.1 to 2.3 lead to similar results, there is a subtle difference in the concept of entropy. In Section 2.1 we defined the entropy for the microcanonical ensemble. The definition was unambiguous in the thermodynamic limit but depended on the tolerance δE for a small system. When we proceeded to the other ensembles in which the energy and particle number could fluctuate, the entropy too became a fluctuating state variable.

In the information-theoretic context, the entropy is a function of the probability distribution in the ensemble and is not fluctuating since it has nothing to do with the state in which the system happens to be. On the other hand, the definition is perfectly unambiguous for systems of any size. Also, there is no restriction to equilibrium situations: The probabilities p_i will be time-dependent if the system evolves dynamically.

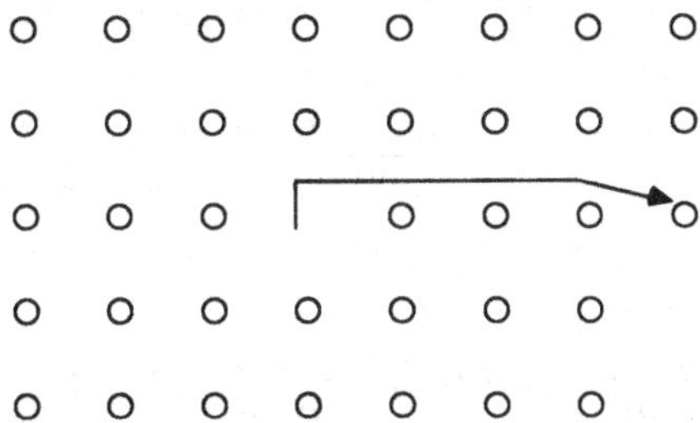

Figure 2.1: An atom migrates to the surface of a crystal, leaving behind a Schottky defect.

2.6 Thermodynamic Variational Principles

We saw in Section 1.3 that the thermodynamic equilibrium state of a system at fixed temperature and volume is the one that minimizes the Helmholtz free energy. The same result followed from maximum-likelihood arguments in Section 2.2 and from the information-theoretic argument of Section 2.5. Similarly, at equilibrium at constant pressure and temperature the Gibbs free energy is at a minimum. These results can be used in statistical calculations. A common approach is to find an approximate expression for the appropriate free energy in terms of certain parameters, and then to determine the equilibrium values of these parameters by minimizing the free energy. This method will be used frequently in our treatment of phase transitions in Chapter 3. At this point we present a simple example that makes use of this method.

2.6.1 Schottky defects in a crystal

Consider a solid in which the atoms are located on a regular lattice. In equilibrium, at nonzero temperature, the crystal will contain a certain number of Schottky defects or vacancies. These can be thought of as atoms which have migrated from interior (bulk) positions to the surface (see Figure 2.1). The problem is to find the equilibrium concentration of vacancies at a given temperature T and pressure P. Let N be the total number of atoms, n the number of vacancies, v_c the volume per unit cell, S_c the configurational entropy of the vacancies, ΔA the change in the free energy of vibration of the crystal due to the presence of a single vacancy, and ϵ the energy cost of creating a vacancy by transferring an atom to the surface. The configurational entropy can be obtained by noting that the number of sites is $N + n$. The number of ways of placing n vacancies on these sites is

$$\frac{(N+n)!}{N!n!}$$

giving

$$S_c = k_B \ln \frac{(N+n)!}{N!n!} \approx Nk_B \ln \frac{N+n}{N} + nk_B \ln \frac{N+n}{n} \,. \tag{2.106}$$

Since the vacancies are in equilibrium at constant pressure, the correct free energy to use is the Gibbs free energy $G = E - TS + PV$. The part of the free energy that depends on n is

$$G_v = n(\epsilon + \Delta A + Pv_c) - TS_c \,.$$

The requirement that G_v be a minimum yields

$$\epsilon + \Delta A + Pv_c = T\frac{\partial S_c}{\partial n} = k_B T \ln \frac{N+n}{n} \approx k_B T \ln \frac{N}{n}$$

and solving for n, we find

$$n = N \exp\{-\beta(\Delta A + Pv_c + \epsilon)\} \,.$$

It is shown in Problem 2.10 that for high temperatures, $e^{-\beta\Delta A}$ approaches a constant typically of the order of 10, while for low T, ΔA approaches a constant. At atmospheric pressures Pv_c is in the range of 10^{-4} to $10^{-5}eV$, which is negligible compared with a typical value of $\epsilon \approx 1eV$. In the laboratory, pressures in the range 10 to 100 kbar are fairly standard and Pv_c then becomes comparable with the vacancy formation energy.

2.7 Problems

2.1 *Ergodic Hypothesis.*

(a) Consider a harmonic oscillator with Hamiltonian

$$H = \frac{1}{2}p^2 + \frac{1}{2}q^2 = \frac{1}{2}(\mathbf{x} \cdot \mathbf{x}) \,.$$

Show that any phase space trajectory $x(t)$ with energy E will, on the average, spend equal time in all regions of the constant energy surface $\Gamma(E)$.

(b) Consider two linearly coupled harmonic oscillators

$$H = \frac{1}{2}({p_1}^2 + {p_2}^2 + {q_1}^2 + {q_2}^2 + q_1 q_2) \ .$$

Express the phase space trajectory $x = (p_1, p_2, q_1, q_2)$ in terms of the initial values of the amplitude and phase of the normal coordinates. Show that there are regions of the constant energy surface which are not visited for any particular trajectory $x(t)$. (If you are not familiar with normal coordinates, a good place to look is Goldstein [113].)

2.2. *Gibbs Paradox.*

(a) Suppose that (2.8) and (2.9) were the correct expressions for the entropy of an ideal gas. Consider two volumes $V_A = V_B = V$, each containing N identical particles with the same mean energy. Show that if the two systems are joined together,

$$S_{A+B} = S_A + S_B + 2Nk_B \ln 2$$

that is, there is an entropy of mixing $2Nk_B \ln 2$.

(b) Show that if the correct expression (2.10) is used,

$$S_{A+B} = S_A + S_B \ .$$

(c) Estimate the entropy of mixing of 1 mol of Ar and 1 mol of Kr.

2.3. *Equipartition.*

(a) A classical harmonic oscillator

$$H = \frac{p^2}{2m} + \frac{Kq^2}{2}$$

is in thermal contact with a heat bath at temperature T. Calculate the partition function for the oscillator in the canonical ensemble and show explicitly that

$$\langle E \rangle = k_B T \qquad \langle (E - \langle E \rangle)^2 \rangle = {k_B}^2 T^2 \ .$$

(b) Consider a system of particles in which the force between the particles is derivable from a potential which is a generalized homogeneous function of degree γ, that is,

$$U(\lambda \mathbf{r}_1, \lambda \mathbf{r}_2, ..., \lambda \mathbf{r}_N) = \lambda^\gamma U(\mathbf{r}_1, \mathbf{r}_2, ..., \mathbf{r}_N) .$$

Show that the equation of state for this system is of the form

$$PT^{-1+3/\gamma} = f\left(\frac{V}{N} T^{-3/\gamma}\right)$$

where $f(x)$ can be calculated (at least in principle) once U is specified.

2.4. *Dielectric Function of a Diatomic Gas.*

Consider a dilute[7] gas made up of molecules which have a permanent electric dipole moment μ (Figure 2.2). The energy H of a molecule in an electric field $\mathbf{E}$ pointing in the z direction can be written

$$H = T_{transl} + T_{rot} - \mu E \cos\theta .$$

Treat the dipoles as classical rods with moment of inertia I. Then

$$T_{rot} = \frac{1}{2} I(\dot{\theta}^2 + \dot{\phi}^2 \sin^2\theta) .$$

Assume that the canonical partition function for the gas can be written

$$Z = \frac{{Z_1}^N}{N!}$$

where the partition function for a single molecule is $Z_1 = Z_{transl} \cdot Z_{rot}$.

(a) Show that

$$Z_{rot} = \frac{2I \sinh \beta E \mu}{\hbar^2 \beta^2 E \mu} .$$

(b) Show that the polarization, P, satisfies

$$P = \frac{N}{V}\langle \mu \cos\theta \rangle = \frac{N}{V}\left(\mu \coth \beta\mu E - \frac{k_B T}{E}\right) .$$

[7] If we are not dealing with a dilute gas, we must make corrections for the fact that the local microscopic ordering field will not be the same as the macroscopic field $\mathbf{E}$. The local field corrections will be different for permanent and induced dipoles [221].

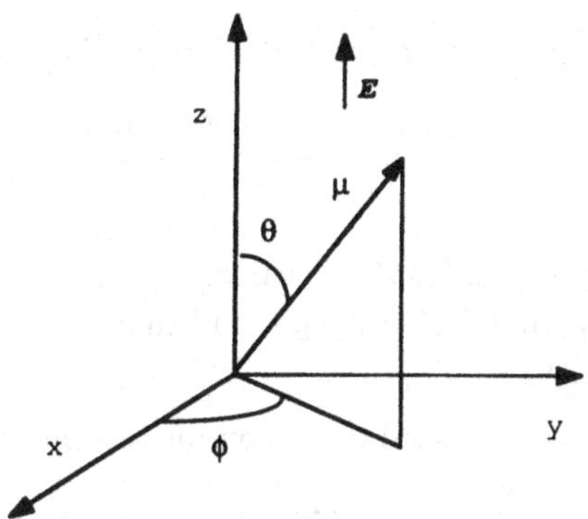

Figure 2.2: Polar molecule with dipole moment μ in an electric field **E**.

(c) Expand the hyperbolic cotangent for small $\beta\mu E$ and show that in the weak-field limit the dielectric constant ϵ, given by

$$\epsilon E = \epsilon_0 E + P$$

satisfies

$$\epsilon = \epsilon_0 + \frac{N\beta\mu^2}{3V} .$$

2.5. *Equivalence of Thermodynamic and Statistical Mechanical Definitions.* Show that in the limit that we are dealing with a very large system

$$k_B T \ln Z_G = \mu\langle N\rangle - \langle A\rangle .$$

Find an estimate for the correction term in analogy with equation (2.41).

2.6. *Classical Ideal Gas in the Canonical and grand Canonical Ensembles.*

(a) Show that, if the kinetic energy of a particle with mass m, momentum p is $E = p^2/2m$, the single particle partition function can be written $Z_1 = V/\lambda^3$ where λ is the thermal wavelength (2.21). The canonical partition function for the ideal gas will then be

$$Z_N = \frac{V^N}{N!\lambda^{3N}} .$$

(b) Use Stirling's approximation to show that in the thermodynamic limit the Helmholtz free energy of an ideal gas is

$$A = -Nk_B T\left[\ln(\frac{V}{N\lambda^3}) + 1\right] .$$

(c) Show that the grand canonical partition function can be written

$$Z_G = \exp\left[e^{\beta\mu}\frac{V}{\lambda^3}\right] .$$

(d) Use the expression for Z_G to calculate the mean values for N, P and S, and show that $PV = Nk_BT$ and that the Gibbs–Duhem relation gives $E = \frac{3}{2}Nk_BT$.

(e) Show that the standard deviation for the energy fluctuations in the ideal gas is

$$\frac{\sqrt{\langle(\Delta E)^2\rangle}}{E} = \sqrt{\frac{2}{3N}} .$$

(f) Show that for an ideal gas $\langle(\Delta N)^2\rangle = \langle N\rangle$ in the grand canonical ensemble.

2.7. *Black–body Radiation.*

(a) The frequency of an electromagnetic mode of wave vector $\mathbf{k}$ is $\omega_{\mathbf{k}} = ck$. In a box of volume V there will be

$$\frac{2Vd^3k}{(2\pi)^3}$$

such modes in a region d^3k surrounding a given wave vector. The Hamiltonian of the system is $\sum \hbar\omega_{\mathbf{k}}n_{\mathbf{k}}$, where $n_{\mathbf{k}}$ is the number of excitations in the mode with wave vector $\mathbf{k}$. Show that the Helmholtz free energy of the electromagnetic radiation in the cavity is

$$A = \frac{Vk_BT}{\pi^2c^3}\int_0^\infty \omega^2 d\omega \ln(1 - e^{-\beta\hbar\omega}) .$$

(b) By comparing the expressions $P = -\partial A/\partial V$ and $E = \partial(\beta A)/\partial\beta$ for the pressure and internal energy, respectively, show that

$$PV = \frac{E}{3} .$$

2.8. *Diatomic Molecule.*

If the energy stored in the rotational and vibrational modes is not too large, we may approximate the Hamiltonian of a diatomic molecule by

$$H = T_{transl} + T_{rot} + T_{vib}$$

neglecting any effect of centrifugal forces on the vibrational modes and the effect of the vibrational distortions on the moment of inertia I. In a dilute gas the density will be low enough that the translational motion can be treated classically. The energy of rotation is

$$T_{rot} = \frac{\hbar^2 j(j+1)}{2I}$$

where j is the rotational quantum number. The state with quantum number j has degeneracy $g_j = 2j+1$ (we have, for simplicity, assumed that the two atoms in the diatomic molecule are different). The vibrational degree of freedom may be taken to be a harmonic oscillator with frequency ω_{vib}.

(a) Find an expression for the specific heat.

(b) Discuss the three cases

$$\begin{array}{ll} (i) & T \ll \theta_{rot} = \hbar^2/(2Ik_B) \ll \hbar\omega_{vib}/k_B = \theta_{vib} \\ (ii) & \theta_{rot} \ll T \ll \theta_{vib} \\ (iii) & \theta_{rot} \ll \theta_{vib} \ll T\,. \end{array}$$

The Euler summation formula

$$\sum_{n=0}^{\infty} f(n) = \int_0^{\infty} dx f(x) + \frac{1}{2}f(0) - \frac{1}{12}f'(0) + \frac{1}{720}f^{(3)}(0) + \cdots$$

may be useful in the solution of part (*ii*).

2.9. *Maximum Entropy Principle for the Grand Canonical Ensemble.*

Construct the Grand canonical ensemble from the maximum entropy principle by maximizing the entropy, subject to the constraints that the mean energy and particle number are fixed.

2.10. *Effect of Lattice Vibrations on Vacancy Formation.*

To establish the effect qualitatively, consider the following crude model. Each atom vibrates as an independent three-dimensional Einstein oscillator of frequency ω_0. Assume further that if a nearest-neighbor site is vacant, the frequency of the mode corresponding to vibration in the direction of the vacancy changes from ω_0 to ω. Let q be the number of nearest neighbors.

(a) Show that in this simple model,

$$\Delta A = nqk_BT \ln \frac{\sinh(\beta\hbar\omega/2)}{\sinh(\beta\hbar\omega_0/2)}$$

where n is the total number of vacancies.

(b) Consider as an example a simple cubic lattice. Each mode then corresponds to the vibration of two springs. If one of them is cut, the simplest assumption one can make is

$$\omega = \frac{\omega_0}{\sqrt{2}} \, .$$

Show that for high temperatures, $\beta\hbar\omega \ll 1$,

$$e^{-\beta\Delta A/n} \approx 8$$

while for low temperatures, $\beta\hbar\omega \gg 1$,

$$\Delta A \approx -\frac{3}{2} n\hbar\omega_0 (2 - \sqrt{2}) \, .$$

2.11. *Partition Function at Fixed Pressure.*

Consider a system of N noninteracting molecules in a container of cross-sectional area $\mathcal{A}$. The bottom of the container (at $z = 0$) is rigid. The top consists of an airtight piston of mass M which slides without friction.

(a) Construct the partition function Z of the $(N + 1)$-particle system (N molecules of mass m, one piston of mass M, cross-sectional area $\mathcal{A}$). You may neglect the effect of gravity on the gas molecules.

(b) Show that the thermodynamic potential $-k_BT \ln Z$ is, in the thermodynamic limit, identical to the Gibbs potential of an ideal gas of N molecules, subject to the pressure $P = Mg/\mathcal{A}$.

2.12. *Stability of a White Dwarf Against Gravitational Collapse.*

It is energetically favorable for a body held together by gravitational forces to be as compact as possible. We take the star to be made up of an approximately equal number N of electrons and protons, since otherwise the Coulomb repulsion would overcome the gravitational interaction. Somewhat arbitrarily we also assume that there is an equal number of neutrons and protons. On earth the gravitational pressure is not large enough to overcome the repulsive forces between atoms and molecules at short distance. Inside the sun, matter does not exist in the form of atoms and molecules, but since it is still burning there is radiation pressure which keeps it from collapsing. Let us consider a burnt out star such as a white dwarf. Assume that the temperature of the star is low enough compared to the electron Fermi temperature that the electrons can be approximated by a $T = 0$ Fermi gas. Because of their large mass the kinetic energy of the protons and neutrons will be small compared to that of the electrons.

(a) Show that, if the electron gas is non-relativistic, the electron mass is m_e, and the radius of the star is R, the electron kinetic energy of the star can be written

$$E_{kin} = \frac{3\hbar^2}{10m_e}\left(\frac{9\pi}{4}\right)^{\frac{2}{3}}\frac{N^{\frac{5}{3}}}{R^2} .$$

(b) The gravitational potential energy is dominated by the neutrons and protons. Let m_N be the nucleon mass. Assume the mass density is approximately constant inside the star. Show that, if there is an equal number of protons and neutrons, the potential energy will be given by

$$E_{pot} = -\frac{12}{5}m_N^2 G\frac{N^2}{R}$$

where G is the gravitational constant ($6.67 \times 10^{-11} Nm^2/kg^2$).

(c) Find the radius for which the potential energy plus kinetic energy is a minimum for a white dwarf with the same mass as the sun ($1.99 \times 10^{30} kg$), in units of the radius of the sun ($6.96 \times 10^8 m$).

(d) If the density is too large the Fermi velocity of the electrons becomes comparable to the velocity of light and we should use the relativistic formula

$$\epsilon(p) = \sqrt{m_e^2c^4 + p^2c^2} - m_ec^2 \tag{2.107}$$

for the relationship between energy and momentum. It is easy to see that, in the ultra relativistic limit ($\epsilon \approx cp$), the electron kinetic energy will be proportional to $N^{3/4}/R$, *i.e.* the R dependence is the same as for the potential energy. Since for large N we have $N^2 \gg N^{\frac{4}{3}}$ we find that if the mass of the star is large enough the potential energy will dominate. The star will then collapse. Show that the critical value of N for this to happen is

$$N_{crit} = \left(\frac{5\hbar c}{36\pi m_N^2 G}\right)^{\frac{3}{2}} \left(\frac{9\pi}{4}\right)^2 .$$

Substituting numbers we find that this corresponds to approximately 1.71 solar masses.

(e) Use the relativistic formula (2.107) to evaluate the kinetic energy numerically and plot the total energy in units of $E_0 = N m_e c^2$ as a function of R when $N = 0.9, 1.0$ and $1.1 \times N_{crit}$.

2.13. *Counting the Number of States of Identical Particles.*

The entropy of a monatomic ideal gas predicted by the Sackur-Tetrode formula becomes negative at very low temparatures, which is quite unphysical.

(a) Find a formula for the temperature at which the Sackur-Tetrode entropy changes sign.

(b) The source of the problem is that the simple $N!$ factor correcting the phase space volume in (2.10) fails when several ideal gas particles are likely to occupy the same state. Suppose each of N identical particles may occupy M states. In how many ways can this be done if there are no restrictions on the number of particles in each state (Bose particles)?

(c) Compute the entropy $k_B \ln \Omega$ in case **(b)**. Will the entropy be extensive in the large N limit if M is constant, independent of N? If $M \propto N$?

(d) How many ways can N fermions occupy M states?

(e) Consider 1000 identical particles , each occupying any one of 2000 single particle states. Calculate the entropy if the particles are *(i)* bosons, *(ii)* fermions, *(iii)* "classical" particles for which the number of available states is given by $M^N/N!$.

Chapter 3

Mean Field and Landau Theory

In this chapter we begin our discussion of the statistical mechanics of systems that display a change in phase as a function of an intensive variable such as the temperature or the pressure. In recent years a great deal of progress has been achieved in our understanding of phase transitions, notably through the development of the renormalization group approach of Wilson, Fisher, Kadanoff, and others. We postpone a discussion of this theory until Chapters 6 and 7. In the present chapter we discuss an older approach known as mean field theory, which generally gives us a qualitative description of the phenomena of interest. We limit ourselves to discussing the most common approaches taken, and postpone our discussion of a number of important applications to Chapter 4. A common feature of mean field theories is the identification of an order parameter. One approach is to express an approximate free energy in terms of this parameter and minimize the free energy with respect to the order parameter (we have used this approach in Section 2.6 in connection with our discussion of Schottky defects in a crystal). Another, often equivalent approach is to approximate an interacting system by a noninteracting system in a self-consistent external field expressed in terms of the order parameter.

To understand the phenomena associated with the sudden changes in the material properties which take place during a phase transition, it has proven most useful to work with simplified models that single out the essential aspects of the problem. One important such model, the Ising model, is introduced in Section 3.1 and discussed in the Weiss molecular field approximation, an example of the self-consistent field approach mentioned earlier. In Section 3.2 we discuss the same model in the Bragg–Williams approximation, which is a free energy minimization approach.

In Section 3.3 we point out that while very useful, mean field theories can give rise to misleading results, particularly in low dimensional problems. In Section 3.4 we discuss an improved version of mean field theory, the Bethe approximation. This method gives better numerical values for the critical temperature and other properties of the system. However, we show in Section 3.5 that the asymptotic critical behavior of mean field theories is always the same.

The most serious fault of mean field theories lies in the neglect of long-range fluctuations of the order parameter. As we shall see, the importance of this omission depends very much on the dimensionality of the problem, and in problems involving one- and two-dimensional systems the results predicted by mean field theory are often qualitatively wrong. In Section 3.6 we illustrate this by discussing properties of the exact solution to the one dimensional Ising model.

Because of its close relation to mean field theory we discuss in Section 3.7 the Landau theory of phase transitions. Symmetry considerations are in general important in determining the order of a transition, and we show in Section 3.8 that the presence of a cubic term in the order paremeter will in general predict that phase transitions or first order (discontinuous).

In Section 3.9 we extend the Landau theory to the case where more than one thermodynamic quantity can be varied independently, and discuss the occurrence of tricritical points. In Section 3.10 we discuss the limitations of mean field theory and derive the Ginzburg criterion for the relevance of fluctuations. We conclude our discussion of Landau theory in Section 3.11 by considering multicomponent order parameters which are needed for a discussion of the Heisenberg ferromagnet and other systems.

An important reference for much of the material in this chapter is Landau and Lifshitz [165]. Many examples are discussed in Kubo et al. [161].

3.1 Mean Field Theory of the Ising Model

We consider here a simple model, known as the Ising model, for a magnetic material. We assume that N magnetic atoms are located on a regular lattice, with the magnetic moments interacting with each other through an exchange interaction of the form

$$H_{Ising} = -J_0 \sum_{\langle ij \rangle} S_{zi} S_{zj} \tag{3.1}$$

where J_0 is a constant and the symbol $\langle ij \rangle$ denotes that the sum is to be carried out over nearest-neighbor pairs of lattice sites. This exchange interaction has a preferred direction, so that the Ising Hamiltonian is diagonal in the representation in which each spin S_{zj} is diagonal. The z component of the spins then takes on the discrete values $-S, -S+\hbar, \ldots, S$. The eigenstates of (3.1) are labeled by the values of S_{zj} on each site, and the model has no dynamics. This makes it easier to work with than the Heisenberg model,

$$H = -J_0 \sum_{\langle ij \rangle} \mathbf{S}_i \cdot \mathbf{S}_j \tag{3.2}$$

where the local spin operators $S_{\alpha j}$ do not commute with H. We specialize to the case $S = \frac{1}{2}\hbar$, and also add a Zeeman term for the energy in a magnetic field directed along the z direction to obtain the final version of the Ising Hamiltonian (which in accordance with the definitions in Chapter 1 should be considered as an enthalpy, but in conformity with common usage will be referred to as an energy),

$$H = -J \sum_{\langle ij \rangle} \sigma_i \sigma_j - h \sum_i \sigma_i \tag{3.3}$$

where $\sigma_i = \pm 1$ and h is proportional to the magnetic field, but has the unit of energy. To obtain an intuitive feeling for the behavior of such a system, consider the limits $T \to 0$ and $T \to \infty$ for the temperature in the case $J > 0$. At $T = 0$ the system will be in its ground state, with all spins pointing in the direction of the applied field. At $T = \infty$ the entropy dominates and the spins will be randomly oriented. In certain cases the two regimes will be separated by a *phase transition*, that is, there will be a temperature T_c at which there is a sudden change from an ordered phase to a disordered phase as the temperature is increased. Suppose that at a certain temperature the expectation value of the magnetization is m, that is,

$$\langle \sigma_i \rangle = m \tag{3.4}$$

for all i. We refer to m as the *order parameter* of the system. Consider the terms in (3.3) which contain a particular spin σ_0. These terms are, with j

restricted to nearest neighbor sites of site 0,

$$\begin{aligned} H(\sigma_0) &= -\sigma_0\left(J\sum_j \sigma_j + h\right) \\ &= -\sigma_0(qJm+h) - J\sigma_0\sum_j(\sigma_j - m) \end{aligned} \tag{3.5}$$

where q is the number of nearest neighbors of site 0. If we disregard the second term on the right hand side in (3.5), we are left with a noninteracting system; that is, each spin is in an effective magnetic field composed of the applied field and an average exchange field due to the neighbors. The magnetization has to be determined self-consistently from the condition

$$m = \langle\sigma_0\rangle = \langle\sigma_j\rangle \ . \tag{3.6}$$

This approximation constitutes a form of mean field theory—the fluctuating values of the exchange field are replaced by an effective average field—and is commonly referred to as the Weiss molecular field theory. We obtain the constitutive equation for m

$$\begin{aligned} m &= <\sigma_0>= \frac{\mathrm{Tr}\,\sigma_0\exp\{-\beta H(\sigma_0)\}}{\mathrm{Tr}\exp\{-\beta H(\sigma_0)\}} \\ &= \tanh[\beta(qJm+h)] \ . \end{aligned} \tag{3.7}$$

To find $m(h,T)$ we must solve (3.7) numerically. However, it is easy to see that $m(h,T) = -m(-h,T)$ and that for each $h \neq 0$ there is at least one solution, and sometimes three. For $h = 0$ there is always the solution $m = 0$ and if $\beta qJ > 1$, two further solutions at $\pm m_0$. We will show in Section 3.2 that the equilibrium state for $T < T_c = qJ/k_B$ is either of the *broken symmetry* states with spontaneous magnetization $\pm m_0(T)$. As $T \to 0$, $\tanh(\beta qJm) \to \pm 1$ for $m \neq 0$, and $m_0 \to \pm 1$. As $T \to T_c$ from below, $|m_0(T)|$ decreases and we may obtain its asymptotic dependence by making a low-order Taylor expansion of the hyperbolic tangent, that is

$$m_0 = \beta qJm_0 - \frac{1}{3}(\beta qJ)^3 m_0^3 + \cdots \tag{3.8}$$

or

$$m_0(T) \approx \pm\sqrt{3}\left(\frac{T}{T_c}\right)^{3/2}\left(\frac{T_c}{T} - 1\right)^{1/2} \ . \tag{3.9}$$

Therefore, we see that the order parameter m_0 approaches zero in a singular fashion as T approaches T_c from below, vanishing asymptotically as

$$m_0(T) \propto \left(\frac{T_c}{T} - 1\right)^{1/2} . \tag{3.10}$$

The exponent for the power law behavior of the order parameter is in general given the symbol β and in more sophisticated theories, as well as in real ferromagnets, is not the simple fraction $\frac{1}{2}$ found here.

3.2 Bragg–Williams Approximation

An alternative approach to mean field theory is to construct an approximate expression for the free energy in terms of the order parameter and apply the condition that its equilibrium value minimizes the free energy. The Hamiltonian for the Ising model of the previous section can be written

$$H = -J(N_{++} + N_{--} - N_{+-}) - h(N_+ - N_-) \tag{3.11}$$

where N_{++}, N_{--}, N_{+-} are the number of nearest neighbor spins that are both $+$, both $-$ or opposite, and the number of spins of each kind are

$$N_+ = \frac{N(1+m)}{2}; \ N_- = \frac{N(1-m)}{2} . \tag{3.12}$$

We now assume that the states of the individual spins are *statistically independent.* This lets us write for the entropy

$$S = -k_B N \left(\frac{N_+}{N} \ln \frac{N_+}{N} + \frac{N_-}{N} \ln \frac{N_-}{N}\right) \tag{3.13}$$

and for the number of pairs

$$N_{++} = q\frac{N_+^2}{2N}; \ N_{--} = q\frac{N_-^2}{2N}; \ N_{+-} = q\frac{N_+ N_-}{N} . \tag{3.14}$$

Substituting (3.14) and (3.12) into (3.11) and (3.13) yields

$$G(h,T) = -\frac{qJN}{2}m^2 - Nhm + Nk_BT\left(\frac{1+m}{2}\ln\frac{1+m}{2} + \frac{1-m}{2}\ln\frac{1-m}{2}\right) .$$

Minimizing with respect to the variational parameter m, we obtain

$$0 = -qJm - h + \frac{1}{2}k_BT \ln\frac{1+m}{1-m} \tag{3.15}$$

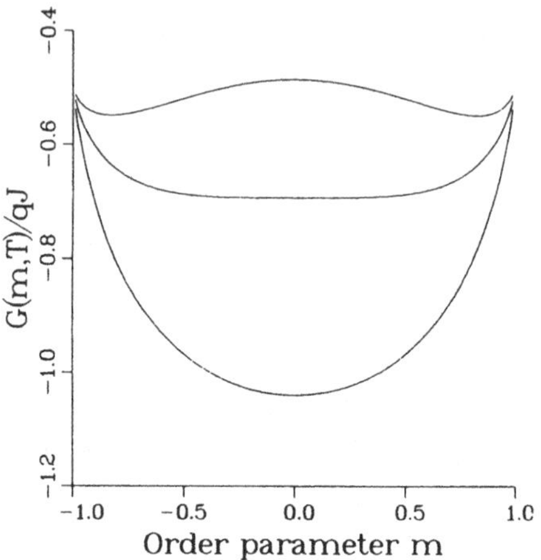

Figure 3.1: Free energy in the Bragg–Williams approximation for the Ising model. The three curves correspond to, respectively, $T = 1.5T_c$, $T = T_c$, and $T = 0.77T_c$.

or

$$m = \tanh[\beta(qJm + h)] \tag{3.16}$$

as in the previous approach (3.7). In the special case $h = 0$ the free energy becomes

$$G(0,T)/N = -\frac{1}{2}qJm^2 + \frac{1}{2}k_BT[(1+m)\ln(1+m) + (1-m)\ln(1-m) - 2\ln 2] \ .$$

For small values of m we may expand in a power series to obtain

$$G(0,T)/N = \frac{m^2}{2}(k_BT - qJ) + \frac{k_BT}{12}m^4 + \cdots - k_BT\ln 2 \tag{3.17}$$

with all higher order terms of even power in m with positive coefficients. The form of $G(0,T)$ is shown for T above and below $T_c = qJ/k_B$ in Figure 3.1. It is clear that the ordered phase is the state of lower free energy when $T < T_c$.

The type of transition seen in this system is known as *continuous*, or second order, since the order parameter increases continuously from zero as a function of $(T_c - T)$ below the transition (Figure 3.2). We shall encounter examples of discontinuous, or first-order, transitions later in this chapter.

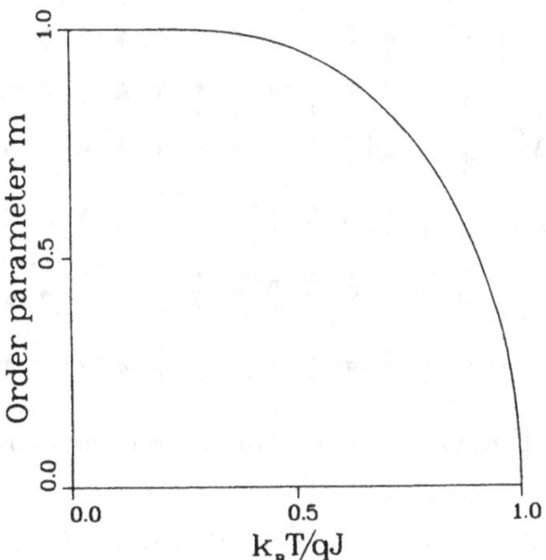

Figure 3.2: Temperature dependence of order parameter below T_c.

3.3 A Word of Warning

The theory presented above is remarkably general. Note that neither the type of lattice nor the spatial dimensionality plays a role in the transition—the sole parameter characterizing the system is the number, q, of nearest neighbors. Therefore, in this approximation, the Ising models on the two-dimensional triangular lattice and the three-dimensional simple cubic lattice have identical properties. This result is quite incorrect and we demonstrate below how one can be misled by mean field arguments.

Consider a one-dimensional chain with free ends. The Hamiltonian in zero field is

$$H = -J \sum_{i=1}^{N-1} \sigma_i \sigma_{i+1} \tag{3.18}$$

with ground state energy $E_0 = -(N-1)J$. Suppose that the system is at a very low temperature and consider the class of excitations defined by $\sigma_i = 1$, $i \leq l$ and $\sigma_i = -1$, $i > l$:

$$\begin{array}{cccccccc} \uparrow & \uparrow & \cdots & \uparrow & \downarrow & \downarrow & \cdots & \downarrow \\ 1 & 2 & & l & l+1 & & & N \end{array}$$

There are $N-1$ such states, all with the same energy $E = E_0 + 2J$. At temperature T the free energy change due to these excitations is

$$\Delta G = 2J - k_B T \ln(N-1)$$

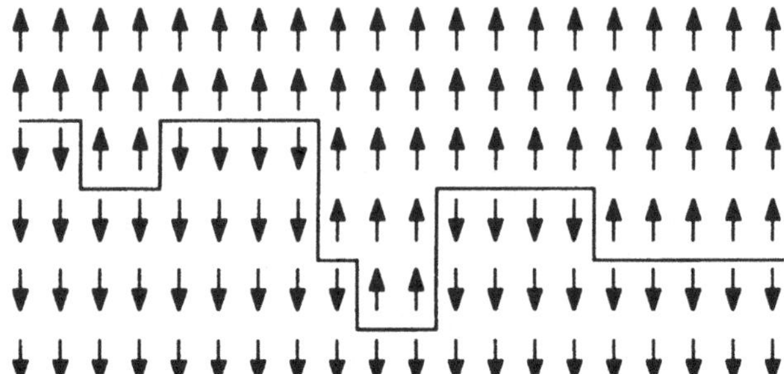

Figure 3.3: Domain wall in a two-dimensional Ising model.

which is less than zero for all $T > 0$ in the limit $N \to \infty$. These excitations disorder the system; the expectation value of the magnetization is zero. Therefore, there cannot be a phase transition to a ferromagnetic state in the one-dimensional Ising model with nearest-neighbor (or with any finite-range) interactions.

A similar argument can be used to give a crude estimate of the transition temperature for the two-dimensional Ising model. Consider an $N \times N$ square lattice with free surfaces. We wish to study the set of excitations of the type shown in Figure 3.3, that is, excitations which divide the lattice into two large domains separated by a wall that extends from one side to the other and has no loops. The energy of the domain wall is

$$\Delta E = 2LJ$$

where L is the number of segments in the wall. If we start the wall from the left there are at least two, sometimes three choices for the direction of the next step if we neglect the possibility of reaching the upper and lower boundaries. The entropy associated with non-looping chains of length L is then at least $k_B \ln 2^L$. There are N possible starting points for the chain. If we assume two choices per site, one of which takes us to the right, the average length of the chain will be $2N$ with a standard deviation $\propto N^{1/2}$, which is small compared to N if N is large enough. The free energy associated with dividing the lattice into two domains is thus approximately

$$\Delta G \approx 4NJ - k_B T \ln(N \times 2^{2N}) .$$

The system is therefore stable against domain formation if

$$T < T_c \approx \frac{2J}{k_B \ln 2} = \frac{2.885J\ldots}{k_B} .$$

This estimate is surprisingly close to the exact result (see Section 6.1) $T_c = 2.269185\ldots J/k_B$.

A more sophisticated version of this type of argument was first devised by Peierls[1] [237] to prove that in two dimensions a phase transition indeed occurs. The one-dimensional argument was presented here to raise a warning flag—mean field arguments, although useful, are not invariably correct. We return to this topic in Section 3.6, where we discuss some exact properties of the Ising model in one dimension.

3.4 Bethe Approximation

In this section we wish to consider an approximation scheme due to Bethe [35]. An extension of the approach, which provides the same results for the order parameter but also yields an expression for the free energy, is due to Fowler and Guggenheim [101].

Consider again the simple Ising model (3.3) of Section 3.1. In our mean field approximation we ignored all correlations between spins and, even for nearest neighbors, made the approximation $\langle\sigma_i\sigma_j\rangle = \langle\sigma_i\rangle\langle\sigma_j\rangle = m^2$. It is possible to improve this approximation in a systematic fashion. We suppose that the lattice has coordination number q and now retain as variables a central spin and its shell of nearest neighbors. The remainder of the lattice is assumed to act on the nearest-neighbor shell through an effective exchange field which we will calculate self-consistently. The energy of the central cluster can be written as

$$H_c = -J\sigma_0\sum_{j=1}^{q}\sigma_j - h\sigma_0 - h'\sum_{j=1}^{q}\sigma_j \;.$$

The situation is depicted in Figure 3.4 for the square lattice. The fluctuating field acting on the peripheral spins $\sigma_1\ldots,\sigma_4$ has been replaced by an effective field h', just as we previously replaced the interaction of σ_0 with its first neighbor shell by a mean energy.

The partition function of the cluster is given by

$$Z_c = \sum_{\sigma_j=\pm1} e^{-\beta H_c} = e^{\beta h}(2\cosh[\beta(J+h')])^q + e^{-\beta h}(2\cosh[\beta(J-h')])^q \;.$$

---

[1] A clear exposition of the arguments is also given in Section 15.4 of [318].

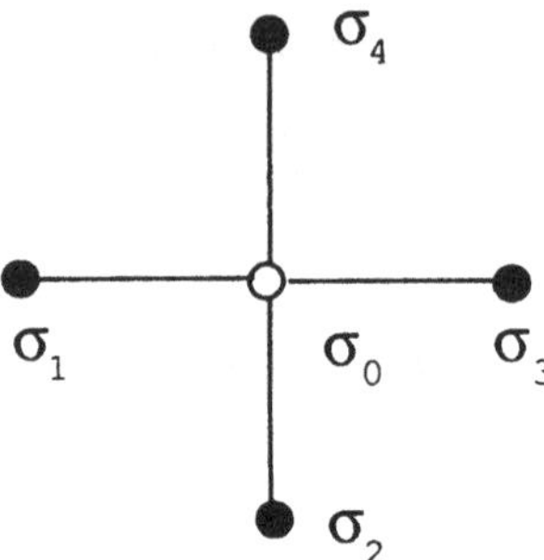

Figure 3.4: Spin cluster used in the Bethe approximation calculation.

The expectation value $\langle\sigma_0\rangle$ is given by

$$\langle\sigma_0\rangle = \frac{1}{Z_c}\left(e^{\beta h}\{2\cosh[\beta(J+h')]\}^q - e^{-\beta h}\{2\cosh[\beta(J-h')]\}^q\right)$$

while for $j = 1, \ldots, q$,

$$\begin{aligned}\langle\sigma_j\rangle = \frac{1}{Z_c}&\left(2e^{\beta h}\sinh[\beta(J+h')]\{2\cosh[\beta(J+h')]\}^{q-1}\right.\\ &\left.- 2e^{-\beta h}\sinh[\beta(J-h')]\{2\cosh[\beta(J-h')]\}^{q-1}\right) \ . \qquad (3.19)\end{aligned}$$

For simplicity we now set $h = 0$. Since the ferromagnet is translationally invariant we must require $\langle\sigma_j\rangle = \langle\sigma_0\rangle$. This yields the equation

$$\begin{aligned}&\cosh^q[\beta(J+h')] - \cosh^q[\beta(J-h')] = \\ &\quad \sinh[\beta(J+h')]\cosh^{q-1}[\beta(J+h')] - \sinh[\beta(J-h')]\cosh^{q-1}[\beta(J-h')]\end{aligned}$$

or

$$\frac{\cosh^{q-1}[\beta(J+h')]}{\cosh^{q-1}[\beta(J-h')]} = e^{2\beta h'} \qquad (3.20)$$

which must be solved for the unknown effective field h'. It is clear that equation (3.20) always has a solution $h' = 0$ corresponding to the disordered high-temperature phase. As $h' \to \infty$ the left side of (3.20) approaches $\exp(2\beta J(q-1))$, *i.e.* a constant, while the right side diverges. Therefore, if the slope of the function on the left at $h' = 0$ is greater than 2β the two functions must intersect again at finite h'. Since (3.20) is invariant under $h' \to -h'$, there will be two further solutions in this case. The critical temperature, below which these solutions exist, is given by

$$\coth \beta_c J = q - 1$$

or

$$\beta_c J = \frac{1}{2} \ln \left(\frac{q}{q-2} \right) \quad . \tag{3.21}$$

On the square lattice this yields $k_B T_c / J = 2.885\ldots$, which may be compared with the exact result $k_B T_c / J = 2.269\ldots$ of Onsager [222] and the prediction $k_B T_c / J = 4$ of the simple mean field theory of the preceding sections. We see that we have achieved a substantial improvement in the prediction of the critical temperature. It is interesting to note that for the one-dimensional Ising model ($q = 2$) the Bethe approximation does not predict a phase transition. This is in agreement with the exact results of Section 3.6.

It is often important to have expressions for the free energy as well as the order parameter. We write the Hamiltonian (3.3) in the form

$$H = H_0 + \lambda V \tag{3.22}$$

where

$$H_0 = -h \sum_i \sigma_i \qquad V = -\sum_{\langle ij \rangle} \sigma_i \sigma_j \ .$$

The physical system we wish to consider has $\lambda = J$ but we can also imagine systems with different values of the exchange coupling strength. The free energy associated with (3.22) is

$$G = -k_B T \ln \mathrm{Tr}\, e^{-\beta H_0 - \beta \lambda V} \ . \tag{3.23}$$

We define

$$G_0 \equiv -k_B T \ln \mathrm{Tr}\, e^{-\beta H_0} \tag{3.24}$$

and

$$\langle \mathcal{O} \rangle_\lambda \equiv \frac{\mathrm{Tr}\, \mathcal{O} e^{-\beta H_0 - \beta \lambda V}}{\mathrm{Tr}\, e^{-\beta H_0 - \beta \lambda V}} \tag{3.25}$$

and see that

$$\frac{\partial G}{\partial \lambda} = \langle V \rangle_\lambda \ . \tag{3.26}$$

Finally,

$$G = G_0 + \int_0^J d\lambda \, \langle V \rangle_\lambda \ . \tag{3.27}$$

Equation (3.27) is exact if the expectation value $\langle V\rangle_\lambda$ can be computed exactly. Useful approximations to the free energy can frequently be obtained by substituting approximate expressions into (3.27). As an example, consider the one-dimensional Ising model ($q = 2$) in zero field. Then

$$G_0 = -Nk_BT\ln 2\ . \tag{3.28}$$

There is no phase transition in this case and $h' = 0$. The Bethe approximation then corresponds to writing

$$\langle\sigma_i\sigma_j\rangle_\lambda = \frac{e^{\beta\lambda} - e^{-\beta\lambda}}{e^{\beta\lambda} + e^{-\beta\lambda}} = \tanh\beta\lambda \tag{3.29}$$

when i, j are nearest neighbors. We obtain

$$G = -Nk_BT\ln 2 - \frac{1}{2}Nq\int_0^J d\lambda\tanh\beta\lambda = -Nk_BT\ln(2\cosh\beta J)\ .$$

As we shall see in Section 3.6, this happens to be exact, whereas in general the Bethe approach produces only approximate free energies.

It should be obvious that still better results can be obtained by considering larger clusters. However, all approximations that depend in an essential way on truncation of correlations beyond a certain distance will break down in the vicinity of a critical point. To show this explicitly, we discuss the critical properties of mean field theories in the next section.

3.5 Critical Behavior of Mean Field Theories

In Section 3.1 we showed that as $T \to T_c$, the order parameter (magnetization) of our Ising model has the asymptotic form (3.10)

$$m(T) \propto (T_c - T)^{1/2}$$

as $T \to T_c$ from below. We now calculate several other thermodynamic functions in the vicinity of the critical point. Consider first the susceptibility per spin,

$$\chi(h,T) = \left(\frac{\partial m}{\partial h}\right)_T\ .$$

From (3.7) we obtain

$$\chi(0,T) = \frac{\beta}{\cosh^2(\beta qJm) - \beta qJ} = \frac{1}{k_B(T - T_c)} \tag{3.30}$$

as $T \to T_c^+$. For $T < T_c$ we use the asymptotic expansion for m to obtain

$$\chi(0,T) \approx \frac{1}{2k_B(T_c - T)} \tag{3.31}$$

and we see that the susceptibility diverges as the critical point is approached from either the low- or high-temperature side. It is conventional to write, for T near T_c,

$$\chi(0,T) \approx A_\pm \left|T - T_c\right|^{-\gamma}$$

and we conclude that in our mean field theory, $\gamma = 1$. The exact solution of the two-dimensional Ising model (6.1) yields $\gamma = 7/4$; for the three-dimensional Ising model, γ is not known exactly, but is approximately 1.25. This failure of mean field theory can be understood in terms of the following exact expression for the susceptibility:

$$\begin{aligned} \chi = \left(\frac{\partial m}{\partial h}\right)_T &= \frac{\partial}{\partial h}\left(\frac{\mathrm{Tr}\ \sigma_0 e^{-\beta H}}{\mathrm{Tr}\ e^{-\beta H}}\right)_T \\ &= \beta \sum_j \left(\langle\sigma_j\sigma_0\rangle - \langle\sigma_j\rangle\langle\sigma_0\rangle\right) . \end{aligned} \tag{3.32}$$

It is clear that χ can diverge only if the *spin-spin correlation function*

$$\Gamma(|\mathbf{r}_j - \mathbf{r}_0|) = \langle\sigma_j\sigma_0\rangle - \langle\sigma_j\rangle\langle\sigma_0\rangle$$

is long-ranged; for example, in three dimensions it must not decay faster than

$$\frac{1}{|\mathbf{r}_j - \mathbf{r}_0|^3}$$

for large separations at $T = T_c$. In our simple mean field approximation, and also in the more sophisticated Bethe approximation, we clearly discarded long-range correlations, and it is therefore not surprising that finite cluster approximations will break down as $T \to T_c$.

Let us next examine the specific heat in both the simple mean field and the Bethe approximations. In zero magnetic field the internal energy in the mean field approximation of Sections 3.1 and 3.2 is given by

$$\begin{aligned} E = \langle H\rangle &= -J\sum_{\langle ij\rangle}\langle\sigma_i\rangle\langle\sigma_j\rangle \\ &= -\frac{N}{2}Jqm^2 \end{aligned}$$

giving

$$C_h = \left(\frac{\partial E}{\partial T}\right)_{h=0} = \begin{cases} -\frac{N}{2}Jq\left(\frac{\partial m^2}{\partial T}\right) \to \frac{3}{2}Nk_B & \text{as } T \to T_c^- \\ 0 & \text{for } T > T_c \end{cases} \tag{3.33}$$

that is, the mean field theory produces a discontinuity at the transition. This behavior is in contrast to more correct theories and experimental results, which yield a power law of the form

$$C_h \approx B_\pm |T - T_c|^{-\alpha}$$

where α is the conventional notation for the specific heat exponent.

The determination of the specific heat singularity in the Bethe approximation is somewhat more tedious. It is easy to show that the correlation function $\langle\sigma_0\sigma_j\rangle$ which determines the internal energy is given by

$$\langle\sigma_0\sigma_j\rangle = \frac{\sinh\beta(J+h')\cosh^{q-1}\beta(J+h') + \sinh\beta(J-h')\cosh^{q-1}\beta(J-h')}{\cosh^q\beta(J+h') + \cosh^q\beta(J-h')}$$

if j is a nearest neighbor of site 0. For $T > T_c$, $h' = 0$ and

$$E = -\frac{N}{2}qJ\langle\sigma_0\sigma_j\rangle = -\frac{N}{2}qJ\tanh\beta J \ .$$

For $h' \neq 0$ we note that $\langle\sigma_0\sigma_j\rangle_{h'} = \langle\sigma_0\sigma_j\rangle_{-h'}$ and we must therefore have

$$\langle\sigma_0\sigma_j\rangle_{h'} = \langle\sigma_0\sigma_j\rangle|_{h'=0} + a(T)h'^2 + \cdots \ .$$

The first piece of this expansion joins continuously with the high temperature form of the internal energy. The second term will yield a discontinuity at T_c if $\partial h'^2/\partial T$ approaches a constant as $T \to T_c$. We leave the explicit demonstration of this as an exercise (Problem 3.4). In a similar way it is possible to show that $m(T) = \langle\sigma_0\rangle \propto |T - T_c|^{1/2}$ in the Bethe approximation. The critical properties of cluster theories thus seem to be in a sense universal and not dependent on the level of sophistication of the approximation.

Another quantity that shows similar behavior in all mean field theories is the critical isotherm $m(T_c, h)$. In the simplest mean field theory we have (3.7)

$$m = \tanh[\beta(qJm + h)] \ .$$

At $T = T_c = qJ/k_B$, we obtain, on expanding the hyperbolic tangent,

$$m \approx m + \beta h - \frac{1}{3}(m + \beta h)^3$$

which gives, near $h = 0$,

$$h \propto |m|^\delta \mathrm{sign}(m)$$

with $\delta = 3$. We again leave it as an exercise for the reader to show that $\delta = 3$ as well in the Bethe approximation. In Section 3.7 we discuss a general theory of phase transitions due to Landau which exhibits the same behavior as mean field and cluster theories near the critical point.

3.6 Ising Chain: Exact Solution

The one-dimensional Ising model is one of a small number of models in statistical mechanics for which one can calculate the partition function exactly. Moreover, the result is simple enough that thermodynamic functions can be evaluated without too much difficulty. Let us first consider a chain of length N with free ends and zero external field:

$$H = -J \sum_{i=1}^{N-1} \sigma_i \sigma_{i+1} .$$

The partition function is given by

$$Z_N = \sum_{\sigma_1=\pm 1} \cdots \sum_{\sigma_N=\pm 1} \exp\left\{ \beta J \sum_{i=1}^{N-1} \sigma_i \sigma_{i+1} \right\} .$$

The last spin occurs only once in the sum in the exponential and we have, independently of the value of σ_{N-1},

$$\sum_{\sigma_N=\pm 1} e^{\beta J \sigma_{N-1} \sigma_N} = 2 \cosh \beta J$$

giving

$$Z_N = [2 \cosh \beta J] Z_{N-1} .$$

We can repeat this process to obtain

$$\begin{aligned} Z_N &= (2 \cosh \beta J)^{N-2} Z_2 \\ Z_2 &= \sum_{\sigma_1=\pm 1} \sum_{\sigma_2=\pm 1} e^{\beta J \sigma_1 \sigma_2} = 4 \cosh \beta J \end{aligned}$$

so that we finally obtain

$$Z_N = 2(2 \cosh \beta J)^{N-1} . \qquad (3.34)$$

The free energy is then

$$G = -k_BT \ln Z_N = -k_BT[\ln 2 + (N-1)\ln(2\cosh\beta J)] \ .$$

In the thermodynamic limit only the term proportional to N is important and

$$G = -Nk_BT \ln(2\cosh\beta J) \ . \tag{3.35}$$

We can also find the free energy in the presence of a magnetic field. To avoid end effects (which do not matter in the thermodynamic limit) we assume periodic boundary conditions, that is, assume that the N'th spin is connected to the first so that the chain forms a ring. Then

$$H = -J\sum_{i=1}^{N}\sigma_i\sigma_{i+1} - h\sum_{i=1}^{N}\sigma_i$$

where the spin labels run modulo N (*i.e.*, $N+i=i$). The Hamiltonian can be rewritten

$$H = -\sum_{i=1}^{N}\left[J\sigma_i\sigma_{i+1} + \frac{h}{2}(\sigma_i+\sigma_{i+1})\right]$$

giving for the partition function

$$\begin{aligned} Z_N &= \sum_{\sigma_1=\pm1}\cdots\sum_{\sigma_N=\pm1}\exp\left\{\beta\sum_{i=1}^{N}\left[J\sigma_i\sigma_{i+1} + \frac{h}{2}(\sigma_i+\sigma_{i+1})\right]\right\} \\ &= \sum_{\sigma_i}\prod_{i=1}^{N}\exp\left\{\beta\left[J\sigma_i\sigma_{i+1} + \frac{h}{2}(\sigma_i+\sigma_{i+1})\right]\right\} \ . \end{aligned}$$

It is convenient to introduce the 2×2 *transfer matrix*

$$\mathbf{P} = \begin{bmatrix} P_{11} & P_{1-1} \\ P_{-11} & P_{-1-1} \end{bmatrix}$$

where

$$\begin{aligned} P_{11} &= e^{\beta(J+h)} \\ P_{-1-1} &= e^{\beta(J-h)} \\ P_{-11} &= P_{1-1} = e^{-\beta J} \ . \end{aligned}$$

We may now use this to express the partition function in terms of a product of these matrices:

$$Z_N = \sum_{\{\sigma_i\}} P_{\sigma_1\sigma_2}P_{\sigma_2\sigma_3}\cdots P_{\sigma_N\sigma_1} = \mathrm{Tr}\,\mathbf{P}^N \ .$$

The matrix **P** can be diagonalized and the eigenvalues λ_1 and λ_2 are the roots of the secular determinant

$$|\mathbf{P} - \lambda \mathbf{I}| = 0 \,. \tag{3.36}$$

Similarly, the matrix $\mathbf{P}^N$ has eigenvalues λ_1^N, λ_2^N and the trace of $\mathbf{P}^N$ is the sum of the eigenvalues:

$$Z_N = \lambda_1^N + \lambda_2^N \,.$$

The solution of (3.36) is

$$\lambda_{1,2} = e^{\beta J} \cosh \beta h \pm \sqrt{e^{2\beta J} \sinh^2 \beta h + e^{-2\beta J}} \,.$$

We note that λ_1, associated with the positive root, is always larger than λ_2. The free energy is

$$\begin{aligned} G &= -k_B T \ln(\lambda_1^N + \lambda_2^N) = -k_B T \left\{ N \ln \lambda_1 + \ln \left[1 + \left(\frac{\lambda_2}{\lambda_1} \right)^N \right] \right\} \\ &\to -N k_B T \ln \lambda_1 \qquad\qquad \text{as } N \to \infty \,. \end{aligned}$$

This gives for the free energy in the thermodynamic limit,

$$G = -N k_B T \ln \left[e^{\beta J} \cosh \beta h + \sqrt{e^{2\beta J} \sinh^2 \beta h + e^{-2\beta J}} \right] \,. \tag{3.37}$$

For the special case $h = 0$ we obtain the previous result (3.35). We may compute the magnetization from

$$m = \langle \sigma_0 \rangle = -\frac{1}{N} \frac{\partial G}{\partial h} = \frac{k_B T}{\lambda_1} \frac{\partial \lambda_1}{\partial h} \,.$$

After some straightforward manipulations we find

$$m = \frac{\sinh \beta h}{\sqrt{\sinh^2 \beta h + e^{-4\beta J}}} \,. \tag{3.38}$$

We see that for $h = 0$ there is no spontaneous magnetization at any nonzero temperature. However, in the limit of low temperatures

$$\sinh^2 \beta h \gg e^{-4\beta J}$$

for any $h \neq 0$ and only a very small field is needed to produce saturation of the magnetization. The zero-field free energy will, in the limit $T \to 0$, approach the value $G(T \to 0) = -NJ$ corresponding to completely aligned spins. We can thus say that we have a phase transition at $T = 0$, while for $T \neq 0$ the free energy is an analytic function of its variables. This behavior contrasts with that of mean field (Section 3.1) or Bragg–Williams approximations (Section 3.2) in which a coexistence line extending from $T = 0$ to $T = T_c$ separates regions of positive and negative order parameter, with a discontinuity of the order parameter across the line. It is interesting to compare in more detail the exact and the mean field solution. For this reason we plot in Figure 3.5 the energy calculated exactly and in the Bragg–Williams approximation for different values of the external field. In Figure 3.6 we plot the susceptibility for different fields, as a function of temperature, in the two approximations, while the specific heat is shown in Figure 3.7. Results from the Bethe approximation are not shown since they agree with the exact ones in this case.

In Sections 3.4 and 3.5 we introduced the pair distribution function

$$g(j) = \langle \sigma_0 \sigma_j \rangle$$

and argued that in mean field theories one neglects long-range correlations between spins. In the simplest mean field theory, one makes the approximation

$$\langle \sigma_0 \sigma_j \rangle = \langle \sigma_0 \rangle \langle \sigma_j \rangle \ .$$

The error introduced by this approximation can be analyzed by studying the spin-spin correlation function

$$\Gamma(j) = \langle \sigma_i \sigma_{i+j} \rangle - \langle \sigma_i \rangle \langle \sigma_{i+j} \rangle$$

which can be calculated quite straightforwardly for the Ising chain. For simplicity we consider only the zero-field case ($h = 0$). Since there is no spin ordering for $T \neq 0$ we have $\langle \sigma_l \rangle = 0$ and $\Gamma(j) = g(j)$ in this case. We assume an Ising chain with free ends and assume that the spins σ_i, σ_{i+j} are far from the ends. We also let the exchange energy between spins l and $l+1$ be a variable J_l which will be set equal to a constant, J, at the end of the calculation. We have

$$\langle \sigma_i \sigma_{i+j} \rangle = \frac{1}{Z_N} \sum_{\{\sigma_l\}} \sigma_i \sigma_{i+j} \exp\left\{ \beta \sum_{l=1}^{N-1} J_l \sigma_l \sigma_{l+1} \right\}$$

and from (3.34)

$$Z_N = 2 \prod_{l=1}^{N-1} (2 \cosh \beta J_l) \ .$$

Since $\sigma_i^2 = 1$,

$$\langle \sigma_i \sigma_{i+j} \rangle = \frac{1}{Z_N} \sum_{\{\sigma_l\}} (\sigma_i \sigma_{i+1})(\sigma_{i+1} \sigma_{i+2}) \cdots (\sigma_{i+j-1} \sigma_{i+j}) \exp\left\{ \beta \sum_{l=1}^{N-1} J_l \sigma_l \sigma_{l+1} \right\}$$

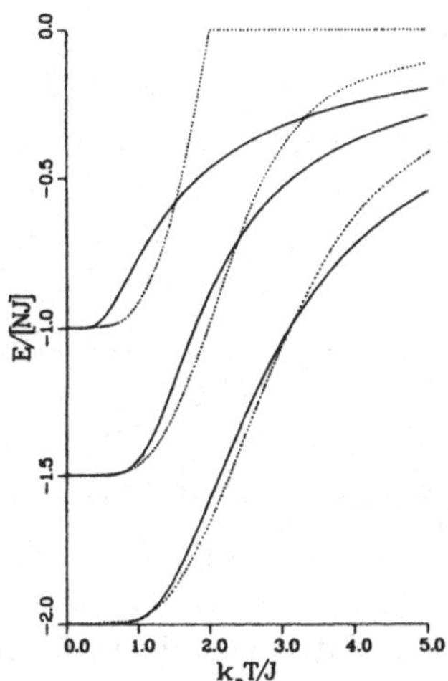

Figure 3.5: Comparison between exact and Bragg–Williams results for the internal energy of the one-dimensional Ising chain. Solid line, exact theory; dotted line, mean field theory. The three sets of curves correspond to $h = 0$, $h = 0.5J$, and $h = J$, respectively.

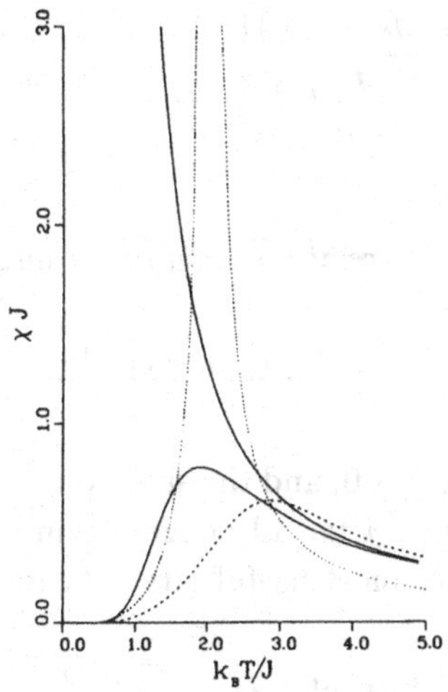

Figure 3.6: Comparison between exact (solid lines) and Bragg–Williams susceptibilities for a one-dimensional Ising chain. The two sets of curves correspond to $h = 0$ and $h = 0.5J$ respectively.

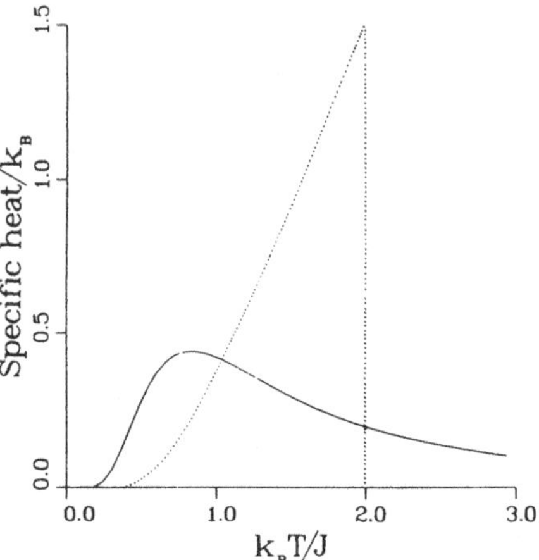

Figure 3.7: Specific heat calculated exactly (solid line) and in the Bragg–Williams approximation (dotted line) for the one-dimensional Ising chain. Only results for $h = 0$ are shown. For $h \neq 0$ the difference between the exact and approximate calculations is similar to that found for the susceptibility shown in Figure 3.6.

$$
\begin{aligned}
&= \frac{1}{Z_N \beta^j} \left. \frac{\partial^j Z_N(J_0 \cdots J_N)}{\partial J_i \cdots \partial J_{i+j-1}} \right|_{J_l = J} \\
&= (\tanh \beta J)^j = e^{-j/\xi} \qquad (3.39)
\end{aligned}
$$

where we have defined the *correlation length* ξ through

$$\xi = -[\ln(\tanh \beta J)]^{-1} .$$

Since $\tanh \beta J < 1$, we have $\xi > 0$, and the spin-spin correlation function decays exponentially with increasing j for all nonzero temperatures. The concept of correlation length will prove most useful later. At low temperatures

$$\ln(\tanh \beta J) \approx -e^{-2\beta J}$$

and we see that the correlation length can become quite large. The divergence of the correlation length at the critical point is a universal feature of continuous phase transitions.

3.7 Landau Theory of Phase Transitions

In 1936, Landau constructed a general theory of phase transitions. The crucial hypothesis is that in the vicinity of the critical point we may expand the free energy in a power series in the order parameter, which we denote by m. The equilibrium value of m is then the value that minimizes the free energy. It is worth pointing out immediately that the basic assumption, that the free energy is an analytic function of m at $m = 0$, is not correct. Nevertheless, Landau theory is of great utility as a qualitative tool and also plays an important role, after suitable generalization, in the modern renormalization theory of Wilson.

We begin by discussing a system in which the Gibbs free energy has the simple symmetry $G(m,T) = G(-m,T)$. We have tacitly assumed that the field h, which is conjugate to the order parameter m, is zero. With this symmetry the most general expansion of $G(m,T)$ is

$$G(m,T) = a(T) + \frac{1}{2}b(T)m^2 + \frac{1}{4}c(T)m^4 + \frac{1}{6}d(T)m^6 + \cdots \,, \tag{3.40}$$

where the fractional coefficients have been introduced in view of later manipulations. We have already encountered this type of expansion in the mean field treatment of the Ising model (Section 3.2), but formula (3.40) is more general than a specific instance of mean field theory.

The coefficients $b(T)$, $c(T)$, $d(T)\cdots$ are at this point unspecified, and we will investigate the consequences of different types of behavior of these functions. The first case we wish to consider is when c, d, e, $\cdots > 0$ and $b(T)$ changes sign at some temperature T_c. We write

$$b(T) = b_0(T - T_c)$$

in the vicinity of $T = T_c$. In this case the function $G(m,T)$ takes the form shown in Figure 3.8 for various values of T.

For $T < T_c$ the point $m = 0$ corresponds to a local maximum of the free energy, and the equilibrium state is one of the two states of spontaneously broken symmetry for which G has an absolute minimum. It is easy to work out the temperature dependence of the order parameter

$$\left(\frac{\partial G}{\partial m}\right)_T = 0 = bm + cm^3 + dm^5 + \cdots \,.$$

Ignoring the term dm^5, we find that

$$m \approx \pm\sqrt{\frac{b_0}{c(T_c)}}\sqrt{T_c - T}$$

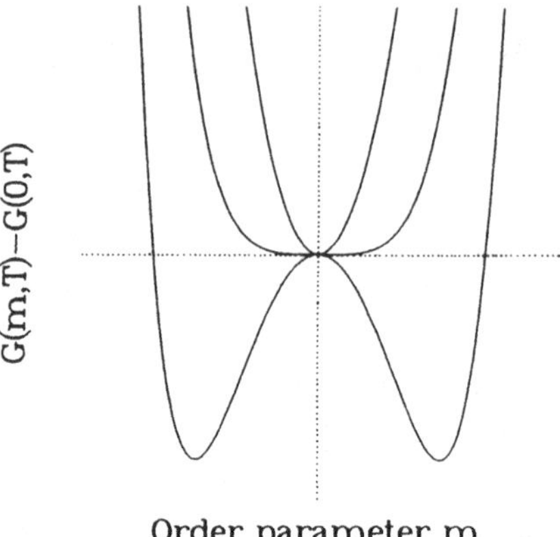

Figure 3.8: Free energy when c, $d \cdots > 0$ and $b(T) = b_0(T - T_c)$.

for $T \to T_c^-$. We may also obtain the behavior of the heat capacity

$$C = T\left(\frac{\partial S}{\partial T}\right) .$$

We let a prime indicate differentiation with respect to T and obtain

$$S = -\frac{\partial G}{\partial T} = -a' - \frac{b'}{2}m^2 - \frac{c'}{4}m^4 - \cdots - \frac{b}{2}(m^2)' - \frac{c}{4}(m^4)' - \cdots .$$

As $T \to T_c^-$

$$C \to -Ta'' - Tb'(m^2)' - \frac{Tc(m^4)''}{4}$$

where

$$\begin{aligned} (m^2)' &\to -\frac{b_0}{c} \\ b' &\to b_0 \\ (m^4)'' &\to \frac{2b_0}{c^2} \end{aligned}$$

giving

$$C \to \begin{cases} -Ta'' + Tb_0^2/2c & T \to T_c^- \\ -Ta'' & T \to T_c^+ \end{cases} .$$

We see that the order parameter and specific heat have the same form that we obtained previously in our mean field treatment of the Ising model.

We now consider a slightly different situation. Assume that c changes sign at some temperature, while $d(T) > 0$ and b is a decreasing function of the

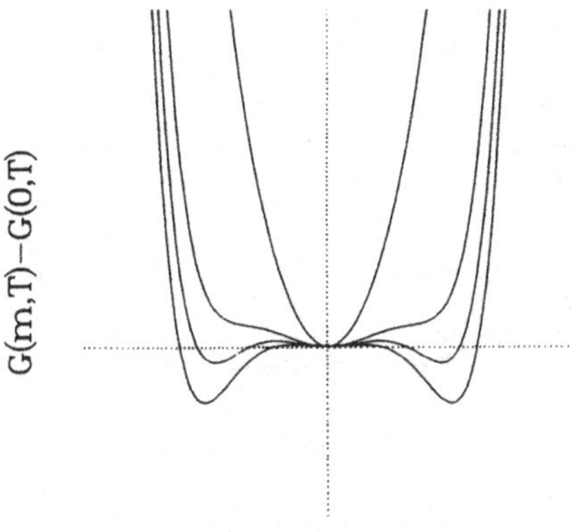

Figure 3.9: Free energy for a sequence of temperatures in the case that $c(T)$ changes sign above the temperature at which $b(T)$ changes sign.

temperature, but is still positive in the region of interest. The free energy in this case will be as shown in Figure 3.9. In this situation a discontinuous jump in the order parameter is expected. To see this, let $m_0 \neq 0$ be the location of a minimum of G. We must show that when $G(m_0, T_c) = G(0, T_c)$, $b(Tc) > 0$; that is, there is a local, rather than global minimum at the point $m = 0$. The equilibrium condition is

$$\left.\frac{\partial G}{\partial m}\right|_{m_0} = 0 = bm_0 + cm_0^3 + dm_0^5 + \cdots .$$

The phase transition occurs when

$$G(m_0) - G(0) = 0 = \frac{b}{2}m_0^2 + \frac{c}{4}m_0^4 + \frac{d}{6}m_0^6 .$$

Solving for the nontrivial value of m_0, we obtain

$$m_0^2 = -\frac{3c(T_c)}{4d} \tag{3.41}$$

and

$$b(T_c) = \frac{3c^2}{16d} > 0 \tag{3.42}$$

which justifies the claim that the first order transition occurs before a continuous transition can take place. The case where b and c approach zero at the same temperature seems at this stage rather unlikely. In Section 3.9 we

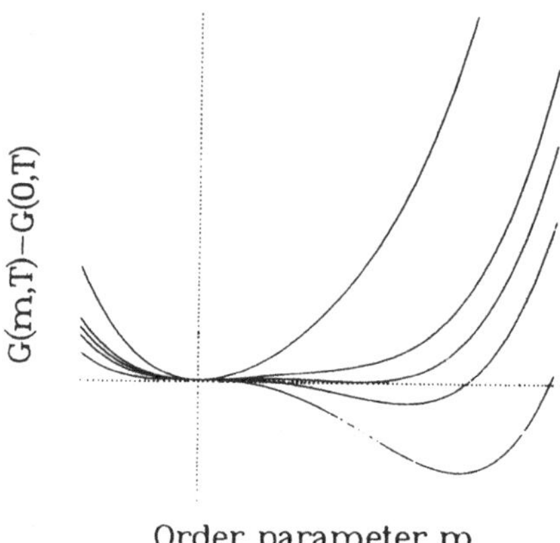

Figure 3.10: Free energy for different temperatures in the case where the Landau expansion contains a cubic term.

shall see that these special points can occur when the coefficients depend on other thermodynamic parameters besides the temperature and when there is competition between different ordered phases.

3.8 Symmetry Considerations

In this section we discuss the situation when $G(m,T) \neq G(-m,T)$; that is, there are terms of odd order in m in the Landau expansion. Below, we shall illustrate this situation by considering the 3-state Potts model. A further example, the Maier Saupe model of nematic liquid crystals, will be considered in Section 4.2, but first consider the general case. Any term in the Landau expansion which is linear in the order parameter can be eliminated by making the transformation $m \to m+\Delta$ and choosing Δ to make the linear term vanish. We therefore assume that the leading term of odd order in m is cubic and write

$$G(m,T) = a(T) + \frac{1}{2}b(T)m^2 - \frac{1}{3}c(T)m^3 + \frac{1}{4}d(T)m^4 + \cdots \ . \tag{3.43}$$

We assume for stability that $d(T) > 0$ and that $c(T) > 0$; $c(T) < 0$ corresponds simply to changing the sign of the order parameter. We also assume as before that $b(T)$ is a decreasing function of T which changes sign at some temperature T^*. With these assumptions the free energy will have the general form shown in Figure 3.10.

As we shall see, a first-order transition will again preempt the second order transition. At the transition point T_c we have

$$G(m_0, T_c) = G(0, T_c)$$

and

$$\left.\frac{\partial G}{\partial m}\right|_{m_0} = 0 = bm_0 - cm_0^2 + dm_0^3 \ .$$

Solving, we obtain

$$m_0 = \frac{2c}{3d}$$

and

$$b(T_c) = \frac{2c^2}{9d} > 0 \ .$$

We see, therefore, that the appearance of a cubic term in the Landau expansion signals a first-order phase transition. This prediction of the theory has been found to hold for three-dimensional systems, but the result turns out to be incorrect in the case of the three-state Potts model in two dimensions (see Sections 3.8.1 and 7.5.2). This is another indication that mean field theory is not reliable for low-dimensional systems.

3.8.1 Potts model

An example that gives rise to a Landau expansion with cubic terms, indicating a first order transition, is the Potts model [248]. Consider a system of N spins, each of which can be in any of q states. Each spin only interacts with nearest neighbor spins of the same type as itself, and the interaction energy is negative. The Hamiltonian is

$$H = -J \sum_{\langle i,j \rangle}^{N} \delta_{S_i, S_j}$$

where $J > 0$.

For $q = 2$ this is just the Ising model. We will restrict our attention to the case $q = 3$, and label the states A, B, C. Let $n_A = N_A/N$, $n_B = N_B/N$ and $n_C = N_C/N$. The free energy in the Bragg–Williams approximation is then

$$A = -\frac{qNJ}{2}[n_A^2 + n_B^2 + n_C^2] + Nk_BT[n_A \ln n_A + n_B \ln n_B + n_C \ln n_C] \ .$$

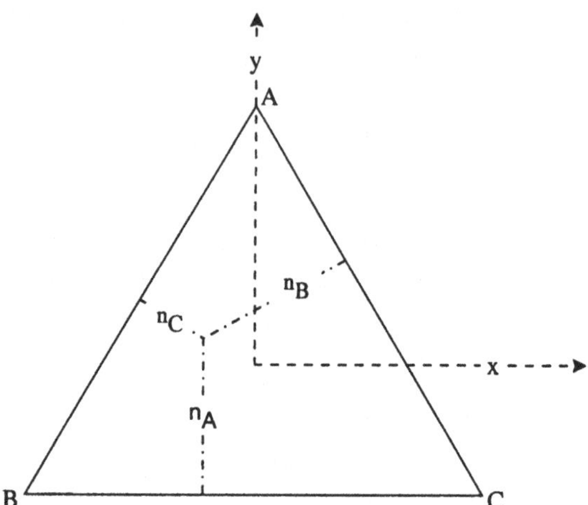

Figure 3.11: The allowed values of x, y are inside the triangle with corners (0,1), $(\sqrt{3}/2, -1/2)$, $(-\sqrt{3}, -1/2)$ corresponding to states of all A, B and C respectively.

In the disordered high temperature phase $n_A = n_B = n_C = 1/3$. In the general case the concentrations are subject to the constraint

$$n_A + n_B + n_C = 1 \ .$$

A possible parametrization is

$$\begin{aligned} n_A &= \tfrac{1}{3}(1+2y) \\ n_B &= \tfrac{1}{3}(1+\sqrt{3}x - y) \\ n_C &= \tfrac{1}{3}(1-\sqrt{3}x - y) \end{aligned} \tag{3.44}$$

with the allowed values restricted to being inside the equilateral triangle of Figure 3.11.

The possible ordered phases have preferential occupation of either the A, B or C state. Because of symmetry the free energy in the three cases will be the same, and to be specific we choose the order parameter to be of the form $x = 0,\ y = m$, with $-1/2 \leq m \leq 1$. We find for the free energy

$$A = -\frac{qNJ}{6}[1+2m^2]+Nk_BT[\frac{2}{3}(1-m)\ln(1-m)+\frac{1}{3}(1+2m)\ln(1+2m)-\ln(3)] \ .$$

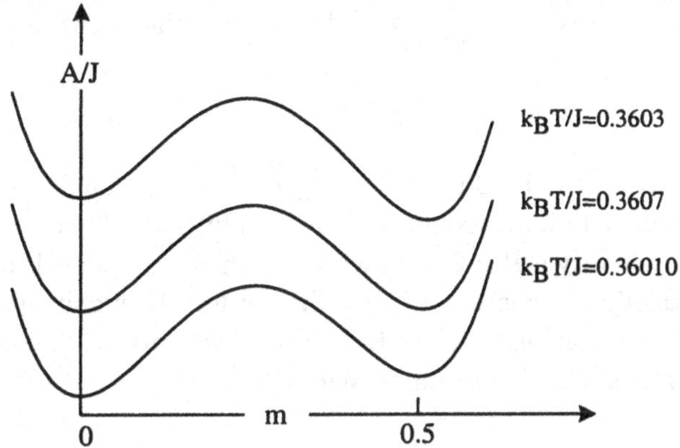

Figure 3.12: Free energy A/J plotted versus order parameter m for $k_BT/J = 0.3603, 0.3607, 0.361$, respectively. The value of y at the transition is seen to be close to 1/2.

We next wish to show that the system, when cooled, will undergo a first-order transition at some temperature T_c as $N \to \infty$ to a phase in which one of the three states A, B or C is selected in preference to the other two states.

The free energy has the power series expansion

$$\frac{A}{N} = -\frac{qJ}{6} - k_BT\ln(3) - (\frac{qJ}{3} - k_BT)m^2 - \frac{k_BT}{3}m^3 + \frac{k_BT}{2}m^4 + \cdots \quad .$$

The presence of a cubic term indicates that the ordering transition will be first order. This result can be confirmed by plotting the free energy vs. the order parameter for some temperatures.

The transition temperature in units of J/k_B can be found numerically. In Figure 3.12 we show the free energy vs. the order parameter for values of T near the transition temperature. The order parameter y was found to be very close to 1/2 at the transition. This result turns out to be exact and we find the transition temperature to be

$$T_C = \frac{J}{4k_B\ln 2} \quad .$$

At the transition we therefore have $n_C = 2/3$, $n_A = n_B = 1/6$.

Since the transition is first order it will be accompanied by latent heat. When $m = 0$ the entropy per spin is $k_B \ln 3$. We find that for $x = 0, y = 1/2$

the entropy per spin is $\frac{k_B}{3}\ln\frac{27}{2}$. We find then that the latent heat

$$L = T\Delta S = \frac{J}{12} \ .$$

We will come back to the 3-state Potts model in Section 7.5.2 where it is employed to model Helium monolayers on graphite at 1/3 coverage. In this case the dimensionality is $d = 2$ and mean field theory gives misleading results; the actual transition is continuous, not first order. In higher dimension the mean field treatment given here will be qualitatively correct. A comprehensive review of the Potts model is given by Wu [330].

3.9 Landau Theory of Tricritical Points

In Section 3.7 we pointed out that it is conceivable that the coefficients b and c in the Landau expansion (3.40) may approach zero simultaneously and that this could lead to new types of critical behavior. This situation is likely to occur when there are more control parameters than just the temperature, and there is more than one order parameter. As an example we consider in Problem 3.13 a simple solvable model in which an elastic field is coupled to an Ising chain. In Section 4.3 we consider another system, the Blume–Emery–Griffiths model for ^{3}He–^{4}He mixtures. Other examples of systems exhibiting tricritical points are the antiferromagnet $FeCl_2$, which undergoes a continuous transition in low applied magnetic fields and a first-order transition to a mixed phase (coexisting antiferromagnetic and ferromagnetic phases) at sufficiently high magnetic fields; the solid NH_4Cl, whose orientational transition changes from second to first order as a function of pressure; the ferroelectric $KDPO_4$; ternary liquid mixtures; and a number of liquid crystal systems. Below we discuss the general Landau approach. For a review of tricritical phenomena see Lawrie and Sarbach [167].

We denote, as is our practice, the order parameter of the system by m while h is the field that couples to m. The field that couples to the subsidiary order parameter x is denoted by Δ. In the case of $FeCl_2$,

$$m = M_Q = \sum_{\mathbf{r}} S_{\mathbf{r}} \exp\{i\mathbf{Q}\cdot\mathbf{r}\}$$

is a staggered magnetization, with $S_{\mathbf{r}}$ the spin at site $\mathbf{r}$, h a staggered magnetic field which is not realizable in the laboratory, x a uniform magnetization, and Δ an applied uniform magnetic field. In the case of the uniaxial-biaxial transition

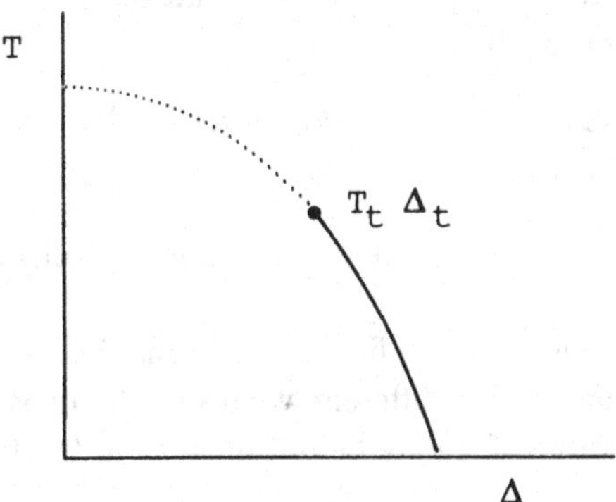

Figure 3.13: Phase behavior near a tricritical point. Solid line, first-order transitions; dotted line, critical points.

in liquid crystals, m and x are the order parameters P and Q of (4.24) and h, Δ two associated fields (*e.g.*, one electric and one magnetic) in two orthogonal directions (see Problem 4.4).

We assume the following form for the free energy:

$$\frac{G(m,\Delta,T)}{N} = a(T,\Delta) + \frac{1}{2}b(T,\Delta)m^2 + \frac{1}{4}c(T,\Delta)m^4 + \frac{1}{6}d(T,\Delta)m^6 \ .$$

The line of critical points is given by $b(T,\Delta) = 0$, which defines a curve in the $T-\Delta$ plane (Figure 3.13). The tricritical point is given by $b(T,\Delta) = c(T,\Delta) = 0$, which, in general, is a unique point (Δ_t, T_t).

We assume that when the temperature is lowered and $\Delta > \Delta_t$, the coefficient c becomes zero at a higher temperature than does b. The usual equations for first-order transitions will then apply. For $\Delta < \Delta_t$, the transition is continuous. We first show that the line of first-order transitions joins the line of critical points in a smooth fashion. The equation for the first-order line is given by (3.42)

$$b(\Delta,T) - \frac{3c^2(\Delta,T)}{16d(\Delta,T)} = 0 \ . \tag{3.45}$$

The equation $b(\Delta,T) = 0$ for the line of critical points yields for the slope of this line at $b = 0$,

$$\left.\frac{d\Delta}{dT}\right|_{crit} = -\frac{\left.\frac{\partial b}{\partial T}\right|_{\Delta}}{\left.\frac{\partial b}{\partial \Delta}\right|_{T}} \equiv -\frac{b_T}{b_\Delta} \ . \tag{3.46}$$

We let the subscript indicate partial differentiation and find from (3.45) for the slope of the first-order line,

$$\left.\frac{d\Delta}{dT}\right|_{first\,order} = -\frac{b_T d + d_T b - \frac{3}{8} c c_T}{b_\Delta d + d_\Delta b - \frac{3}{8} c c_\Delta} . \tag{3.47}$$

As $(T, \Delta) \rightarrow (T_t, \Delta_t)$, $c \rightarrow 0$, $b \rightarrow 0$, and equations (3.46) and (3.47) become the same.

It is also easy to see that the first-order transition for $\Delta > \Delta_t$, implies coexistence of two phases with different values of the density x. We suppose that the expectation value of x may be obtained from the free energy through

$$x = -\frac{1}{N} \left.\frac{\partial G}{\partial \Delta}\right|_T \tag{3.48}$$

where the minus sign implies a suitable sign convention for Δ. The first-order transition occurs when

$$G(m, T, \Delta) = G(0, T, \Delta) \tag{3.49}$$

and

$$\left.\frac{\partial G}{\partial m}\right|_{T,\Delta} = 0 . \tag{3.50}$$

We leave it as an exercise to prove that (3.48)–(3.50) yield an equation for the discontinuity in x. As the tricritical point is approached along the first-order line, the discontinuity in x takes the form

$$\delta x = -\frac{3}{8d}\,(b_\Delta c + c_\Delta b) + \mathcal{O}(c^2) \tag{3.51}$$

where we have used (3.41). Therefore, in the $T - x$ plane the phase diagram has the shape shown in Figure 3.14.

It is also of interest to calculate the asymptotic behavior of the order parameter and specific heat as the tricritical point is approached. Solving (3.50) we obtain

$$m^2 = \sqrt{\frac{c^2}{4d^2} - \frac{b}{d}} - \frac{c}{2d} . \tag{3.52}$$

There are two different asymptotic forms of this function, depending on how the tricritical point is approached. Since both b and c approach zero linearly, we expect that in most cases

$$\left|\frac{b}{d}\right| \gg \frac{c^2}{4d^2} .$$

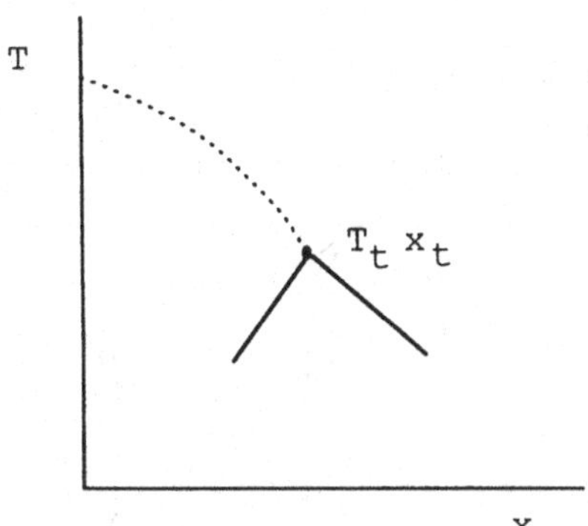

Figure 3.14: Phase diagram in the $T - x$ plane near the tricritical point.

If $b(T, \Delta) = b_0(T - T_t)$, we obtain

$$m(T) \approx \left[\frac{b_0}{d(T_t)}\right]^{1/4} (T_t - T)^{1/4} \ .$$

The exponent β at a tricritical point is therefore $\frac{1}{4}$ rather than the value $\frac{1}{2}$ found near a critical point. There is, however, a narrow region in the $T - \Delta$ plane given by

$$\left|\frac{b}{d}\right| < \frac{c^2}{4d^2}$$

in which the asymptotic behavior given above does not hold. In this region all terms in (3.52) are proportional to $T_t - T$ and

$$m \propto \sqrt{T_t - T} \ .$$

A rough sketch of the critical and tricritical regions is shown in Figure 3.15.

The exponents for a path of approach that lies in the tricritical region are often subscripted with a t, those for a path in the critical region with a u. Thus $\beta_t = \frac{1}{4}$, $\beta_u = \frac{1}{2}$. We leave it as an exercise (Problem 3.12) to show that $\gamma_u = 2$, $\gamma_t = 1$, $\alpha_t = \frac{1}{2}$, $\alpha_u = -1$. It is a remarkable fact that in three dimensions the predictions regarding the tricritical exponents are exact (to within logarithmic corrections [320]). In Section 3.10 we present self-consistency arguments which indicate the reasons for this result and which also show that the Landau–Ginzburg theory of critical points will be correct for spatial dimensionality $d \geq 4$.

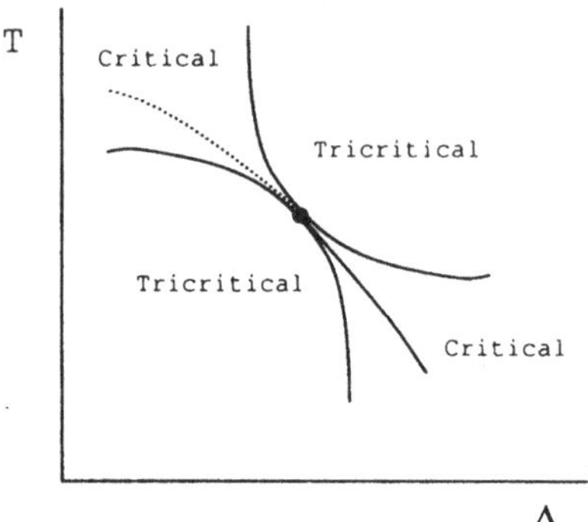

Figure 3.15: Critical and tricritical regions near a tricritical point.

3.10 Landau–Ginzburg Theory for Fluctuations

We have pointed out in Section 3.5 that the failure of mean field theories at a critical point is due to the neglect of long-range correlations. It is possible to generalize the Landau theory to incorporate fluctuations at least in an approximate fashion. Instead of the Gibbs free energy[2] which we have used throughout this chapter, we now make a Legendre transformation to a Helmholtz free energy which in the homogeneous case is $A(M,T) = G + hM$ with $dA = -SdT + hdM$. We will allow the independent variable to depend on position

$$M = \int d^3r \; m(\mathbf{r})$$

and assume that the free energy can be written

$$\begin{aligned} A(\{m(\mathbf{r})\},T) &= \int d^3r \left\{ a(T) + \frac{b(T)}{2} m^2(\mathbf{r}) + \frac{c(T)}{4} m^4(\mathbf{r}) \right. \\ &\left. + \frac{d(T)}{6} m^6(\mathbf{r}) + \cdots + \frac{f}{2} [\nabla m(\mathbf{r})]^2 \right\} . \end{aligned} \quad (3.53)$$

[2]We could equally well use the Gibbs potential

$$G(\{h(\mathbf{r})\},T,\{m(\mathbf{r})\}) = A - \int d^3r h(\mathbf{r}) m(\mathbf{r})$$

and treat $m(\mathbf{r})$ as a variational parameter. The resulting expressions (3.54)–(3.59) are identical.

The first three terms are a simple generalization of (3.40); the last term expresses the fact that the free energy is increased when the order parameter is not constant in space. The coefficient f can thus be assumed to be positive. We have in (3.53) expressed the free energy as a volume integral over the free-energy density. This is valid even for discrete systems, such as spins on a lattice, as long as $m(\mathbf{r})$ varies significantly only over sufficiently large distances that we may "coarse grain" dynamical variables. Near a critical point this approximation will certainly be valid. However, the fundamental objection to the Landau theory of Section 3.7, namely that the free energy is not necessarily an analytic function of the order parameter, applies equally to the inhomogeneous form (3.53). In the homogeneous case we have

$$h = \left.\frac{\partial A}{\partial M}\right|_T .$$

In the inhomogeneous case the generalization is the functional derivative

$$h(\mathbf{r}) = \frac{\delta A}{\delta m(\mathbf{r})} .$$

We construct the variation of A,

$$\delta A = \int d^3r \{\delta m(\mathbf{r})[bm(\mathbf{r}) + cm^3(\mathbf{r}) + dm^5(\mathbf{r}) + \cdots] + f\nabla\delta m(\mathbf{r})\cdot\nabla m(\mathbf{r})\} .$$

The last term can be simplified by carrying out an integration by parts and demanding that $\delta m(\mathbf{r}) = 0$ at the surface of the sample. We then obtain

$$h(\mathbf{r}) = bm(\mathbf{r}) + cm^3(\mathbf{r}) + dm^5(\mathbf{r}) + \cdots - f\nabla^2 m(\mathbf{r}) . \qquad (3.54)$$

From this equation we may recover the results of the homogeneous Landau theory by letting $h(\mathbf{r}) = 0$ and $\nabla m(\mathbf{r}) = 0$. Near a second-order transition the (uniform) order parameter then obeys the equation

$$m_0^2 = -\frac{b}{c} \qquad T < T_c \qquad (3.55)$$

which is familiar from Section 3.7.

Imagine now that a localized perturbation $h_0\delta(\mathbf{r})$ is applied to the material. Equation (3.54) allows us to calculate the effect of this perturbation throughout the system. Let $m(\mathbf{r}) = m_0(T) + \phi(\mathbf{r})$. Neglecting nonlinear terms in ϕ we write $m^3(\mathbf{r}) = m_0^3 + 3m_0^2\phi(\mathbf{r})$. With these approximations we obtain

$$\nabla^2\phi(\mathbf{r}) - \frac{b}{f}\phi(\mathbf{r}) - 3m_0^2\frac{c}{f}\phi(\mathbf{r}) - \frac{b}{f}m_0 - \frac{c}{f}m_0^3 = -\frac{h_0}{f}\delta(\mathbf{r}) . \qquad (3.56)$$

With $m_0 = 0$ for $T > T_c$, and given by (3.55) for $T < T_c$, we find

$$\begin{aligned}\nabla^2\phi - \frac{b}{f}\phi &= -\frac{h_0}{f}\delta(\mathbf{r}) \qquad T > T_c \\ \nabla^2\phi + 2\frac{b}{f}\phi &= -\frac{h_0}{f}\delta(\mathbf{r}) \qquad T < T_c\ .\end{aligned} \tag{3.57}$$

In three dimensions, these equations are easily solved in spherical coordinates:

$$\phi = \frac{h_0}{4\pi f}\frac{e^{-r/\xi}}{r} \tag{3.58}$$

with

$$\begin{aligned}\xi(T) &= \sqrt{\frac{f}{b(T)}} \qquad T > T_c \\ \xi(T) &= \sqrt{-\frac{f}{2b(T)}} \qquad T < T_c\ .\end{aligned} \tag{3.59}$$

The function $\xi(T)$ is the correlation length, and with

$$b(T) = b'(T - T_c)$$

we see that it diverges as $T \to T_c$ from both the low- and high-temperature sides. In this theory

$$\xi(T) \propto |T - T_c|^{-1/2}\ .$$

Experimentally, and in more exact theories,

$$\xi(T) \propto |T - T_c|^{-\nu}$$

with the critical exponent ν dependent on the model and the dimensionality.

We may relate the function $\phi(\mathbf{r})$ to a correlation function. Assuming that a term

$$-\int d^3r\, m(\mathbf{r})h(\mathbf{r})$$

is included in the Hamiltonian, we have

$$\langle m(\mathbf{r})\rangle = \frac{\mathrm{Tr}\, m(\mathbf{r}) \exp\{-\beta[H_0 - \int d^3r'\ h(\mathbf{r}')m(\mathbf{r}')]\}}{\mathrm{Tr}\ \exp\{-\beta[H_0 - \int d^3r'\ h(\mathbf{r}')m(\mathbf{r}')]\}}$$

where H_0 refers to the part of H which is independent of $h(\mathbf{r})$. We see that

$$\frac{\delta\langle m(\mathbf{r})\rangle}{\delta h(0)} = \phi(\mathbf{r})/h_0 = \beta\left(\langle m(\mathbf{r})m(0)\rangle - \langle m(\mathbf{r})\rangle\langle m(0)\rangle\right) = \beta\Gamma(\mathbf{r})\ . \tag{3.60}$$

The function $\phi(\mathbf{r})$ is thus proportional to the order parameter-order parameter correlation function. The susceptibility (a similar expression holds for the compressibility in the case of a fluid) is given by

$$\chi = \beta \int d^3r \; \Gamma(\mathbf{r})$$

and it is easily seen that the usual mean field result

$$\chi \propto |T - T_c|^{-1}$$

is recovered.

The results obtained above allow us to establish a self-consistency criterion for mean field (or Landau) theories known as the Ginzburg criterion. We first generalize the analysis to systems of spatial dimensionality d. Equations (3.57) and (3.58) are quite general—one simply replaces the operator ∇^2 by the analogous d-dimensional operator and the δ-function by the appropriate d-dimensional δ-function. The solutions to (3.57)–(3.58) are generally not of the simple form (3.59). However, one can show that in arbitrary dimension, d, for $r \ll \xi$, $\phi \propto r^{-d+2}$, while for $r \gg \xi, \phi \propto e^{-r/\xi}$. For the purpose of order of magnitude estimates we can thus write

$$\phi(\mathbf{r}) \approx \frac{e^{-r/\xi}}{r^{d-2}} \; .$$

In mean field theories we always crudely approximate the correlation functions (3.60) at large distances. Therefore, one might expect that such approximations would be valid if the ratio

$$\frac{\int_{\Omega(\xi)} d^dr[\langle m(\mathbf{r})m(0)\rangle - \langle m(\mathbf{r})\rangle\langle m(0)\rangle]}{\int_{\Omega(\xi)} d^dr \; m_0^2} \ll 1 \tag{3.61}$$

where the integral is carried out over a d-dimensional hypersphere of radius ξ. It is of interest to estimate the dimensionality d at which Landau theory correctly describes the critical behavior of the system. To do this we substitute the asymptotic form, as calculated from the Landau model, for the various functions appearing in (3.61). Substitution of

$$m_0^2 \approx |T - T_c|^{2\beta}$$

and

$$\langle m(\mathbf{r})m(0)\rangle - \langle m(\mathbf{r})\rangle\langle m(0)\rangle \approx \frac{\exp\{-r/\xi\}}{r^{d-2}}$$

and carrying out the integration in spherical coordinates, we obtain the condition

$$\frac{Bd\int_0^\xi drr^{d-1}e^{-r/\xi}r^{-(d-2)}}{B\xi^d\,|T-T_c|^{2\beta}} \ll 1 \tag{3.62}$$

where Br^d is the volume of a d-dimensional sphere of radius r. Letting $r = \xi x$ in the numerator produces

$$\left(d\int_0^1 dx\ xe^{-x}\right)|T-T_c|^{d\nu-2\beta-2\nu} \ll 1\ .$$

The first factor is simply a constant of order unity and the inequality will be satisfied as $T \to T_c$, if and only if $d\nu - 2\beta - 2\nu > 0$, or

$$d > 2 + \frac{2\beta}{\nu}\ . \tag{3.63}$$

At critical points the Landau theory yields $\beta = \frac{1}{2}$, $\nu = \frac{1}{2}$, and we obtain $d_c \geq 4$. At tricritical points we have $\beta_t = \frac{1}{4}$, $\nu_t = \frac{1}{2}$, and hence $d_t \geq 3$. The borderline values $d_c = 4$ and $d_t = 3$ are called *upper critical dimensionalities* and play an important role in the development of the renormalization group approach to critical phenomena. At these marginal dimensionalities there are small corrections to the Landau critical exponents. The Landau theory of tricritical points thus provides an excellent representation of the correct cooperative effect in three dimensions.

Another application of the Ginzburg criterion is that estimates of the correlation length can be used to determine the range of temperatures near T_c where critical fluctuations play an important role [146]. We return to this question in Section 11.3.4 where we argue that in the case of the BCS theory of superconductivity the temperature range is too small to be significant. On the other hand in the recently discovered high temperature superconductors fluctuations are quite significant in the critical region [147]. A further example of Landau–Ginzburg theory is given in Section 5.4 where we study properties of liquid-vapor interfaces.

3.11 Multicomponent Order Parameters: n-Vector Model

In many cases of physical interest the ground state of the system has a degeneracy which is greater than the twofold degeneracy of the zero-field Ising model.

An example of such a system is the Heisenberg model with the Hamiltonian

$$H = -\sum_{i<j} J_{ij} \mathbf{S}_i \cdot \mathbf{S}_j \tag{3.64}$$

in the absence of an applied field. The dynamical variables are the three-dimensional spin operators

$$\mathbf{S}_i = (S_{xi}, S_{yi}, S_{zi})$$

obeying the usual angular momentum commutation relations. The Hamiltonian (3.64) favors the parallel alignment of neighboring spins if $J_{ij} > 0$, and it is easily shown that the ground state has all the spins aligned in the same direction, which we label z, $S_{zi} = S$. The Hamiltonian (3.64) is rotationally invariant in spin space, and the z direction may be taken to be any direction. The application of a magnetic field breaks this symmetry, but the nature of the correlated fluctuations that determine the critical behavior of the system depend essentially on the existence of rotational symmetry. In this section we only wish to demonstrate the appropriate generalizations of Landau theory to take such symmetries into account. The equilibrium state of the Heisenberg model may be expressed in terms of the three thermal expectation values

$$\begin{aligned} m_x &= \frac{1}{N}\left\langle \sum_i S_{xi} \right\rangle \\ m_y &= \frac{1}{N}\left\langle \sum_i S_{yi} \right\rangle \\ m_z &= \frac{1}{N}\left\langle \sum_i S_{zi} \right\rangle . \end{aligned}$$

The rotational symmetry of (3.64) can then be incorporated into the Landau theory by constructing an expansion that is invariant under arbitrary rotations of the vector $\mathbf{m}$, that is,

$$G = a + \frac{1}{2} b(T)(m_x^2 + m_y^2 + m_z^2) + \frac{1}{4} c(T)(m_x^2 + m_y^2 + m_z^2)^2 + \cdots .$$

The general n-vector model, in which the three-component order parameter of the Heisenberg model is replaced by an n-component order parameter, will thus have its Landau expansion in terms of the quantity

$$m^2 = \sum_{\alpha=1}^{n} m_\alpha^2 .$$

Similarly, in the case of a nematic liquid crystal where the order parameter is the symmetric and traceless tensor $Q_{\alpha\beta}$ defined by (4.21), the Landau expansion must be expressible in terms of the two invariants which can be formed from such a tensor, namely

$$\sum_{\alpha,\beta} Q_{\alpha\beta}Q_{\beta\alpha}$$

$$\sum_{\alpha,\beta,\gamma} Q_{\alpha\beta}Q_{\beta\gamma}Q_{\gamma\alpha} \ .$$

It is clear from the foregoing that specific forms of symmetry breaking may easily be incorporated in the Landau free energy. For example, a magnet in a cubic lattice is in general subject to a crystal field of cubic symmetry. For a Heisenberg model on a cubic lattice the appropriate form of the Landau free energy is

$$\begin{aligned} G(\{m\},T) &= a + \frac{b}{2}(m_x^2 + m_y^2 + m_z^2) + \frac{c}{4}(m_x^2 m_y^2 + m_x^2 m_z^2 + m_y^2 m_z^2) \\ &+ \frac{d}{4}(m_x^4 + m_y^4 + m_z^4) + \cdots \end{aligned} \tag{3.65}$$

where in the absence of crystal fields $c(T) = 2d(T)$. The equilibrium behavior of the system is obtained for an n-component order parameter from the equations

$$\frac{\partial G}{\partial m_\alpha} = 0 \qquad \alpha = 1, 2, \ldots, n \ .$$

Other examples of systems with a multicomponent order parameter include superfluid ^{4}He (two components), superconductors (two components), and the q-state Potts model ($q - 1$ components) which describes the critical behavior of a number of two- and three-dimensional materials [see, *e.g.*, Sections 3.8.1 7.5.2].

3.12 Problems

3.1. *Ising Model with Long Range Interactions.* Consider a long chain of spins $\sigma_i = 1$ *or* -1. The interaction between the spins is not just between nearest neighbors, but long range

$$H = -\sum_{i=1}^{N}\sum_{j<i} \frac{J}{|i-j|^\alpha}\sigma_i\sigma_j$$

where α is a constant between 1 and 2, and $J > 0$ (ferromagnetic coupling). Assume periodic boundary conditions.

(a) Make a mean field approximation for the system in the limit $N \to \infty$ and estimate the transition temperature for ferromagnetic order, and the order parameter $m = \langle \sigma \rangle$ below the transition temperature. You will need to approximate the sum over neighboring spins by an integral.

(b) If the interaction between the spins had been between nearest neighbors only

$$H = -\sum_{i=1}^{N} \sigma_i \sigma_{i+1}$$

it was argued in the text that the energy associated with creating a "domain wall" separating regions with up and down spins would have been $2J$ while the entropy associated with creating two domains would have been proportional to $\ln N$. Since $\ln N > 2J$ in the limit $N \to \infty$ the system would be unstable against splitting up into domains and ferromagnetic ordering impossible. Show that this argument has to be modified if the interaction is long range and that ordering at a non-zero temperature is possible if $\alpha < 2$.

3.2. *One-Dimensional Ising Model in Bethe Approximation.*

Calculate the magnetization for the one-dimensional Ising model in a magnetic field in the Bethe approximation and compare with the exact result (3.38).

3.3. *Critical Exponents.*

(a) Fill in the missing steps to obtain equations (3.30)–(3.31).

(b) Show that the specific heat C_h at $h = 0$ is discontinuous in the Bethe approximation at $T = T_c$ for $q > 2$.

(c) Show that in the Bethe approximation to the Ising model $m(h = 0) \propto |T - T_c|^{1/2}$ near T_c.

(d) Show that the exponent for the critical isotherm in the Bethe approximation for the Ising model satisfies $\delta = 3$.

3.4. *Cluster Approximation for the Two-Dimensional Ising Model.*

The Bethe approximation can be modified to treat clusters of a more

Figure 3.16: Clusters for the two-dimensional Ising model on the square and triangular lattices.

general type. Consider as an example the two-dimensional Ising model on the square and triangular lattices. Divide the lattice into blocks of four and three spins as shown in Figure 3.16. Treat the interactions within a block exactly, while using the molecular field approximation for the interactions between spins in different blocks. Calculate the critical temperature for **(a)** the triangular ($\triangle$) lattice and **(b)** the square ($\square$) lattice and compare with the exact values

$$\frac{J}{k_B T_c} = 0.441\ldots(\square) \quad \frac{J}{k_B T_c} = 0.275\ldots(\triangle)$$

and with results from the simplest molecular field theory.

3.5. *Application of the One-Dimensional Ising Model to a Polymer Problem.* Use the one-dimensional Ising model to describe the following observation: The length l of molecules in a dilute solution of long chain-like polymer molecules is found to change with the temperature T as shown in Figure 3.17.

3.6. *One-Dimensional Ising Model with Spin 1.* Calculate the internal energy of the one-dimensional Ising model defined by the Hamiltonian

$$H = -J\sum_i \sigma_i\sigma_{i+1} \quad \sigma_i = 0, \pm 1, \quad J > 0 \,.$$

The solution of this problem requires differentiation of the root of a cubic equation. You may wish to do this numerically.

3.7. *Generalized Random Walk Problem.* Use the transfer matrix formalism to solve the following generalized

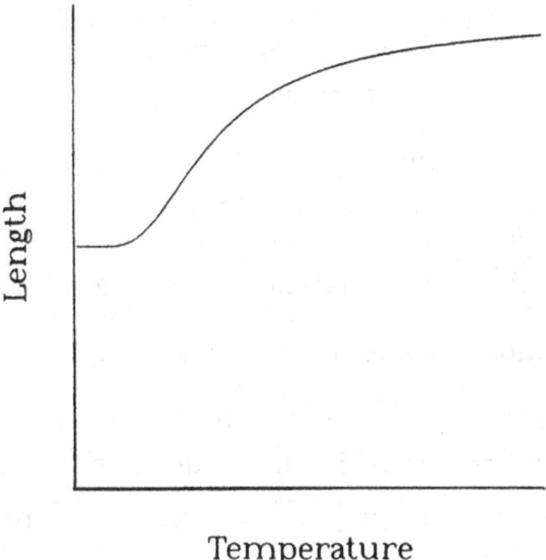

Figure 3.17: Temperature dependence of the average length l of long chain-like polymer molecules.

random walk problem: By observing drunks, one may notice that the completely random walk is only a crude first approximation in describing the motion. Inertia plays an important role in determining which way to take the next step. For this reason the next step will take place with greater probability in the direction of the previous one. A simple model for the motion can be constructed by assuming that there is a correlation between nearest neighbor steps, namely that the next step will be in the same direction as the previous one with probability p and in the opposite direction with probability $(1-p)$.

Calculate the mean-square displacement after N steps. The motion can be assumed to be one-dimensional.

3.8. *Clusters of Spins for the Ising Chain.*

(a) Consider the one-dimensional Ising model in a magnetic field h subject to periodic boundary conditions. Suppose that spin j is in a specific state, either up or down. Using the transfer matrix of Section 3.6, calculate the probability that spins $j+1, j+2, \cdots, j+n$ will be in the same state as spin j and that spin $j+n+1$ will be in the opposite state.

(b) Remove the restriction on spin $j+n+1$ and again calculate the probability.

3.9. *Latent Heat of a First-Order Transition.*

Consider the Landau free energy

$$G(m,T) = a(T) + \frac{b}{2}m^2 + \frac{c}{4}m^4 + \frac{d}{6}m^6$$

and assume that $b > 0, c < 0$, so that a first-order transition takes place. Derive an expression for the latent heat of transition.

3.10. *Asymptotic Behavior near a Tricritical Point.*

(a) Derive the result (3.51) for the discontinuity of the order parameter x across the first-order line near the tricritical point.

(b) Show that $\gamma_t = 1$, $\alpha_t = \frac{1}{2}$ in the tricritical region.

(c) Show that in the critical region the exponents predicted by Landau theory are $\gamma_u = 2$, $\alpha_u = -1$.

3.11. *Heisenberg Model in a Crystal Field.*

In Section 3.11 the Landau free energy in the presence of a cubic crystal field was given by equation (3.65). Assuming that the coefficients of higher than fourth order in m are all positive, determine the nature of the ordered phase. You may assume that the system will order in a (100), (111), or (110) preferred spin orientation and minimize the free energy with respect to a simple amplitude. In which situations will the transition be discontinuous?

3.12. *Alben Model.*

The symmetry breaking aspect of second order phase transitions can be nicely illustrated in a simple mechanical model [9]. An airtight piston of mass m is inside a tube of cross sectional area a. The tube is bent into a semicircular shape (see Figure 3.18) of radius R. The system is kept at temperature T. On each side of the piston there is an ideal gas consisting of N atoms. The volume to the right of the piston is $aR(\frac{\pi}{2} - \phi)$, while the volume to the left is $aR(\frac{\pi}{2} + \phi)$. Using the formula derived in problem 2.6 for the Helmholtz free energy of an ideal gas we find for the free energy of the system

$$A = MgR\cos\phi - Nk_BT\left[\ln\frac{aR(\frac{\pi}{2}+\phi)}{N\lambda^3} + \ln\frac{aR(\frac{\pi}{2}-\phi)}{N\lambda^3} + 2\right] .$$

(a) Show by minimizing the free energy that the system undergoes a symmetry breaking phase transition ($\phi \neq 0$) at a temperature

$$T_c = \frac{MgR\pi^2}{8Nk_B} .$$

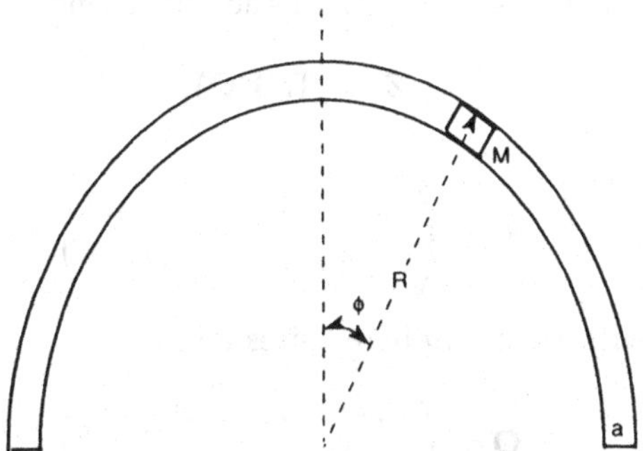

Figure 3.18: Alben model.

(b) Plot the "order parameter" ϕ vs. T/T_c for $T < T_c$.

(c) Describe what happens to the phase transition if the number of atoms on the left and right side of the piston is $N(1-\delta)$ and $N(1+\delta)$, respectively.

(d) At a certain temperature the right chamber (containing $N(1+\delta)$ molecules) is found to contain a puddle of liquid coexisting with its vapor. Which of the following statements may be true at equilibrium:

(i) The left chamber will contain a liquid in coexistence with its vapor.

(ii) The left chamber contains only vapor.

(iii) The left chamber contains only liquid.

3.13. *Solvable Model with Tricritical Point.*

Consider an Ising chain with N spins $\sigma_i = \pm 1$ and periodic boundary conditions. The chain is coupled to an elastic field ϵ. Nonzero values of ϵ causes a dimerization of the chain, *i.e.* alternating bonds are strengthened (or weakened). The dimensionless Hamiltonian for the system can be written (a similar model which includes a magnetic field has been described by Zaspel [333]).

$$H = -\sum_{i=1}^{N}[1 - \epsilon(-1)^i]\sigma_i\sigma_{i+1} + N\omega\epsilon^2 \ .$$

The partition function associated with the summation over spins can be computed by the transfer matrix method as discussed in Section 3.6

$$Z_\sigma = \mathrm{Tr}(\mathbf{PQ})^{\frac{N}{2}}$$

where

$$\mathbf{P} = \begin{pmatrix} e^{\beta(1+\epsilon)} & e^{-\beta(1+\epsilon)} \\ e^{-\beta(1+\epsilon)} & e^{\beta(1+\epsilon)} \end{pmatrix}$$

is associated with even numbered sites and

$$\mathbf{Q} = \begin{pmatrix} e^{\beta(1-\epsilon)} & e^{-\beta(1-\epsilon)} \\ e^{-\beta(1-\epsilon)} & e^{\beta(1-\epsilon)} \end{pmatrix}$$

corresponds to odd numbered sites. Let λ be the largest eigenvalue of the transfer matrix **PQ**. The partition function for the whole system can then be written as $N \to \infty$

$$Z_{tot} = \int_{-\infty}^{\infty} d\epsilon e^{-\beta N g(\epsilon)}$$

where

$$g(\epsilon) = -\frac{k_B T}{2} \ln \lambda(\epsilon) + \omega\epsilon^2 \ .$$

If $g(\epsilon)$ has an absolute minimum at ϵ_0 we find that $\epsilon = \epsilon_0$ at equilibrium, and that the free energy per spin is $g(\epsilon_0)$.

(a) Show that the largest eigenvalue of the transfer matrix is

$$\lambda = 2[\cosh(2\beta) + \cosh(2\beta\epsilon)] \ .$$

(b) There will be no phase transition if $\omega > 0.25$. Show that if $\omega = 0.20$ the system will undergo a second order phase transition to a dimerized state $\epsilon \neq 0$. Estimate the value of β at the transition.

(c) Show that if $\omega = 0.24$ the system will undergo a first order transition to a dimerized state. Estimate β at the transition (*e.g.* by plotting the free energy as a function of epsilon for a few temperatures).

(d) Estimate the values of ω and β at the tricritical point.

3.14. *The Potts Chain.* Consider a chain of N spins, each of which can be in any of three spin states. The system is subject to periodic boundary conditions and an external field with different components in the direction of each state. If two neighboring spins are in the same state, the energy of interaction between them is $-J$, otherwise the interaction is zero:

$$H = -\sum_{i=1}^{N}\left(J\delta_{S_i,S_{i+1}} + \sum_{\alpha=1}^{3} H_\alpha \delta_{S_i,\alpha}\right)$$

where

$$\delta_{i,j} = \begin{cases} 0 & i \neq j \\ 1 & i = j \end{cases} .$$

(a) Construct a transfer matrix for the system.

(b) Calculate the free energy per spin of the system in the thermodynamic limit $N \to \infty$ for the special case $H_1 = H, H_2 = H_3 = 0$.

(c) Plot the "magnetization" $m = \langle S_1 \rangle$ vs. H/k_BT for $J/k_BT = 0.1, 1, 4$ respectively.

3.15. *Mean Field Theory for q-state Potts Model.* Use the method of Section 3.8.1 to analyze the general case of the q-state Potts model, with $q \geq 3$.

(a) Show that there is a first order phase transition when the temperature is

$$k_BT = \frac{J(q-2)}{2(q-1)\ln(q-1)}$$

with order parameter jump to

$$m = \frac{q-2}{q-1} .$$

(b) Show that the latent heat of the transition is

$$L = \frac{J(q-2)^2}{2q(q-1)} .$$

Chapter 4

Applications of Mean Field Theory

We now continue our discussion of mean field theories. While the previous chapter concentrated on presenting the various techniques available, we now wish to give some typical examples of the very wide range of applications allowed by the method. In Section 4.1 we show that the usefulness of the Ising model is not limited to the magnetic problems for which it was originally intended, and apply it to a problem involving order-disorder transitions in alloys. Further examples are given as problems.

In Section 3.8 we showed that the presence of cubic terms in the Landau expansion led to a prediction of first order (or discontinuous) transitions and illustrated the argument by a discussion of the 3-state Potts model. In Section 4.2 we give an important further example, the Maier Saupe model for the isotropic to nematic transitions in liquid crystals. The discussion of tricritical points of Section 3.9 is applied in Section 4.3 to the problem of phase separation in ^{3}He–^{4}He mixtures at low temperatures.

The van der Waals theory of fluids of Section 4.4 was the first modern mean field theory of phase transitions. It also led to the theory of "corresponding states", hinting at the idea of universality, which we will discuss further in Chapters 6 and 7. In Section 4.5 we extend mean field theory to a problem of population biology and consider a model of insect infestations that is mathematically very similar to the van der Waals model. We argue that it is possible frame the problem as one of equilibrium phase transitions. We return to the

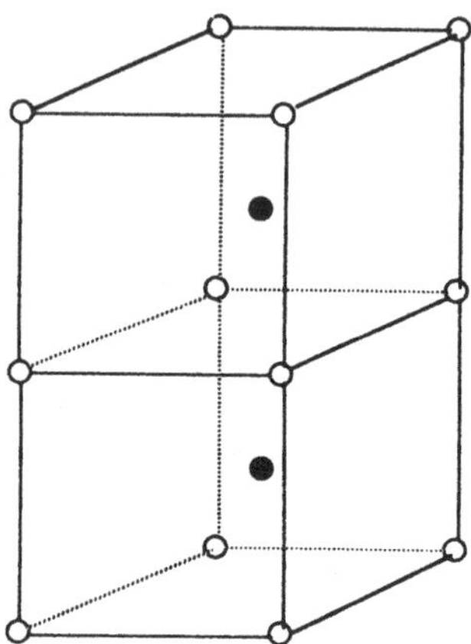

Figure 4.1: Ordered structure for β' brass: open circles, Zn filled circles Cu.

insect infestation problem in Section 8.2 where we calculate the equilibrium fluctuations of the model. Finally in Section 4.6 we consider a more explicitly non-equilibrium model. Again the model exhibits phase transitions similar to those of equilibrium theory.

4.1 Order–Disorder Transition

In the binary alloy of Cu and Zn (brass), at compositions in a narrow range around 50 atomic % Cu, 50 atomic % Zn, the atoms occupy the sites of a body-centered cubic (bcc) lattice forming β-brass. The distribution of atoms on these sites is disordered above a temperature T_c which is close to $740K$. Below T_c there is ordering with atoms of each kind preferentially distributed on one of the two simple cubic sublattices of the bcc lattice (β' phase, see Figure 4.1).

The simplest model that can account for the low-temperature structure is one in which the energy of nearest-neighbor pairs depends on what kind of pair it is. We define the quantities N_{AA}, N_{BB}, N_{AB} to be the number of nearest-neighbor pairs of the Cu-Cu, Zn-Zn, and Cu-Zn type, respectively, and take the energy of the configuration to be

$$E = N_{AA}e_{AA} + N_{AB}e_{AB} + N_{BB}e_{BB} \tag{4.1}$$

where e_{AA}, e_{AB}, and e_{BB} are, respectively, the energies of an AA, AB, and BB bond.

Let N be the numbers of lattice sites and N_A, N_B the number of Cu and Zn atoms, respectively. Referring to Figure 4.1, we introduce the occupation numbers for each of the two simple cubic sublattices: N_{A1} and N_{B1} are the number of atoms of each type on sublattice 1, N_{A2} and N_{B2} the number of atoms of each type on sublattice 2. We have

$$\begin{aligned} N_{A1} + N_{A2} &= N_A = c_A N \\ N_{B1} + N_{B2} &= N_B = c_B N \\ N_{A1} + N_{B1} &= \frac{1}{2} N \\ N_{A2} + N_{B2} &= \frac{1}{2} N \, . \end{aligned} \tag{4.2}$$

For the sake of definiteness we let $N_A \leq N_B$ and define the order parameter

$$m = \frac{N_{A1} - N_{A2}}{N_A} \, . \tag{4.3}$$

With this definition $-1 \leq m \leq 1$ and

$$\begin{aligned} N_{A1} &= \tfrac{1}{2} N_A (1+m) \quad & N_{B1} &= \tfrac{1}{2}(N_B - N_A m) \\ N_{A2} &= \tfrac{1}{2} N_A (1-m) \quad & N_{B2} &= \tfrac{1}{2}(N_B + N_A m) \, . \end{aligned} \tag{4.4}$$

Up to this point our treatment is exact. We now make the crucial approximations

$$\begin{aligned} N_{AA} &= q \frac{N_{A1} N_{A2}}{\frac{1}{2} N} \qquad\qquad N_{BB} = q \frac{N_{B1} N_{B2}}{\frac{1}{2} N} \\ N_{AB} &= q \left(\frac{N_{A1} N_{B2}}{\frac{1}{2} N} + \frac{N_{A2} N_{B1}}{\frac{1}{2} N} \right) \end{aligned} \tag{4.5}$$

where q is the number of nearest neighbors surrounding each atomic site. The mean energy is obtained by substituting (4.5) into (4.1), while the entropy can be evaluated using the method of Section 3.2. The appropriate free energy is

$$A = E - TS \tag{4.6}$$

with

$$E = \frac{1}{2} q N (e_{AA} c_A^2 + 2 e_{AB} c_A c_B + e_{BB} c_B^2) - q N c_A^2 \epsilon \, m^2 \tag{4.7}$$

where

$$\epsilon = \frac{1}{2}(e_{AA} + e_{BB}) - e_{AB} \tag{4.8}$$

and

$$
\begin{aligned}
S &= -k_B\left(N_{A1}\ln\frac{2N_{A1}}{N}+N_{A2}\ln\frac{2N_{A2}}{N}+N_{B1}\ln\frac{2N_{B1}}{N}+N_{B2}\ln\frac{2N_{B2}}{N}\right)\\
&= -\frac{1}{2}Nk_B\Big(c_A(1+m)\ln[c_A(1+m)]+c_A(1-m)\ln[c_A(1-m)]\\
&\quad +(c_B-mc_A)\ln[c_B-mc_A]+(c_B+mc_A)\ln[c_B+mc_A]\Big)\ . \qquad (4.9)
\end{aligned}
$$

The quantities c_A and c_B are held fixed while m must be adjusted to minimize the free energy. Differentiation of A with respect to m gives

$$0 = -2qc_A\epsilon\, m + \frac{1}{2}k_BT\ln\frac{(1+m)(c_B+c_Am)}{(1-m)(c_B-c_Am)}\ . \qquad (4.10)$$

For low temperature and $\epsilon > 0$ there are three solutions to (4.10): a trivial solution $m = 0$, and two nonzero solutions symmetric about $m = 0$. At high temperatures only the trivial solution exists. As can easily be verified by differentiating (4.10) with respect to m, the trivial solution yields a minimum of the free energy for

$$T > T_c = \frac{2q\epsilon c_Ac_B}{k_B}\ .$$

This situation is very similar to that of Section 3.2; if we were to plot the free energy (4.6) as a function of the order parameter m, the resulting figure would look similar to Figure 3.1. In the special case $c_A = c_B = \frac{1}{2}$, (4.10) can be written

$$0 = -q\epsilon m + k_BT\ln\frac{1+m}{1-m} \qquad (4.11)$$

which is equivalent to (3.15) if in that equation we take $h = 0$ and $J = \epsilon/2$. The nature of the phase transition in the alloy system is thus identical to the phase transition of the Ising ferromagnet. This last conclusion is independent of the mean field approximation, as we now show.

We introduce the variables n_{iA}, n_{iB}, where $n_{iA} = 1$ if an atom of type A occupies site i and $n_{iA} = 0$ otherwise. Similarly, $n_{iB} = 1 - n_{iA}$. These variables can be expressed in terms of Ising spin variables

$$
\begin{aligned}
n_{iA} &= \tfrac{1}{2}(1+\sigma_i)\\
n_{iB} &= \tfrac{1}{2}(1-\sigma_i)
\end{aligned} \qquad (4.12)
$$

with $\sigma_i = \pm 1$. With $\epsilon = 2J$, the energy (4.1) becomes

$$H = J\sum_{\langle ij\rangle}\sigma_i\sigma_j + \frac{q}{4}(e_{AA}-e_{BB})\sum_i\sigma_i + \frac{q}{8}N(e_{AA}+e_{BB}+2e_{AB})\ . \qquad (4.13)$$

Remembering that we have a system with a fixed total number of particles of each kind, we see that the last two terms are constant and therefore irrelevant. We obtain the equivalent Hamiltonian

$$H = J\sum_{\langle ij\rangle} \sigma_i\sigma_j \ . \tag{4.14}$$

If $J > 0$, that is, if it is energetically favorable for atoms of opposite kind (spin) to be nearest neighbors, this Hamiltonian represents an antiferromagnet.

We can also show that under certain circumstances the ferromagnetic and antiferromagnetic Ising models are equivalent. Consider a crystal structure that can be divided into two *sublattices*, so that the nearest neighbors to the sites on one of the sublattices belong to the other (*e.g.*, the square, honeycomb, simple cubic, and body-centered cubic lattices, but not the triangular, face-centered cubic, or hexagonal close-packed lattices). We may then make the transformation $\sigma_i = -\tau_i$ for i on one of the sublattices and $\sigma_i = +\tau_i$ for i on the other sublattice and have

$$H = -J\sum \tau_i\tau_j \ . \tag{4.15}$$

Since $\tau_l = \pm 1$, the partition function for the Hamiltonian (4.15) is the same as that of the Ising ferromagnet with $h = 0$. Thus the two systems have identical thermodynamic properties at all temperatures.

In the derivations given above, we allowed the concentrations of the two components of the alloy to vary freely. In practice, β-brass occurs only in a fairly narrow concentration range[1] around 50% Cu, 50% Zn. At other stoichiometries the face-centered cubic, more complex cubic structures, or the hexagonal closed packed structure may be thermodynamically stable, or the system may be in a mixture of different phases. In general, there is no guarantee that a particular choice of lattice structure, division into sublattices, or selection of order parameter is the correct one. One should therefore be guided by physical intuition in trying out a number of different alternatives, selecting the one with lowest free energy.

The homogeneous phase with the lowest free energy may also be unstable with respect to *phase separation.* The difference in concentration of each species is analogous to the magnetization in the Ising model, and since we are dealing with a system with fixed magnetization (rather than external field), we use the symbol A in preference to G for the free energy. Consider a sample with concentration $c_A = c_0$ and let the minimum single-phase free energy be

[1] see *e.g.* [156]

$A(c_0)$. If the sample were to split into two phases, one with a fraction y of the total number of sites, the constraint that the overall concentration of A-atoms is c_0 is expressed through the *lever rule*,

$$yc_1 + (1 - y)c_2 = c_0 \tag{4.16}$$

where c_1 and c_2 are the concentrations of A atoms in the two phases. The homogeneous phase is stable against phase separation if for all c_1 and c_2

$$yA(c_1) + (1 - y)A(c_2) > A(c_0) \ . \tag{4.17}$$

Geometrically, (4.17) corresponds to requiring that $A(x)$ be a *convex function.* If $A(x)$ can be differentiated twice, (4.17) is equivalent to the condition that

$$\frac{\partial^2 A}{\partial c^2} > 0 \quad \text{for all } c \ . \tag{4.18}$$

When the convexity requirement is violated, phase separation will occur. The resulting free energy lies on the convex envelope of $A(c)$. The equilibrium concentrations c_1, and c_2 are given by the lever rule (4.16) and by the condition $\partial A/\partial c|_{c_1} = \partial A/\partial c|_{c_2}$ *i.e.*, that the chemical potential is the same in the two phases. This constitutes the double tangent construction of Figure 4.2. A simple model for phase separation is found in Problem 4.1.

4.2 Maier–Saupe Model

An example of a model that gives rise to a Landau expansion with a cubic term is the Maier–Saupe model for the isotropic to nematic transition in liquid crystals[2] [186][187]. We consider a system of anisotropic molecules with a symmetry axis. The center of the i'th molecule is taken to be at $\mathbf{r}_i$ and the unit vector pointing in the direction of the symmetry axis is denoted by $\hat{\mathbf{n}}_i$. We further assume that the directions $\hat{\mathbf{n}}_i$ and $-\hat{\mathbf{n}}_i$ are equivalent. The interaction between the molecules is represented by a pair potential $W(\mathbf{r}_j - \mathbf{r}_i, \hat{\mathbf{n}}_j, \hat{\mathbf{n}}_i)$. We assume that the number of molecules ρ per unit volume is constant and let $f(\hat{\mathbf{n}})$ be the probability density that a molecule is oriented along $\hat{\mathbf{n}}$, and define $\mathbf{r}_{ji} = \mathbf{r}_j - \mathbf{r}_i$. We introduce the pair distribution function $g(\mathbf{r}_{ji}, \hat{\mathbf{n}}_j, \hat{\mathbf{n}}_i)$ as the conditional probability that there is a molecule at $\mathbf{r}_i$ with orientation $\hat{\mathbf{n}}_i$ given that there is a molecule at $\mathbf{r}_j$ with orientation $\hat{\mathbf{n}}_j$. In analogy with

[2]For an introduction to the properties of liquid crystals, we recommend de Gennes and Prost [71], Priestley et al. [251], Stephens and Straley [288] and Chandrasekhar [59].

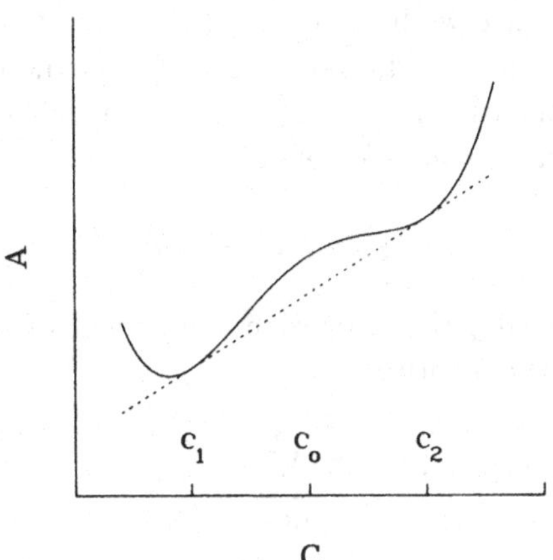

Figure 4.2: Phase separation. Material of initial concentration c_0 splits up into two phases with concentrations c_1 and c_2. The amount of material in each phase is given by the double tangent construction.

our discussion of the Weiss molecular field in Section 3.1 we write for the *pseudopotential* or the term in the average total energy which depends on the orientation $\hat{\mathbf{n}}_i$:

$$\epsilon(\hat{\mathbf{n}}_i) = \text{const.} + \rho \int d^3 r_{ji} \int d\Omega_j W(\mathbf{r}_{ji}, \hat{\mathbf{n}}_i, \hat{\mathbf{n}}_j) f(\hat{\mathbf{n}}_j) g(\mathbf{r}_{ji}, \hat{\mathbf{n}}_j, \hat{\mathbf{n}}_i)$$

where the integration over Ω_j extends over the solid angle of $\hat{\mathbf{n}}_j$. Apart from the mean field approach, a fundamental approximation of the Maier–Saupe model consists of ignoring the dependence of the pair distribution function on orientation. This allows us to write

$$\epsilon(\hat{\mathbf{n}}_i) = \text{const.} + \rho \int d\Omega_j f(\hat{\mathbf{n}}_j) \int d^3 r_{ji} W(\mathbf{r}_{ji}, \hat{\mathbf{n}}_i, \hat{\mathbf{n}}_j) g(\mathbf{r}_{ji}) \ . \tag{4.19}$$

We can expand the second integral in (4.19) in a Legendre series in $\hat{\mathbf{n}}_i \cdot \hat{\mathbf{n}}_j$ to obtain

$$\epsilon(\hat{\mathbf{n}}_i) = \text{const.} + \rho \int d\Omega_j f(\hat{\mathbf{n}}_j) \left(\gamma - 2U P_2(\hat{\mathbf{n}}_i \cdot \hat{\mathbf{n}}_j) + \mathcal{O}[P_4(\hat{\mathbf{n}}_i \cdot \hat{\mathbf{n}}_j)] \cdots\right) \tag{4.20}$$

where γ and U are constants and $U > 0$ corresponds to the situation where it is energetically advantageous for the molecules to parallel align. In (4.20)

$P_2(x) = \frac{1}{2}(3x^2 - 1)$ is the second Legendre polynomial, and P_4 is the fourth polynomial. Note that since we have assumed that the directions $\hat{\mathbf{n}}$ and $-\hat{\mathbf{n}}$ are equivalent, there will be no odd-index Legendre polynomials in the series (4.20). The final approximation in the Maier–Saupe model is to stop at second order in the Legendre expansion. We define

$$\sigma_{i\alpha\beta} = \frac{1}{2}(3n_{i\alpha}n_{i\beta} - \delta_{\alpha\beta})$$

where α, $\beta = x, y, z$ and $n_{i\alpha}$ is a Cartesian component of $\hat{\mathbf{n}}_i$ and $\delta_{\alpha\beta}$ is the Kronecker delta. Using the identity

$$P_2(\hat{\mathbf{n}}_j \cdot \hat{\mathbf{n}}_i) = \frac{2}{3}\sum_{\alpha,\beta=1}^{3} \sigma_{i\alpha\beta}\sigma_{j\alpha\beta}$$

and defining

$$Q_{\alpha\beta} = \langle\sigma_{j\alpha\beta}\rangle = \int d\Omega_j \sigma_{j\alpha\beta} f(\hat{\mathbf{n}}_j) \tag{4.21}$$

we can rewrite (4.20) to obtain

$$\epsilon(\hat{\mathbf{n}}_i) = -\frac{4}{3}\rho U \sum_{\alpha\beta} Q_{\alpha\beta}\, \sigma_{i\beta\alpha}$$

where we have omitted terms that do not depend on particle orientation. Taking into account the double counting that occurs when we sum the pair potential over all molecules, we find for the orientational contribution to the internal energy,

$$E = -\frac{2}{3}\rho U N \sum_{\alpha\beta} Q_{\alpha\beta} Q_{\beta\alpha} \ . \tag{4.22}$$

The orientational contribution to the entropy is given by

$$S_{or} = -Nk_B \int d\Omega f(\hat{\mathbf{n}}) \ln f(\hat{\mathbf{n}}) \ .$$

It then follows from the argument used in Section 2.5 that the single-particle distribution function

$$f(\hat{\mathbf{n}}) = \frac{\exp\{-\beta\epsilon(\hat{\mathbf{n}})\}}{\int d\Omega \exp\{-\beta\epsilon(\hat{\mathbf{n}})\}}$$

will minimize the free energy $G = E - TS_{or}$, and the order parameter $Q_{\alpha\beta}$ can then be determined from the self-consistency criterion

$$Q_{\alpha\beta} = \int d\Omega\, \sigma_{\alpha\beta} f(\hat{\mathbf{n}}) \ . \tag{4.23}$$

In analogy with the Weiss molecular field theory there will always be a solution $Q_{\alpha\beta} = 0$ and we identify such solutions with the isotropic high-temperature phase. At low temperatures nonzero solutions of (4.23) will appear. Each such solution corresponds to a preferential orientation of the *director field* $\hat{\mathbf{n}}$ and we call this phase *nematic*. The nematic phase does not exhibit long range spatial order, and we distinguish it from the more complicated *smectic* phases, which exhibit varying degrees of translational ordering. The nonzero solutions of (4.23) will not be unique, because of the overall rotational symmetry of the system. However, since $Q_{\alpha\beta}$ is real and symmetric, there will always be a principal-axis coordinate system in which $Q_{\alpha\beta}$ is diagonal. We let θ and ϕ be the polar angles of $\hat{\mathbf{n}}$ in such a system

$$\begin{aligned} n_x &= \sin\theta\cos\phi \\ n_y &= \sin\theta\sin\phi \\ n_z &= \cos\theta \ . \end{aligned}$$

Defining

$$\begin{aligned} p &= \tfrac{3}{2}\sin^2\theta\cos 2\phi; \quad & q &= \tfrac{1}{2}(3\cos^2\theta - 1) \\ P &= \langle p\rangle; & Q &= \langle q\rangle \end{aligned} \tag{4.24}$$

we find after some algebra

$$\mathbf{Q} = \begin{bmatrix} -\frac{1}{2}(Q-P) & 0 & 0 \\ 0 & -\frac{1}{2}(Q+P) & 0 \\ 0 & 0 & Q \end{bmatrix}$$

$$\begin{aligned} \epsilon(\hat{\mathbf{n}}) &= -2\rho U(Qq + \tfrac{1}{3}Pp) \\ E &= -N\rho U(Q^2 + \tfrac{1}{3}P^2) \ . \end{aligned}$$

If we choose the z-axis to be in the direction of the eigenvector belonging to the largest eigenvalue of $\mathbf{Q}$, we will, for molecules with a symmetry axis, have $P = 0$, and with $\mu = \cos\theta$

$$g = \frac{G}{N} = -k_BT\ln\left(4\pi\int_0^1 d\mu\exp\{\rho U\beta[(3\mu^2-1)Q - Q^2]\}\right) \ . \tag{4.25}$$

It is now a straightforward matter to obtain a Taylor expansion of (4.25) in powers of Q, and after some algebra we obtain the expansion

$$\begin{aligned} g &= -k_BT\ln(4\pi) + \rho UQ^2\left(1 - \frac{2}{5}\beta\rho U\right) \\ &\quad - \frac{8}{105}\beta^2\rho^3U^3Q^3 + \frac{4}{175}\beta^3\rho^4U^4Q^4 + \cdots \ . \end{aligned} \tag{4.26}$$

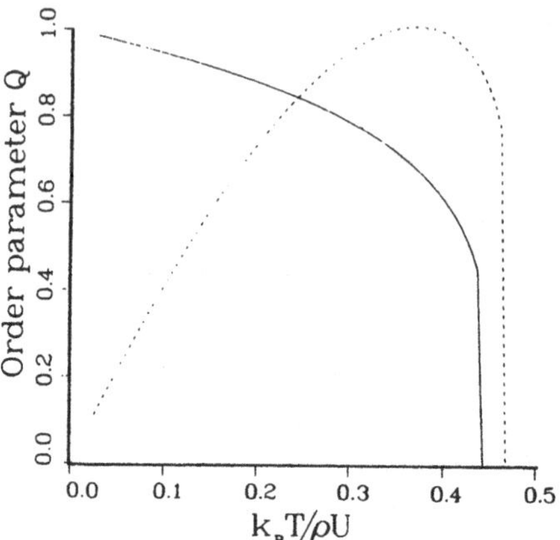

Figure 4.3: Order parameter as a function of temperature. The dashed line was obtained from the Landau expansion (4.26) and the solid line from the self-consistent equation (4.23).

This expansion is indeed of the form (3.43). We leave it as an exercise for the reader (Problem 4.2) to work out further details of the model.

In Figure 4.3 we plot the order parameter obtained by minimizing (4.26) as a function of the temperature. We also plot the order parameter resulting from the self-consistent equation (4.23). In the principal-axis frame with the molecules aligned preferentially along the z axis, this equation takes the form

$$Q = \frac{1}{2}\langle 3\mu^2 - 1\rangle = \frac{\int_0^1 d\mu \frac{1}{2}(3\mu^2 - 1)\exp\{3\beta\rho U Q\mu^2\}}{\int_0^1 d\mu \exp\{3\rho\beta Q U\mu^2\}} \ . \tag{4.27}$$

Equation (4.27) is most easily solved by choosing a value for $x = 3\beta\rho U Q$, evaluating Q numerically and then using the calculated value of Q to obtain the temperature.

Figure 4.3 illustrates that while the Landau theory gives a correct qualitative picture, there are difficulties associated with using the Landau expansion for quantitative purposes in the case of a first-order transition. The problem is that the jump in the order parameter is not necessarily small and the expansion (3.43) to low order may not be accurate.

We note that the simple mean field theory (4.27) predicts that the discontinuity in the order parameter Q at the transition should be 0.43. Both smaller and larger values have been observed experimentally. When steric effects are included in the theory, as in the van der Waals theory of Section 4.4, one tends to get a more strongly first-order transition. On the other hand, when one takes into account the fact that actual molecules do not have cylindrical symmetry (see, *e.g.*, Straley [291]) the discontinuity in the order parameter tends to be smaller than in the Maier–Saupe theory. It is then also possible, in principle, to obtain biaxial phases [59].

It is of interest to extend the Maier–Saupe theory to include the effect of a magnetic (or electric) field. The magnetic susceptibility will in general be anisotropic and we let $\chi_{\parallel}$ and $\chi_{\perp}$ correspond to orientations of the field parallel and perpendicular to the molecular axis. The energy of a molecule in the field is then

$$-\chi_{\parallel}(\hat{\mathbf{n}}\cdot\mathbf{H})^2-\chi_{\perp}[H^2-(\hat{\mathbf{n}}\cdot\mathbf{H})^2] = -\frac{1}{3}H^2\{(\chi_{\parallel}+2\chi_{\perp})+(\chi_{\parallel}-\chi_{\perp})[3(\hat{\mathbf{n}}\cdot\hat{\mathbf{H}})^2-1]\} . \tag{4.28}$$

We assume for simplicity that $\Delta\chi = (\chi_{\parallel}-\chi_{\perp}) > 0$ (*i.e.*, there is a tendency for the molecules to align parallel to the field). The field direction then becomes the preferred axis. We drop the first term in (4.28), which is independent of the molecular orientation, and write for the molecular pseudopotential in the presence of a field,

$$\epsilon_H(\hat{\mathbf{n}}) = -2\rho U(Q+\gamma)q \tag{4.29}$$

where

$$\gamma = \frac{\Delta\chi H^2}{3\rho U} . \tag{4.30}$$

The self-consistent equation now becomes

$$Q = \frac{\int_0^1 d\mu\, \frac{1}{2}(3\mu^2-1)\exp\{3\rho U\beta(Q+\gamma)\mu^2\}}{\int_0^1 d\mu\, \exp\{3\rho U\beta(Q+\gamma)\mu^2\}} . \tag{4.31}$$

The order parameter Q can easily be evaluated numerically for a given value of $x = 3\rho U\beta(Q+\gamma)$; once Q is determined, we can solve for the effective temperature. The results are depicted in Figure 4.4.

We see that for γ below a critical value γ_c there is a narrow temperature range for which Q is a triple-valued function of T. Not all these values correspond to a minimum in the free energy. This can be seen by plotting g and T as the parameter x is varied. The resulting curves are plotted in Figure 4.5. The region that is not a minimum of the free energy is the loop in the top curve of Figure 4.5. This region corresponds to the dashed curve in Figure 4.4. A first-order transition occurs at the intersection point. The point T_c, γ_c at which the loop has degenerated into a point is an ordinary critical point where

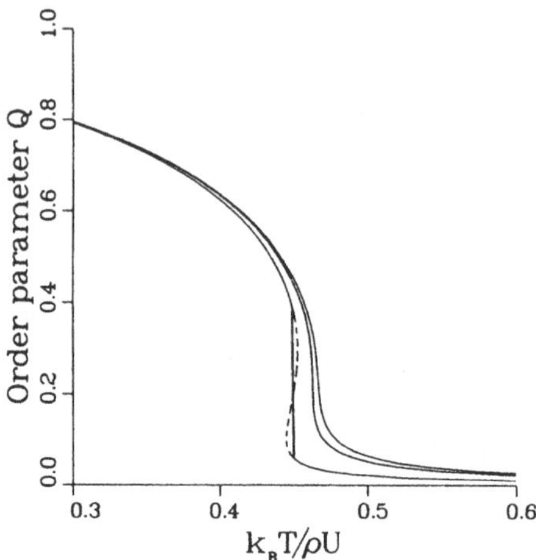

Figure 4.4: The order parameter as a function of temperature for different fields when $\Delta\chi > 0$. Solid curves, $\gamma = 0.005$, $\gamma = 0.011 = \gamma_c$, and $\gamma = 0.013$. The dashed curve for the lowest γ value corresponds to the unphysical region discussed in the text.

the transition is second order. We refer the interested reader to Wojtowicz and Sheng [329] and Palffy–Muhoray and Dunmur [229] for further details (the latter authors also discuss the interesting case $\gamma < 0$, for which there is a tricritical point and a biaxial phase occurs). The full phase diagram is given by Frisken et al. [104].

4.3 Blume–Emery–Griffiths Model

We next consider, as an example of a system exhibiting a tricritical point, (see also Section 3.9) a mixture of ^{3}He and ^{4}He in the liquid phase. When the temperature is lowered, pure ^{4}He undergoes a transition to the superfluid state (see Section 11.2). This transition is continuous, and is known as the λ transition, because of the λ-like shape of the specific heat singularity. If ^{3}He is added to the system, the transition temperature is lowered and the mixture remains homogeneous for low ^{3}He concentrations both below and above the critical temperature. At concentrations greater than $x_t = n_3/(n_3 + n_4) \approx 0.670$, the transition is discontinuous and accompanied by phase separation. One of the coexisting phases is a ^{4}He-rich superfluid, the other a ^{3}He-rich

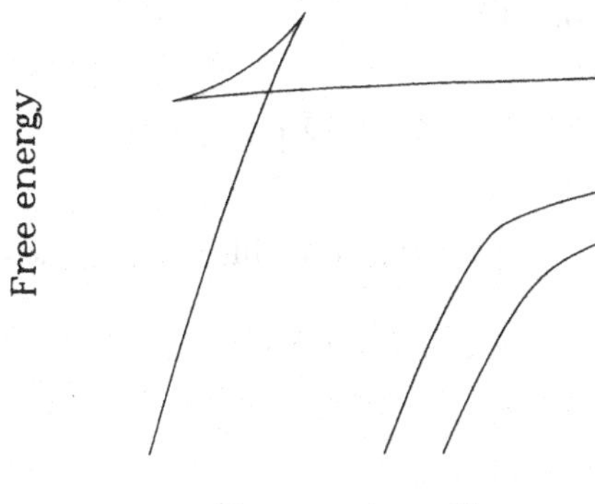

Figure 4.5: $g = G/N$ as a function of T for different values of γ.

normal fluid. The dividing point x_t, T_t, is the tricritical point alluded to above. A simple model exhibiting a tricritical point is found in Problem 3.13. Here we discuss a model for the ^{3}He$-^4$He system described above. This model, known as the Blume–Emery–Griffiths model [45], is a classical lattice gas model which ignores the quantum-statistical nature of the λ transition (Section 11.2) but does take into account the effect of non-ordering impurities (^{3}He) on the transition. The simplest version of the BEG model has the Hamiltonian

$$H = -J\sum_{\langle ij\rangle} S_i S_j + \Delta \sum_i S_i^2 - \Delta N$$

where $S_i = 0$ or ± 1 and the spins occupy a three-dimensional lattice with coordination number q. The connection between this "magnetic" Hamiltonian and the ^{3}He–^{4}He system is made by identifying $S_i = \pm 1$ with a ^{4}He atom on site i (or in cell i) and $S_i = 0$ with a ^{3}He atom. The parameter Δ controls the number of ^{3}He atoms and represents the difference $\mu_3 - \mu_4$ in chemical potentials. The concentration of ^{3}He atoms is given by

$$x = 1 - \langle S_i^2 \rangle$$

and it is clear that $x \to 0$ as $\Delta \to -\infty$ and $x \to 1$ as $\Delta \to +\infty$. The normal to superfluid transition is modeled by the transition from a high-temperature paramagnetic phase, $\langle S_i \rangle = 0$, to an ordered ferromagnetic phase, $\langle S_i \rangle = m$, and we now proceed to construct a mean field theory for this transition. The expectation value $m = \langle S_i \rangle$ is the order parameter in this theory. Let $p_i(S_i)$ be the probability that the spin on site i takes on the value S_i and assume, in the spirit of mean field theory, that

$$\tilde{p}(S_1, S_2, \cdots S_N) = \prod_{i=1}^{N} p_i(S_i) \ .$$

By translational invariance the probabilities will be the same on all sites

$$p_i(S_i) = p(S) \ .$$

We then have

$$\frac{G(T,\Delta)}{N} = -\frac{qJ}{2}\left(\sum_S p(S)S\right)^2 + \Delta\sum_S p(S)S^2 + k_BT\sum_S p(S)\ln p(S) - \Delta \ . \tag{4.32}$$

Minimizing (4.32) subject to the constraint $\sum_S p(S) = 1$ as in Section 2.5, we obtain

$$p(S) = \frac{\exp\{\beta(qJmS - \Delta S^2)\}}{1 + 2e^{-\beta\Delta}\cosh(\beta qJm)} \ . \tag{4.33}$$

Substituting (4.33) into (4.32), we obtain the approximate free energy

$$\begin{aligned}\frac{G(T,\Delta,m)}{N} &= -\frac{1}{2}qJm^2 + \Delta\langle S_i^2\rangle + k_BT\left[\frac{e^{\beta(qJm-\Delta)}}{D}\ln\frac{e^{\beta(qJm-\Delta)}}{D}\right.\\ &\left. + \frac{e^{-\beta(qJm+\Delta)}}{D}\ln\frac{e^{-\beta(qJm-\Delta)}}{D} - \frac{1}{D}\ln D\right] - \Delta\end{aligned} \tag{4.34}$$

where

$$D = 1 + 2e^{-\beta\Delta}\cosh\beta qJm$$

and

$$\langle S_i^2\rangle = \frac{2e^{-\beta\Delta}\cosh\beta qJm}{D} \ .$$

The expression for G may be simplified by decomposing the logarithms and we find

$$\frac{G(T,\Delta,m)}{N} = \frac{1}{2}qJm^2 - k_BT\ln(1 + 2e^{-\beta\Delta}\cosh\beta qJm) - \Delta \ . \tag{4.35}$$

This function must still be minimized with respect to the parameter m in order to obtain the equilibrium state for each (Δ, T). We construct the Landau expansion for the free energy

$$\frac{G(T,\Delta,m)}{N} = a(T,\Delta) + \frac{1}{2}b(T,\Delta)m^2 + \frac{1}{4}c(T,\Delta)m^4 + \frac{1}{6}d(T,\Delta)m^6 \ .$$

By comparing terms, one finds

$$
\begin{aligned}
a(T,\Delta) &= -k_BT\ln\left(1+2e^{-\beta\Delta}\right)-\Delta \\
b(T,\Delta) &= qJ\left(1-\frac{qJ}{\delta k_BT}\right) \\
c(T,\Delta) &= \frac{qJ}{2\delta^2}(\beta qJ)^3\left(1-\frac{\delta}{3}\right)
\end{aligned}
$$

where $\delta = 1+\frac{1}{2}e^{\beta\Delta}$. In the disordered phase $m=0$ and

$$x(T,\Delta) = \frac{1}{1+2e^{-\beta\Delta}} = \frac{\delta-1}{\delta} \ .$$

Combining this with $b(T_c(\Delta))=0$ we find

$$\frac{T_c(x)}{T_c(0)} = 1-x \tag{4.36}$$

and using $c(T_t,\Delta_t)=0$ we have $x_t=\frac{2}{3}$.

This value of the tricritical concentration is in remarkable agreement with the experimental value $x_t \approx 0.670$. The predicted linear dependence of the transition temperature on concentration is not observed. The actual transition temperature varies as $(1-x)^{2/3}$ for small x and the discrepancy is a consequence of the use of classical rather than quantum statistics in our model. The tricritical temperature in the BEG model $[T_t/T_c(0)=\frac{1}{3}]$ is for this reason somewhat lower than the observed value of 0.4.

The nomenclature "tricritical point" is a consequence of the fact that at this point three lines of critical points meet. In our treatment of the BEG model we have not considered the effect of the field h that couples to the order parameter m and have therefore found only the line of λ transitions. The other two lines emerge in a symmetrical fashion from the tricritical point in the $\pm h$ directions. The interested reader is encouraged to consult the original paper [45] for a more general treatment of the model that makes the full structure of the critical surface apparent.

4.4 Mean Field Theory of Fluids: van der Waals Approach

With the exception of the Maier–Saupe model, our examples of mean field theories to this point have been lattice models. We now turn to the case of a fluid consisting of particles interacting via a pair potential which contains a hard core, preventing the particles from overlapping, and a weak attractive tail. We wish to obtain an approximate equation of state for such a system in the spirit of mean field theory. Various modifications of the ideal gas law have been put forward to take into account the effect of interparticle interaction. One approach, with considerable physical appeal, was put forward by van der Waals, more than a hundred years ago. The van der Waals equation of state can be derived through many different routes; perhaps the simplest approach is through the following observations:

1. The internal energy of the ideal gas is purely kinetic in origin and independent of the volume. The entropy (2.15) can be written

$$S = Nk_B \ln V + \text{terms independent of volume.}$$

   The Helmholtz free energy is thus

$$A = -Nk_BT \ln \frac{V}{N} + \text{terms independent of volume.}$$

   This form of the free energy can be used to derive the equation of state for the pressure

$$P = -\left(\frac{\partial A}{\partial V}\right)_T = \frac{Nk_BT}{V}\,. \qquad (4.37)$$

2. In a first approximation, the attraction between the particles reduces the internal energy per particle by an amount proportional to the average number of surrounding particles (*i.e.*, to the density). This allows us to approximate the volume-dependent part of the internal energy

$$E = -a\left(\frac{N}{V}\right)N$$

   where a is a constant that depends on molecular properties.

3. Short distance repulsion prevents particles from approaching each other too closely. This has no direct effect on the internal energy, but reduces the free volume available to each particle. Let b be the excluded volume per particle. The total free volume is thus $V_f = V - Nb$. It is in the spirit of the derivation of the expression (2.15) for the entropy to interpret the volume-dependence as being due to the free volume, while the energy in (2.15) is the kinetic energy. With this interpretation we obtain for the free energy

$$A = -a\frac{N^2}{V} - Nk_BT \ln \frac{V - Nb}{N} + \text{terms independent of volume.}$$

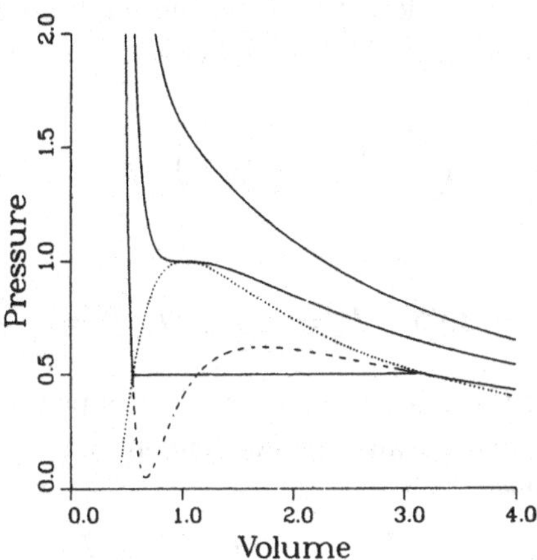

Figure 4.6: Isotherms and coexistence region according to van der Waals theory. Solid line: isotherm; dashed line: unphysical part of isotherm in coexistence region; dotted line: coexistence region. Pressure and volume in units of P_c and V_c.

The van der Waals equation of state follows by differentiation as in (4.37). After rearranging terms, we have

$$\left[P + a\left(\frac{N}{V}\right)^2\right](V - Nb) = Nk_BT \,. \tag{4.38}$$

This equation crudely describes the condensation of a gas into a liquid. For an extremely dilute gas $N/V \to 0$, $Nb \ll V$ and (4.38) reduces to the ideal gas equation of state. We are concerned here with lower temperatures and higher densities.

In Figure 4.6 we plot the behavior predicted by the van der Waals equation of state in the $P-V$ plane. The critical isotherm is characterized by an infinite compressibility at the critical temperature, that is,

$$\left.\frac{\partial P}{\partial V}\right|_{N,T} = 0$$

at $T = T_c$, $V = V_c$. The isotherms predicted by (4.38) will, below $T = T_c$, have both a maximum and a minimum, that is, for certain values of the pressure below the critical point there will be three real roots when solving for the volume. As $T \to T_c$ from below, the maximum and minimum of the isotherm merge and we obtain an inflection point. The critical point is therefore given by

$$\left(\frac{\partial P}{\partial V}\right)_T = \left(\frac{\partial^2 P}{\partial V^2}\right)_T = 0 . \tag{4.39}$$

These equations yield

$$V_c = 3Nb \quad P_c = \frac{a}{27b^2} \quad k_B T_c = \frac{8a}{27b} . \tag{4.40}$$

Using these values of the critical parameters, it is possible to rewrite the van der Waals equation in a parameter-independent way. Defining the reduced dimensionless variables

$$v = \frac{V}{V_c} \quad p = \frac{P}{P_c} \quad t = \frac{T}{T_c}$$

and substituting the reduced quantities into the van der Waals equations gives the *law of corresponding states*,

$$\left(p + \frac{3}{v^2}\right)\left(v - \frac{1}{3}\right) = \frac{8t}{3} . \tag{4.41}$$

We must now deal with the coexistence region. For $t < 1$, the system undergoes a first-order phase transition from the gas to the liquid phase. The unphysical behavior of the isotherms given by (4.41) in this region (mechanical stability does not allow $\partial p/\partial v > 0$) is a characteristic of mean field theory. There is a simple method known as the equal-area or Maxwell construction for removing the unphysical regions. The coexisting regions must be at the same pressure and on the same isotherm for reasons of mechanical and thermal equilibrium. Consider the chemical potential (or Gibbs free energy per particle)

$$\mu = \frac{G}{N} = \frac{A + PV}{N} \tag{4.42}$$

$$d\mu = -\frac{S}{N}dT + \frac{V}{N}dP . \tag{4.43}$$

Two coexisting phases must have the same chemical potential and along an isotherm we have

$$\int_1^2 d\mu = \mu(2) - \mu(1) = \frac{1}{N}\int V dP = 0 \tag{4.44}$$

where the coexisting phases have been labeled 2 and 1. In Figure 4.7 we have exchanged the axes of the plot of Figure 4.6 and it is easy to see that (4.44)

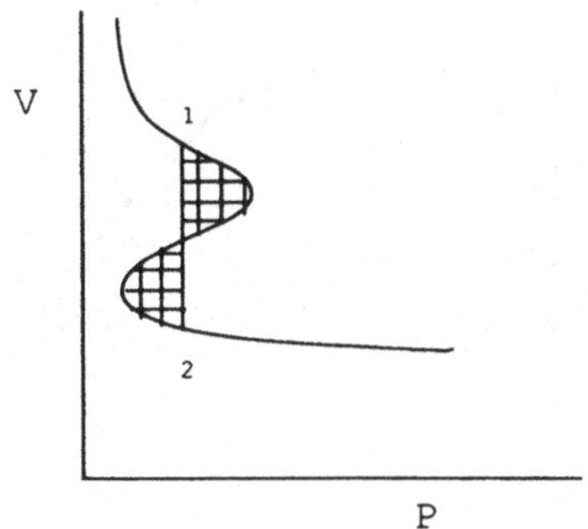

Figure 4.7: Maxwell construction.

implies that the areas of the two shaded regions must be the same—hence the name "equal-area construction".

We note that for $T < T_c$, the equation $(\partial P/\partial v)_T = 0$ defines a curve known as the spinodal. Van der Waals suggested that the states between the coexistence curve and the spinodal are metastable single-phase states. In the case of phase separation discussed in Section 4.1, the spinodal is given by $(\partial^2 A/\partial c^2)_T = 0$. Spinodals also occur in the Maier–Saupe model of Section 4.2. They are a general feature of first order transitions in mean field theory.

Since (4.41) does not contain any free parameters, the law of corresponding states implies that when expressed in terms of the reduced variables, all fluids should exhibit similar behavior (*i.e.*, the coexistence region in reduced units should look the same for all fluids). Experimental evidence (Figure 4.8) indicates that the law of corresponding states is a valid concept, but that the van der Waals equation of state does not provide a good quantitative approximation to it.

The van der Waals theory outlined above can be generalized in a number of different ways. Different phenomenological equations of state which are parameterized in terms of the van der Waals constants a and b, and which give better agreement with observed corresponding states than the original van der Waals theory have been put forward by a number of authors [123], [300], [179].

The van der Waals approach can also be generalized to more complicated systems. Consider first an isotropic mixture containing N_i molecules of species i. We can obtain a van der Waals theory of mixing by writing, in the spirit of the Bragg–Williams approximation in Section 3.2, for the configurational part

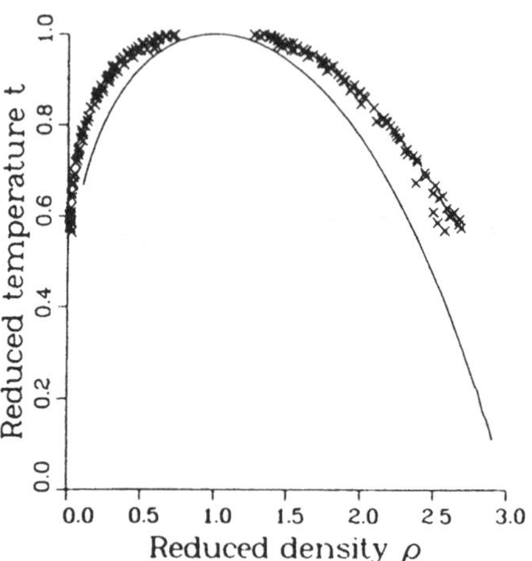

Figure 4.8: Density versus temperature in reduced units. Curve from van der Waals law of corresponding states. Data points are experimental results for several fluids quoted in [124].

of the internal energy per particle

$$E = \frac{1}{2}\sum_{ij} N_i \rho_j e_{ij}$$

where $\rho_j = N_j/V$ is the number density of species j and the e_{ij}'s are constants. If we write for the entropy of mixing

$$S_m = -k_B \sum_i N_i \ln \frac{N_i}{V_f^i}$$

and for the free volume of particles of species i,

$$V_f^i = V\left(1 - \frac{1}{2}\rho_j v_{ij}\right)$$

we may approximate the Helmholtz free energy by

$$A = E + k_B T \sum_i N_i \ln \frac{N_i}{V_f^i} \ .$$

It is now straightforward to compute the pressure from $P = -\partial A/\partial V$ to obtain the equation of state. Depending on the values of the parameters, this theory produces a rich variety of possible phase diagrams. Phase coexistence curves can be obtained by imposing the condition of mechanical and thermodynamic stability (*i.e.*, that the pressure and chemical potential $\mu_i = \partial A/\partial N_i$ of each species be the same in each phase). For a discussion of this type of calculation, we refer the interested reader to Hicks and Young [130].

It is also possible to construct van der Waals theories for anisotropic systems such as nematics in a similar spirit [98], [99], [228]. The van der Waals theory for interfaces is discussed in Section 5.4.

4.5 Spruce Budworm Model

Some time ago Ludwig et al. [181] proposed a simple ecological model intended to describe spruce budworm infestations. The model has three components *trees*, which are parasitized by *budworms*, who in turn are prey to *birds*. Changes in the respective populations take place on different timescales, with the budworm timescale being the fastest. In the simplest version of the model the tree and bird population are taken to be constant external parameters. The time derivative of the number N of insects is governed by the ordinary differential equation

$$\frac{dN}{dt} = r_B N(1 - \frac{N}{K_B}) - \frac{BN^2}{A^2 + N^2} + \Delta \ . \tag{4.45}$$

Here r_B is the net population growth rate (birth rate – death rate in the absence of predation). The constant K_B, commonly referred to as "the carrying capacity", takes into account that even in the absence of predators the insect population will saturate. It is proportional to the number of trees available for infestation. The constant B is proportional to the number of predators (birds) present. If the number of budworms is too low the birds will prefer other insects species, hence the proportionality to N^2 rather than N in the predation term. The denominator is intended to describe a saturation effect — there are only so many budworms that a bird can eat. Because of its simplicity the model serves as an important toy model in population biology [210]. The last term on the right hand side of (4.45) represents *immigration*. This term was not present in the original model, but it is necessary to include such a term in the presence of fluctuations (to be discussed in Section 8.2). The reason for this is that in the absence of immigration the zero population state is an *absorbing state*, meaning that once such a state is reached there is no escape. Since there is also a finite, albeit extremely small probability that the system will evolve towards the absorbing state the insect population will strictly speaking eventually go extinct if $\Delta = 0$. We note that equation (4.45) is mean-field-like in the following sense: A more complete description of the system would retain as dynamical variable at least the budworm populations on individual trees rather than simply the total number N.

Fick's law for diffusion assumes that the particle current is proportional to the gradient of the chemical potential. It is tempting to generalize this law and assume that the rate of approach of an order parameter to thermal equilibrium is proportional to the deviation of the appropriate free energy from its equilibrium value. In the context of Landau theory for phase transitions this approach is referred to as the time dependent Landau approximation. We then consider N to be the *order parameter* of the system:

$$\frac{dN}{dt} \propto -\frac{\partial G}{\partial N} \ . \tag{4.46}$$

The stable steady states of (4.46) are then the ones for which the *free energy* G has a local minimum. This assumption cannot hold for an arbitrary system, *e.g.*, if the model is extended to more species that evolve dynamically, (*e.g.*, trees and birds of prey). The problem is that we cannot in general find an exact differential [3]

$$dG = \sum_i \frac{\partial G}{\partial N_i} dN_i \tag{4.47}$$

yielding transport near the steady state for

$$\dot{N}_i \propto -\frac{\partial G}{\partial N_i} \ . \tag{4.48}$$

Following the notation of [210], we introduce the reduced variables

$$u = \frac{N}{A}; \; r = \frac{Ar_B}{B}, \; q = \frac{K_B}{A}, \; \delta = \frac{\Delta}{B} \tag{4.49}$$

[3] This issue has led to a controversy in theoretical economics which dates back to J. Willard Gibbs according to the book *More Heat than Light* by Mirowski [206]. Often attempts are made to exploit an analogy between thermodynamics and economics in which *utility* plays the role of a free energy and with quantities such as price given by derivatives in a manner similar to what we did for the *pressure* and *chemical potential* (for a recent example see e.g. [277]). Gibbs's objection, that realistic utilities will not have exact differentials, was according to Mirowski never satisfactorily answered nor even understood, hence the title of Mirowski's book.

in terms of which (4.45) becomes

$$\frac{A}{B}\frac{du}{dt} = ru(1 - \frac{u}{q}) - \frac{u^2}{1+u^2} + \delta \ . \qquad (4.50)$$

Integrating (4.50) with respect to u we find that for fixed values of the parameters r, q, δ the reduced "free energy" will be proportional to

$$g = -\frac{ru^2}{2} + \frac{ru^3}{3q} + (u - \arctan u) - \delta u \ . \qquad (4.51)$$

Of the parameters describing the mean field theory (4.45) we expect K_B, A, B to be proportional to system size (extensive) while r_B is intensive (independent of system size). Similarly the dependent variable N is extensive, while the reduced variables u, r, q, g, δ are all intensive.

The *steady states* are obtained by putting the right hand side of (4.50) to zero. This gives rise, in the limit $\delta \to 0$, to a cubic equation somewhat analogous to the law of corresponding states of the van der Waals theory of fluids (Section 4.4)

$$r(1+u^2)(1-\frac{u}{q}) - u^2 = 0 \ . \qquad (4.52)$$

In Figure 4.9 we plot a few "isoqs" (analogous to the isotherms of van der Waals theory). The system has a *critical point* $r = r_c$, $q = q_c$ for which $u = u_c$ is a triple root of (4.52). We have in the limit $\delta \to 0$

$$r_c = \frac{3\sqrt{3}}{8}; \ q_c = 3\sqrt{3}; \ u_c = \sqrt{3} \ . \qquad (4.53)$$

For $q < q_c$ (4.52) has only a single real root while for $q > q_c$ there is a region $r_1 < r < r_2$ for which (4.52) has three real roots $u_g < u_i < u_l$. We refer to the two curves $r_1(q)$ and $r_2(q)$ as the *spinodals* (again exploiting the analogy with van der Waals theory). Of the three roots u_l and u_g are locally stable while the intermediate root is always unstable.

In Figure 4.10 we plot the phase diagram of the system. If we allow the interpretation of $g(u)$ as a free energy, the coexistence line will separate the regions where $g(u_l) < g(u_g)$ and $g(u_l) > g(u_g)$. Below this line we expect the low density phase (g) to be globally stable and the high density phase (l) to be *metastable*, while it is the other way around immediately above the coexistence line. This behavior corresponds to a first order (or discontinuous transition). In Section 8.2 we show how one can construct a different coexistence line from a microscopic stochastic version of the model.

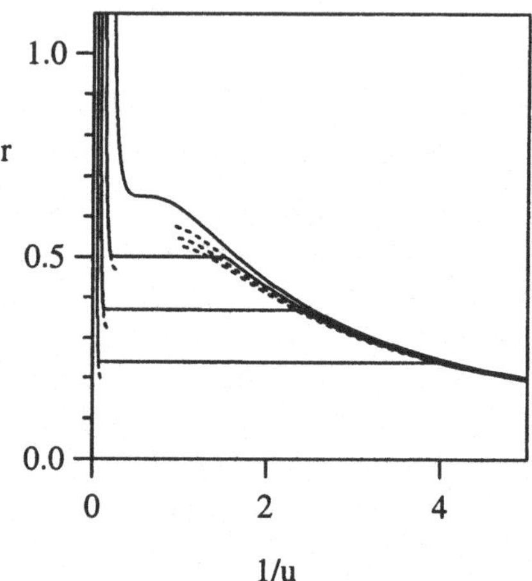

Figure 4.9: "Isoq"-curves. From top to bottom: $q = q_c$, $q = 8$, $q = 12$, $q = 20$. Dotted curves: metastable (local) minimum in free energy. Solid curves global (equilibrium) minimum in free energy.

Figures 4.9 and 4.10 suggest that $1/q$ is a temperature-like variable and that r is analogous to the pressure and u is analogous to the density in van der Waals theory.[4] We will return to the spruce budworm model in Section 8.2 where we discuss fluctuations.

4.6 A Non-Equilibrium System: Two Species Asymmetric Exclusion Model

We argued in Section 3.3 that a one-dimensional system in equilibrium, with short range (local) interactions, cannot undergo a symmetry breaking phase transition at nonzero temperature. The argument was based on an estimate of

[4]There is a certain arbitrariness in this assignment, but this need not be of concern to us, since even in van der Waals theory the two-phase coexistence line is not parallel to either the pressure or temperature axis. Hence the pressure and temperature are not the scaling fields (see Section 7.2) in van der Waals theory.

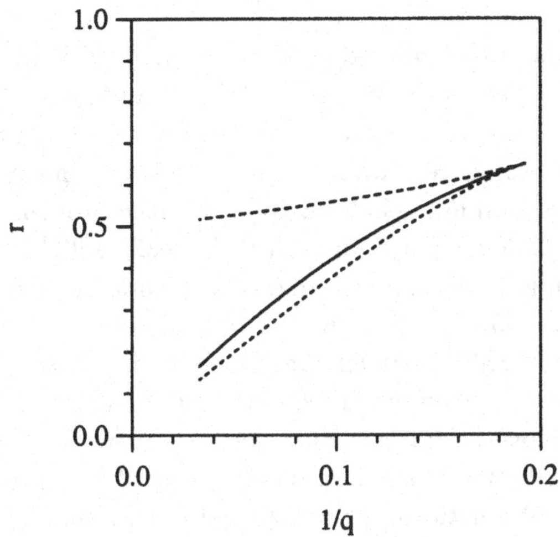

Figure 4.10: Phase diagram of spruce budworm model. Solid curve coexistence line, dotted curve spinodals (see text).

the free energy of a domain wall separating two symmetry-equivalent phases. If a system is kept out of equilibrium, the defect probability may not follow the equilibrium distribution, and the argument need not hold. Maintaining the system out of equilibrium will, however, require a supply of free energy. In a biological system this energy is provided by the metabolism of the organism. In the case of an electric current in a conductor the energy is provided by a battery or a power supply. In these cases the non-equilibrium nature of the system manifests itself by violation of the principle of *detailed balance* to be discussed in Sections 9.2.2 and 12.1.

In this section we wish to show that a symmetry breaking phase transition can take place in a one-dimensional *driven* diffusive system with short range interactions. The model has been studied by Evans et al. [86] and we refer to the original articles for some of the details and will here only summarize the results.

The model consists of a one-dimensional lattice. Each site on this lattice is either empty, or occupied by a positive charge, or by a negative charge. Positive charges can only move from left to right, while negative charges only move from right to left never in reverse[5] (this is the non-equilibrium feature of the model).

The model should be regarded as a "toy model" without any direct application, although the exclusion models were originally motivated by *superionic conductors*, materials that conduct electricity at elevated temperatures by ionic hopping under the influence of an applied field between empty sites in a lattice. The one species version has also been applied to the problem of pumping ions through pores in biological membranes and in zeolites [61].

Another possible motivation is *molecular motors*: certain proteins such as myosin, dynein and kinesin move within cells along filaments such as microtubulin and actin. These filament are *polar* in the sense that the different motor proteins only move in one direction along the filaments although different proteins can move in different directions (just like the two charges in our model). These small motors (their size is typically of the order 10nm) perform important biological functions. They transport chemicals across the cells, *e.g.* neurotransmitters along the axons, they help move chromosomes around during cell division, and they power muscles. Molecular motors are combustion engines; the energy for the unidirectional motion violating detailed balance is provided by the hydrolysis and reaction of adenosine triphosphate (ATP) $\rightarrow$ adenosine diphosphate (ADP) +P.

Returning to our toy model: the time evolution of the model is specified by the following rules:

1. $+$ particles jump to an empty site to the right at unit rate (*i.e.*, the particle will move with probability dt to the right in the time interval dt).

2. $-$ particles jump to an empty site to the left at unit rate.

3. A $+$ particle immediately to the left of a $-$ particle will exchange position with the $-$ particle ($+- \rightarrow -+$) at rate q.

4. If the left-most site is empty it will become occupied by a $+$ particle in a time interval dt with probability αdt (*i.e.*, the rate is α).

5. An empty right-most site will become occupied by a $-$ particle with probability αdt in a time interval dt.

[5] For this reason the model is called *asymmetric*. Since only one charge can occupy a given site at any time we are dealing with an *exclusion* model. Hence the name *Two species asymmetric exclusion model.*

6. A $-$ particle occupying the left-most site jumps off the chain at the rate β.

7. A $+$ particle occupying the right-most site jumps off the chain at the same rate β.

Note the symmetry between $+$ and $-$ particles. We expect that when the system has been operating under the above rules for some time a *steady state* will establish itself with approximately constant currents j_+ and j_- of the two types of particles. From the symmetry one would expect the two currents to be the same. Somewhat surprisingly this turns out not always to be the case. The mechanism is as follows: If the rate α is large compared to β there is a tendency for the particles to pile up and cause a "traffic jam" as will happen on a highway if too many vehicles enter the road. Suppose now that a fluctuation causes there to be more $+$ than $-$ particles in the system. The $+-$ exchange rule makes it possible for the $-$ particles and some of the $+$ particles to move, but some of the $+$ particles will have $+$ neighbors and be stuck in the traffic. This effect may cause an amplification of the fluctuation and result in symmetry breaking, somewhat in analogy with the Alben model of Problem 3.12.

It is relatively straightforward to set up an approximate mean field theory for the steady state in a chain of length N. In the steady state the current must be constant along the chain. Let p_i be the probability that site i from the left is occupied by a $+$ particle and let m_i be the corresponding probability for $-$ particles, while the probability that site i is empty is given by $1 - p_i - m_i$. This allows us to write for the two currents

$$j_+ = p_i(1 - p_{i+1} - m_{i+1}) + qm_{i+1} \tag{4.54}$$

$$j_- = m_i(1 - p_{i-1} - m_{i-1}) + qp_{i-1} \tag{4.55}$$

where $i \neq 1, N$. The currents will be subject to the boundary conditions

$$j_+ = \alpha(1 - p_1 - m_1) = \beta p_N \tag{4.56}$$

$$j_- = \beta m_1 = \alpha(1 - p_N - m_N) \ . \tag{4.57}$$

In the special case $q = 1$ the two equations (4.54) and (4.55) decouple and it is possible to obtain analytic results. We here only summarize some of these and refer the interested reader to Evans et al. [86] for the details.

1: *Maximum current phase.* If

$$\frac{\alpha\beta}{\alpha + \beta} > \frac{1}{2}$$

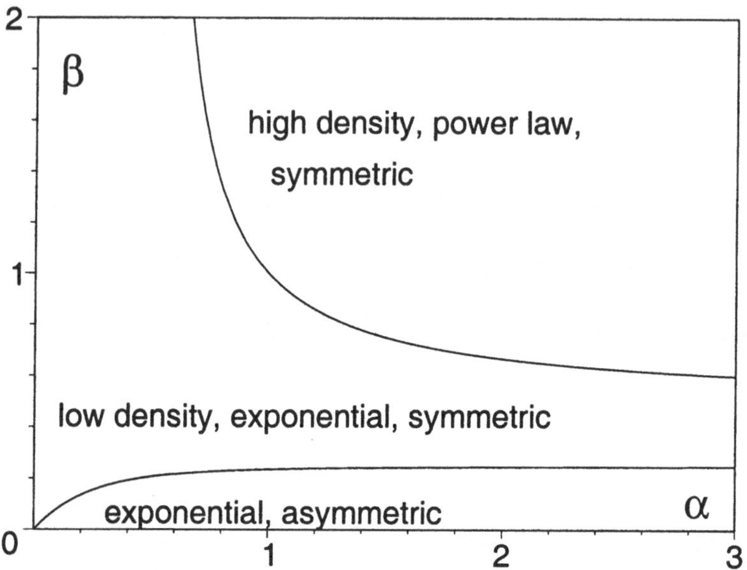

Figure 4.11: Mean field phase diagram for the 2-species asymmetric exclusion model of Evans et al. [86].

the particle densities slowly approach the constant value $\frac{1}{2}$ in the middle of the chain. This approach follows a power law and there is $\pm$ symmetry. The current takes on its maximum value $j\pm = 1/4$.

2: *Low density* $\pm$ *symmetric phase.* If

$$\frac{1}{2} > \frac{\alpha\beta}{\alpha+\beta} > \frac{1}{2} - \frac{\alpha\beta}{\alpha-\beta}$$

the particle density in the middle of the chain is less than $1/2$. There is a pile-up of particles near the exit end and the approach to the asymptotic density in the middle of the chain is exponential (fast).

3: *Asymmetric phase.* For

$$\frac{\alpha\beta}{\alpha+\beta} < \frac{1}{2} - \frac{\alpha\beta}{\alpha-\beta}$$

the symmetry between + and − particles is broken. For a narrow range of parameter values the density of both types of particles is low, but in most of this region the density is low for one species and high for the other.

The overall phase diagram for $q = 1$ is depicted in Figure 4.11. The phase transitions between the different phases outlined above are believed to be continuous. The qualitative features of the mean field phase diagrams has been confirmed by simulations. If $q \neq 1$, the mean field equations can only be solved numerically. One still finds the same phases as before, but some of the phase transitions are now discontinuous (first order).

4.7 Problems

4.1. *Solid–Solid Solutions.*

A crystalline solid is composed of constituents A and B. The energies associated with nearest-neighbor pairs of different types are, respectively, e_{AA}, e_{BB} and e_{AB}. Assume that

$$\epsilon = \frac{1}{2}e_{AA} + \frac{1}{2}e_{BB} - e_{AB} < 0$$

and that each site has q nearest neighbors.

(a) Calculate the Helmholtz free energy in the Bragg–Williams approximation for a homogeneous system in which the concentration of type A atoms is c_A and that of B atoms is $c_B = 1 - c_A$.

(b) Show that the system will phase separate when $c_A = c_B = \frac{1}{2}$ below the temperature

$$k_B T_c = \frac{1}{2} q|\epsilon| \ .$$

(c) For $c_A \neq c_B$, show that phase separation will occur at a lower temperature than the T_c of part **(b)**.

(d) Find the coexistence curve numerically and plot the result in the $k_B T/q|\epsilon|$, c_A plane.

4.2. *Maier–Saupe Model of a Liquid Crystal.*

(a) Using the Maier–Saupe expression (4.25) for the free energy, derive the Landau expansion (4.26).

(b) Find the transition temperature and the discontinuity in the order parameter at the transition predicted by the expansion found in part **(a)**.

(c) Evaluate numerically the transition temperature and the discontinuity of the order parameter and entropy at the transition in the Maier–Saupe model. (One of the weaknesses of the Maier–Saupe approximation is that the predicted latent heat is usually too large.)

4.3. *Critical Point of Liquid Crystal with Positive Susceptibility Anisotropy* Consider a liquid crystal described by the Maier–Saupe model with a pseudopotential of the form (4.29) and $\gamma > 0$. As the field strength parameter γ is increased from zero the ordering transition will become more weakly first order until a critical point is reached. Find the critical value of the inverse temperature β and γ at the critical point in units where $\rho U = 1$.

4.4. *Tricritical Point of Liquid Crystals with Negative Susceptibility Anisotropy* Consider a liquid crystal for which the anisotropy constant γ given by (4.29) is *negative.* It is now energetically favorable for the molecules to orient perpendicular to the magnetic field. If we choose the field to be the z-axis the order parameter Q will then be negative and the biaxial order parameter P will be non-zero in the low temperature phase and we can write for the pseudopotential in the Maier–Saupe model

$$\epsilon(\hat{n}) = 2\rho U[(Q+\gamma)q + \frac{1}{3}Pp] \ .$$

The model will exhibit a tricritical point for a certain value of β and γ. The problem is to locate this point. It is convenient to work in units of the inverse temperature parameter for which $\rho U = 1$. A possible way to proceed is as follows.

(i) Find expressions for the coefficients a and b in the Landau expansion

$$G = a(\beta,\gamma) + \frac{1}{2}b(\beta,\gamma) + \frac{1}{4}c(\beta,\gamma) + \cdots$$

assuming that Q has been determined selfconsistently from (4.31).

(ii) For given values of $x = 3\beta(Q+\gamma)$ and γ Q can be determined numerically and from the definition of x the order parameter Q.

(iii) Adjust β so that $b(\beta,\gamma)$ vanishes and evaluate $c(\beta,\gamma)$.

(iv) Modify the value of *gamma* and repeat steps **ii** and **iii** until b and c vanish simultaneously. The resulting values of β and γ locate the tricritical point!

4.5. *Lotka Volterrra Model.* Big fish sometimes eat little fish. If they don't find little fish they starve, but if they find them the population prospers. The little fish reproduce at a certain rate, but when too many are eaten by the big fish the population declines. Vito Volterra modeled this by the differential equations

$$\frac{dN_B}{dt} = -aN_B + bN_B N_L$$

$$\frac{dN_L}{dt} = cN_L - dN_B N_L \ . \tag{4.58}$$

(a) Find the steady state values of N_L and N_B and discuss the stability of the steady state. Describe the nature of the solutions to (4.58) for arbitrary starting states.

(b) If one or both populations become very small immigration may become important. Modify (4.58) to

$$\frac{dN_B}{dt} = -a(N_B - \lambda) + bN_B N_L$$

$$\frac{dN_L}{dt} = c(N_L + \gamma) - dN_B N_L \tag{4.59}$$

and assume that λ and γ are small compared to the steady state values of the populations found in **a**. How do the properties of the steady state change?

(c) If the big fish disappear (4.58) allows the population of little fish to grow without limit. Modify the differential equation so that

$$\frac{dN_B}{dt} = -aN_B + bN_B N_L$$

$$\frac{dN_L}{dt} = cN_L(1 - \frac{N_L}{K_L}) - dN_B N_L \tag{4.60}$$

where K_L is the *prey carrying capacity* of the environment. Assume that K_L is large compared with the steady state value of N_L found in **a**. Discuss how the steady state properties change and stability of the steady states.

4.6. *Paper-Scissor-Stone Model.* Consider a set of reactions

$$A + B \Rightarrow 2B; \quad B + C \Rightarrow 2C; \quad C + A \Rightarrow 2A$$

described by the differential equations

$$\begin{aligned}\frac{dN_A}{dt} &= \beta N_A(N_C - N_B)\\ \frac{dN_B}{dt} &= \beta N_B(N_A - N_C)\\ \frac{dN_C}{dt} &= \beta N_C(N_B - N_A)\ .\end{aligned} \tag{4.61}$$

(a) Show that the system of equations admits the conservation laws

$$N_A + N_B + N_C = N = \text{constant}$$

$$N_A N_B N_C = p = \text{constant}\ .$$

(b) Use the conservation laws to describe the family of solutions. Paramet the concentrations $n_A = N_A/N$, $n_B = N_B/N$; $n_C = N_C/N$ in the same way as we did for the 3-state Potts model in (3.44) and plot a family of curves inside the triangle of Figure 3.11 using different values of the constant p.

4.7. *Mean Field Theory for the Transverse Field Ising Model.* Consider a system of half integer Ising spins with the Hamiltonian

$$H = -J\sum_{<ij>}\sigma_{iz}\sigma_{jz} - \Gamma\sum_i \sigma_{ix}$$

where

$$\sigma_{ix} = \begin{pmatrix} 0 & 1\\ 1 & 0\end{pmatrix}_i ; \quad \sigma_{ix} = \begin{pmatrix} 1 & 0\\ 0 & -1\end{pmatrix}_i .$$

This model was originally proposed by de Gennes [69] and has recently been much studied as a model for "quantum phase transitions" in systems such as $LiHoF_4$ (see *e.g.* Rönnow et al. [257] and references therein).

In the spirit of mean field theory we approximate the magnetization components by

$$\begin{aligned}m_z \equiv \langle\sigma_{iz}\rangle &= \frac{\mathrm{Tr}\,\sigma_z \exp(K\sigma_z + h\sigma_x)}{\mathrm{Tr}\exp(K\sigma_z + h\sigma_x)}\\ &= \frac{\partial}{\partial K}\left(\ln \mathrm{Tr}\exp(K\sigma_z + h\sigma_x)\right)\end{aligned} \tag{4.62}$$

$$m_x = \frac{\partial}{\partial h}\left(\ln \mathrm{Tr}\exp(K\sigma_z + h\sigma_x)\right) \tag{4.63}$$

where q is the number of neighbors to each spin, $K = \beta q J m_z$ and $h = \beta\Gamma$.

(a) Evaluate the trace by diagonalizing

$$K\sigma_z + h\sigma_x = \begin{pmatrix} K & h \\ h & -K \end{pmatrix}$$

and show that

$$m_z = \frac{K}{\sqrt{K^2 + h^2}} \tanh(\sqrt{K^2 + h^2})$$

$$m_x = \frac{h}{\sqrt{K^2 + h^2}} \tanh(\sqrt{K^2 + h^2}) \ .$$

(b) Show that the result in **(a)** gives rise to a phase transition between a high temperature paramagnetic phase and a low temperature ferromagnetic phase at a temperature $k_B T_c = 1/\beta_c$ given by

$$\tanh(\beta_c \Gamma) = \frac{\Gamma}{qJ} \ .$$

Chapter 5

Dense Gases and Liquids

In this chapter we discuss selected topics in the theory of nonideal gases and liquids, a subject with a lengthy history and one in which a considerable degree of understanding of the basic phenomena has been attained. One of the earliest theories of dense gases is the well-known van der Waals equation which we discussed in Chapter 3 as an example of mean field theory. We shall not return to the van der Waals theory of bulk liquids in this chapter but rather concentrate on more general theories of gas and liquid phases. In the case of atomic and molecular gases and fluids we may, except in the case of very light constituents such as hydrogen or helium, safely neglect quantum effects and concentrate on the evaluation of the classical partition function

$$Z_c = \frac{1}{N!h^{3N}} \int d^{3N}p\, d^{3N}r\, e^{-\beta H} \tag{5.1}$$

where

$$H = \sum_i \frac{p_i^2}{2m} + \sum_{i<j} U\left(|\mathbf{r}_i - \mathbf{r}_j|\right) \tag{5.2}$$

in the case of a simple atomic gas or liquid. At this point it is worth pointing out that even for a rare-gas system, such as liquid argon, the Hamiltonian (5.2) is not complete. Three-body interactions, $V(\mathbf{r}_1, \mathbf{r}_2, \mathbf{r}_3)$, play an important role in the thermodynamic properties of these systems. In molecular fluids the potential energy is generally a function of the relative orientation of the molecules as well as of their separation. The determination of an appropriate intermolecular potential is a difficult, and only partially solved problem in quantum chemistry. The reader is referred to the review article by Barker and Henderson [28] for a discussion of the role and parametrization of intermolecular potentials and to the classic monograph of Hirshfelder *et al.* [131] or the newer book by Maitland *et al.* [188] for a more detailed treatment of this topic.

In this chapter we invariably assume that our system can be described by the Hamiltonian (5.2) with only central two-body forces between the constituents. The calculation of Z_c can then be reduced, after an integration over the momentum variables, to

$$Z_c = \lambda^{-3N} \frac{1}{N!} \int d^{3N}r \exp\left\{ -\beta \sum_{i<j} U(r_{ij}) \right\} \tag{5.3}$$

where $\lambda = [h^2/(2m\pi k_B T)]^{1/2}$ is the thermal wavelength introduced in (2.21). The remaining integral will be denoted by $Q_N(V,T)$ and is called the *configuration integral*:

$$Q_N(V,T) = \frac{1}{N!} \int d^{3N}r \exp\left\{ -\beta \sum_{i<j} U(r_{ij}) \right\} . \tag{5.4}$$

The evaluation of this expression is the central problem in the theory of dense gases and liquids. We shall describe, in the following sections, a number of different approaches that have been devised for this purpose. In Section 5.1 we discuss the virial expansion for $Q_N(V,T)$. In Section 5.2 we focus on the reduced distribution functions and summarize some of the more successful approximation schemes for the solution of the Ornstein–Zernike equation. In Section 5.3 we discuss perturbation theories, and in Section 5.4 we turn to the topic of inhomogeneous fluids. In this section we finally return to van der Waals theory when we construct a Landau–Ginzburg theory of the liquid-vapor interface. We conclude with a brief introduction to density-functional methods in Section 5.5.

During the last thirty years computer simulations have played an important part in developing our intuition for and understanding of condensed matter systems, including liquids. This technique is of such great current interest and importance that we devote a separate Chapter 9 to this topic.

There are a number of excellent general references for the material of this chapter. Among them are the aforementioned review of Barker and Henderson [28] and the book by Hansen and McDonald [125].

5.1 Virial Expansion

The ideal gas equation of state provides a reasonable approximation to the properties of interacting atoms or molecules only in the dilute limit. A systematic approach to the effects of increasing density, or lower temperature, is the virial expansion in which one expands the pressure in a power series in the density

$$\frac{P}{k_B T} = \frac{N}{V}\left(1 + B_2(T)\frac{N}{V} + B_3(T)\left(\frac{N}{V}\right)^2 + \cdots\right) . \tag{5.5}$$

The coefficients B_j are known as virial coefficients. The most elegant method of deriving the virial coefficients utilizes the grand canonical ensemble, rather than the canonical, and is due to J. E. Mayer. The pressure is given by

$$\frac{P}{k_B T} = \frac{1}{V}\ln Z_G = \frac{1}{V}\ln\left[\sum_{N=0}^{\infty} e^{\beta\mu N}\lambda^{-3N} Q_N(V,T)\right] \tag{5.6}$$

with Q_N given by equation (5.4). The potential $U(r_{ij})$ which appears in (5.4) depends on the particular system in question, but typically, for a neutral system, will be sharply repulsive at short distances due to overlap of electronic wave functions and will be weakly attractive at larger separations. A potential frequently used to describe rare gases and fluids is the Lennard–Jones or 6–12 potential, which is of the form

$$U(r) = 4\epsilon\left[\left(\frac{\sigma}{r}\right)^{12} - \left(\frac{\sigma}{r}\right)^{6}\right] . \tag{5.7}$$

This potential has a minimum value of $-\epsilon$ at $r = 2^{1/6}\sigma$. For argon, appropriate values of ϵ and σ are $\epsilon/k_B = 120$ K, $\sigma = 0.34$ nm.

The function $e^{-\beta U(r)}$ which appears in the configuration integral has the undesirable property of approaching unity rather than zero as r goes to infinity. To construct an expansion in powers of the density, we need a function of the potential which is significant only if groups of atoms are close to each other. Such a function is the Mayer function

$$f_{ij}(r) = \exp\{-\beta U(r_{ij})\} - 1 \tag{5.8}$$

which is sketched in Figure 5.1. In terms of this function the configuration integral becomes

$$Q_N(V,T) = \frac{1}{N!}\int d^{3N}r \prod_{j<m}(1 + f_{jm}) . \tag{5.9}$$

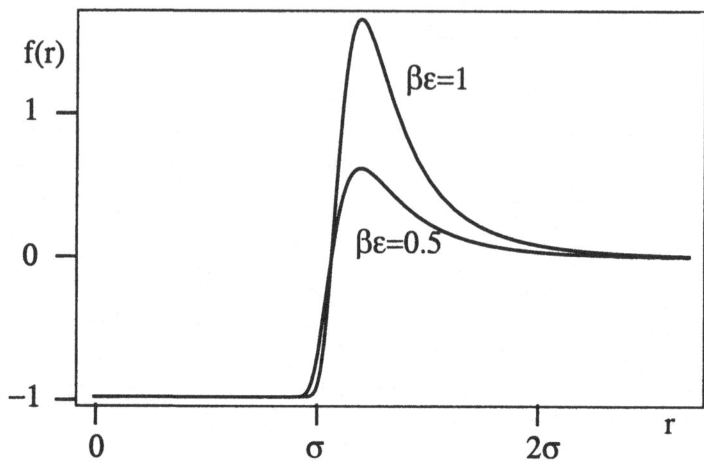

Figure 5.1: The function $f(r)$ for the Lennard–Jones potential for two values of $\beta\epsilon$.

The expansion of the product (5.9) results in a series

$$Q_N(V,T) = \frac{1}{N!}\int d^{3N}r \left(1 + \sum_{j<m} f_{jm} + \sum_{j<m,r<s} f_{jm}f_{rs} + \cdots\right) . \qquad (5.10)$$

The evaluation of the various terms in (5.10) is greatly facilitated by a graphical notation. We identify a graph with each term. A particle is denoted by a heavy dot, the function f_{jm} by a line connecting the dots corresponding to particles j and m. Thus each term in (5.10) consists of a graph with N dots and a variable number of lines joining pairs of dots. Some simple graphs with the corresponding term in the integrand of (5.10) are given in Table 5.1.

The integrations over different disconnected pieces of a graph can clearly be carried out independently. We therefore focus on the different connected subgraphs and define a cluster integral for each topologically distinct connected subgraph. In view of later simplifications, we define the *cluster integral* $b_j(T)$ for graph j to be

$$b_j(T) = \frac{1}{n_j! V}\sum \int d^3r_1 d^3r_2 \ldots d^3r_{n_j} \left(\prod_{i,j} f_{ij}\right) . \qquad (5.11)$$

In formula (5.11), n_j is the number of vertices (particles) in graph j and the product over Mayer functions is the appropriate combination determined by

Table 5.1:

Integrand	Graph
1	•
f_{12}	 1 •———• 2
$f_{12}f_{23}$	 1 •—•(2)—• 3

the connectivity of the graph. The summation in (5.11) is a sum over distinct permutations of the labels on the vertices which occur in the expansion (5.10). This requires some comment and we first illustrate the notion with a few simple examples. The three-particle graphs

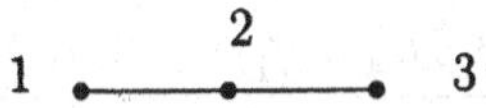

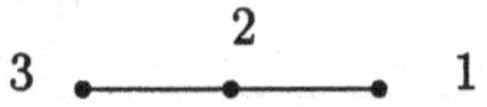

1 •—•(2)—• 3 3 •—•(2)—• 1

are not distinct from each other but *are* distinct from

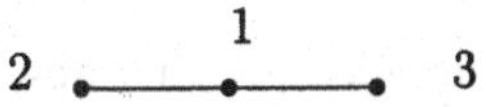

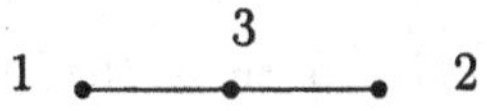

2 •—•(1)—• 3 1 •—•(3)—• 2

The four-particle graph

has the three distinct assignments of labels (1234), (1324), (1243) (counterclockwise from the lower left-hand corner), and no others. Any other assignment of labels, such as (1432), corresponds to the same product of Mayer functions as one of the three previous assignments and this product appears only once in the configuration integral (5.10). We note that since the particle coordinates are integrated out in (5.11), the cluster integral depends only on the topology of the graph and we may henceforth drop the labeling of the vertices. Moreover, since the function f is short ranged, one of the vertices of a connected graph, or equivalently the center of mass coordinate, can be freely integrated over the volume. Thus $b_j(T)$ is independent of the volume. The cluster integrals of the first few graphs are listed in Table 5.2.

The most general term in (5.10), expressed in terms of cluster integrals, is

$$\frac{1}{N!}\left\{\frac{N!}{\prod_j m_j!(n_j!)^{m_j}}\prod_s [Vn_s!b_s(T)]^{m_s}\right\} . \tag{5.12}$$

Table 5.2: Some elementary graphs and the corresponding cluster integrals.

j	Graph	$b_j(T)$
1		1
2		$\frac{1}{2}\int d^3x f(x)$
3		$\frac{1}{2}\int d^3x d^3y f(x) f(y) = 2b_2^2(T)$
4		$\frac{1}{3!}\int d^3x d^3y f(x) f(y) f(\lvert\mathbf{x}-\mathbf{y}\rvert)$
5		$\frac{1}{4!V}\int d^3r_1 \ldots d^3r_4 [f_{12}f_{23}f_{14}f_{34}$ $f_{13}f_{14}f_{23}f_{24} + f_{12}f_{24}f_{34}f_{13}]$

In equation (5.12), m_j is the number of times that the unlabeled graph j appears in the given term, n_j is the number of vertices of graph j, and $Vn_s!b_s(T)$ is the contribution of graph s to Q_N when a specific set of particles is assigned to the vertices of the graph. The combinatorical factor

$$\frac{N!}{\prod_j m_j!(n_j!)^{m_j}}$$

is the number of ways that the N particles can be assigned to the set of disconnected graphs and $N = \sum_j m_j n_j$.

We show explicitly that this formula produces the correct contribution in a specific simple case. One term in the configuration integral is graph 4 in Table 5.2. The corresponding term in the configuration integral is

$$\frac{1}{N!}\sum_{i,j,k}\int d^{3N}r f_{ij} f_{jk} f_{ki}$$
$$= \frac{N(N-1)(N-2)}{3!}\frac{V^{N-3}}{N!}\int d^3r_1 d^3r_2 d^3r_3 f(r_{12}) f(r_{13}) f(r_{23})$$

by direct counting. In expression (5.12), we have $m_4 = 1$, $m_1 = N-3$, $n_1 = 1$, and $n_4 = 3$ and we obtain

$$\frac{N(N-1)(N-2)V^{N-3}}{N!} V b_4(T)$$

which is the same as the previous expression. The reader is encouraged to check a few other simple cases.

We now return to expression (5.6) for the pressure:

$$\begin{aligned} \frac{P}{k_B T} &= \frac{1}{V} \ln \sum_{N=0}^{\infty} z^N \lambda^{-3N} Q_N(V,T) \\ &= \frac{1}{V} \ln \sum_{N=0}^{\infty} z^N \lambda^{-3N} \sum_{\{m_j\}} \prod_{j=1}^{\infty} \frac{[V b_j(T)]^{m_j}}{m_j!} \end{aligned} \tag{5.13}$$

where $z = e^{\beta\mu}$ and $\sum_{\{m\}}$ indicates a sum over all possible combinations of graphs subject to the restriction $\sum m_j n_j = N$. The sum (5.13) may be decomposed into a product of sums over m_j. Using $N = \sum m_j n_j$, we obtain

$$\begin{aligned} \frac{P}{k_B T} &= \frac{1}{V} \ln \left\{ \prod_j \sum_{m_j=0}^{\infty} \frac{(z\lambda^{-3})^{n_j} V b_j(T)]^{m_j}}{m_j!} \right\} \\ &= \frac{1}{V} \ln \prod_j \exp[(z\lambda^{-3})^{n_j} V b_j(T)] = \sum_j (z\lambda^{-3})^{n_j} b_j(T) \; . \end{aligned} \tag{5.14}$$

To complete the virial expansion we must still express the chemical potential in terms of the density $n = N/V$ and the temperature. This is accomplished by constructing an expansion of the density

$$N = z \left(\frac{\partial}{\partial z} \ln Z_G \right)_{T,V} = V z \frac{\partial}{\partial z} \left(\sum_{j=1}^{\infty} (z\lambda^{-3})^{n_j} b_j(T) \right)_{T,V}$$

or

$$n = \frac{N}{V} = \sum_{j=1}^{\infty} n_j (z\lambda^{-3})^{n_j} b_j(T) \; . \tag{5.15}$$

Substituting $z = a_1 n + a_2 n^2 + a_3 n^3 + \cdots$ and solving for the a_j's, one finally obtains, on substituting into (5.14), the virial expansion

$$\frac{P}{k_B T} = n + B_2(T) n^2 + B_3(T) n^3 + \cdots \; . \tag{5.16}$$

The completion of this task is left as an exercise. We simply quote the result:

$$
\begin{aligned}
B_2(T) &= -\frac{1}{2}\int d^3r f(r) \\
B_3(T) &= -\frac{1}{3V}\int d^3r_1 d^3r_2 d^3r_3 f_{12} f_{13} f_{23} \ .
\end{aligned}
\qquad (5.17)
$$

The reader will note that only graph 4 from Table 5.2 contributes to the third virial coefficient B_3. The contributions from graph 3 have canceled in the process of eliminating z. This result is a particular manifestation of a general theorem: The virial expansion can be expressed to all orders in terms of cluster integrals of *stars*. A star graph is a graph that cannot be separated into disjoint pieces by cutting through a single vertex. The difference between stars and other graphs is that the cluster integral of non stars can be expressed as the product of cluster integrals of the separable pieces. Such is not the case with stars. An example of this decomposition has already appeared in Table 5.2 in the case of graph 3 and a general proof is straightforward. The general expression of the virial coefficients in terms of star graph cluster integrals is derived in the book by Mayer and Mayer [195] and can also be found in that of Uhlenbeck and Ford [307].

It is clear that the evaluation of even the third virial coefficient presents computational difficulties for realistic potentials. For the hard-sphere potential the virial coefficients up to B_7 have been computed, either analytically or numerically. For the 6–12 potential the virial coefficients up to B_5 have been calculated. In Figure 5.2 we show the equations of state for a system of hard spheres obtained from the virial expansion using more and more terms. The dots represent the results of computer simulations. We see that the agreement is rather good except at higher densities. The data are taken from Barker and Henderson [27]–[28].

From the virial series for the 6–12 potential one can obtain a series of estimates for the critical temperature by requiring that $(\partial P/\partial V)$ and $(\partial^2 P/\partial V^2)$ both be zero at $T = T_c$. The results are given in Table 5.3 for the dimensionless critical temperature $(k_B T_c/\epsilon)$ (from Temperley *et al.* [299]). Argon, which is thought to be a good example of a Lennard-Jones system, has an experimental value of 1.26 for this parameter. We see that the virial expansion does seem to be converging although rather slowly.

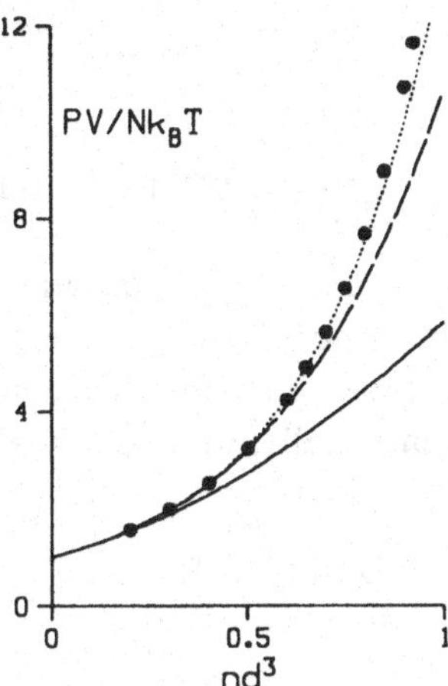

Figure 5.2: Virial equation of state for hard spheres of diameter d. Solid curve, two virial coefficients. Dashed curve, four coefficients; dotted curve, six coefficients. (Filled circles are Monte Carlo data of Barker and Henderson [27]).

5.2 Distribution Functions

5.2.1 Pair correlation function

One of the most useful approaches to the theory of liquids has been the study of reduced distribution functions and, in particular, the calculation by a number of sophisticated approximation schemes of the pair correlation function.

Table 5.3: Critical temperature of a 6–12 fluid obtained from the virial expansion.

	k_BT_c/ϵ
B_3	1.445
B_4	1.300
B_5	1.291

Consider the function

$$P(\mathbf{r}_1,\mathbf{r}_2\ldots,\mathbf{r}_N)=\frac{1}{N!Q_N(V,T)}\exp\{-\beta W(\mathbf{r}_1,\mathbf{r}_2\ldots,\mathbf{r}_N)\} \tag{5.18}$$

with $W(\mathbf{r}_1,\mathbf{r}_2,\ldots,\mathbf{r}_N)=\sum_{i<j}U(\mathbf{r}_i-\mathbf{r}_j)$. This function is the probability density that the N particles are at positions $\mathbf{r}_1,\ldots,\mathbf{r}_N$. The function P provides far more information than is necessary for the calculation of thermodynamic functions. To proceed systematically, we define a sequence of reduced distribution functions:

$$n_1(\mathbf{x})=\sum_{i=1}^{N}\langle\delta(\mathbf{x}-\mathbf{r}_i)\rangle \tag{5.19}$$

$$n_2(\mathbf{x}_1,\mathbf{x}_2)=\sum_{i\neq j}\langle\delta(\mathbf{x}_1-\mathbf{r}_i)\delta(\mathbf{x}_2-\mathbf{r}_j)\rangle \tag{5.20}$$

and, in general,

$$n_s(\mathbf{x}_1,\mathbf{x}_2,\cdots,\mathbf{x}_s)=\sum_{i\neq j\neq\ldots m}\langle\delta(\mathbf{x}_1-\mathbf{r}_i)\delta(\mathbf{x}_2-\mathbf{r}_j)\cdots\delta(\mathbf{x}_s-\mathbf{r}_m)\rangle\,. \tag{5.21}$$

In a homogeneous system the reduced distribution function n_1 is simply the density:

$$n_1(\mathbf{r}_1)=N\frac{\int d^3r_2\cdots d^3r_N\exp\left\{-\beta\sum_{i<j}U(\mathbf{r}_i-\mathbf{r}_j)\right\}}{\int d^3r_1\cdots d^3r_N\exp\left\{-\beta\sum_{i<j}U(\mathbf{r}_i-\mathbf{r}_j)\right\}} \tag{5.22}$$

which is easily seen by letting $\mathbf{r}_j=\mathbf{r}_1+\mathbf{x}_j$ for $j\neq 1$, integrating over $\mathbf{x}_j$ and noting that the integrand in the denominator becomes independent of $\mathbf{r}_1$. The two-particle distribution

$$n_2(\mathbf{x}_1,\mathbf{x}_2)=N(N-1)\frac{\int d^3r_3d^3r_4\ldots d^3r_N\exp\{-\beta W(\mathbf{x}_1,\mathbf{x}_2,\mathbf{r}_3\ldots\mathbf{r}_N)\}}{\int d^3r_1d^3r_2\ldots d^3r_N\exp\{-\beta W(\mathbf{r}_1,\mathbf{r}_2,\ldots\mathbf{r}_N)\}} \tag{5.23}$$

is the probability that two particles occupy the positions $\mathbf{x}_1$ and $\mathbf{x}_2$ and, as $|\mathbf{x}_1-\mathbf{x}_2|\to\infty$, approaches the limiting value $N(N-1)/V^2$. It is easy to see that the expectation value of the interaction energy may be expressed in terms of $n_2(\mathbf{x}_1,\mathbf{x}_2)$:

$$\begin{aligned}\langle U\rangle &= \sum_{i<j}\langle U(\mathbf{r}_i-\mathbf{r}_j)\rangle\\ &= \frac{1}{2}\sum_{i\neq j}\int d^3x_1d^3x_2\langle U(\mathbf{x}_1-\mathbf{x}_2)\delta(\mathbf{x}_1-\mathbf{r}_i)\delta(\mathbf{x}_2-\mathbf{r}_j)\rangle\end{aligned}$$

$$= \frac{1}{2}\sum_{i\neq j}\int d^3x_1 d^3x_2 U(\mathbf{x}_1-\mathbf{x}_2)\langle\delta(\mathbf{x}_1-\mathbf{r}_i)\delta(\mathbf{x}_2-\mathbf{r}_j)\rangle \quad (5.24)$$

$$= \frac{1}{2}\int d^3x_1 d^3x_2 U(\mathbf{x}_1-\mathbf{x}_2) n_2(\mathbf{x}_1,\mathbf{x}_2)\ .$$

In a homogeneous system $n_2(\mathbf{x}_1,\mathbf{x}_2) = n_2(|\mathbf{x}_1-\mathbf{x}_2|)$. It is conventional to define another function $g(|\mathbf{x}_1-\mathbf{x}_2|)$ called the pair distribution function through

$$n_2(|\mathbf{x}_1-\mathbf{x}_2|) = \left(\frac{N}{V}\right)^2 g(|\mathbf{x}_1-\mathbf{x}_2|)\ . \quad (5.25)$$

In the thermodynamic limit $N\to\infty$, $N/V=$ constant, we have $N(N-1)\approx N^2$ and as the separation becomes large $g(|\mathbf{x}_1-\mathbf{x}_2|)\to 1$. Therefore, from (5.24) and (5.25), we have

$$\langle U\rangle = \frac{N^2}{2V}\int d^3r\, U(r) g(r)\ . \quad (5.26)$$

The Fourier transform of the pair distribution function is intimately related to the *static structure factor* $S(\mathbf{q})$ which we define as follows:

$$S(\mathbf{q}) - 1 = \frac{N}{V}\int d^3r\,\{g(r)-1\}\, e^{i\mathbf{q}\cdot\mathbf{r}}\ . \quad (5.27)$$

By substituting the definitions (5.20) and (5.25), we find

$$S(\mathbf{q}) = \frac{1}{N}\left\langle\sum_{i,j}\exp\{i\mathbf{q}\cdot(\mathbf{r}_i-\mathbf{r}_j)\}\right\rangle - N\delta_{\mathbf{q},0} \quad (5.28)$$

where $\delta_{\mathbf{q},0}$ is the three-dimensional Kronecker delta:

$$\delta_{\mathbf{q},0} = \frac{1}{V}(2\pi)^3\delta(\mathbf{q})\ .$$

Equation (5.28) can be used as an alternative definition of the structure factor. The last term is sometimes not included, but removes an uninteresting singularity at $\mathbf{q}=0$. The structure factor plays an important role in the interpretation of elastic scattering experiments employing *e.g.* neutrons or light. To see this, consider the situation in which an incoming beam can be described in terms of plane waves

$$\Psi_{\mathbf{k}}(\mathbf{r}) = \frac{1}{\sqrt{V}}e^{i\mathbf{k}\cdot\mathbf{r}}\ .$$

One then detects an elastically scattered outgoing wave with wave vector $\mathbf{k}' = \mathbf{k}+\mathbf{q}$,

$$\Psi_{\mathbf{k}'}(\mathbf{r}) = \frac{1}{\sqrt{V}}e^{i\mathbf{k}'\cdot\mathbf{r}}\ .$$

We assume that the interaction between the probe and the particles of the system can be expressed as a sum of contributions from the individual particles:

$$\sum_i u(\mathbf{r}-\mathbf{r}_i) \ .$$

Let the Fourier transform of this potential be given by

$$u(\mathbf{q}) = \int d^3r \, u(\mathbf{r}) e^{-i\mathbf{q}\cdot\mathbf{r}} \ .$$

The golden rule transition rate for elastic scattering to the state $\mathbf{k}' = \mathbf{k}+\mathbf{q}$ is given by

$$W_{\mathbf{k}\to\mathbf{k}'} = \frac{2\pi}{\hbar}\left|\langle \mathbf{k}+\mathbf{q}|\sum_i u(\mathbf{r}-\mathbf{r}_i)|\mathbf{k}\rangle\right|^2 \delta(\epsilon(\mathbf{k})-\epsilon(\mathbf{k}')) \ . \tag{5.29}$$

The thermal average of the squared matrix element in (5.29), for $\mathbf{q} \neq 0$, is given by

$$\frac{N}{V^2}|u(\mathbf{q})|^2 S(\mathbf{q})$$

where we have used (5.28). We thus have, for the scattered intensity,

$$I(\mathbf{k}'-\mathbf{k}=\mathbf{q}) \propto f(\mathbf{q})S(\mathbf{q})I_0 \tag{5.30}$$

where I_0 is the intensity of the incoming beam and the *form factor* $f(\mathbf{q}) = |u(\mathbf{q})|^2$. In the special case of neutron scattering, the potential $u(\mathbf{r}-\mathbf{r}_i)$ will be essentially a δ-function potential and the form factor will therefore be a slowly varying function of $\mathbf{q}$. Hence $S(\mathbf{q})$ will be proportional to the intensity of the scattered beam. The functions g and S have the general appearance shown in Figure 5.3 for a dense fluid.

Knowledge of the pair distribution function thus allows us to predict the internal energy (5.26) and the results of elastic scattering experiments. We next show that the pair distribution function is also intimately related to the compressibility $K_T = -1/V\,(\partial V/\partial P)_T$.

For a homogeneous system with a fixed number, N, of particles, we have

$$\int d^3x_1 n_2(|\mathbf{x}_1-\mathbf{x}_2|) = \frac{N(N-1)}{V} \ . \tag{5.31}$$

If the number of particles is allowed to fluctuate, as in the grand canonical ensemble, the right-hand side will be replaced by $\langle N(N-1)\rangle/V$ and the definition (5.25) changes to

$$g(|\mathbf{x}_1-\mathbf{x}_2|) = \frac{V}{\langle N\rangle^2} n_2(|\mathbf{x}_1-\mathbf{x}_2|) \ .$$

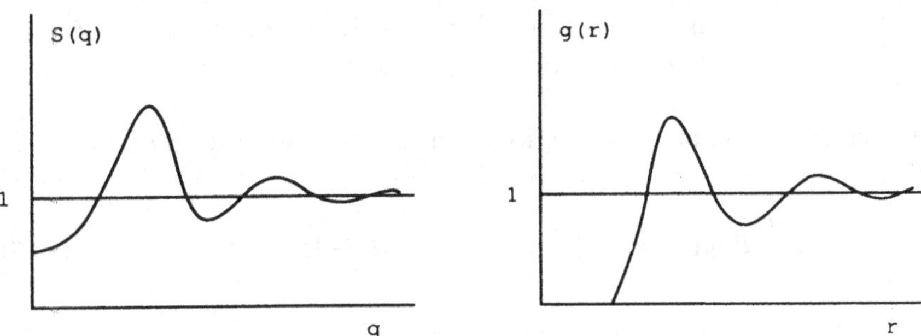

Figure 5.3: Sketch of $g(r)$ and $S(q)$ for a typical liquid.

We therefore have

$$\int d^3r\,(g(r)-1)) = \frac{V}{\langle N\rangle^2}\langle N(N-1)\rangle - V = V\frac{(\Delta N)^2}{\langle N\rangle^2} - \frac{V}{\langle N\rangle} \tag{5.32}$$

where

$$(\Delta N)^2 = \langle (N-\langle N\rangle)^2\rangle \ .$$

If we now recall the relationship (2.65) between the particle fluctuation ΔN and the compressibility, we find the *fluctuation* or *compressibility* equation of state.

$$n\int d^3r\,[g(r)-1] = nk_BTK_T - 1 \ . \tag{5.33}$$

The quantity

$$h(r) = g(r) - 1 \tag{5.34}$$

is commonly referred to as the *pair correlation function*.

The use of the grand canonical definition of $g(r)$ deserves some comment. A mechanical measurement of the compressibility would be carried out at fixed N. However, as we have shown above, the pair correlation function is related to the intensity of elastically scattered radiation and is normally obtained from such a scattering experiment. In a scattering experiment the beam samples a fraction of the total volume and in this subvolume the number of particles, while macroscopic, fluctuates. Thus the use of the grand canonical ensemble is appropriate.

The pair correlation function provides a measure of the distance over which particles are correlated. For an ideal gas, $g(r) = 1$ or $h(r) = 0$ and the fluctuation equation (5.33) gives the compressibility directly:

$$K_T = \frac{V}{Nk_BT} \ . \tag{5.35}$$

The same result can be obtained from the ideal gas equation of state $PV = Nk_BT$ and the definition of the isothermal compressibility: $K_T = -1/V\,(\partial V/\partial P)_{N,T}$.

Consider next an ideal solid in which the particles sit at fixed positions $\mathbf{R}_i$. In this case

$$n\int d^3r g(\mathbf{r}) = \frac{1}{N}\sum_{i\neq j}\int d^3r\,\delta(\mathbf{r}-\mathbf{R}_i+\mathbf{R}_j) = N-1\ . \tag{5.36}$$

Thus

$$n\int d^3r\,[g(\mathbf{r})-1] = -1$$

and the fluctuation equation of state gives $K_T = 0$. Physically, this means that if the atoms are not allowed to vibrate about their equilibrium positions, the compressibility is zero.

In self-condensed systems, such as solids and liquids far from the critical point, the compressibility will be much smaller than that of an ideal gas

$$0 < nk_BTK_T \ll 1\ .$$

Thus for nearly incompressible systems

$$n\int d^3r\,[g(r)-1] \approx -1\ .$$

Another situation in which the fluctuation equation of state offers valuable insight is in the discussion of the critical region (*i.e.*, the region of the phase diagram near the liquid-vapor critical point). At the critical point $\partial P/\partial V = 0$, so that $K_T = \infty$. This means that

$$\int d^3r\,[g(r)-1] = \int d^3r\,h(r) \to \infty \tag{5.37}$$

as the system approaches the critical point. The divergence is due to a long tail in $h(r)$. Conversely, the structure factor $S(q)$ becomes very large for small q. This feature is responsible for the phenomenon of critical opalescence observed in light-scattering studies near critical points.

The reader will note that the left-hand side of (5.33) is equal to

$$\lim_{q\to 0}[S(q)-1]\ .$$

In the canonical ensemble this limit is equal to -1, but as we have already pointed out above, a scattering experiment invariably samples a fluctuating number of particles and we have indicated in Figure 5.3 that $S(q)$ approaches a nonzero limit as q becomes small.

Having demonstrated the central role of the pair correlation function, we now discuss methods of calculating this function.

5.2.2 BBGKY hierarchy

We proceed to derive a set of equations for the reduced distribution functions introduced at the beginning of this section. This hierarchy is an equilibrium version of the BBGKY (Born, Bogoliubov, Green, Kirkwood, Yvon) hierarchy for the evolution of time-dependent distribution functions (see *e.g.* Balescu, [23]). Consider the function $\nabla n_1(\mathbf{x})$:

$$\begin{aligned}
\nabla n_1(\mathbf{x}) &= \frac{N}{N!Q_N}\nabla\int d^3r_2 d^3r_3\cdots d^3r_N \\
&\quad\times\exp\left\{-\beta\left(\sum_{i\neq 1}U(\mathbf{x}-\mathbf{r}_i)+\sum_{1\neq i<j}U(\mathbf{r}_i-\mathbf{r}_j)\right)\right\} \\
&= -\frac{\beta N(N-1)}{N!Q_N}\int d^3r_2\,\nabla_x U(\mathbf{x}-\mathbf{r}_2)\int d^3r_3\cdots d^3r_N \\
&\qquad\times\exp\{-\beta W(\mathbf{x}_1,\mathbf{r}_2,\ldots,\mathbf{r}_N)\} \\
&= -\beta\int d^3r_2\,[\nabla_x U(\mathbf{x}-\mathbf{r}_2)]\,n_2(\mathbf{x},\mathbf{r}_2) \qquad (5.38)
\end{aligned}$$

where $W(\mathbf{r}_1,\mathbf{r}_2,\ldots,\mathbf{r}_N)=\sum_{i<j}U(\mathbf{r}_{ij})$.

For a homogeneous system both sides of equation (5.38) are zero and this derivation serves only to indicate how to obtain a coupled set of integrodifferential equations for the reduced distribution functions. Proceeding in a similar fashion, we find

$$\begin{aligned}
\nabla_1 n_2(\mathbf{x}_1-\mathbf{x}_2) &= \frac{N(N-1)}{N!Q_N}\nabla_1 \\
&\quad\times\int d^3r_3\ldots d^3r_N\,\exp\{-\beta W(\mathbf{x}_1,\mathbf{x}_2,\ldots,\mathbf{r}_N)\} \\
&= -\beta\,[\nabla_1 U(\mathbf{x}_1-\mathbf{x}_2)]\,n_2(\mathbf{x}_1,\mathbf{x}_2) \\
&\quad -\beta\int d^3r_3\,[\nabla_1 U(\mathbf{x}_1-\mathbf{r}_3)]\,n_3(\mathbf{x}_1,\mathbf{x}_2,\mathbf{r}_3)\,. \qquad (5.39)
\end{aligned}$$

Converting to the pair and triplet distribution functions

$$n_2(\mathbf{x}_1,\mathbf{x}_2)=\left(\frac{N}{V}\right)^2 g(\mathbf{x}_1,\mathbf{x}_2) \qquad n_3(\mathbf{x}_1,\mathbf{x}_2,\mathbf{x}_3)=\left(\frac{N}{V}\right)^3 g_3(\mathbf{x}_1,\mathbf{x}_2,\mathbf{x}_3)$$

we have

$$-k_B T \nabla_1 g(\mathbf{x}_1, \mathbf{x}_2) = [\nabla_1 U(\mathbf{x}_1 - \mathbf{x}_2)]\, g(\mathbf{x}_1, \mathbf{x}_2) + n \int d^3x_3\, [\nabla_1 U(\mathbf{x}_1 - \mathbf{x}_3)]\, g_3(\mathbf{x}_1, \mathbf{x}_2, \mathbf{x}_3). \quad (5.40)$$

Equation (5.40) is the first of an infinite series of equations known as the BBGKY hierarchy. These equations link low-order distribution functions to functions of successively higher order and may be solved to jth order by approximating g_{j+1} in some fashion. The best known of these approximations is the Kirkwood superposition approximation. In this theory one writes $g_3(\mathbf{x}_1, \mathbf{x}_2, \mathbf{x}_3) = g(\mathbf{x}_1, \mathbf{x}_2) g(\mathbf{x}_1, \mathbf{x}_3) g(\mathbf{x}_2, \mathbf{x}_3)$. This converts equation (5.40) into a closed nonlinear equation for the pair function which is known as the Born–Green–Yvon (BGY) equation. This equation has been solved for a number of systems. In particular, in the case of hard spheres the results are in good agreement with numerical simulations at low density. It is clear from the nature of the decoupling that the superposition approximation can be valid only at low density. This is born out by a calculation of the virial coefficients (5.1) resulting from the BGY equation. The first two terms in the density expansion of $g(r)$ are correct; the higher coefficients are approximate. We shall not discuss the BBGKY hierarchy further. The reader is referred to Barker and Henderson [28] for a discussion of the merits of this type of approach.

5.2.3 Ornstein–Zernike equation

An equation that has been much used to develop approximate theories of dense gases and fluids is the Ornstein–Zernike equation. One defines a function, the *direct correlation function*, $C(\mathbf{r}_1, \mathbf{r}_2)$, by demanding that it be a solution of the integral equation

$$h(\mathbf{r}_1, \mathbf{r}_2) = C(\mathbf{r}_1, \mathbf{r}_2) + n \int d^3r_3 h(\mathbf{r}_1, \mathbf{r}_3) C(\mathbf{r}_3, \mathbf{r}_2) \quad (5.41)$$

where $h(\mathbf{r}_1, \mathbf{r}_2) = g(\mathbf{r}_1, \mathbf{r}_2) - 1$. The origin of the term "direct correlation" function is clear from this equation—in the limit of low density, $C(\mathbf{r}_1, \mathbf{r}_2)$ is precisely the correlation function for particles at positions $\mathbf{r}_1$ and $\mathbf{r}_2$ and is simply the Mayer function $f(|\mathbf{r}_1 - \mathbf{r}_2|)$. The second term on the right-hand side of (5.41) contains the effect of three or more particles on the function h. Equation (5.41) obviously cannot be solved since it contains two unknown functions. It can, however, be closed by expressing C in terms of h on the basis of some physically appealing approximation. One of the most useful of these approximations is the Percus–Yevick (PY) approximation, which we now briefly discuss. The pair distribution function is given by

$$g(\mathbf{r}_1,\mathbf{r}_2) = V^2 \frac{\int d^3r_3 \dots d^3r_N \exp\{-\beta W(\mathbf{r}_1,\mathbf{r}_2,\dots,\mathbf{r}_N)\}}{\int d^{3N}r \exp\{-\beta W(\mathbf{r}_1,\mathbf{r}_2,\dots,\mathbf{r}_N)\}} . \tag{5.42}$$

This equation can be used to derive a virial expansion for g. The first term in the expansion is simply $\exp\{-\beta U(r_{12})\}$ and we define

$$y(\mathbf{r}_{12}) \equiv \exp\{\beta U(\mathbf{r}_1 - \mathbf{r}_2)\}\, g(\mathbf{r}_1,\mathbf{r}_2) = 1 + \sum_{j\geq 1} y_j(\mathbf{r}_1 - \mathbf{r}_2) n^j \tag{5.43}$$

where the y_j's can be expressed in terms of cluster integrals by the same methods that we used for the pressure in Section 5.1. We leave the derivation of $y_1(\mathbf{r}_{12})$ as an exercise and simply quote the result

$$y_1(\mathbf{r}_1 - \mathbf{r}_2) = \int d^3r_3\, f(\mathbf{r}_1 - \mathbf{r}_3) f(\mathbf{r}_3 - \mathbf{r}_2) \tag{5.44}$$

where, as usual, $f(r) = \exp\{-\beta U(r)\} - 1$. Noting that

$$h(r) = f(r) + \sum_{j\geq 1} n^j y_j(r)\, [1 + f(r)]$$

and substituting in (5.41), we obtain

$$C(r) = \sum_{j=0}^{\infty} n^j C_j(r)$$

with

$$\begin{aligned} C_0(r) &= f(r) \\ C_1(r) &= f(r) y_1(r) . \end{aligned} \tag{5.45}$$

We now make the approximation, correct to first order in the density, that

$$C(r) = f(r) y(r) = \left(1 - e^{\beta U(r)}\right) g(r) , \tag{5.46}$$

which when substituted in equation (5.41) provides a nonlinear integral equation known as the Percus–Yevick equation (see, *e.g.* [23] for further details). This equation can be solved analytically in three dimensions for hard spheres [322], [300] and by numerical methods for arbitrary interaction potentials. The importance of this equation lies in the fact that it provides an excellent representation of the pair correlation function for hard spheres which are a simple

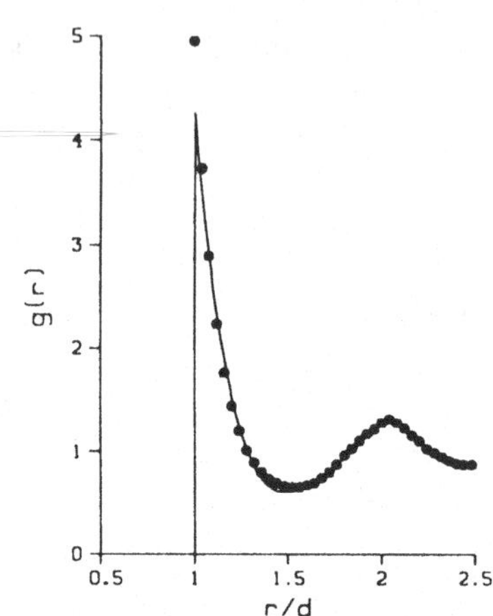

Figure 5.4: Comparison of $g(r)$ from the PY equation with computer simulations for $n\pi d^3/6 = 0.463$, where n is the density and d the hard-sphere diameter. (Data points from Alder and Hecht [10].)

and remarkably successful model for real liquids [313]. In Figure 5.4 we compare the pair distribution functions for hard spheres obtained from the PY equation and from molecular dynamics calculations. The agreement between the two is excellent except at $r/d = 1$. For more realistic interatomic potentials such as the Lennard–Jones potential, the Percus–Yevick equation is not quite as successful, particularly when it comes to predictions of thermodynamic properties. Nevertheless, the attractive part of the real interatomic potential is, in many approximations, seen to provide only a small perturbation and it is this insight that led to the successful modern perturbation theories discussed in Section 5.3.

There are a number of other approximate closures of the Ornstein–Zernike equation that have been developed and a thorough review of these theories may be found in [28]. We mention only the mean spherical approximation and the hypernetted chain approximation. The mean spherical approximation consists of the ansatz

$$\begin{aligned} g(r) &= 0 & r &< d \\ C(r) &= -\beta U(r) & r &> d \end{aligned} \tag{5.47}$$

where d is the hard-sphere diameter for a potential consisting of a hard core and a long-range piece given by $U(r)$, and $C(r)$ the direct correlation function. This approximation, when substituted into the Ornstein–Zernike equation, provides an exactly solvable equation, both for charged hard spheres [315], [316] and dipolar hard spheres [323].

Another frequently used closure of the Ornstein–Zernike equation is the hypernetted chain approximation (HNC). In this scheme one writes

$$C(r) = h(r) - \beta U(r) - \ln[1 + h(r)] \ . \tag{5.48}$$

When analyzed in terms of a virial expansion, the HNC seems, at first glance, to be a more satisfactory approximation than the PY equation. However, it turns out that for hard spheres and for other short-range potentials, the PY equation gives better results. For systems, such as electrolytes or the classical Coulomb gas, in which the interparticle potential is of long range, the HNC approximation is preferable.

5.3 Perturbation Theory

In this section we outline the ideas on which the modern perturbation theories of liquids are based and display some of the results obtained by such methods. The basic physical idea is that in systems of atoms or molecules interacting through a potential like the 6–12 potential, the short-range repulsive piece of the interaction is responsible for most of the structure seen in the pair correlation function. One should therefore be able to use the hard-sphere system as a reference system or unperturbed system and to treat the corrections due to the attractive part of the potential perturbatively. A 6–12 potential has a soft core rather than a hard core, but this does not present any essential difficulties. We decompose the pair potential into two pieces:

$$U(r_{ij}, \lambda) = U_0(r_{ij}) + \lambda U_1(r_{ij}) \tag{5.49}$$

where

$$\begin{aligned} U_0(r_{ij}) &= 0 \quad \text{for } r_{ij} > \sigma \\ U_1(r_{ij}) &= 0 \quad \text{for } r_{ij} < \sigma \ . \end{aligned} \tag{5.50}$$

In equation (5.49) the case $\lambda = 1$ corresponds to the original potential. The logarithm of the configuration integral is given by

$$\ln Q_N(V, T, \lambda) = \ln \frac{1}{N!} \int d^{3N}r \, \exp\left\{ -\beta \sum_{i<j} [U_0(r_{ij}) + \lambda U_1(r_{ij})] \right\} \ . \tag{5.51}$$

We now expand this function in powers of λ:

$$\ln Q_N(V,T,\lambda) = \ln Q_N(V,T,0) + \lambda \frac{\partial}{\partial \lambda} \ln Q_N(V,T,\lambda)|_{\lambda=0} + O(\lambda^2) \quad (5.52)$$

with

$$\begin{aligned} &\frac{\partial}{\partial \lambda} \ln Q_N(V,T,\lambda)|_{\lambda=0} \\ &= -\beta \frac{N(N-1)}{2} \int d^{3N}r \, U_1(r_{12}) \frac{\exp\left\{-\beta \sum_{i<j} U_0(r_{ij})\right\}}{N! Q_N(V,T,0)} \\ &= -\frac{\beta}{2} \int d^3r_1 d^3r_2 U_1(r_{12}) n_2^{(0)}(\mathbf{r}_1, \mathbf{r}_2) \\ &= -\frac{\beta}{2} \left(\frac{N}{V}\right)^2 \int d^3r_1 d^3r_2 U_1(r_{12}) g^{(0)}(\mathbf{r}_1, \mathbf{r}_2) \end{aligned}$$

where $g^{(0)}$ is the pair distribution function of the system for $\lambda = 0$ (*i.e.*, for the *reference system*). Therefore, we obtain, for the Helmholtz free energy,

$$A(V,T) = A_0(V,T) + \frac{N^2}{2V} \int d^3r \, U_1(r) g^{(0)}(r) + \cdots . \quad (5.53)$$

The higher-order terms can be similarly expressed as integrals over three- and higher-particle correlation functions of the reference system. Ideally, one would like to take the hard-sphere system as the reference system, partly because its properties are well known and partly because it would provide a common starting point for a number of different systems whose potentials are of the same general form but which may not have exactly the same reference potential $U_0(r)$. Barker and Henderson [26] have provided a method of achieving this goal. We shall not repeat their argument. The result is that one can indeed replace $g^{(0)}(r)$ by a hard-sphere correlation function provided that the hard-sphere diameter, d, is taken to be temperature dependent. Specifically,

$$d = \int_0^\sigma dr \left[1 - \exp\{-\beta U_0(r)\}\right] . \quad (5.54)$$

With this choice of effective hard-sphere diameter very good agreement with computer simulations and experiment is obtained. This is illustrated in Figure 5.5, where the pair distribution function $g(r)$ obtained from zeroth and first-order perturbation theory is compared with the results of computer simulations. The agreement in first-order perturbation theory is extremely good, indicating that the ideas behind this approach are indeed correct.

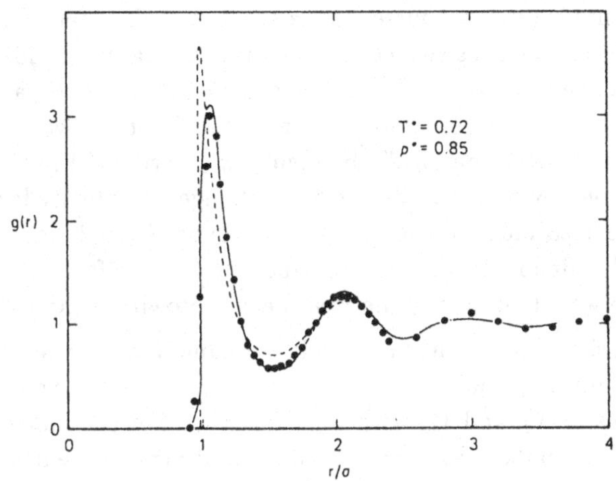

Figure 5.5: Pair distribution function of a Lennard–Jones fluid for $n = 0.85\sigma^{-3}$ and $T = 0.72\epsilon/k_B$ (near the triple point). Points are the results of simulations of Verlet [313]. The dashed and solid curves correspond to zeroth- and first-order Barker–Henderson perturbation theory. (From Barker and Henderson [28].)

In this short discussion we have only mentioned one of the perturbation theories which are useful for real liquids. Several other versions have been developed and the reader is referred to [28] and the original articles cited therein.

5.4 Inhomogeneous Liquids

In this section we discuss some of the properties of inhomogeneous liquids. Section 5.4.1 is devoted to the van der Waals theory of the liquid vapor interface and of the surface tension and in Section 5.4.2 we construct a simple theory of the normal modes—the capillary waves—of a free liquid surface.

5.4.1 Liquid–vapor interface

In most elementary physics courses the concept of surface tension of liquids is introduced. In the present subsection we wish to relate this quantity to the statistical mechanics of the two distinct coexisting phases separated by the surface, namely the bulk fluid and the vapor. In Chapter 1 we introduced the surface tension thermodynamically by including a term $\sigma d\mathcal{A}$ in the expression for the work done by a system in a general change of state. Before beginning our mean field treatment of the interface we expand on the thermodynamic treatment and define more carefully the appropriate interfacial parameters. In this discussion we follow closely the treatment of Rowlinson and Widom [258].

Consider a single-component system in a volume V at temperature T. Suppose that two phases, liquid and vapor, coexist and let the volume occupied by these phases be V_L and V_G with $V_L + V_G = V$. Let the molecular density well inside the two bulk phases be n_L and n_G and $n(\mathbf{r})$ be the density at point $\mathbf{r}$. We assume that the interface between the two phases is planar and perpendicular to the z direction. The assignment of volumes V_L and V_G to liquid and gas is somewhat arbitrary. The density will vary between the liquid and gas densities over some distance, d, and the boundary between liquid and gas is, therefore, not sharp. We shall see that there is a convenient choice of the "dividing surface" which makes the subsequent calculations easier.

We now define the surface energy, the number of particles in the surface region, and the surface Helmholtz free energy through the equations

$$\begin{aligned} V_L n_L + V_G n_G + N_S &= N \\ V_L e_L + V_G e_G + E_S &= E \\ V_L a_L + V_G a_G + A_S &= A \end{aligned} \tag{5.55}$$

where the quantities N, E, A are the total particle number, energy, and Helmholtz free energy of the system and the lower-case symbols refer to the corresponding bulk densities. If the boundary between liquid and vapor were mathematically sharp, and coincided with our choice of dividing surface, the number of particles N_S in the surface would be zero.

We take the temperature of the system to be fixed. A slight generalization of the derivation of the Gibbs–Duhem equation (1.39) yields

$$A = -PV + \sigma\mathcal{A} + \mu N \tag{5.56}$$

where $\mathcal{A}$ is the area of the interface. Similarly,

$$\begin{aligned} A_L &= -PV_L + \mu N_L \\ A_G &= -PV_G + \mu N_G \end{aligned} \tag{5.57}$$

and from (5.55),

$$\begin{aligned} A_S = \mathcal{A} a_S &= \sigma \mathcal{A} + \mu(N - N_L - N_G) \\ &= \sigma \mathcal{A} + \mu N_S \ . \end{aligned} \tag{5.58}$$

From equation (5.58) we see that if we choose the location of the Gibbs dividing surface (at $z = 0$) to satisfy the equation

$$N_S = \mathcal{A} \int_{-\infty}^{0} dz[n(z) - n_L] + \mathcal{A} \int_{0}^{\infty} dz[n(z) - n_G] = 0 \tag{5.59}$$

the surface tension will be related to the excess Helmholtz free energy per unit area through

$$\sigma = a_S \ . \tag{5.60}$$

We now construct an approximate theory of the interfacial region following the ideas of van der Waals [308] and Cahn and Hilliard [53]. We are considering a system for which the total volume (gas + liquid) is held fixed at constant temperature. Thus the appropriate free energy to minimize with respect to any variational parameters is the Helmholtz free energy. We suppose that there exists a *local* Helmholtz free energy per unit volume, $\Psi(\mathbf{r})$, which, in our geometry, can only be a function of z, $\Psi(z)$. The surface tension (5.60) is related to this function through

$$\sigma = \int_{-\infty}^{0} dz \, [\Psi(z) - \Psi_L] + \int_{0}^{\infty} dz \, [\Psi(z) - \Psi_G] \tag{5.61}$$

where $\Psi_L = A_L/V_L$, $\Psi_G = A_G/V_G$. This free-energy density is, at fixed temperature, a functional of the local density, that is, $\Psi(z) = \Psi[n(z)]$. We will assume a phenomenological free-energy density which is composed of two parts, a term $\Gamma(dn/dz)^2/2$, where Γ is a positive constant, and a term that results from analytic continuation of a mean field free energy into the region of unphysical single-phase density. In Section 4.4 this unphysical region was avoided by means of the Maxwell construction. We assume that the Helmholtz free-energy density resembles the curve shown in Figure 5.6 for $T < T_c$. The straight-line section connecting the bulk densities is the result of the Maxwell construction (or equivalently the double tangent construction of Section 4.1).

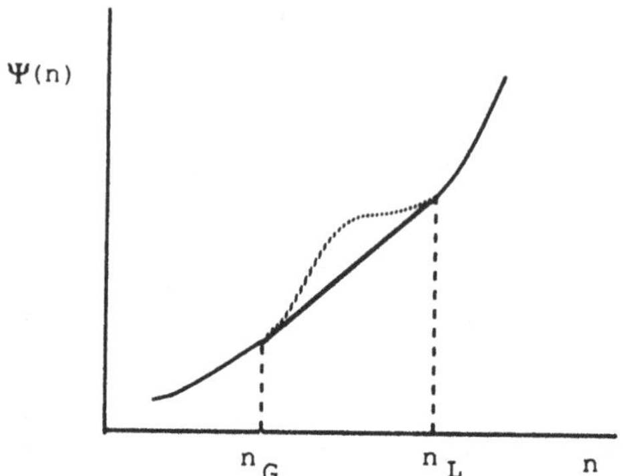

Figure 5.6: Phenomenological free energy density.

The dotted curve represents a mean field free-energy density resulting, for example, from a van der Waals equation of state (Section 4.4). Therefore, we have

$$\sigma = \int_{-\infty}^{\infty} dz \left\{ \frac{\Gamma}{2} \left(\frac{dn}{dz} \right)^2 + [\Psi_n([n(z)]) - \Psi_e(z)] \right\} \tag{5.62}$$

where Ψ_n is the nonequilibrium free-energy density (dotted curve in Figure 5.6) and Ψ_e is the equilibrium bulk-liquid or gas free energy density, depending on whether z is greater or less than zero. The density profile is determined by minimizing (5.62) with respect to $n(z)$. The first term in (5.62) is a typical Landau–Ginzburg term which reduces the fluctuations in the free energy.

To make further progress, we must find an approximate expression for the term

$$\Delta\Psi = \Psi_n\left([n(z)]\right) - \Psi_e(z)$$

in the transition region. We define $n_0 = (n_L + n_G)/2$, $\lambda = (n_L - n_G)/2$ and assume that $\Delta\Psi$ has an expansion of the form

$$\Delta\Psi = \sum_j \alpha_j \left[n(z) - n_0\right]^j = \sum_j \alpha_j \zeta^j(z) \lambda^j \tag{5.63}$$

where the reduced variable

$$\zeta(z) = \frac{2[n(z) - n_0]}{n_L - n_G} \tag{5.64}$$

takes on the value $\zeta = -1$ when $n(z) = n_G$ and $\zeta = 1$ when $n(z) = n_L$. We must require that $\Delta\Psi(n_G) = \Delta\Psi(n_L) = 0$. Furthermore, since Ψ is the free energy density, $d\Psi(n,T)/dn = \mu$, where μ is the chemical potential, which must be the same in the two bulk phases. Also, since by the tangent construction in Figure 5.6, Ψ must match smoothly to the bulk solution, we have

$$\left.\frac{d(\Delta\Psi)}{d\zeta}\right|_{-1} = \left.\frac{d(\Delta\Psi)}{d\zeta}\right|_{+1} = 0\,. \tag{5.65}$$

If we truncate (5.63) at the fourth-order term, we find

$$\Delta\Psi = \alpha_4\lambda^4\left[1-\zeta^2(z)\right]^2 \tag{5.66}$$

where the coefficient α_4 is undetermined, but positive. Minimizing (5.62) with respect to $n(z)$ or, equivalently, $\zeta(z)$, we find

$$\Gamma\frac{d^2\zeta}{dz^2} + 4\alpha_4\lambda^2\zeta(1-\zeta^2) = 0\,. \tag{5.67}$$

A first integral of this differential equation is easily obtained by substituting $\phi = d\zeta/dz$ and noting that $d\phi/dz = \phi d\phi/d\zeta$. Thus we find

$$\frac{d\phi^2}{d\zeta} = -\frac{8\alpha_4\lambda^2}{\Gamma}\zeta(1-\zeta^2) \tag{5.68}$$

and integrating from -1 to ζ yields

$$\phi^2(\zeta) - \phi^2(-1) = \frac{2\alpha_4\lambda^2}{\Gamma}(\zeta^4 - 2\zeta^2 + 1)\,.$$

Because we expect that

$$\phi(-1) = \left.\frac{d\zeta}{dz}\right|_{\infty} = 0$$

we have

$$\frac{d\zeta}{dz} = -\sqrt{\frac{2\alpha_4\lambda^2}{\Gamma}}(1-\zeta^2)$$

where the minus sign on the right has been chosen in order to have liquid at $z = -\infty$. Integrating, we find that

$$\zeta(z) = -\tanh\left(\sqrt{\frac{2\alpha_4\lambda^2}{\Gamma}}z\right) \tag{5.69}$$

and the density profile in the liquid-vapor interface is

$$n(z) = \frac{n_L + n_G}{2} - \frac{n_L - n_G}{2}\tanh\left(\sqrt{\frac{2\alpha_4\lambda^2}{\Gamma}}z\right)\,. \tag{5.70}$$

The quantity $\sqrt{\Gamma/(2\alpha\lambda^2)}$ also appears in the theory of superconductivity where it is commonly referred to as the Landau–Ginzburg coherence length. In Section 11.3.4 we use an argument, very similar in spirit to the derivation above, to discuss superconductivity. In the theory of the liquid-vapor interface presented above, it plays the role of an interfacial width and we see from (5.70) that if the coefficient Γ in (5.62) were taken to be zero, the profile would be infinitely sharp. We may now substitute (5.70) into (5.62) to find the surface tension:

$$\sigma = 2\alpha_4\lambda^4 \int_{-\infty}^{\infty} dz\,\mathrm{sech}^4\sqrt{\frac{2\alpha_4\lambda^2}{\Gamma}}z = \sqrt{2\alpha_4\Gamma}\lambda^3 \int_{-\infty}^{\infty} dy\,\mathrm{sech}^4 y\ . \qquad (5.71)$$

As the system approaches the liquid-vapor critical point

$$\lambda = \frac{n_L - n_G}{2} \sim (T_c - T)^{1/2}$$

and we see in (5.71) the classical scaling form of the surface tension

$$\sigma \sim (T_c - T)^{\mu}$$

with $\mu = \frac{3}{2}$.

A generalization of the theory given above for the liquid-vapor interface which incorporates the scaling theory of critical phenomena (Chapter 6) has been provided by Fisk and Widom [97] but is beyond the scope of this book. Our treatment has been very much a phenomenological one, containing two parameters α_4 and Γ, which at this level we cannot determine microscopically, as well as some questionable assumptions about the form of the free energy. Other more sophisticated mean field theories based on the inhomogeneous Ornstein–Zernike equation are parameter free in the sense that the interparticle potential, temperature, and bulk density completely determine the density profile [245]. These theories, as well as density functional theories, are capable of producing surface tensions to within roughly 10% of the experimental values except very close to the critical point where mean field theories fail in general.

The liquid-vapor interface has also been extensively studied by computer simulations (Chapela *et al.*, [60], and references therein). For some time there was considerable controversy concerning the possible existence of small-amplitude oscillations superimposed on a smooth density variation such as that given by (5.70). For molecular liquids, such as condensed rare gases, it is now generally accepted that there are no oscillations in the density profile and that the general form (5.70), with the phenomenological constant $(\Gamma/2\alpha_4\lambda^2)^{1/2}$ replaced by a function $d(T)$ of the temperature, provides a very good fit to the numerical results.

Experimental measurements of the density profile can be carried out in a number of ways. Beaglehole [31] measured the thickness of the interface in liquid argon at 90K and 120K by ellipsometry. At 90K he found a thickness of roughly 0.8 nm (defined as the distance over which the density goes from 10% to 90% of its final value). Other techniques, such as the use of synchrotron radiation incident at glancing angles [12], hold out the promise that much more of the detailed structure of the interfacial region will become known in the near future.

An essential assumption in our derivation of the density $n(z)$ was that the interface is planar. If we think of the interface as a stretched membrane, we see that there will be thermally excited normal modes of vibration which will distort the interface and broaden it further. A simple treatment of these "capillary waves" forms the subject matter of the next subsection.

5.4.2 Capillary waves

Let us imagine that the position of the interface between the liquid and vapor phases can be specified by a function, $z(x, y)$, which may refer either to the location of an infinitely sharp dividing surface or to the midpoint of a more diffuse profile such as (5.70). If the energy of a flat surface at $z = 0$ is taken to be zero, the added energy due to distortion and displacement is given by

$$\Delta E = \int dxdy \left[\frac{n_L - n_G}{2} g z^2(x, y) + \sigma \left\{ \sqrt{1 + \left(\frac{\partial z}{\partial x}\right)^2 + \left(\frac{\partial z}{\partial y}\right)^2} - 1 \right\} \right]$$

where σ is the surface tension, g the acceleration due to gravity and n_L, n_G the *mass* densities in the liquid and gas phases. The derivation of the first term, the change in gravitational potential energy due to a displacement of the dividing surface, is left as an exercise. The second term represents the change in area of the interface due to distortion. Assuming that the distortion is small (*i.e.*, $|\partial z/\partial x|$, $|\partial z/\partial y| \ll 1$), we may expand the second term to obtain

$$\Delta E = \int dxdy \left[\frac{n_L - n_G}{2} g z^2(x, y) + \frac{\sigma}{2} \left\{ \left(\frac{\partial z}{\partial x}\right)^2 + \left(\frac{\partial z}{\partial y}\right)^2 \right\} \right] . \tag{5.72}$$

If we consider a liquid in a cubical container of length L on each side and apply periodic boundary conditions in the x and y directions,

$$z(x + L, y) = z(x, y + L) = z(x, y)$$

we may express the energy (5.72) in terms of the energies of a set of normal modes. We let

$$z(x,y) = \frac{1}{L}\sum_{\mathbf{q}} \hat{z}_{\mathbf{q}} e^{i\mathbf{q}\cdot\mathbf{r}} \tag{5.73}$$

with $\mathbf{q} = 2\pi(n_x, n_y)/L$ and n_x, $n_y = 0, \pm1, \pm2, \ldots$. Substituting, we obtain

$$\Delta E = \frac{\sigma}{2}\sum_{\mathbf{q}} \left[\frac{(n_L - n_G)g}{\sigma} + q^2\right] \hat{z}_{\mathbf{q}}\hat{z}_{-\mathbf{q}} \; . \tag{5.74}$$

The quantity $a^2 = 2\sigma/\{(n_L - n_G)g\}$ has the dimension of (length)^2 and its square root is given the name "capillary length". Modes with a wavelength much larger than the capillary length are called gravity waves, while modes with wavelength less than $2\pi a$ are known as capillary waves.

Decomposing $\hat{z}_{\mathbf{q}}$ into its real and imaginary parts and using the equipartition theorem of classical statistics we find, for the thermal expectation value $\langle \hat{z}_{\mathbf{q}}\hat{z}_{-\mathbf{q}}\rangle$,

$$\langle \hat{z}_{\mathbf{q}}\hat{z}_{-\mathbf{q}}\rangle = \frac{2k_BT/\sigma}{2a^{-2} + q^2} \; . \tag{5.75}$$

The broadening of the interface due to thermal excitation of these modes is conveniently expressed in terms of the quantity

$$\xi^2 = \frac{1}{L^2}\int dxdy \left\langle [z(x,y) - \bar{z}]^2 \right\rangle \tag{5.76}$$

where

$$\bar{z} = \frac{1}{L^2}\int dxdy\, z(x,y) \; .$$

Substituting (5.73) into (5.76), we see that

$$\xi^2 = \frac{1}{L^2}\sum_{|\mathbf{q}|>0} \langle \hat{z}_{\mathbf{q}}\hat{z}_{-\mathbf{q}}\rangle = \frac{2k_BT}{\sigma\mathcal{A}}\sum_{|\mathbf{q}|>0} \frac{1}{q^2 + 2a^{-2}} \; . \tag{5.77}$$

Converting the sum over $\mathbf{q}$ to an integral ($\sum_{\mathbf{q}} = \mathcal{A}/(2\pi)^2 \int d^2\mathbf{q}$) and carrying out the integration in cylindrical coordinates, we obtain

$$\xi^2 = \frac{k_BT}{\pi\sigma}\int_{2\pi/L}^{q_{max}} dq \frac{q}{q^2 + 2a^{-2}} = \frac{k_BT}{2\pi\sigma}\ln\frac{q_{max}^2 a^2/2 + 1}{2\pi^2 a^2/L^2 + 1} \; . \tag{5.78}$$

The cutoff at q_{max} has been introduced because wave vectors greater than $2\pi/a_0$, where a_0 is a molecular diameter, are essentially meaningless as the interface is not a true continuum. Taking $a_0 = .34$ nm (appropriate for argon) and the surface tension to be the experimental value for argon at its triple point $\sigma = 1.51 \times 10^{-4}$ N/m, we find a capillary length of 1.5 mm and from equation (5.78) $\xi \approx .64$ nm for $L = 5$ cm. The dependence of ξ on L is quite weak due to the slow variation of the logarithm.

The divergence of the interfacial width in zero gravity due to the long-wavelength modes is found in other physical situations as well. Simple models of crystal growth, such as the solid-on-solid model [319], display roughening of the surface due to excitation of the analogous modes. Similarly, domain walls in the three-dimensional Ising model are rough in the same way.

5.5 Density-Functional Theory

Density-functional methods were first developed in the context of quantum many-body theory of electrons in atoms, molecules and solids and they have had a great impact on these fields. The fundamental theorem that the ground-state energy of a many-electron system is a unique functional of the electronic density, achieving its minimum at the physical density, was later shown to hold for the Helmholtz free energy and grand potential of classical liquids. Applications to bulk and inhomogeneous fluids soon followed and these methods are now among the most successful of the 'analytic' theories of dense liquids. In this section we will develop the underlying concepts of density-functional theories and show how some of the theories of the previous two sections can be expressed in this language. We will not review the numerous approximate schemes that have been developed in this context, but rather refer the reader to the two review articles that we have found most helpful [87] [88].

5.5.1 Functional differentiation

Before discussing this formalism, we digress briefly to review some of the fundamentals of functional differentiation. A more extensive discussion can be found in [231]. The notion of a functional will be to some extent familiar from Hamilton's principle of classical mechanics which states that

$$\delta I[q(t)] = \delta \int_{t_0}^{t_1} L(q(t), \dot{q}(t), t)dt = 0$$

i.e., that the path $q(t)$ of a particle between times t_0 and t_1 is stationary with respect to variations of the integral I of the Lagrangian L between times t_0 and t_1. Here I is a functional of $q(t)$ and the operation δ gives the change in I when an *arbitrary* continuous infinitesimal function $\delta q(t)$ is added to $q(t)$ subject to $\delta q(t_0) = \delta q(t_1) = 0$. We can express this equation in the alternative form

$$\delta I[q(t)] = \int_{t_0}^{t_1} \frac{\delta I}{\delta q(t)} \delta q(t) dt = 0 \ .$$

This form is the natural generalization of the differential $dF = \sum_{j=1}^{N} (\partial F/\partial x_j) dx_j$ for a function $F(x_1, x_2, ..., x_N)$ of a finite number N of independent variables.

We may obtain an expression for the functional derivative $\delta I/\delta q(t)$ in the following way. Imagine adding an arbitrary function $\epsilon \tilde{q}(t)$ to $q(t)$ and constructing the limit

$$\lim_{\epsilon \to 0} \frac{I[q(t) + \epsilon \tilde{q}(t)] - I[q(t)]}{\epsilon} = \frac{d}{d\epsilon} I[q(t) + \epsilon \tilde{q}(t)] \bigg|_{\epsilon=0} = \int_{t_0}^{t_1} \frac{\delta I}{\delta q(t)} \tilde{q}(t) dt \ .$$

By choosing $\tilde{q}(t) = \delta(t - t')$ we then obtain the expression

$$\frac{\delta I[q(t)]}{\delta q(t')} = \lim_{\epsilon \to 0} \frac{I[q(t) + \epsilon \delta (t - t')] - I[q(t)]}{\epsilon} \tag{5.79}$$

which allows an explicit evaluation of the quantity $\delta I[q]/\delta q(t)$. Before discussing some other properties of functional derivatives we present a simple example. In the van der Waals theory of the liquid-vapor interface we encountered the free energy functional (5.62)

$$\sigma = \int_{-\infty}^{\infty} dz \left\{ \frac{\Gamma}{2} \left(\frac{dn}{dz} \right)^2 + [\Psi_n[n(z)] - \Psi_e(z)] \right\} \ .$$

We use the definition (5.79) to construct the functional derivative:

$$\frac{\delta \sigma[n(z)]}{\delta n(z')} = -\Gamma \frac{d^2 n(z')}{dz'^2} + \Psi_n'[n(z')]$$

where the first term is obtained by integrating by parts the expression

$$\Gamma \int_{-\infty}^{\infty} dz \left(\frac{dn}{dz} \right) \epsilon \frac{d}{dz} \delta (z - z')$$

in order to remove the derivative from the δ function. The resulting expression is in principle still a functional of $n(z)$ so that higher functional derivatives such as $\delta^2 \sigma / \delta n(z) \delta n(z')$ are defined in strict analogy with the first derivative. We also note that the definition (5.79) has the obvious corollary:

$$\frac{\delta q(t)}{\delta q(t')} = \delta (t - t') \ . \tag{5.80}$$

It also follows from the basic definition (5.79) that many of the familiar rules of ordinary differentiation hold as well for functional derivatives. Examples are

$$\frac{\delta}{\delta q(t)}\left(aF[q(t)]+bG[q(t)]\right)=a\frac{\delta F}{\delta q(t')}+b\frac{\delta G}{\delta q(t')}$$

$$\frac{\delta}{\delta q(t')}F[q(t)]G[q(t)]=F[q(t)]\frac{\delta G[q(t)]}{\delta q(t')}+G[q(t)]\frac{\delta F[q(t)]}{\delta q(t')}\;.$$

We may also derive a chain-rule for functional derivatives. Suppose that $q(t)$ is itself a functional of $s(t)$, i.e. $q(t)=q[s(t),t]$. Then

$$\delta F=\int\frac{\delta F}{\delta q(t)}\delta q(t)dt=\int dt\int dt'\frac{\delta F}{\delta q(t)}\frac{\delta q(t)}{\delta s(t')}\delta s(t')$$

and

$$\frac{\delta F}{\delta s(t')}=\int dt\frac{\delta F}{\delta q(t)}\frac{\delta q(t)}{\delta s(t')}$$

which is the functional derivative form of the chain rule of ordinary differentiation $dF/dx=(dF/dz)(dz/dx)$. Finally, we note that it is possible to define a formal inverse of a functional derivative. If q is a functional of s, then s is itself a functional of q (although the inversion of a functional relation is not necessarily trivial). Then

$$\delta q(t)=\int dt'\frac{\delta q(t)}{\delta s(t')}\delta s(t')=\int dt'\int dt''\frac{\delta q(t)}{\delta s(t')}\frac{\delta s(t')}{\delta q(t'')}\delta q(t'')\;.$$

Using (5.80) we have

$$\int dt''\frac{\delta q(t)}{\delta s(t'')}\frac{\delta s(t'')}{\delta q(t')}=\delta\left(t-t'\right)\tag{5.81}$$

which is a formal statement of the inverse relation between $\delta q(t)/\delta s(t')$ and $\delta s(t')/\delta q(t'')$.

To this point our functionals have depended on functions of a single variable t or z. The generalization to higher dimensions is straightforward. If $F=F[n(\mathbf{r})]$ then

$$\frac{\delta F}{\delta n(\mathbf{r})}=\lim_{\epsilon\to 0}\frac{F[n(\mathbf{r}')+\epsilon\delta\left(\mathbf{r}'-\mathbf{r}\right)]-F[n(\mathbf{r}')]}{\epsilon}\tag{5.82}$$

and

$$\frac{\delta n(\mathbf{r})}{\delta n(\mathbf{r}')}=\delta\left(\mathbf{r}-\mathbf{r}'\right)\;.$$

The other relations given above have the same obvious generalizations.

5.5.2 Free-energy functionals and correlation functions

We now wish to show that the Helmholtz free energy (or equivalently the grand potential) is a *unique* functional of the *density* $n(\mathbf{r})$. This nontrivial result is the generalization to finite temperature of the Kohn–Hohenberg theorems [231] which establish the same result for the quantum mechanical ground-state energy of a many-body system. Our discussion of this topic follows closely the treatment of [87] where references to the original articles may also be found.

We consider a system of N particles interacting through two-body interactions as in (5.2) but subject to a position-dependent single-particle potential $v(\mathbf{r})$ so that

$$\begin{aligned} H &= \sum_i \frac{p_i^2}{2m} + \sum_{i<j} U\left(|\mathbf{r}_i - \mathbf{r}_j|\right) + \sum_i v(\mathbf{r}_i) \\ &= H_0 + \mathcal{V}(\mathbf{r}_1, \ldots, \mathbf{r}_N) \end{aligned} \tag{5.83}$$

where $\mathcal{V} = \sum_i v(\mathbf{r}_i)$. The canonical partition function

$$Z_c = e^{-\beta A} = \frac{1}{N!h^{3N}} \int d^{3N}p\, d^{3N}r\, e^{-\beta\{H_0 + \mathcal{V}\}} \tag{5.84}$$

is clearly a functional of the 'external' potential $v(\mathbf{r})$, as is the Helmholtz free energy $A = A[N, V, T; v(\mathbf{r})]$. Similarly, the density $n(\mathbf{r})$ (5.19)

$$n(\mathbf{r}) = \sum_i \langle \delta\left(\mathbf{r} - \mathbf{r}_i\right) \rangle = n[\mathbf{r}; v(\mathbf{r})] \tag{5.85}$$

is a functional of $v(\mathbf{r})$. This implies, in turn, that $v(\mathbf{r})$ is a functional of n. What we wish to prove is that this relation is unique up to a constant. More precisely, to a given density $n(\mathbf{r})$ there corresponds, for a given T, a single function $v(\mathbf{r})$.

Consider a phase-space probability density $\rho(\mathbf{r}_i, \mathbf{p}_i)$ such that

$$\mathrm{Tr}\, \rho(\mathbf{r}_i, \mathbf{p}_i) \equiv \frac{1}{N!h^{3N}} \int d^{3N}r d^{3N}p\, \rho(\mathbf{r}_i, \mathbf{p}_i) = 1 \tag{5.86}$$

where we have defined the operation Tr (the 'classical trace') to condense the notation. The canonical equilibrium probability density

$$\rho_0(\mathbf{r}_i, \mathbf{p}_i) = \frac{1}{Z} e^{-\beta H} \tag{5.87}$$

is one example of such a function. We now define a Helmholtz free energy functional

$$\mathcal{A}[\rho] = \mathrm{Tr}\, \rho \left\{H + k_B T \ln \rho\right\} \ . \tag{5.88}$$

The equilibrium value of this free energy functional $A = \mathcal{A}[\rho_0]$ is obtained when the canonical probability density ρ_0 is used in (5.88). We first show that for *any* N-particle $\rho \neq \rho_0$, we have

$$\mathcal{A}[\rho] > \mathcal{A}[\rho_0] \; . \tag{5.89}$$

Noting from (5.87) that $H = -k_BT \ln \rho_0 - k_BT \ln Z = -k_BT \ln \rho_0 + \mathcal{A}[\rho_0]$ we have

$$\mathcal{A}[\rho] = \mathrm{Tr}\,\rho \left\{ k_BT \ln \frac{\rho}{\rho_0} + \mathcal{A}[\rho_0] \right\} \tag{5.90}$$

or

$$\begin{aligned} \beta \left\{ \mathcal{A}[\rho] - \mathcal{A}[\rho_0] \right\} &= \mathrm{Tr} \left\{ \rho \ln \frac{\rho}{\rho_0} - \rho + \rho_0 \right\} \\ &= \mathrm{Tr}\,\rho_0 \left\{ \frac{\rho}{\rho_0} \ln \frac{\rho}{\rho_0} - \left(\frac{\rho}{\rho_0} - 1 \right) \right\} \end{aligned} \tag{5.91}$$

where, in the first step, we have used the fact that both ρ and ρ_0 are normalized. The proof now follows from the inequality $x \ln x \geq x - 1$ where the equality holds only at the point $x = 1$. Therefore, since $\rho_0 \geq 0$ we have shown that the canonical probability density minimizes the Helmholtz free energy functional, a result obtained by other methods in Section 2.5.

We now assume that we may express the external potential $v(\mathbf{r})$ as a functional of the density $n(\mathbf{r})$. We will have shown that the free energy is a unique functional of $n(\mathbf{r})$ if we can demonstrate that v is uniquely determined by n. To do this we assume the contrary and demonstrate that a contradiction ensues. Suppose that there are two distinct external potentials $v(\mathbf{r})$ and $v'(\mathbf{r})$ that produce the same equilibrium density $n(\mathbf{r})$. There are then two distinct normalized *equilibrium* probability densities corresponding to these two external potentials with associated Helmholtz free energy functionals $\mathcal{A}[\rho]$ and $\mathcal{A}[\rho']$. Let $H = H_0 + \mathcal{V}$ and $H' = H_0 + \mathcal{V}'$. Then

$$\begin{aligned} \mathcal{A}[\rho'] &= \mathrm{Tr}\rho' \left\{ H' + k_BT \ln \rho' \right\} \\ &< \mathrm{Tr}\rho \left\{ H' + k_BT \ln \rho \right\} = \mathcal{A}[\rho] + \mathrm{Tr}\rho \left\{ \mathcal{V}' - \mathcal{V} \right\} \; . \end{aligned}$$

Since the external potential is a sum of single-particle energies, the last term in this equation can be written in the form

$$\mathrm{Tr}\rho \left\{ \mathcal{V}' - \mathcal{V} \right\} = \int d^3r \, n(\mathbf{r})(v'(\mathbf{r}) - v(\mathbf{r}))$$

and we have the final form

$$\mathcal{A}[\rho'] < \mathcal{A}[\rho] + \int d^3r \, n(\mathbf{r})(v'(\mathbf{r}) - v(\mathbf{r})) \; . \tag{5.92}$$

Since both ρ and ρ' are equilibrium distributions, we can carry out the same manipulations beginning with H instead of H' to obtain

$$\mathcal{A}[\rho] < \mathcal{A}[\rho'] + \int d^3r\, n(\mathbf{r})(v(\mathbf{r}) - v'(\mathbf{r})) \ . \tag{5.93}$$

Adding equations (5.92) and (5.93) we have the contradiction that $\mathcal{A}[\rho] + \mathcal{A}[\rho'] < \mathcal{A}[\rho'] + \mathcal{A}[\rho]$ and thus the assumption that $v \neq v'$ which implies that $\rho \neq \rho'$ must have been false. We have therefore established that

$$\mathcal{A}[\rho] = \mathcal{A}[n(\mathbf{r})] \ . \tag{5.94}$$

In the derivation of (5.89) we explicitly required the distributions to be N-particle distributions. When we re-express (5.89) in terms of a functional derivative with respect to $n(\mathbf{r})$ we must incorporate the constraint $\int d^3r\, n(\mathbf{r}) = N$ by means of a Lagrange multiplier that, of course, turns out to be the chemical potential μ. The resulting equation is

$$\left.\frac{\delta \mathcal{A}}{\delta n(\mathbf{r})}\right|_{n_0(\mathbf{r})} - \mu = 0 \tag{5.95}$$

where $n_0(\mathbf{r})$ is the equilibrium density.

We note also that the foregoing results could equally have been obtained in the grand canonical ensemble. Using the grand Hamiltonian $K = H_0 + \mathcal{V} - \mu N$ instead of H, we may show that

$$\Omega[\rho] = \Omega[n(\mathbf{r})] \qquad \left.\frac{\delta \Omega}{\delta n(\mathbf{r})}\right|_{n_0(\mathbf{r})} = 0 \tag{5.96}$$

where

$$\Omega[\rho] = \mathrm{Tr}\, \rho\, (H - \mu N + k_B T \ln \rho) \tag{5.97}$$

and where the classical trace now also includes a sum over N. The explicit demonstration is left as an exercise. The functionals $\mathcal{A}[n]$ and $\Omega[n]$ are simply related:

$$\Omega[n] = \mathcal{A}[n] - \mu \int d^3r\ n(\mathbf{r}) \ . \tag{5.98}$$

In the case of an ideal gas subject to the external potential $v(\mathbf{r})$ it is possible to write down an explicit form for the equilibrium free energy functional:

$$\Omega_{ideal}[n] = k_B T \int d^3r \left[n(\mathbf{r}) \ln \left\{n(\mathbf{r})\lambda^3\right\} - n(\mathbf{r})\right] + \int d^3r\, n(\mathbf{r}) v(\mathbf{r}) - \mu \int d^3r\, n(\mathbf{r}) \tag{5.99}$$

where $\lambda = [h^2/(2m\pi k_B T)]^{1/2}$ is the thermal wavelength. Taking the functional derivative with respect to $n(\mathbf{r})$ and using (5.96), we have

$$\left.\frac{\delta\Omega_{ideal}}{\delta n(\mathbf{r})}\right|_{n_0(\mathbf{r})} = k_B T \ln n\lambda^3 - \mu + v(\mathbf{r}) = 0$$

or

$$n_0(\mathbf{r}) = \frac{1}{\lambda^3}\exp\{-\beta(v(\mathbf{r}) - \mu)\} \tag{5.100}$$

which can be recognized as the generalization of (2.20) to an inhomogeneous system.

In the case of interacting particles we may express the grand potential as a sum of two terms, one of which contains the contributions due to the pair-potential:

$$\begin{aligned}\Omega[n(\mathbf{r})] &= \Omega_{ideal}[n(\mathbf{r})] - \Phi[n(\mathbf{r})] \\ &= k_B T \int d^3r\, \left[n(\mathbf{r}) \ln\left\{n(\mathbf{r})\lambda^3\right\} - n(\mathbf{r})\right] \\ &\quad + \int d^3r n(\mathbf{r}) u[\mathbf{r}; n(\mathbf{r})] - \Phi[n(\mathbf{r})]\end{aligned} \tag{5.101}$$

where we have defined the $u[\mathbf{r}; n(\mathbf{r})] = v(\mathbf{r}) - \mu$ and explicitly recognized that it is a functional of $n(\mathbf{r})$. From (5.96) we therefore have

$$\begin{aligned}0 = \left.\frac{\delta\Omega}{\delta n(\mathbf{r})}\right|_{n_0} &= k_B T \ln n_0(\mathbf{r})\lambda^3 + u(\mathbf{r}) - \left.\frac{\delta\Phi}{\delta n(\mathbf{r})}\right|_{n_0} \\ &= k_B T \ln n_0(\mathbf{r})\lambda^3 + u(\mathbf{r}) - k_B T C_1[\mathbf{r}; n_0(\mathbf{r})]\end{aligned} \tag{5.102}$$

where we have defined $C_1[\mathbf{r}; n_0(\mathbf{r})] \equiv \beta\delta\Phi/\delta n_0$. Thus the density is given implicitly by

$$n_0(\mathbf{r})\lambda^3 = \exp\left\{-\beta u(\mathbf{r}) + C_1[\mathbf{r}; n_0(\mathbf{r})]\right\} . \tag{5.103}$$

To this point these expressions are entirely formal, since we have not yet shown how the function C_1 can be determined. Before discussing this aspect, we show that the functional derivative of C_1 is the direct correlation function that appeared in the Ornstein–Zernike equation (5.41).

We first note that from the second equation in (5.102)

$$C_2[\mathbf{r}, \mathbf{r}'; n_0] \equiv \frac{\delta C_1[\mathbf{r}; n_0]}{\delta n_0(\mathbf{r}')} = \frac{\delta(\mathbf{r} - \mathbf{r}')}{n_0(\mathbf{r})} + \beta\frac{\delta u[\mathbf{r}; n_0(\mathbf{r})]}{\delta n_0(\mathbf{r}')} . \tag{5.104}$$

Consider now the equilibrium grand potential. We define the density operator $\hat{n}(\mathbf{r}) = \sum_i \delta(\mathbf{r} - \mathbf{r}_i)$. Then

$$\begin{aligned} \Omega[n_0] &= -k_B T \ln \sum_N \frac{1}{N!} \int d^{3N} r \, \lambda^{-3N} \\ & \quad \times \exp\left\{ -\beta \sum_{i<j} U(\mathbf{r}_i - \mathbf{r}_j) - \beta \int d^3 r \, u(\mathbf{r}) \hat{n}(\mathbf{r}) \right\} . \end{aligned} \tag{5.105}$$

Clearly,

$$\frac{\delta \Omega}{\delta u(\mathbf{r})} = n_0(\mathbf{r}) \tag{5.106}$$

and

$$-k_B T \frac{\delta^2 \Omega}{\delta u(\mathbf{r}) \delta u(\mathbf{r}')} = \langle \hat{n}(\mathbf{r}) \hat{n}(\mathbf{r}') \rangle - n_0(\mathbf{r}) n_0(\mathbf{r}') . \tag{5.107}$$

Noting that

$$\begin{aligned} \langle \hat{n}(\mathbf{r}) \hat{n}(\mathbf{r}') \rangle &= \left\langle \sum_{i \neq j} \delta(\mathbf{r} - \mathbf{r}_i) \, \delta(\mathbf{r}' - \mathbf{r}_j) + \sum_i \delta(\mathbf{r} - \mathbf{r}_i) \, \delta(\mathbf{r}' - \mathbf{r}_i) \right\rangle \\ &= n_2(\mathbf{r}, \mathbf{r}') + n_0(\mathbf{r}) \delta(\mathbf{r} - \mathbf{r}') \end{aligned} \tag{5.108}$$

where $n_2(\mathbf{r}, \mathbf{r}')$ is the two-particle reduced distribution function defined in (5.20). We obtain,

$$n_2(\mathbf{r}, \mathbf{r}') - n_0(\mathbf{r}) n_0(\mathbf{r}') + n_0(\mathbf{r}) \delta(\mathbf{r} - \mathbf{r}') = -k_B T \frac{\delta n_0(\mathbf{r})}{\delta u(\mathbf{r}')} = -k_B T \left[\frac{\delta u(\mathbf{r}')}{\delta n_0(\mathbf{r})} \right]^{-1} \tag{5.109}$$

where the inverse functional derivative is formally defined in (5.81). Using this equation and (5.104) we have

$$\begin{aligned} & \int d^3 r'' \left[n_2(\mathbf{r}, \mathbf{r}'') - n_0(\mathbf{r}) n_0(\mathbf{r}'') + n_0(\mathbf{r}) \delta(\mathbf{r} - \mathbf{r}'') \right] \\ & \qquad \times \left[-\frac{\delta(\mathbf{r}'' - \mathbf{r}')}{n_0(\mathbf{r}'')} + C_2(\mathbf{r}'', \mathbf{r}') \right] = -\delta(\mathbf{r} - \mathbf{r}') . \end{aligned} \tag{5.110}$$

We now define $n_2(\mathbf{r}, \mathbf{r}') \equiv n_0(\mathbf{r}) n_0(\mathbf{r}')[h(\mathbf{r}, \mathbf{r}') + 1]$, where h is the pair correlation function (5.34) in the case of an inhomogeneous fluid. Rearranging (5.110) we obtain the inhomogeneous Ornstein–Zernike equation

$$h(\mathbf{r}, \mathbf{r}') = C_2(\mathbf{r}, \mathbf{r}') + \int d^3 r'' n_0(\mathbf{r}'') h(\mathbf{r}, \mathbf{r}'') C_2(\mathbf{r}'', \mathbf{r}') \tag{5.111}$$

which shows that C_2 is indeed the direct correlation function.

Finally, we note that we may obtain an alternate expression for the pair correlation function for the case of particles interacting through a two-body potential $U(\mathbf{r}_{ij})$ as in (5.83). We can write this potential energy in the form

$$\sum_{i<j} U(\mathbf{r}_i - \mathbf{r}_j) = \frac{1}{2} \int d^3r\, d^3r' \,(\hat{n}(\mathbf{r})\hat{n}(\mathbf{r}') - \hat{n}(\mathbf{r})\delta(\mathbf{r}-\mathbf{r}'))\, U(\mathbf{r},\mathbf{r}') \,.$$

Regarding the grand potential or Helmholtz free energy as a functional of U we have

$$\frac{1}{2} n_2(\mathbf{r},\mathbf{r}') = \frac{\delta\Omega}{\delta U(\mathbf{r},\mathbf{r}')} \,. \tag{5.112}$$

5.5.3 Applications

We now proceed to sketch how one can apply the formalism developed above to bulk and inhomogeneous liquids. Consider first equations (5.101) and (5.102).

$$\frac{\delta}{\delta n(\mathbf{r})} \{\Omega[n] - \Omega_{ideal}[n]\} = -k_B T C_1[n(\mathbf{r})] \,.$$

Suppose now that we know the grand potential at some reference density $n_i(\mathbf{r})$ and imagine slowly increasing the density from $n_i(\mathbf{r})$ to its final value $n(\mathbf{r})$. We can do this by increasing a parameter λ from 0 to 1 if

$$n_\lambda(\mathbf{r}) = n_i(\mathbf{r}) + \lambda\,[n(\mathbf{r}) - n_i(\mathbf{r})] = n_i(\mathbf{r}) + \lambda\Delta n(\mathbf{r}) \,.$$

When $\lambda \to \lambda + d\lambda$, $\delta n_\lambda(\mathbf{r}) = d\lambda\Delta n(\mathbf{r})$ and the change in grand potential is

$$\delta\,\{\Omega[n] - \Omega_{ideal}[n]\} = -k_B T d\lambda \int d^3r\, \Delta n(\mathbf{r}) C_1[n_\lambda(\mathbf{r})] \,.$$

Therefore, with $\Omega_{exc} \equiv \Omega - \Omega_{ideal}$, the 'excess' free energy,

$$\Omega_{exc}[n(\mathbf{r})] = \Omega_{exc}[n_i(\mathbf{r})] - k_B T \int_0^1 d\lambda \int d^3r\, \Delta n(\mathbf{r}) C_1[n_\lambda(\mathbf{r})] \,. \tag{5.113}$$

It is more common to approximate the direct correlation function C_2 than the function C_1 and we therefore carry out one further step. Using the definition (5.104) and carrying out exactly the same procedure as above, we may express C_1 in terms of a functional integral of C_2:

$$C_1[\mathbf{r}; n_\lambda(\mathbf{r})] = C_1[\mathbf{r}; n_i(\mathbf{r})] + \int_0^\lambda d\lambda' \int d^3r'\, \Delta n(\mathbf{r}') C_2[\mathbf{r},\mathbf{r}'; n_{\lambda'}] \,.$$

Substituting this expression into (5.113) we find

$$\begin{aligned}\Omega_{exc}[n(\mathbf{r})] &= \Omega_{exc}[n_i(\mathbf{r})] - k_BT\int d^3r\,\Delta n(\mathbf{r})C_1[\mathbf{r};n_i] \\ &\quad -k_BT\int_0^1 d\lambda\int_0^\lambda d\lambda'\int d^3r\int d^3r'\Delta n(\mathbf{r})\Delta n(\mathbf{r}')C_2[\mathbf{r},\mathbf{r}';n_{\lambda'}] \\ &= \Omega_{exc}[n_i(\mathbf{r})] - k_BT\int d^3r\,\Delta n(\mathbf{r})C_1[\mathbf{r};n_i] \\ &\quad -k_BT\int_0^1 d\lambda(1-\lambda)\int d^3r\int d^3r'\Delta n(\mathbf{r})\Delta n(\mathbf{r}')C_2[\mathbf{r},\mathbf{r}';n_\lambda]\end{aligned} \tag{5.114}$$

where, in the last step, we have reversed the order of integration over λ and λ'. Before discussing this rather complicated formula further, we simplify it for a special case, a homogeneous bulk fluid at density n and use as a reference fluid the ideal gas. Then $C_1[\mathbf{r};n_i]=0$, $\Delta n(\mathbf{r})=n$ and

$$\Omega(n) = \Omega_{ideal}(n) + k_BTV\int_0^1 d\lambda\,(\lambda-1)\int d^3r\,C_2[r;\lambda n]\,. \tag{5.115}$$

From this formula we see some of the difficulties of this approach. In order to carry out the integral on the right-hand side, we need to know the direct correlation function $C_2[r;n']$ for *all* densities in the range $0<n'\le n$. This is clearly not possible. Nevertheless, there have been a number of quite impressively successful approximation schemes that make use of formulas (5.115) and even of (5.114) in the case of an inhomogeneous fluid such as one confined between closely spaced walls. The review by Evans [88] contains a thorough discussion of these specialized topics. Equation (5.114) is also the starting point for theories of freezing. In that case, the reference fluid is generally the bulk homogeneous fluid (see [128] for a review).

We can also make contact with the perturbation theories briefly described in Section 5.3. To do this, we use (5.112) and imagine turning on a pair-potential in the same way as we increased the density above. We write

$$U(\mathbf{r},\mathbf{r}') = U_0(\mathbf{r},\mathbf{r}') + \lambda U_1(\mathbf{r},\mathbf{r}')\,. \tag{5.116}$$

Then, in strict analogy with (5.113) we have

$$\begin{aligned}\Omega[U] &= \Omega[U_0] + \int_0^1 d\lambda\int d^3r\int d^3r'\,U_1(\mathbf{r},\mathbf{r}')\frac{\delta\Omega}{\delta U(\mathbf{r},\mathbf{r}')} \\ &= \Omega[U_0] + \frac{1}{2}\int_0^1 d\lambda\int d^3r\int d^3r'\,U_1(\mathbf{r},\mathbf{r}')n_2[\mathbf{r},\mathbf{r}';\lambda]\,.\end{aligned} \tag{5.117}$$

In the case of a homogeneous bulk fluid, $n_2[\mathbf{r}, \mathbf{r}'; \lambda] = (N/V)^2 g[r; \lambda]$ and since the pair-potential depends only on the distance between the particles we obtain the simpler version

$$\Omega = \Omega_0 + \frac{N^2}{2V} \int_0^1 d\lambda \int d^3r\, U_1(r) g[r; \lambda] \; . \tag{5.118}$$

If we decide to approximate the pair distribution $g[r; \lambda] \approx g[r; 0]$ we recover the grand canonical version of equation (5.53) of Section 5.3.

Finally, we have already used a version of density functional theory to construct the van der Waals theory of the liquid-vapor interface. The particular form of the free energy functional chosen in that section is a particular case of a functional Taylor expansion of a free energy density, valid if the density is not too rapidly varying. For a systematic development of this type of approximation, we refer again to [87], [88].

5.6 Problems

5.1. *Tonks Gas.*

Consider a one-dimensional gas of particles of length a confined to a strip of length L. The particles interact through the potential

$$\begin{aligned} U(x_i - x_j) &= \infty \text{ for } |x_i - x_j| < a \\ &= 0 \text{ for } |x_i - x_j| > a \end{aligned}$$

(a) Calculate the partition function and equation of state exactly.

(b) Evaluate the virial coefficients B_2 and B_3 and show that your results are consistent with part (a).

5.2. *Decomposition of Reducible Graphs.*

Consider an arbitrary graph g with $n = n_1 + n_2 - 1$ vertices which can be cut at a point into two disjoint pieces g_1 and g_2 with n_1 and n_2 vertices, respectively. Find an expression for the cluster integral $b(g)$ in terms of the cluster integrals $b(g_1)$ and $b(g_2)$.

5.3. *Virial Expansion for the Pair Distribution Function.*

Generalize the method of Section 5.1 to the pair distribution function. In particular, find an expression for the function $y_1(\mathbf{r}_{12})$ of equation (5.43).

5.4. *Inversion Temperature of Argon.*

In the Joule–Thompson process a gas is forced from a high-pressure chamber through a porous plug into a lower-pressure chamber. The process takes place at constant enthalpy and the change in temperature of the gas, for a small pressure difference between the two chambers, is given by $dT = \mu_J dP$, where μ_J is the Joule–Thompson coefficient:

$$\mu_J = \left(\frac{\partial T}{\partial P}\right)_{H,N} = \frac{1}{C_P}\left[T\left(\frac{\partial V}{\partial T}\right)_{P,N} - V\right] .$$

The locus $\mu_J = 0$ is called the Joule–Thompson inversion curve. Use the first three terms of the virial equation of state

$$P = \frac{Nk_BT}{V}\left(1 + B_2(T)\frac{N}{V} + B_3(T)\left(\frac{N}{V}\right)^2 + \ldots\right)$$

to obtain an equation for the inversion curve in the $P - T$ plane.

Chapter 6

Critical Phenomena I

In this chapter and in Chapter 7, we discuss continuous phase transitions in some detail. We have already developed, in Chapter 3, several approximate methods for dealing with strongly interacting, or highly correlated systems, in particular the mean field approach, and its extensions, as well as the Landau theory of phase transitions. In that chapter we also demonstrated the inherent limitation of mean field theories, namely the fact that the correlation function becomes very long ranged near a critical point. For this reason, theories based on an exact treatment of small clusters cannot be expected to produce the correct critical behavior of a system.

We will commence by following the historical development of the field. In the present chapter, we begin in Section 6.1 with a derivation of the Onsager solution of the two-dimensional Ising model. We then discuss, in Section 6.2, the series expansion methods developed primarily during the 1950s and 1960s to determine the critical properties of model systems. Section 6.3 contains a discussion of the scaling theory of Widom [324], Domb and Hunter [79], and Kadanoff *et al.* [146]. The material of Section 6.4 extends the scaling theory to systems that have one or more finite physical dimensions. In Section 6.5 we concern ourselves with the universality hypothesis and examine the theoretical and experimental evidence in its favor. Finally, in Section 6.6, we present a short and qualitative discussion of the Kosterlitz–Thouless mechanism for phase transitions in two-dimensional systems with continuous symmetry. At that point we shall be in a position to undertake a study of the renormalization group approach to critical phenomena developed by Wilson and others in the 1970s. This theory, which has provided a firm theoretical basis for both scaling and universality, forms the subject matter of Chapter 7.

6.1 Ising Model in Two Dimensions

In a landmark paper, Onsager [222] exactly calculated the free energy of the two-dimensional ferromagnetic Ising model in zero magnetic field on the rectangular lattice. This calculation provided the first exact solution of a model that displays a phase transition. Onsager's original derivation is mathematically complex. Since his original paper, a number of more transparent solutions of the problem have appeared. Below, we present a brief account of one of these, namely that of Schultz *et al.* [270]. Our motivation for including this calculation is twofold. Most of this book is concerned with approximation techniques, but we feel that it is worthwhile to exhibit an exact calculation in statistical physics as a counterpoint to the examples of mean field theory and approximate renormalization group calculations. Second, we frequently quote the exact results of Onsager for the specific heat and order parameter and feel that some readers may not feel comfortable with these results without the evidence of a derivation. Those readers not interested in the technical details may skip ahead to Section 6.1.4.

6.1.1 Transfer matrix

We have already solved the one-dimensional Ising model in Section 3.6 by use of the transfer matrix approach and will also apply this method in two dimensions. We first formulate the one-dimensional problem in a slightly different way. Consider, again, the Hamiltonian

$$H = -J\sum_{i=1}^{N}\sigma_i\sigma_{i+1} - h\sum_i \sigma_i \ . \tag{6.1}$$

The partition function is

$$Z = \sum_{\{\sigma\}}\left(e^{\beta h\sigma_1}e^{K\sigma_1\sigma_2}\right)\left(e^{\beta h\sigma_2}e^{K\sigma_2\sigma_3}\right)\ldots\left(e^{\beta h\sigma_N}e^{K\sigma_N\sigma_1}\right) \tag{6.2}$$

where we have grouped the factors somewhat differently from (3.36) and where $K = \beta J$.

We now introduce two orthonormal basis states $|+1\rangle$ and $|-1\rangle$ and Pauli operators, which in this basis have the representation

$$\sigma_Z = \begin{pmatrix} 1 & 0 \\ 0 & -1 \end{pmatrix} \quad \sigma^+ = \begin{pmatrix} 0 & 1 \\ 0 & 0 \end{pmatrix} \quad \sigma^- = \begin{pmatrix} 0 & 0 \\ 1 & 0 \end{pmatrix} \tag{6.3}$$

with $\sigma_X = \sigma^+ + \sigma^-$ and $\sigma_Y = -i(\sigma^+ - \sigma^-)$. It is now easy to see that the Boltzmann weight $\exp\{\beta h \sigma_i\}$ can be expressed as a diagonal matrix, $\mathbf{V}_1$, in this basis:

$$\langle +1|\mathbf{V}_1| + 1\rangle = e^{\beta h}, \quad \langle -1|\mathbf{V}_1| - 1\rangle = e^{-\beta h}$$

or

$$\mathbf{V}_1 = \exp\{\beta h \sigma_Z\} . \tag{6.4}$$

Similarly, we define the operator $\mathbf{V}_2$ corresponding to the nearest-neighbor coupling by its matrix elements in this basis:

$$\langle +1|\mathbf{V}_2| + 1\rangle = \langle -1|\mathbf{V}_2| - 1\rangle = e^K$$

$$\langle +1|\mathbf{V}_2| - 1\rangle = \langle -1|\mathbf{V}_2| + 1\rangle = e^{-K} .$$

Therefore,

$$\mathbf{V}_2 = e^K \mathbf{1} + e^{-K}\sigma_X = A(K)\exp\{K^* \sigma_X\} \tag{6.5}$$

where in the second step we have used the fact that $(\sigma_X)^{2n} = \mathbf{1}$. The constants $A(K)$ and K^* are determined from the equations

$$\begin{aligned} A\cosh K^* &= e^K \\ A\sinh K^* &= e^{-K} \end{aligned} \tag{6.6}$$

or $\tanh K^* = \exp\{-2K\}$, $A = \sqrt{2\sinh 2K}$. Using these results, we write the partition function as follows:

$$\begin{aligned} Z &= \sum_{\{\mu=+1,-1\}} \langle\mu_1|\mathbf{V}_1|\mu_2\rangle\langle\mu_2|\mathbf{V}_2|\mu_3\rangle\langle\mu_3|\mathbf{V}_1|\mu_4\rangle \cdots \langle\mu_{2N}|\mathbf{V}_2|\mu_1\rangle \\ &= \mathrm{Tr}(\mathbf{V}_1\mathbf{V}_2)^N = \mathrm{Tr}(\mathbf{V}_2^{1/2}\mathbf{V}_1\mathbf{V}_2^{1/2})^N = \lambda_1^N + \lambda_2^N \end{aligned} \tag{6.7}$$

where λ_1 and λ_2 are the two eigenvalues of the Hermitian operator

$$\mathbf{V} = (\mathbf{V}_2^{1/2}\mathbf{V}_1\mathbf{V}_2^{1/2}) = \sqrt{2\sinh 2K}\, e^{K^*\sigma_X/2} e^{\beta h \sigma_Z} e^{K^*\sigma_X/2} . \tag{6.8}$$

In arriving at this symmetric form of the transfer matrix $\mathbf{V}$ we have used the invariance of the trace of a product of matrices under a cyclic permutation

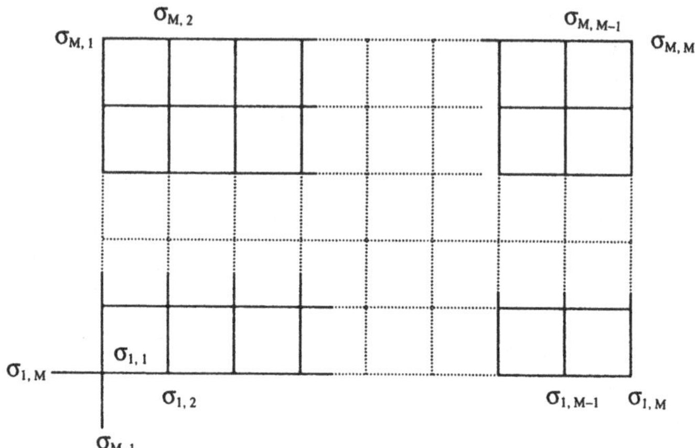

Figure 6.1: $M \times M$ square lattice with periodic boundary conditions.

of the factors. Clearly, in the case $h = 0$, the two eigenvalues are given by $\lambda_1 = A \exp\{K^*\}$, $\lambda_2 = A \exp\{-K^*\}$ and we recover our previous result (3.37).

We note, in passing, that in this procedure a one-dimensional problem in classical statistics has been transformed into a zero-dimensional (only one "site") quantum-mechanical ground-state problem (largest eigenvalue). This result is quite general. There exists a correspondence between the ground state of quantum Hamiltonians in $d-1$ dimensions and classical partition functions in d dimensions which can sometimes be exploited, for example in numerical simulations of quantum-statistical models [296], [297].

We now generalize this procedure to the two-dimensional Ising model and consider an $M \times M$ square lattice with periodic boundary conditions (see Figure 6.1) and the Hamiltonian

$$H = -J \sum_{r,c} \sigma_{r,c}\sigma_{r+1,c} - J \sum_{r,c} \sigma_{r,c}\sigma_{r,c+1} \tag{6.9}$$

where the label r refers to rows, c to columns, and $\sigma_{r+M,c} = \sigma_{r,c+M} = \sigma_{r,c}$.

The first term in (6.9) contains only interactions in column c and is, in this sense, analogous to the magnetic field term in (6.1). The second term in (6.9) is the coupling between neighboring columns and will lead to a non-diagonal factor in the complete transfer matrix.

In analogy with the one-dimensional case, we now introduce the 2^M basis states

$$|\mu\rangle \equiv |\mu_1, \mu_2, \cdots, \mu_M\rangle \equiv |\mu_1\rangle|\mu_2\rangle \cdots |\mu_M\rangle \tag{6.10}$$

with $\mu_j = \pm 1$ and M sets of Pauli operators $(\sigma_{jX}, \sigma_{jY}, \sigma_{jZ})$ which act on the jth state in the product (6.10), that is,

$$\begin{aligned}
\sigma_{jZ} \quad |\mu_1, \mu_2, \ldots \mu_j, \ldots, \mu_M\rangle &= \mu_j \qquad |\mu_1, \mu_2, \ldots \mu_j, \ldots, \mu_M\rangle \\
\sigma_j^+ \quad |\mu_1, \mu_2, \ldots \mu_j, \ldots, \mu_M\rangle &= \delta_{\mu_j,-1} \quad |\mu_1, \mu_2, \ldots \mu_j + 2, \ldots, \mu_M\rangle \\
\sigma_j^- \quad |\mu_1, \mu_2, \ldots \mu_j, \ldots, \mu_M\rangle &= \delta_{\mu_j,1} \quad |\mu_1, \mu_2, \ldots \mu_j - 2, \ldots, \mu_M\rangle \ .
\end{aligned} \tag{6.11}$$

Moreover, we impose the commutation relations $[\sigma_{j\alpha}, \sigma_{m\beta}] = 0$ for $j \neq m$. For $j = m$ the usual Pauli matrix commutation relations apply.

If we think of the index μ_i as the orientation of the ith spin in a given column, we see immediately that the Boltzmann factors $\exp\{K \sum_r \sigma_{r,c}\sigma_{r+1,c}\}$ are given by the matrix elements of the operator $\mathbf{V}_1 = \exp\{K \sum_j \sigma_{jZ}\sigma_{j+1,Z}\}$. Similarly, the matrix element

$$\begin{aligned}
\langle\{\mu\}|\mathbf{V}_2|\{\mu'\}\rangle &= \langle \mu_M, \mu_{M-1}, \ldots, \mu_1| \prod_{j=1}^{M} \left(e^K \mathbf{1} + e^{-K}\sigma_{jX}\right) |\mu_1', \mu_2', \ldots, \mu_M'\rangle \\
&= \exp\{(M-2n)K\}
\end{aligned} \tag{6.12}$$

where n of the indices $\{\mu'\}$ differ from the corresponding entries in $\{\mu\}$. Thus the partition function of the two-dimensional Ising model, in zero magnetic field, is, as can easily be verified, given by

$$\begin{aligned}
Z &= \sum_{\{\mu_1\},\{\mu_2\},\ldots,\{\mu_M\}} \langle\mu_1|\mathbf{V}_1|\mu_2\rangle\langle\mu_2|\mathbf{V}_2|\mu_3\rangle\langle\mu_3|\mathbf{V}_1|\mu_4\rangle \cdots \langle\mu_M|\mathbf{V}_2|\mu_1\rangle \\
&= \mathrm{Tr}(\mathbf{V}_1\mathbf{V}_2)^M = \mathrm{Tr}(\mathbf{V}_2^{1/2}\mathbf{V}_1\mathbf{V}_2^{1/2})^M \ .
\end{aligned} \tag{6.13}$$

In (6.13) the sum over each $\{\mu_j\}$ is, of course, over the entire set of 2^M basis states. Using (6.5) and (6.6), we may write

$$\mathbf{V}_2 = (2\sinh 2K)^{M/2} \exp\left\{K^* \sum_{j=1}^{M} \sigma_{jX}\right\} \tag{6.14}$$

and we have reduced the calculation of the partition function to the determination of the largest eigenvalue of the Hermitian operator

$$\begin{aligned}
\mathbf{V} &= \mathbf{V}_2^{1/2}\mathbf{V}_1\mathbf{V}_2^{1/2} \\
&= (2\sinh 2K)^{M/2} \exp\left\{\frac{K^*}{2} \sum_{j=1}^{M} \sigma_{jX}\right\}
\end{aligned}$$

$$\times \exp\left\{K\sum_{j=1}^{M}\sigma_{jZ}\sigma_{j+1,Z}\right\}\exp\left\{\frac{K^*}{2}\sum_{j=1}^{M}\sigma_{jX}\right\} \qquad (6.15)$$

which is still a nontrivial task since the factors in (6.15) do not commute with each other and, since the matrix $\mathbf{V}$ becomes infinite dimensional in the thermodynamic limit.

6.1.2 Transformation to an interacting fermion problem

It is convenient for what follows to perform a rotation of the spin operators and to let $\sigma_{jZ} \to -\sigma_{jX}$, $\sigma_{jX} \to \sigma_{jZ}$ for all j. These rotations, of course, leave the eigenvalues invariant. Using $\sigma_{jZ} = 2\sigma_j^+\sigma_j^- - \mathbf{1}$ and $\sigma_{jX} = \sigma_j^+ + \sigma_j^-$, we arrive at the forms

$$\begin{aligned}\mathbf{V}_1 &= \exp\left\{K\sum_{j=1}^{M}(\sigma_j^+ + \sigma_j^-)(\sigma_{j+1}^+ + \sigma_{j+1}^-)\right\} \\ \mathbf{V}_2 &= (2\sinh 2K)^{M/2}\exp\left\{2K^*\sum_{j=1}^{M}(\sigma_j^+\sigma_j^- - \frac{1}{2}\mathbf{1})\right\} . \end{aligned} \qquad (6.16)$$

Schultz *et al.* [270] showed that these operators can be simplified by a series of transformations. The first of these is the Jordan–Wigner transformation which converts the Pauli operators to fermion operators (see the Appendix for a discussion of second quantization). This step is useful because of subsequent canonical transformations that are not possible for angular momentum operators. One writes

$$\begin{aligned}\sigma_j^+ &= \exp\left\{\pi i\sum_{m=1}^{j-1}c_m^\dagger c_m\right\}c_j^\dagger \\ \sigma_j^- &= c_j\exp\left\{-\pi i\sum_{m=1}^{j-1}c_m^\dagger c_m\right\} = \exp\left\{\pi i\sum_{m=1}^{j-1}c_m^\dagger c_m\right\}c_j \end{aligned} \qquad (6.17)$$

where the operators c, $c^\dagger$ obey the commutation relations

$$[c_j, c_m^\dagger]_+ \equiv c_jc_m^\dagger + c_m^\dagger c_j = \delta_{jm}$$

$$[c_j, c_m]_+ = [c_j^\dagger, c_m^\dagger]_+ = 0 .$$

The operator $c_m^\dagger c_m$ is the fermion number operator for site m with integer eigenvalues 0 and 1. Since $e^{i\pi n} = e^{-i\pi n}$ the last step of (6.17) follows. To see that the spin commutation relations are preserved under this transformation consider, for $n > j$,

$$[\sigma_j^-, \sigma_n^+] = \exp\left\{\pi i \sum_{m=j+1}^{n-1} c_m^\dagger c_m\right\}\left(c_j e^{\pi i c_j^\dagger c_j} c_n^\dagger - c_n^\dagger e^{\pi i c_j^\dagger c_j} c_j\right) .$$

Noting that $\exp\left\{\pi i c_j^\dagger c_j\right\} c_j = c_j$ and $c_j \exp\left\{\pi i c_j^\dagger c_j\right\} = -c_j$, we have $[\sigma_j^-, \sigma_n^+] = 0$ for $n \neq j$. We also immediately see that the on-site anticommutator

$$[\sigma_j^-, \sigma_j^+]_+ = [c_j, c_j^\dagger]_+ = 1 .$$

The verification of further commutation relations and the derivation of the inverse of the transformation (6.17) is left as an exercise. Using (6.17), we can express the operators $\mathbf{V}_1$, and $\mathbf{V}_2$ in terms of the fermion operators. The operator $\mathbf{V}_2$ presents no difficulties and is immediately given by

$$\mathbf{V}_2 = (2\sinh 2K)^{M/2} \exp\left\{2K^* \sum_{j=1}^{M}\left(c_j^\dagger c_j - \frac{1}{2}\right)\right\} . \tag{6.18}$$

In the case of $\mathbf{V}_1$, there is a slight difficulty due to the periodic boundary conditions. We first note that for $j \neq M$ the term

$$(\sigma_j^+ + \sigma_j^-)(\sigma_{j+1}^+ + \sigma_{j+1}^-) = c_j^\dagger c_{j+1}^\dagger + c_j^\dagger c_{j+1} + c_{j+1}^\dagger c_j + c_{j+1} c_j .$$

For the specific case $j = M$,

$$\begin{aligned}
(\sigma_M^+ + \sigma_M^-)(\sigma_1^+ + \sigma_1^-) &= \exp\left\{\pi i \sum_{j=1}^{M-1} c_j^\dagger c_j\right\} c_M^\dagger (c_1^\dagger + c_1) \\
&\quad + \exp\left\{\pi i \sum_{j=1}^{M-1} c_j^\dagger c_j\right\} c_M (c_1^\dagger + c_1) \\
&= \exp\left\{\pi i \sum_{j=1}^{M} c_j^\dagger c_j\right\} \left[e^{\pi i c_M^\dagger c_M}(c_M^\dagger + c_M)(c_1^\dagger + c_1)\right] \\
&= (-1)^n (c_M - c_M^\dagger)(c_1^\dagger + c_1)
\end{aligned}$$

where $n = \sum_j c_j^\dagger c_j$ is the total fermion number operator. The operator n commutes with $\mathbf{V}_2$ but not with $\mathbf{V}_1$. On the other hand, $(-1)^n$ commutes with both $\mathbf{V}_1$, and $\mathbf{V}_2$ as the various terms in $\mathbf{V}_1$ change the total fermion number by 0 or ± 2. Thus if we consider separately the subspaces of even and odd total number of fermions, we may write $\mathbf{V}_1$ in a simple universal way, that is,

$$\mathbf{V}_1 = \exp\left\{ K \sum_{j=1}^{M} (c_j^\dagger - c_j)(c_{j+1}^\dagger + c_{j+1}) \right\} \tag{6.19}$$

where

$$\begin{array}{ll} c_{M+1} \equiv -c_1,\ c_{M+1}^\dagger \equiv -c_1^\dagger & \text{for } n \text{ even} \\ c_{M+1} \equiv c_1,\ c_{M+1}^\dagger \equiv c_1^\dagger & \text{for } n \text{ odd}\ . \end{array} \tag{6.20}$$

With this choice of boundary condition on the fermion creation and annihilation operators, we have recovered translational invariance and now carry out the *canonical* transformation

$$\begin{aligned} a_q &= \frac{1}{\sqrt{M}} \sum_{j=1}^{M} c_j e^{-iqj} \\ a_q^\dagger &= \frac{1}{\sqrt{M}} \sum_{j=1}^{M} c_j^\dagger e^{iqj} \end{aligned} \tag{6.21}$$

with inverse

$$\begin{aligned} c_j &= \frac{1}{\sqrt{M}} \sum_q a_q e^{iqj} \\ c_j^\dagger &= \frac{1}{\sqrt{M}} \sum_q a_q^\dagger e^{-iqj}\ . \end{aligned} \tag{6.22}$$

To reproduce the boundary conditions (6.20), we take $q = j\pi/M$ with

$$\begin{array}{ll} j = \pm 1, \pm 3, \ldots, \pm(M-1) & \text{for } n \text{ even} \\ j = 0, \pm 2, \pm 4, \ldots, \pm(M-2), M & \text{for } n \text{ odd} \end{array}$$

and where we have also assumed, without loss of generality, that M is even. It is easy to see that the operators a_q, $a_q^\dagger$ obey fermion commutation relations, that is, $[a_q, a_{q'}^\dagger]_+ = \delta_{q,q'}$ and $[a_q, a_{q'}]_+ = [a_q^\dagger, a_{q'}^\dagger]_+ = 0$ for all q and q'. Substituting into (6.18) and (6.19), we find for n even,

$$\begin{aligned} \mathbf{V}_2 &= (2\sinh 2K)^{M/2} \exp\left\{ 2K^* \sum_{q>0} (a_q^\dagger a_q + a_{-q}^\dagger a_{-q} - 1) \right\} \\ &= (2\sinh 2K)^{M/2} \prod_{q>0} \mathbf{V}_{2q} \end{aligned} \tag{6.23}$$

and

$$\begin{aligned}\mathbf{V}_1 &= \exp\left\{2K\sum_{q>0}[\cos q(a_q^\dagger a_q + a_{-q}^\dagger a_{-q}) - i\sin q(a_q^\dagger a_{-q}^\dagger + a_q a_{-q})]\right\}\\ &= \prod_{q>0}\mathbf{V}_{1q} \qquad (6.24)\end{aligned}$$

where, in (6.23) and (6.24), we have combined the terms corresponding to q and $-q$, and recognized in writing the resulting operators as products, that bilinear operators with different wave vectors commute. This is a great simplification since the eigenvalues of the transfer matrix can now be written as a product of eigenvalues of, as we shall see, at most 4×4 matrices. For the case of odd n we also need the operators $\mathbf{V}_{1q}$ and $\mathbf{V}_{2q}$ for $q = \pi$ and $q = 0$. These are given by

$$\begin{array}{ll}\mathbf{V}_{10} = \exp\left\{2Ka_0^\dagger a_0\right\} & \mathbf{V}_{20} = \exp\left\{2K^*(a_0^\dagger a_0 - \frac{1}{2})\right\}\\ \mathbf{V}_{1\pi} = \exp\left\{-2Ka_\pi^\dagger a_\pi\right\} & \mathbf{V}_{2\pi} = \exp\left\{2K^*(a_\pi^\dagger a_\pi - \frac{1}{2})\right\}\end{array} \qquad (6.25)$$

which are already in diagonal form and, of course, commute with each other.

6.1.3 Calculation of eigenvalues

We proceed to calculate the eigenvalues of the operator

$$\mathbf{V}_q = \mathbf{V}_{2q}^{1/2}\mathbf{V}_{1q}\mathbf{V}_{2q}^{1/2}$$

for $q \neq 0$ and $q \neq \pi$. Since we are dealing with fermions, we have only four possible states: $|0\rangle$, $a_q^\dagger|0\rangle$, $a_{-q}^\dagger|0\rangle$, and $a_q^\dagger a_{-q}^\dagger|0\rangle$, where $|0\rangle$ is the zero particle state defined by $a_q|0\rangle = a_{-q}|0\rangle = 0$. These states are already eigenstates of $\mathbf{V}_2$, and since the operator $\mathbf{V}_1$ has nonzero off-diagonal matrix elements only between states that differ by two in fermion number, the problem reduces to finding the eigenvalues of $\mathbf{V}_q$ in the basis $|0\rangle$ and $|2\rangle = a_q^\dagger a_{-q}^\dagger|0\rangle$. We note that

$$\mathbf{V}_{1q}a_{\pm q}^\dagger|0\rangle = \exp\{2K\cos q\}\, a_{\pm q}^\dagger|0\rangle \qquad (6.26)$$

and

$$\begin{aligned}\mathbf{V}_{2q}^{1/2}|0\rangle &= \exp\{-K^*\}\,|0\rangle\\ \mathbf{V}_{2q}^{1/2}|2\rangle &= \exp\{K^*\}\,|2\rangle\ .\end{aligned} \qquad (6.27)$$

To obtain the matrix elements of $\mathbf{V}_{1q}$ in the basis $|0\rangle$, $|2\rangle$, we let

$$\mathbf{V}_{1q}|0\rangle = \alpha(K)|0\rangle + \beta(K)|2\rangle\ .$$

Differentiating this expression with respect to K, we obtain

$$\begin{aligned}\frac{d\alpha}{dK}|0\rangle + \frac{d\beta}{dK}|2\rangle &= 2\left[\cos q\left\{a_q^\dagger a_q + a_{-q}^\dagger a_{-q}\right\}\right.\\ &\quad\left. - i\sin q\left\{a_q^\dagger a_{-q}^\dagger + a_q a_{-q}\right\}\right]\{\alpha\,|0\rangle + \beta\,|2\rangle\}\\ &= 2i\beta\sin q\,|0\rangle + [4\beta\cos q - 2i\alpha\sin q]|2\rangle \end{aligned} \tag{6.28}$$

or

$$\begin{aligned}\frac{d\alpha}{dK} &= 2i\beta(K)\sin q\\ \frac{d\beta}{dK} &= 4\beta(K)\cos q - 2i\alpha(K)\sin q\,. \end{aligned} \tag{6.29}$$

We solve these equations subject to the boundary conditions $\alpha(0) = 1$, $\beta(0) = 0$. The result is

$$\begin{aligned}\langle 0|\mathbf{V}_{1q}|0\rangle &= \alpha(K) = e^{2K\cos q}(\cosh 2K - \sinh 2K\cos q)\\ \langle 2|\mathbf{V}_{1q}|0\rangle &= \beta(K) = -ie^{2K\cos q}\sinh 2K\sin q\,. \end{aligned} \tag{6.30}$$

By the same method we can find the matrix elements $\langle 2|\mathbf{V}_{1q}|2\rangle$ and $\langle 0|\mathbf{V}_{1q}|2\rangle = \langle 2|\mathbf{V}_{1q}|0\rangle^*$ and obtain the matrix

$$\mathbf{V}_{1q} = e^{2K\cos q}\begin{bmatrix}\cosh 2K - \sinh 2K\cos q & i\sinh 2K\sin q\\ -i\sinh 2K\sin q & \cosh 2K + \sinh 2K\cos q\end{bmatrix} \tag{6.31}$$

and

$$\mathbf{V}_q = \begin{bmatrix}\exp\{-K^*\} & 0\\ 0 & \exp\{K^*\}\end{bmatrix}[\mathbf{V}_{1q}]\begin{bmatrix}\exp\{-K^*\} & 0\\ 0 & \exp\{K^*\}\end{bmatrix}. \tag{6.32}$$

The eigenvalues of this matrix are easily determined. Since we wish, eventually, to take the logarithm of the largest eigenvalue of the complete transfer matrix in order to calculate the free energy, we write the eigenvalues in the form

$$\lambda_q^{\pm} = \exp\{2K\cos q \pm \epsilon(q)\} \tag{6.33}$$

and after a bit of algebra, we obtain the equation

$$\cosh\epsilon(q) = \cosh 2K\cosh 2K^* + \cos q\sinh 2K\sinh 2K^* \tag{6.34}$$

for $\epsilon(q)$. By convention we choose $\epsilon(q) \geq 0$. We see that the minimum of the right-hand side of (6.34) occurs as $q \to \pi$ and that, for all q,

$$\epsilon(q) > \epsilon_{min} = \lim_{q\to\pi}\epsilon(q) = 2|K - K^*| \tag{6.35}$$

and also note that

$$\lim_{q\to 0} \epsilon(q) = 2(K + K^*) . \tag{6.36}$$

We are now in a position to combine all this information. Consider first the subspace in which all states contain an even number of fermions. In this case the allowed wave vectors do not include $q = 0$ or $q = \pi$, and comparing (6.33) and (6.26), we see that the largest eigenvalue of $\mathbf{V}_q$ for each q is λ_q^+. Thus the largest eigenvalue in this subspace, Λ_e, is given by

$$\begin{aligned} \Lambda_e &= (2\sinh 2K)^{M/2} \prod_{q>0} \lambda_q^+ \\ &= (2\sinh 2K)^{M/2} \exp\left\{ \sum_{q>0} [2\cos q + \epsilon(q)] \right\} \\ &= (2\sinh 2K)^{M/2} \exp\left\{ \frac{1}{2} \sum_q \epsilon(q) \right\} \end{aligned} \tag{6.37}$$

where, in the last step, we have used $\sum_q \cos q = 0$ and have also extended the summation over the entire range $-\pi < q < \pi$.

The other subspace must be examined more carefully. For $q \neq 0$ and $q \neq \pi$ the maximum possible eigenvalue is λ_q^+. The corresponding eigenstates are all states with $(-1)^n = -1$. To make the overall state have $(-1)^n = -1$, we occupy the $q = 0$ state and leave the $q = \pi$ state empty and obtain a contribution of $(2\sinh 2K)^{M/2} \exp\{2K\}$ to the eigenvalue Λ_o. Therefore the largest eigenvalue in the odd subspace is

$$\Lambda_o = (2\sinh 2K)^{M/2} \exp\left\{ 2K + \frac{1}{2} \sum_{q\neq 0,\pi} \epsilon(q) \right\} . \tag{6.38}$$

Since the wave vectors in the two subspaces are not identical, a direct comparison between the two largest eigenvalues is somewhat complicated. However, we note that

$$\begin{aligned} \frac{1}{2}\lim_{q\to 0} \epsilon(q) + \frac{1}{2}\lim_{q\to\pi} \epsilon(q) &= |K - K^*| + (K + K^*) \\ &= 2K \qquad \text{for } K > K^* \\ &= 2K^* \qquad \text{for } K^* > K . \end{aligned}$$

Thus if $K > K^*$, it is quite plausible, and can be shown rigorously in the thermodynamic limit $M \to \infty$, that Λ_o and Λ_e are degenerate. A little reflection will convince the reader that unless such a degeneracy exists, the order parameter $m_0(T)$ will be strictly zero. Therefore, the critical temperature of the two-dimensional Ising model is given by the equation $K = K^*$, or using the identity [from (6.6)]

$$\sinh 2K \sinh 2K^* = 1 \tag{6.39}$$

by the more usual expression

$$\sinh \frac{2J}{k_B T_c} = 1 \tag{6.40}$$

or $k_B T_c/J = 2.269185\ldots$.

The degeneracy of the two largest eigenvalues of the transfer matrix contributes only an additive term of $\ln 2$ to the dimensionless free energy and is thus negligible. Therefore, at any temperature the free energy is given by

$$\begin{aligned}\frac{\beta G(0,T)}{M^2} &= \beta g(0,T) = -\frac{1}{2}\ln(2\sinh 2K) - \frac{1}{2M}\sum_q \epsilon(q) \\ &= -\frac{1}{2}\ln(2\sinh 2K) - \frac{1}{4\pi}\int_{-\pi}^{\pi} dq\, \epsilon(q)\end{aligned} \tag{6.41}$$

where we have converted the sum over wave vectors to an integral.

6.1.4 Thermodynamic functions

With a bit more algebra, we can simplify the expression (6.41) for the zero-field free energy. Using (6.39) and $\cosh 2K^* = \coth 2K$, which follows from (6.6), we have

$$\cosh\{\epsilon(q)\} = \cosh 2K \coth 2K + \cos q\ . \tag{6.42}$$

Consider, now, the function

$$f(x) = \frac{1}{2\pi}\int_0^{2\pi} d\phi \ln(2\cosh x + 2\cos\phi)\ . \tag{6.43}$$

Differentiating with respect to x and evaluating the resulting integral by contour integration, we find

$$\frac{df(x)}{dx} = \mathrm{sign}\,(x) \quad \text{or} \quad f(x) = |x|\ . \tag{6.44}$$

Taking $x = \epsilon(q)$ we obtain the integral representation:

$$\epsilon(q) = \frac{1}{\pi}\int_0^{\pi} d\phi \ln(2\cosh 2K \coth 2K + 2\cos q + 2\cos\phi)\ . \tag{6.45}$$

We define

$$I = \frac{1}{2\pi}\int_0^\pi dq\epsilon(q) \;=\; \frac{1}{2\pi^2}\int_0^\pi dq \int_0^\pi d\phi \ln[2\cosh 2K \coth 2K + 2\cos q + 2\cos\phi] \; . \quad (6.46)$$

Using the trigonometric identity

$$\cos q + \cos\phi = 2\cos\frac{q+\phi}{2}\cos\frac{q-\phi}{2}$$

and changing variables of integration to

$$\omega_1 = \frac{q-\phi}{2} \qquad \omega_2 = \frac{q+\phi}{2}$$

we have

$$I = \frac{1}{\pi^2}\int_0^\pi d\omega_2 \int_0^{\pi/2} d\omega_1 \ln[2\cosh 2K\coth 2K + 4\cos\omega_1\cos\omega_2] \; . \quad (6.47)$$

The integration over ω_2 is almost in the form (6.43) and we can put it into this form by writing

$$\begin{aligned} I &= \frac{1}{\pi^2}\int_0^\pi d\omega_2 \int_0^{\pi/2} d\omega_1 \ln(2\cos\omega_1) \\ &\quad + \frac{1}{\pi^2}\int_0^{\pi/2} d\omega_1 \int_0^\pi d\omega_2 \ln\left(\frac{\cosh 2K\coth 2K}{\cos\omega_1} + 2\cos\omega_2\right) \\ &= \frac{1}{\pi}\int_0^{\pi/2} d\omega_1 \ln(2\cos\omega_1) \\ &\quad + \frac{1}{\pi}\int_0^{\pi/2} d\omega_1 \cosh^{-1}\frac{\cosh 2K\coth 2K}{2\cos\omega_1} \; . \end{aligned} \quad (6.48)$$

But $\cosh^{-1} x = \ln[x + \sqrt{x^2 - 1}]$ and hence

$$I = \frac{1}{2}\ln(2\cosh 2K\coth 2K) + \frac{1}{\pi}\int_0^\pi d\theta \ln\frac{1+\sqrt{1-q^2(K)\sin^2\theta}}{2} \quad (6.49)$$

where

$$q(K) = \frac{2\sinh 2K}{\cosh^2 2K} \; . \quad (6.50)$$

Substituting in (6.41), we finally arrive at the form

$$\beta g(0,T) = -\ln(2\cosh 2K) - \frac{1}{\pi}\int_0^{\pi/2} d\theta \ln\frac{1+\sqrt{1-q^2\sin^2\theta}}{2} \quad (6.51)$$

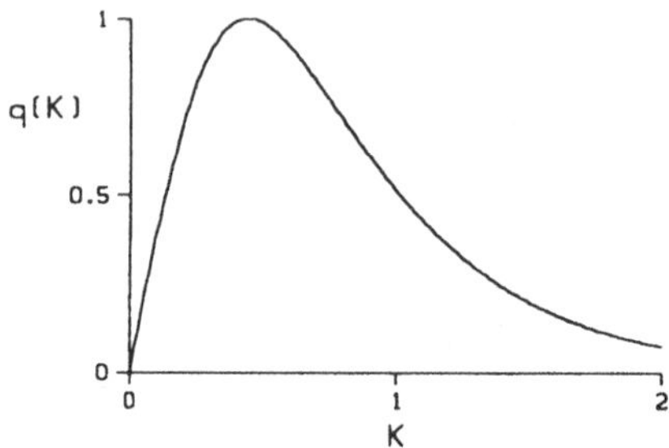

Figure 6.2: The function $q(K)$ defined in (6.50).

for the free energy per spin.

The function $q(K)$, defined in (6.50), has the form shown in Figure 6.2. It takes on a maximum value, $q = 1$, at $\sinh 2K = 1$, and it is clear that the integral on the right-hand side of equation (6.51) can only be nonanalytic at that point since the term inside the square root cannot vanish for $q < 1$. The internal energy per spin of the system is given by

$$u(T) = \frac{d}{d\beta}\left[\beta g(T)\right] = -J \coth 2K \left[1 + \frac{2}{\pi}\left(2\tanh^2 2K - 1\right) K_1(q)\right] \qquad (6.52)$$

where

$$K_1(q) = \int_0^{\pi/2} \frac{d\phi}{\sqrt{1 - q^2 \sin^2 \phi}}$$

is the complete elliptic integral of the first kind. As $q \to 1$, the term $(2\tanh^2 2K - 1) \to 0$, and the internal energy is continuous at the transition. The specific heat per spin $c(T)$ can be obtained by differentiating once more with respect to temperature. Some analysis (Problem 6.2) shows that

$$\begin{aligned}\frac{1}{k_B} c(T) &= \frac{4}{\pi}(K \coth 2K)^2 \Big\{ K_1(q) - E_1(q) \\ &\quad -(1 - \tanh^2 2K)\left[\frac{\pi}{2} + (2\tanh^2 2K - 1) K_1(q)\right]\Big\} \end{aligned} \qquad (6.53)$$

where

$$E_1(q) = \int_0^{\pi/2} d\phi \sqrt{1 - q^2 \sinh^2 \phi}$$

is the complete elliptic integral of the second kind. Near T_c the specific heat (6.53) is given, approximately, by

$$\frac{1}{k_B} c(T) \approx -\frac{2}{\pi}\left(\frac{2J}{k_B T_c}\right)^2 \ln\left|1 - \frac{T}{T_c}\right| + \text{const.} \qquad (6.54)$$

The internal energy and specific heat are shown in Figure 6.3.

The difference between the exact specific heat and that obtained in Chapter 3 from mean field and Landau theories is striking. Instead of a discontinuity in $c(T)$, we find a logarithmic divergence. In modern theories of critical phenomena, the form assumed for the specific heat is

$$c(T) \sim \left|1 - \frac{T}{T_c}\right|^{-\alpha} . \tag{6.55}$$

Onsager's result is a special case of this power law behavior. The limiting form of the function

$$\lim_{\alpha \to 0} \frac{1}{\alpha} \left(X^{-\alpha} - 1\right) = -\ln X .$$

The formula (6.54) is thus seen to be a special case of the power law singularity with $\alpha = 0$.

The calculation of the spontaneous magnetization is a nontrivial extension of the present derivation and may be found in Schultz *et al.* [270]. The result is

$$\begin{aligned} m_0(T) &= -\lim_{h\to 0} \tfrac{\partial}{\partial h} g(h,T) \\ &= \left[1 - \tfrac{(1-\tanh^2 K)^4}{16 \tanh^4 K}\right]^{1/8} \quad T < T_c \qquad (6.56) \\ &= 0 \quad T > T_c . \qquad (6.57) \end{aligned}$$

As $T \to T_c$ from below, the limiting form of the spontaneous magnetization is given by

$$m_0(T) \approx (T_c - T)^{1/8} \equiv (T_c - T)^{\beta} .$$

As in mean field theories, the order parameter has a power law singularity at the critical point but the exponent $\beta = \frac{1}{8}$, not $\frac{1}{2}$ as obtained from mean field and Landau theories. The derivation of (6.56) was first published by Yang [331], but Onsager had previously announced the result at a conference. The asymptotic form as $T \to T_c$ of the zero-field susceptibility is also known [198]:

$$\chi(0,T) = \lim_{h\to 0} \frac{\partial m(h,T)}{\partial h} \sim |T - T_c|^{-7/4} = |T - T_c|^{-\gamma} . \tag{6.58}$$

The exponent $\gamma = \frac{7}{4}$ in (6.58) again is to be compared with the classical value $\gamma = 1$. It is clear from the exact results described above that the form of the free energy near a critical point is quite different from that postulated in the Landau theory.

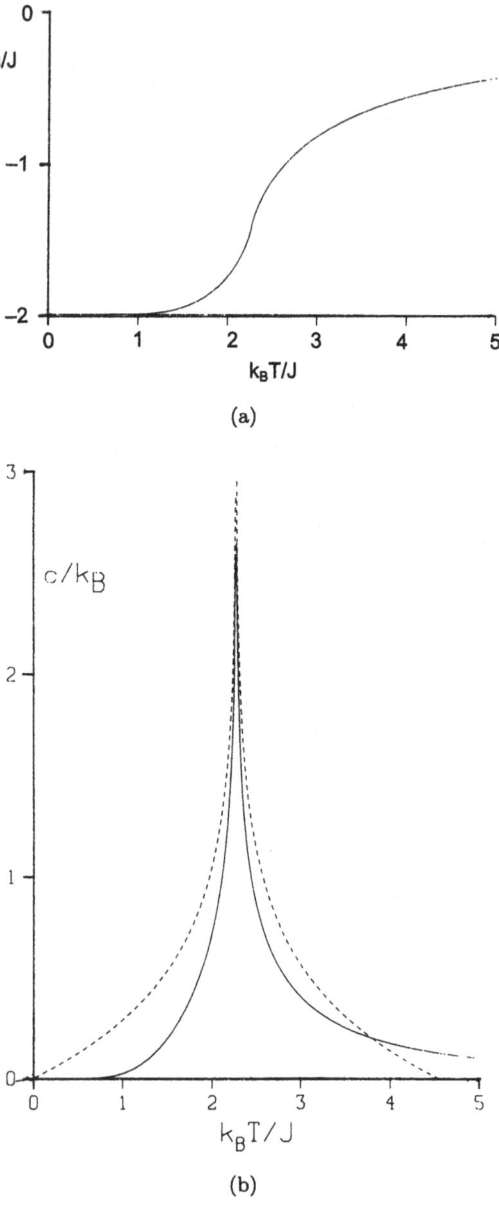

Figure 6.3: The internal energy (a) and specific heat (b) of the two-dimensional Ising model on the square lattice. The dotted curve corresponds to the approximation (6.54).

6.1.5 Concluding remarks

The reader who has worked through the details of the preceding subsections will appreciate the difficulty of calculating even the zero-field free energy exactly. One can easily write down the transfer matrix of the two-dimensional Ising model in a finite magnetic field and arrive at a generalization of (6.15). However, the subsequent transformation to fermion operators yields a transfer matrix which is not bilinear in fermion operators and which cannot be diagonalized, at least by presently known techniques.

Similarly, one can construct the transfer matrix of the three-dimensional Ising model. In this case the matrix $\mathbf{V}$ is of dimension $2^L \times 2^L$, where $L = M^2$ if the lattice is an $M \times M \times M$ simple cubic lattice. The reader can verify that the difficulty here is not this increase in the dimensionality of the transfer matrix but rather that the Jordan–Wigner transformation (6.17) does not produce a bilinear form in fermion operators.

Since Onsager's solution appeared, a small number of other two-dimensional problems have been solved exactly. The reader is referred to the book by Baxter [30] for an account of this work. The exact solution for the Ising model on a fractal is presented in Section 7.3. Since exact results near the critical point were so elusive, workers in the field devised various approximate techniques to probe the critical behavior of strongly interacting systems. We first discuss the method of series expansions which initially provided the greatest amount of information on critical behavior.

6.2 Series Expansions

The method of series expansions was first introduced by Opechowski [225] and has proved to be, with the help of modern computers, a powerful tool for the study of critical phenomena. To motivate the approach, let us consider first a simple function $f(z)$ and its power series expansion about $z = 0$:

$$f(z) = \left(1 - \frac{z}{z_c}\right)^{-\gamma} = \sum_{n=0}^{\infty} \binom{\gamma}{n} \left(\frac{z}{z_c}\right)^n \tag{6.59}$$

where

$$\binom{\gamma}{n} = \frac{\gamma(\gamma+1)(\gamma+2)\cdots(\gamma+n-1)}{n!} .$$

The power series (6.59) converges for $|z| < |z_c|$. Now suppose that we have available a certain number of terms in the power series of an unknown function.

Can we infer the function from this limited information? The answer to this question is, of course, no. However, if we have reason to believe that the unknown function has a specific structure, such as the power law singularity of the function $f(z)$ in (6.59), we may be able to determine the parameters of interest (z_c and γ) from the available information. Let us write, in general,

$$f(z) = \sum_{n=0}^{\infty} a_n z^n \ . \tag{6.60}$$

For the case (6.59), the ratio of successive coefficients, a_n, takes the form

$$r(n) = \frac{a_n}{a_{n-1}} = \frac{\gamma + n - 1}{n} \frac{1}{z_c} = \frac{1}{z_c}\left(1 + \frac{\gamma - 1}{n}\right) \ . \tag{6.61}$$

A plot of this ratio as function of $1/n$ for a number of integers thus yields a discrete set of points which fall on a straight line with intercept $1/z_c$ at $1/n = 0$ and with slope $(\gamma - 1)/z_c$. The reason for concentrating on the form (6.59) is that we believe, because of Onsager's exact solution, that thermodynamic functions have precisely this type of structure (with $z = 1/T$, $z_c = 1/T_c$) near the critical point. Of course, real thermodynamic functions are more complicated than the simple form (6.59), but it may still be possible for us to extract the information of interest, namely the critical temperature and exponent, from a finite number of terms in the power series expansion. In general, the coefficient a_n, in the power series of a function is determined, at least for large n, primarily by the nearest singularity in the complex plane to the point about which the expansion is carried out. A function of temperature, such as the specific heat, will, in general, have singularities at various points in the complex $1/T$ plane (only real $1/T$ is of interest), but if the closest of these singularities is the physical one at $1/T_c$, then we may expect a power series about $T = \infty$ (*i.e.*, $1/T = 0$) to have a simple structure similar to that of the function (6.59), at least after the first few terms. In Figure 6.4 we show a plot of $r(n)$ versus $1/n$ for the function $f(z) = \exp\{z\}(1 - z)^{-1.5}$. The convergence of the ratios $r(n)$ to the point $z_c = 1$ is evident. This is the basic idea behind series expansions and we will return to the analysis of series after briefly discussing the generation of such expansions.

6.2.1 High-temperature expansions

The derivation of a high-temperature expansion is, in principle, straightforward. Suppose that we wish to calculate the expectation value of an operator

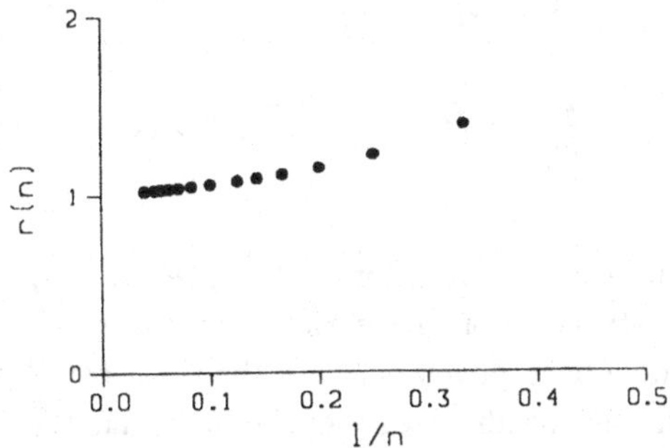

Figure 6.4: Plot of the ratios $r(n) = a_n/a_{n-1}$ for n up to 20 for the simple function $f(z) = (1-z)^{-1.5}\exp(z)$. Some points have been omitted for clarity.

O with respect to the canonical distribution for a system with Hamiltonian H. This expectation value is given by

$$\langle O\rangle = \frac{\mathrm{Tr}\, Oe^{-\beta H}}{\mathrm{Tr} e^{-\beta H}} = \frac{\mathrm{Tr}\, O\sum_{j=0}^{\infty}(-\beta H)^j/j!}{\mathrm{Tr}\sum_{j=0}^{\infty}(-\beta H)^j/j!} . \tag{6.62}$$

For a system with a finite number, Z_0, of discrete states, such as a spin system, it is convenient to divide the numerator and denominator of (6.62) by $Z_0 = \mathrm{Tr}\,\mathbf{1}$ and to define

$$\overline{O} = \frac{\mathrm{Tr}\, O}{\mathrm{Tr}\,\mathbf{1}} .$$

Note that Z_0 is the partition function at $T = \infty$. $\overline{O}$ is thus the ensemble average of O at infinite temperature. At finite temperatures, we have

$$\begin{aligned}\langle O\rangle &= \frac{\overline{O} - \beta\overline{OH} + \beta^2\overline{OH^2}/2! + \cdots}{1 - \beta\overline{H} + \beta^2\overline{H^2}/2! \cdots} \\ &= \overline{O} - \beta(\overline{OH} - \overline{O}\,\overline{H}) \\ &\quad + \frac{\beta^2}{2!}(\overline{OH^2} - 2\overline{OH}\,\overline{H} + 2\overline{O}\,\overline{H}^2 - \overline{O}\,\overline{H^2}) + \cdots .\end{aligned} \tag{6.63}$$

Therefore, if

$$\langle O\rangle = \sum_n \frac{a_n}{T^n} \tag{6.64}$$

we obtain

$$a_0 = \overline{O}$$

$$\begin{aligned} a_1 &= -\frac{1}{k_B}(\overline{OH} - \overline{O}\,\overline{H}) \qquad (6.65) \\ a_2 &= \frac{1}{2!k_B^2}(\overline{OH^2} - 2\overline{OH}\,\overline{H} + 2\overline{O}\,\overline{H}^2 - \overline{O}\,\overline{H^2}) \\ &\cdots \end{aligned}$$

As in the case of the Mayer expansion of Chapter 4, graphical methods are very useful in the construction of such series expansions. A number of different graphical methods have been developed and the most convenient, or powerful, technique depends on the specific details of the problem. We derive below a few terms in the high-temperature series of the Ising model on the square lattice and simple cubic lattices and leave a similar derivation for the Heisenberg model as an exercise (Problem 6.3).

Consider, once again, the model

$$H = -J\sum_{\langle ij\rangle}\sigma_i\sigma_j - h\sum_i \sigma_i \qquad (6.66)$$

with $\sigma_i = \pm 1$ and where the sum over i, j is over nearest-neighbor pairs on a lattice which, at present, we do not specify. The zero-field susceptibility per spin in the disordered phase is given by (3.32)

$$\begin{aligned} k_BT\chi(T) &= \frac{1}{N}\frac{\sum_{i,j}\mathrm{Tr}\,\sigma_i\sigma_j\exp\{-\beta H_0\}}{\mathrm{Tr}\exp\{-\beta H_0\}} \\ &= 1 + \frac{1}{N}\frac{\sum_{i\neq j}\mathrm{Tr}\,\sigma_i\sigma_j\exp\{-\beta H_0\}}{\mathrm{Tr}\exp\{-\beta H_0\}} \qquad (6.67) \end{aligned}$$

with $H_0 = -J\sum_{\langle ij\rangle}\sigma_i\sigma_j$. It is convenient to make use of the identity

$$\begin{aligned} \exp\{\beta J\sigma_i\sigma_j\} &= \cosh\beta J + \sigma_i\sigma_j\sinh\beta J \\ &= \cosh\beta J\,(1 + v\sigma_i\sigma_j) \qquad (6.68) \end{aligned}$$

where $v = \tanh\beta J$. The identity (6.68) can be easily demonstrated by expanding the left side in a power series and using $\sigma_i^2 = 1$. The variable v, which approaches zero as T goes to infinity, can be used as an expansion parameter instead of $1/T$ and we shall construct our expansion in powers of v. Equation (6.67) may now be written as

$$k_BT\chi = 1 + \frac{1}{N}\frac{\sum_{i\neq j}\mathrm{Tr}\,\sigma_i\sigma_j\prod_{\langle nm\rangle}(1 + v\sigma_n\sigma_m)}{\mathrm{Tr}\prod_{\langle jm\rangle}(1 + v\sigma_j\sigma_m)}\,. \qquad (6.69)$$

The structure of both numerator and denominator of (6.69) is similar to that of the Mayer expansion of the configuration integral that appeared in Chapter 4. We now carry out an analogous graphical expansion. We associate a line between nearest neighbors j, m with the term $v\sigma_j\sigma_m$. We also divide the top and bottom by 2^N and note that

$$2^{-N}\mathrm{Tr}\,\mathbf{1} = 1 = 2^{-N}\mathrm{Tr}\,\sigma_i^2$$

since the trace is carried out over the 2^N states of the system. Consider now the expansion of the numerator in (6.69):

$$\mathrm{Num} = 2^{-N}\sum_{i\neq j}\mathrm{Tr}\,\sigma_i\sigma_j\left(1 + v\sum_{\langle nm\rangle}\sigma_n\sigma_m + v^2\sum_{\langle nm\rangle\neq\langle rt\rangle}\sigma_n\sigma_m\sigma_r\sigma_t + \cdots\right).$$

Since $\mathrm{Tr}\,\sigma_i = 0$, all terms that do not contain even powers (including zero) of any spin variable will yield zero. We do not draw a line connecting the two distinct, but otherwise unrestricted, points i, j. The first few terms of the numerator can be graphically represented as follows:

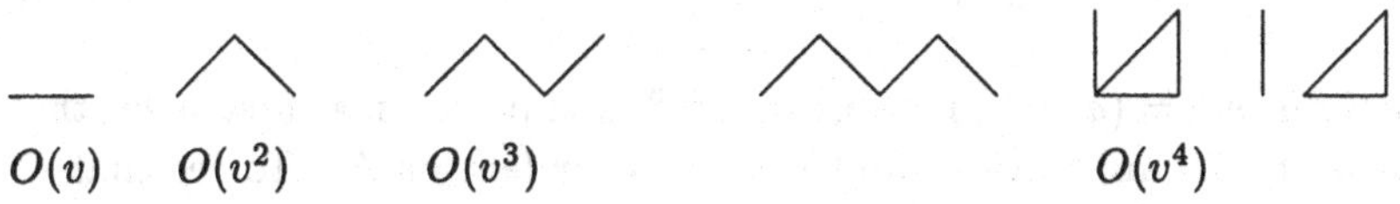

In certain types of lattices, such as the square, simple cubic, and body-centered cubic lattices, diagrams involving triangles cannot occur. We suppose that the lattice is of this type. We must now count the number of times that a particular diagram will appear. This number, called the lattice constant of graph g, will be denoted by (g). The sum over i, j in (6.69) is not restricted to nearest neighbors but $i \neq j$. Therefore, we fix i to be a specific lattice point and sum over $j \neq i$. Since v connects only nearest neighbors, we quickly obtain the lattice constants of the first three graphs appearing in the numerator

$$(g_1) = (\ \text{——}\ ) = Nq$$

$$(g_2) = (\ \wedge\ ) = Nq(q-1)$$

$$(g_3) = \left(\ \vee\wedge\ \right) = Nq(q-1)^2$$

where q is the coordination number of the lattice. For graph 4, the combinatorical factor is not simply $q(q-1)^3$, as this number contains contributions from configurations in which the last point j is identical to the first point i. A more careful calculation yields

$$(g_4) = \left(\wedge\!\!\vee\!\!\wedge \right) \begin{array}{l} = 100N \text{ on the square lattice} \\ \\ = 726N \text{ on the simple cubic lattice.} \end{array}$$

Now consider the denominator. The first nonzero contribution comes from the square $g_5 = \square$. This graph appears N times on the square lattice, $3N$ times on the simple cubic lattice. To this point, our expression for the susceptibility is

$$\begin{aligned} k_B T\chi &= 1 + \frac{1}{N}\frac{(g_1)v + (g_2)v^2 + (g_3)v^3 + (g_4)v^4 + \cdots}{1 + (g_5)v^4 + \cdots} \\ &= 1 + \frac{1}{N}\left[(g_1)v + (g_2)v^2 + (g_3)v^3 + (g_4)v^4 + \cdots\right] \\ &\quad \times \left[1 - (g_5)v^4 \cdots\right] . \end{aligned} \tag{6.70}$$

The term $(g_6) = (g_1)(g_5)$ is of order N^2v^5, and it seems, at first sight, that the expansion will approach a limit that is dependent on N. To see that this is not the case, we must add the fifth-order contribution to the numerator. This consists of three graphs:

$g_7 =$ | $\square$ $g_8 =$ /$\square$ $g_9 =$ $\wedge\!\vee\!\wedge\!\vee$

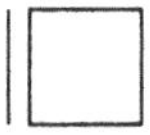

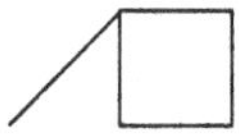

The graph g_8 will enter with a factor of 2 which comes from the fact that the square can be attached to the line either at vertex i or vertex j. The complete term of order v^5 is therefore

$$(g_7) + 2(g_8) + (g_9) - (g_1)(g_6) .$$

The lattice constant of the disconnected graph g_7 can be related to the product of the lattice constants in the last term. Since neither of the two ends of the line are allowed to touch the square, we have

$$(g_7) = (g_1)(g_6) - 2(g_8) - (g_{10})$$

where the new graph g_{10} is

$$g_{10} = $$

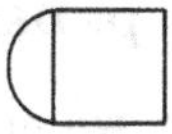

The fifth-order term thus reduces to

$$[(g_9) - (g_{10})]\, v^5 \quad . \tag{6.71}$$

Since these graphs are both connected, we see that the contribution is of order N. The cancellation of terms proportional to higher powers of N is quite general and occurs in all orders. It is clear from this example that the calculation of high-order terms in the series can be quite complicated and very sophisticated techniques have been developed to push these calculations as far as possible. Evaluating the lattice constants in (6.71), we obtain, through order v^5, the series

$$k_B T\chi = 1 + 4v + 12v^2 + 36v^3 + 100v^4 + 276v^5 + \cdots \tag{6.72}$$

and

$$k_B T\chi = 1 + 6v + 30v^2 + 150v^3 + 726v^4 + 33510v^5 + \cdots \tag{6.73}$$

on the square and simple cubic lattices. To obtain (6.72) and (6.73), we have used the following easily derived results: $(g_9) = 284N$ (square lattice) or $3534N$ (simple cubic lattice) and $(g_{10}) = 8N$ (square lattice) or $24N$ (simple cubic lattice).

The length of series that can be derived in any finite time clearly depends on the lattice structure, as more graphs with nonzero lattice constants exist on close-packed lattices such as the triangular and face-centered cubic lattices than on open lattices such as the simple cubic lattice. The susceptibility expansion is known through order v^{15} on the fcc lattice [200], through order v^{22} on the *bcc* lattice [216]. We do not list the series here but will use all available terms when we discuss analysis of such series in Section 6.2.3. While the example above focuses on the susceptibility, it is clear that similar series can be (and have been) constructed for other thermodynamic functions and for other lattice models. The power of the expansion method, for lattice models, comes from the fact that the trace over spin variables is essentially trivial. The corresponding step in the Mayer expansion, the evaluation of the cluster integral, is the limiting factor that prevents the method from being as useful in the case of liquids.

6.2.2 Low-temperature expansions

If the ground state of a system is known and if the excitations from this state can be classified in a simple way, it is possible to construct a series expansion that is complementary to the high-temperature series. Applications of this method have been primarily to Ising models ([78]). Models like the Heisenberg model or the XY model cannot be treated in the same way because their elementary excitations (known for both the classical and quantum Heisenberg model and not well known for the XY model) form a continuum. In the case of the Heisenberg model, a short low-temperature series has been constructed by Dyson [82] by a different approach.

Consider, once again, the Ising model in a magnetic field at $T = 0$. The ground state is the completely aligned ferromagnetic state. Excitations from this state consist of clusters of overturned spins. The cost in energy of flipping a single spin is $2qJ + 2h$, where q is the coordination number of the lattice. The partition function is, therefore,

$$Z = \left[1 + Nu^q w + \frac{N(N-q-1)}{2}u^{2q}w^2 + \frac{qN}{2}u^{2q-2}w^2 + \cdots\right] e^{-\beta E_0}$$

where $u = e^{-2\beta J}$, $w = e^{-2\beta h}$ and where E_0 is the ground-state energy. The third and fourth terms correspond to two flipped spins. The third term arises from pairs of spins that are not nearest neighbors, the fourth from nearest-neighbor pairs of spins. Clearly, u and w are suitable variables for a power series expansion. As in the case of high-temperature series, the logarithm of Z contains only terms proportional to N. The generation of lengthy low-temperature series is a highly specialized art and we refer the reader to the review of Domb [78] for a discussion of the details.

6.2.3 Analysis of series

We discuss, in this subsection, some of the simpler methods of series analysis and some of the results that have been obtained from analysis of high-temperature series. In the introduction to this topic, we showed that if a function has a simple power law singularity of the form

$$\chi(v) = (v - v_c)^{-\gamma} \tag{6.74}$$

the ratio of successive coefficients in the power series is

$$\frac{a_n}{a_{n-1}} = \frac{1}{v_c}\left(1 + \frac{\gamma - 1}{n}\right) .$$

Let us construct the sequence

$$r_n = n\frac{a_n}{a_{n-1}} - (n-1)\frac{a_{n-1}}{a_{n-2}} \; .$$

This sequence should approach $1/v_c$ faster than the simple ratio a_n/a_{n-1} since the term of order $1/n$ is eliminated. If χ is of the form (6.74), without other singularities or analytic correction terms, then the approximants r_n should be simply $1/v_c$, where v_c is the critical value of $\tanh \beta J$ in the case of the Ising model. If χ contains a term of the form (6.74), then we may hope that as n becomes large enough, r_n will approach a limit that we will interpret as $1/v_c$. In Table 5.1 we list these approximants for the square and simple cubic lattices using the currently available terms in the expansion. The numbers were calculated from Table I of Domb [78].

We see that the even approximants (r_n for $n = 2, 4, 6, \ldots$) increase uniformly and the odd approximants (r_n for $n = 3, 5, 7, \ldots$) decrease and that they seem to approach a common limit. Crudely extrapolating the even and odd approximants to their intersection with the $1/n = 0$ axis, we arrive at an estimate $1/v_c = 2.4151$ corresponding to $k_B T_c/J = 2.2701$ for the square lattice. The exact Onsager result is $2.269185\ldots$ and it is clear that the series expansion gives an excellent estimate of the critical temperature.

The corresponding estimate of T_c for the simple cubic lattice is $k_B T_c/J = 4.515$. The method of series analysis used above is very unsophisticated, and there are far more powerful methods based on Padé approximants to the series [106]. Such methods can provide extremely well-converged estimates of the critical temperatures. We note, in passing, that the oscillation of the approximants r_n for the square and simple cubic lattices is characteristic of lattices that can be divided into two sublattices with nearest neighbors of atoms on one sublattice lying on the other sublattice. In such a situation there is a competing singularity at $-v_c$ which represents the phase transition for the nearest-neighbor Ising antiferromagnet. On close-packed lattices, such as the triangular or face-centered cubic lattices, the antiferromagnetic transition occurs at a lower value of T ($T_c = 0$ on the triangular lattice). Thus the antiferromagnetic singularity is further from the origin in the complex v plane and the aforementioned oscillations in the approximants are absent.

To obtain a sequence of estimates for the exponent, γ, which characterizes the singularity, we can construct, for example, the sequence of approximants

$$S_n = 1 + n\left(\frac{a_n}{a_{n-1}}v_c - 1\right)$$

Table 5.1

n	r_n (Square)	r_n (Simple Cubic)	S_n (Square)	S_n (Simple Cubic)
2	2	4		
3	3	5	1.7265	1.2687
4	2.1111	4.3600	1.6007	1.2188
5	2.6889	4.8136	1.7140	1.2677
6	2.2870	4.3905	1.6610	1.2245
7	2.5671	4.7368	1.7239	1.2567
8	2.3277	4.4553	1.6877	1.2276
9	2.4962	4.6966	1.7213	1.2510
10	2.3714	4.4870	1.7031	1.2288
11	2.4625	4.6716	1.7227	1.2468
12	2.3857	4.5060	1.7106	1.2287
13	2.4500	4.6556	1.7250	1.2432
14	2.3914	4.5191	1.7152	1.2280
15	2.4424	4.6445	1.7265	1.2401
16	2.3959	4.5287	1.7186	1.2270
17	2.4365	4.6363	1.7274	1.2373
18	2.3996		1.7221	
19	2.4322		1.7281	
20	2.4021		1.7227	
21	2.4292		1.7285	

which should tend to γ as $n \to \infty$. These approximants are biased in the sense that an estimate of v_c is first required. As in the determination of T_c, far better tools exist and we use this method only as a demonstration of the utility of high-temperature series. The approximants S_n are also listed in Table 5.1 for the square and simple cubic series. Bearing in mind that we are using a rough estimate of v_c to bias the approximants, we see that the internal convergence of each sequence is remarkably good. The two-dimensional approximants increase as function of n and are certainly consistent with the exact value (see Section 5.A) of $\gamma = \frac{7}{4}$. The approximants for the simple cubic lattice are also well converged and indicate a value of γ near 1.25. For a number of years, many scientists speculated that this result (*i.e.*, $\gamma = \frac{5}{4}$) might be exact but recent longer series ([216]) indicate that a slightly smaller value $\gamma = 1.239$ is, in fact, closer to the truth.

An interesting question, at this point, is whether the nature of the singularity in χ is sensitive to details such as the type of lattice, the range of the interaction, or the magnitude of the spin. In reference [78], the coefficients a_n are tabulated for a number of two- and three-dimensional lattices and the interested reader may wish to carry out the straightforward analysis demonstrated above. It turns out that the exponent γ is quite insensitive to the lattice structure, whereas quantities such as T_c or v_c are not. The inclusion of a second-neighbor or longer-range, interaction also has no effect on the critical exponents, and the size of the spin also plays no role. The fact that such details are *irrelevant* in the language of the renormalization group theory, is now understood to be a particular manifestation of "universality" , a topic that we discuss in Section 6.5.

Series for other thermodynamic quantities such as the specific heat, the magnetization (low-temperature series), the second derivative of the susceptibility, $\partial^2\chi/\partial h^2$ and a number of other quantities have been derived. Analysis of these series reveals that in general thermodynamic functions have power law singularities at the critical point with exponents which are, again, independent of lattice structure and of other details.

$$\begin{aligned} \chi(0,T) &\sim A_\pm \left|T - T_c\right|^{-\gamma} \\ C(0,T) &\sim E_\pm \left|T - T_c\right|^{-\alpha} \\ m(0,T) &\sim B\,(T_c - T)^{\beta} \\ m(h,T_c) &\sim |h|^{1/\delta}\,\mathrm{sign}\,(h)\ . \end{aligned} \tag{6.75}$$

For the Ising model in three dimensions, estimates of the exponents were, until quite recently, consistent with the following values: $\gamma = \frac{5}{4}$, $\alpha = \frac{1}{8}$, $\beta = \frac{5}{16}$ and $\delta = 5$. We quote these numbers to illustrate certain simple relations between the exponents. For example,

$$\begin{aligned} \alpha + 2\beta + \gamma = 2 \\ \gamma = \beta(\delta - 1)\ . \end{aligned} \tag{6.76}$$

Relations of this type are called *scaling laws* and it is believed that they are exact. We discuss them further in Section 6.3.

Before discussing scaling theory in detail, we briefly review the results of series expansions for some other models. For the Heisenberg model

$$H = -J\sum_{\langle ij\rangle} \mathbf{S}_i \cdot \mathbf{S}_j$$

only high-temperature series of any substantial length have been derived and exponents such as β and δ can therefore not be determined without use of the scaling laws (6.76). The series for the Heisenberg model are shorter than those for the Ising model and the available critical exponents are substantially different from the corresponding Ising exponents. Again, within the margins of error of series analysis, no dependence on lattice structure or size of spin has been found. The susceptibility exponent is found to be $\gamma = 1.43 \pm 0.01$ for the spin-$\frac{1}{2}$ Heisenberg model in three dimensions and $\gamma = 1.425 \pm 0.02$ for the spin-∞ (*i.e.*, classical spin) Heisenberg model (see Camp and van Dyke [56] for a critical discussion).

The specific heat series are difficult to analyze and the exponent α is not very accurately known for the Heisenberg model. The weight of the evidence indicates that α is small and negative, corresponding to a cusp in the specific heat, rather than a divergence. The dependence of the critical behavior on spatial dimensionality is even more pronounced than for the Ising model. Indeed, in two dimensions, the Heisenberg model does not order at any finite temperature. This result can be proven rigorously (Mermin and Wagner, [203]).

Another model that has been studied extensively is the XY model, which has the Hamiltonian

$$H = -J \sum_{\langle ij \rangle} \left(S_i^x S_j^x + S_i^y S_j^y \right) \ .$$

This model is thought to be a good model for the critical behavior of liquid ^{4}He at the normal fluid to superfluid transition. As in the case of the Heisenberg model, only high temperature series are available. The best estimates [36] of the critical exponents γ, α in three dimensions are $\gamma = \frac{4}{3}$, $\alpha = 0$ (logarithmic divergence). These exponents differ from both the Ising and Heisenberg exponents. As for the Heisenberg model in two dimensions, one can rigorously prove that the order parameter of the XY model is zero at any finite temperature in two dimensions. Nevertheless, the XY model undergoes a transition, known as a Kosterlitz–Thouless transition, at a finite temperature in two dimensions (see Section 6.6). On the experimental side, thin films of liquid helium seem to show the same transition.

Finally, we mention the results obtained for the n-vector model (sometimes referred to as the D-vector model). This model is a generalization of the Heisenberg model to an n-dimensional spin space $S_j = (S_1, S_2, \ldots, S_n)$ and has the Hamiltonian

$$H = -J \sum_{\langle ij \rangle} \sum_{\alpha=1}^{n} S_{i\alpha} S_{j\alpha} \ .$$

The susceptibility exponent for this model, which has the Ising, XY, and Heisenberg models as the special cases $n = 1, 2, 3$ has, in three dimensions, the approximate dependence on n

$$\gamma(n) = \frac{2n + 8}{n + 7}$$

independent of lattice [283]. This expression is purely phenomenological and without theoretical basis.

6.3 Scaling

In Section 6.2 we noted that the analysis of high- and low-temperature series suggested a number of simple relationships between critical exponents of the type (6.76). Experimental data on many different materials were also, within experimental error, consistent with these scaling laws [314]. In this section we pursue the matter from different points of view and show that these scaling laws indicate a particular type of structure for the free energy near a critical point.

6.3.1 Thermodynamic considerations

We first note that considerations of thermodynamic stability allow us to derive inequalities for the critical exponents. One of the simplest of these is due to Rushbrooke [261]. Consider a magnetic system. The specific heats at constant field, C_H, and constant magnetization, C_M, satisfy the relationship (Problem 1.4)

$$\chi_T \left(C_H - C_M \right) = T \left(\frac{\partial M}{\partial T} \right)_H^2 \tag{6.77}$$

where $\chi_T = \left(\frac{\partial M}{\partial H}\right)_T$ is the isothermal susceptibility. Thermodynamic stability requires that χ_T, C_H, and C_M all be greater than or equal to zero. Therefore,

$$C_H > T \chi_T^{-1} \left(\frac{\partial M}{\partial T} \right)_H^2 \ . \tag{6.78}$$

We now consider a system in zero field at a temperature below the critical temperature, T_c, but close enough to it that we may use (6.75). We obtain

$$(T_c - T)^{-\alpha} > \text{const. } (T_c - T)^{\gamma + 2(\beta - 1)} \tag{6.79}$$

which leads to the Rushbrooke inequality:

$$\alpha + 2\beta + \gamma \geq 2 \ . \tag{6.80}$$

A number of similar inequalities have been derived and we refer to the book by Stanley [282] for further details. The intriguing feature of these inequalities is that they seem to hold as equalities, and that therefore there are only a small number of independent critical exponents.

6.3.2 Scaling hypothesis

To be specific, let us assume that we have a system for which the appropriate free energy is a function of two independent thermodynamic variables [*e.g.*, $E(S,M)$, $A(T,M)$, $G(T,H)$]. At the critical point of the system, these variables have the values T_c, H_c, M_c, and so on. We introduce the relative variables

$$\begin{aligned} h &= H - H_c \\ m &= M - M_c \\ t &= \frac{T - T_c}{T_c} \end{aligned} \tag{6.81}$$

and consider the quantities

$$\begin{aligned} \chi(t, h=0) = \left(\tfrac{\partial m}{\partial h}\right)_t &\sim (-t)^{-\gamma'} & t<0 \\ &\sim t^{-\gamma} & t>0 \\ C_h(t,0) = -\tfrac{T}{T_0^2}\left(\tfrac{\partial^2 G}{\partial t^2}\right)_h &\sim (-t)^{-\alpha'} & t<0 \\ &\sim t^{-\alpha} & t>0 \\ m(t,0) = -\left(\tfrac{\partial G}{\partial h}\right)_t &\sim (-t)^{\beta} & t<0 \end{aligned} \tag{6.82}$$

$$m(0,h) \sim |h|^{1/\delta}\,\text{sign}\,(h) \ .$$

Using the Helmholtz free energy, we have for the equation of state of the system,

$$h = \left(\frac{\partial A}{\partial m}\right) \ . \tag{6.83}$$

A number of authors (see, e.g., Griffiths [117] for references and a more complete discussion) asked themselves what the functional form of the free energy or the equation of state must be in order to produce the correct critical exponents. We saw in Section 3.G that if the free energy is assumed to be analytic at the critical point,

$$A(t,m) = a_0 + \frac{1}{2}a_2 m^2 + \frac{1}{4}a_4 m^4 + \cdots \tag{6.84}$$

and with $a_2 \approx at$ near $t = 0$, we have

$$h \approx am\left(t + \frac{a_4}{a}m^2 + \cdots\right)$$

for the equation of state. From this equation of state we automatically obtain the classical critical exponents $\alpha = 0$, $\beta = \frac{1}{2}$, $\gamma = 1$ and $\delta = 3$. If, however, we modify the equation of state to read

$$h \approx m\left(t + cm^{1/\beta}\right)$$

we can have an arbitrary value for β. This equation is still not satisfactory since if we differentiate this equation at constant t, we find

$$\frac{\partial h}{\partial m} \sim t$$

giving $\gamma = 1$ for the susceptibility exponent. The situation can be improved if, instead, we assume the equation of state

$$h \approx m\left(t + cm^{1/\beta}\right)^{\gamma}$$

or, more generally,

$$h = m\psi(t, m^{1/\beta})$$

where ψ is an arbitrary homogeneous function of degree γ, that is,

$$\psi(\lambda^{1/\gamma}t, \lambda^{1/\gamma}m^{1/\beta}) = \lambda\psi(t, m^{1/\beta}) \ . \tag{6.85}$$

To be more systematic, let us assume that the singular part of the free energy, $G(h,t)$, near the transition is dominated by a term that changes under a change of scale [1] according to

$$G(t,h) = \lambda^{-d}G(\lambda^y t, \lambda^x h) \ , \tag{6.86}$$

[1] The traditional form of (6.86) is $G(t,h) = \eta G(\eta^s t, \eta^r h)$. Since η is arbitrary, the choice $\eta = \lambda^{-d}$ leads to the present form with $x = -s/d, y = -r/d$.

where d is the spatial dimensionality. This form of the free energy implies that $m = -\frac{\partial G}{\partial h}$ will scale according to

$$m(t,h) = \lambda^{-d+x} m(\lambda^y t, \lambda^x h) \tag{6.87}$$

while

$$\begin{aligned} \chi(t,h) &= \lambda^{-d+2x} \chi(\lambda^y t, \lambda^x h) \\ C_h(t,h) &= \lambda^{-d+2y} C_h(\lambda^y t, \lambda^x h) \ . \end{aligned} \tag{6.88}$$

We now consider the special cases $h = 0$, $\lambda = |t|^{-1/y}$, and $t = 0$, $\lambda = |h|^{-1/x}$ to obtain

$$\begin{aligned} m(t,0) &= (-t)^{(d-x)/y} m(-1,0) \\ m(0,h) &= |h|^{(d-x)/x}\, m(0,\pm 1) \\ \chi(t,0) &= |t|^{-(2x-d)/y}\, \chi(\pm 1,0) \\ C_h(t,0) &= |t|^{-(2y-d)/y}\, C_h(\pm 1,0) \ . \end{aligned} \tag{6.89}$$

In (6.89) the coefficients of the leading power law term are the thermodynamic functions evaluated at points far from the singularity ($t = 0$, $h = 0$) and therefore are simply finite constants. We see that when critical exponents exist on both sides of the critical point [see (6.82)] they will have to be identical,

$$\begin{aligned} \alpha &= \alpha' \\ \gamma &= \gamma' \end{aligned} \tag{6.90}$$

but the prefactor of the power law will in general be different, that is, there is no reason for $C_h(1,0)$ or $\chi(1,0)$ to be the same as $C_h(-1,0)$ or $\chi(-1,0)$. By comparing with (6.82), we find

$$\begin{aligned} \alpha &= \frac{2y-d}{y} \\ \beta &= \frac{d-x}{y} \\ \gamma &= \frac{2x-d}{y} \\ \delta &= \frac{x}{d-x} \ . \end{aligned} \tag{6.91}$$

We see that there are only two independent critical exponents and that the scaling relations

$$\begin{aligned} \alpha + 2\beta + \gamma &= 2 \\ \beta\,(\delta - 1) &= \gamma \end{aligned} \tag{6.92}$$

follow directly from (6.91). As we shall see, in the next section, further scaling relations appear if an assumption similar to (6.85) is made for the behavior of the pair correlation function near T_c.

6.3.3 Kadanoff block spins

The agreement of the results of the preceding section with those of experiments (see Figure 6.6) and analyses of series is very satisfactory, but our formulation of the scaling law offers no physical justification for the basic assumption (6.86). An important development occurred when Kadanoff produced an intuitively appealing plausibility argument for the scaling form of the free energy. (For a very readable review of these ideas, as they were presented at the time, see [146].)

Let us consider an Ising model on a d-dimensional hypercubic lattice with nearest-neighbor separation a_0. The common property of all systems near a critical point is that the correlation length $\xi \gg a_0$ and, exactly at T_c, $\xi = \infty$. Two parameters characterize the state of the system: $t = (T - T_c)/T_c$ and h. Consider now a small block of spins. In Figure 6.5 neighboring groups of nine spins on a square lattice have been combined into such blocks and the blocks themselves form a square lattice with nearest-neighbor spacing $3a_0$. Each block can be in one of 2^9 states, or, in general 2^n states with $n = L^d$, where L is the linear dimension of a block. In the following analysis we assume that $\xi/a_0 \gg L$. If this is the case, many of these states will be suppressed because of the strong correlation over short distances.

Following Kadanoff, we now assume that each block can be characterized by an Ising spin, σ_J, for the Jth block and that this new dynamical variable also takes on the values ± 1, as do the original "site spins" σ_i. Heuristically, we expect that the state of the block spin system can be described in terms of effective parameters $\tilde{t}$, $\tilde{h}$ which measure the distance of the block spin system from criticality. The parameters $\tilde{t}$, $\tilde{h}$ depend on t and h as well as on the linear dimension, L, of the block. The simplest possible relation between these variables consistent with the symmetry requirements $\tilde{h} \to -\tilde{h}$ when $h \to -h$ and $\tilde{t} \to \tilde{t}$ when $h \to -h$, as well as with the condition $\tilde{t} = \tilde{h} = 0$ when $t = h = 0$ is

$$\begin{aligned} \tilde{h} &= hL^x \\ \tilde{t} &= tL^y \end{aligned} \tag{6.93}$$

where x and y are unspecified except that they must be positive so as to insure that the block spin system is further from criticality than the site spin system.

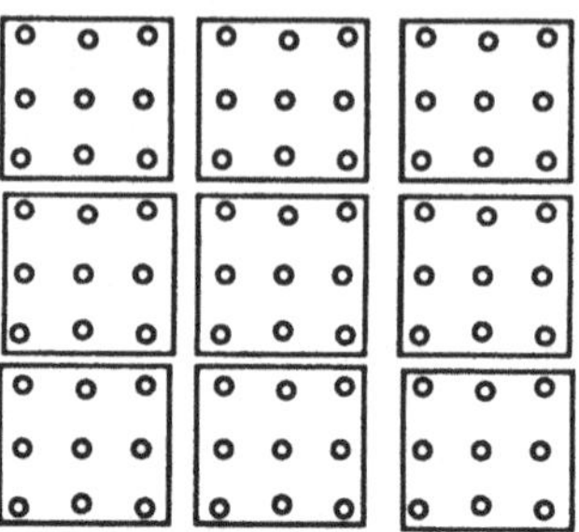

Figure 6.5: Grouping of site spins into nine-spin blocks on a square lattice.

The singular part of the free energy of the block spin system must be the same function of $\tilde{t}$ and $\tilde{h}$ as the singular piece of the free energy of the site spin system is of t and h. Since there are L^d site spins per block spin, we have

$$G(t,h) = L^{-d}G(L^y t, L^x h) \tag{6.94}$$

which completes the derivation of the scaling form of the free energy. Equation (6.94) is of the same form as (6.86). The quantity λ in the latter equation is now identified as a "dilation" parameter. Since the block spin separation is L times the site spin separation, we also expect that the correlation length will be reduced by the same factor:

$$\xi(t,h) = L\xi(L^y t, L^x h)\ . \tag{6.95}$$

The last equation leads to new predictions. Near the critical point the correlation length diverges according to the form

$$\xi(t,0) \sim |t|^{-\nu}$$

and we may now relate the exponent ν to the unspecified exponent y. Assuming that equations (6.94) and (6.95) hold for any value of L and letting $L = |t|^{-1/y}$ in (6.94) and (6.95), as well as $h = 0$, we obtain

$$G(t,0) = |t|^{d/y}\, G(\pm 1, 0) \tag{6.96}$$

$$\xi(t,0) = |t|^{-1/y}\, \xi(\pm 1, 0)\ . \tag{6.97}$$

The exponent d/y is related to the specific heat exponent through $d/y = 2 - \alpha = d\nu$, and we have therefore obtained a scaling relation that explicitly involves the spatial dimensionality

$$d\nu = 2 - \alpha \tag{6.98}$$

called a *hyperscaling* equation. If we use only the scaling form of the free energy, we easily reproduce the results (6.85)–(6.92).

To this point we have only obtained the scaling form of the correlation length. We can also find a scaling equation for the correlation function $\Gamma(r,t,h) = \langle\sigma_r\sigma_0\rangle - \langle\sigma_r\rangle\langle\sigma_0\rangle$ by introducing local fields h_r, $\tilde{h}_r$ and demanding that the variation of the free energy with respect to changes in the local field be the same in the site and block spin picture:

$$\Gamma(r) = \frac{\delta^2 G}{\delta h(r)\delta h(0)}$$

and

$$\delta^2 G = \Gamma(r)\delta h(0)\delta h(r) = L^{-2d}\Gamma\left(\frac{r}{L}\right)\delta\tilde{h}(0)\delta\tilde{h}\left(\frac{r}{L}\right) . \tag{6.99}$$

Using $\tilde{h} = L^x h$, we have

$$\Gamma(r,t,h) = L^{2(x-d)}\Gamma\left(\frac{r}{L}, L^y t, L^x h\right) . \tag{6.100}$$

At the critical point the correlation function $\Gamma(r,0,0)$ has the asymptotic form $\Gamma(r,0,0) \sim r^{-(d-2+\eta)}$. We let $L = r$ and find $2(x-d) = 2-d-\eta$, or

$$2-\eta = 2x - d . \tag{6.101}$$

The susceptibility exponent, γ, may be obtained from (6.94):

$$\begin{aligned} \gamma &= (2x-d)/y = (2x-d)\nu \\ \gamma &= (2-\eta)\nu . \end{aligned} \tag{6.102}$$

The derivation of further scaling relations is left to the problem section.

The verification of the scaling laws is difficult. All experiments of which the authors are aware are at least consistent with the exact equalities derived from the scaling form of the free energy. [2] In Figure 6.6 we show an experimental plot of the function $h/|t|^{\gamma+\beta}$ as function of the scaled magnetization $m/|t|^{\beta}$ for the ferromagnet $CrBr_3$. Using the scaling form of the order parameter

$$m(t,h) = |t|^{\beta}\, m\left(\pm 1, \frac{h}{|t|^{\beta+\gamma}}\right) = |t|^{\beta}\,\phi_{\pm}\left(\frac{h}{|t|^{\beta+\gamma}}\right) \tag{6.103}$$

which is easily obtained from (6.89), we expect that the data should fall on at most two universal curves, one for $t < 0$ corresponding to the function ϕ_- and

[2] Hyperscaling will not hold for systems with mean field exponents, which are independent of dimensionality.

one for $t > 0$ corresponding to ϕ_+. We see that this is indeed the case for a range of temperatures near T_c.

A more accurate check of the scaling relations can be carried out for the two-dimensional Ising model for which $\alpha = 0$, $\beta = \frac{1}{8}$, $\gamma = \frac{7}{4}$, $\eta = \frac{1}{4}$ and $\nu = 1$ are all known exactly. Clearly, all scaling relations are found to hold. For the three-dimensional Ising model the results of series expansions were for many years consistent with those scaling relations not involving the correlation length exponent ν. It seemed, however, that the hyperscaling relation (6.98) might not be satisfied. More recent work of Nickel [216] found no violation of hyperscaling, at least at the present level of accuracy of the series results. For a detailed discussion of experimental tests of the scaling laws, the reader is referred to the article by Vincentini-Missoni [314].

We have emphasized Kadanoff's heuristic derivation of the scaling laws because the concept of replacing a group of dynamic variables by an effective single dynamic variable can be made into a concrete calculational tool. Indeed, it is clearly the correct way to attack problems with large correlation lengths (*i.e.*, where dynamic variables are strongly coupled over large distances). By successively removing intermediate dynamical variables we can hope to arrive at a problem which is simple (*i.e.*, one for which the correlation length relative to the new unit of length, La_0, is much smaller). Critical behavior is then seen to be a property of the rescaling of block spin interactions (t, h) under successive coarse graining. This is the program of the renormalization group approach that is discussed in detail in Chapter 7.

6.4 Finite-Size Scaling

In this section we begin our discussion of finite-size effects in the vicinity of a critical point. This is an important topic both from a conceptual and a practical point of view. On the conceptual side, we know that a finite system cannot have a true singularity at a non-zero temperature. This is very easy to understand in the case of discrete models such as the Ising model in any dimension. For a finite set of Ising spins on a lattice, the partition function consists of a long but finite polynomial in $v = \tanh K$ (Section 6.2) and this polynomial is a smooth function of v. The same holds for all other thermodynamic functions. We have also already referred to this fact in Section 6.1 when we discussed the degeneracy of the two largest eigenvalues of the transfer matrix for the two-dimensional Ising model. We asserted that this degeneracy occurs only in the thermodynamic limit $M \to \infty$ and that without such degeneracy, there cannot

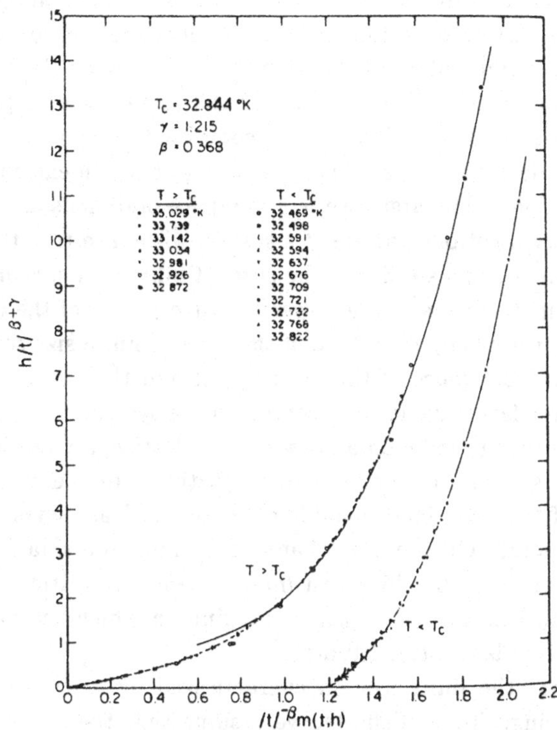

Figure 6.6: Magnetization of $CrBr_3$ in the critical region plotted in a scaled form (see the text). (From Ho and Litster [132].)

be a spontaneous magnetization. Experimental systems, of course, are finite albeit very large. For all practical purposes, true phase transitions do occur in these finite systems and we have to reconcile this fact with the rigorous proof that no phase transition can occur in any finite system. This is one of the roles of finite size scaling theory.

As we discuss elsewhere (Chapter 9), the simulation of finite systems by Monte Carlo and molecular dynamics techniques is one of the most powerful techniques in statistical physics. Simulations of finite numbers of particles or spins are plagued to a much larger extent than experiments by finite size effects and a proper analysis of numerical data, in particular near the critical point of the corresponding infinite system, requires an understanding of finite-size scaling. Finally, one of the most powerful renormalization group techniques for two-dimensional systems has its origins in finite-size scaling theory. This topic will be discussed in Section 7.6. An excellent reference for the material in the present section and for Section 7.6 is the review by Barber [24].

We begin, following [24], by discussing the role of sample geometry. It is clear that finite-size effects will occur in a somewhat different way if the sample is of finite volume than if it is infinite in at least two dimensions and finite in the other dimension. The first case, *e.g.* a lattice of dimension $L \times L \times L$ falls into the category mentioned above of a system without a true thermodynamic phase transition. In this situation, the susceptibility and specific heat remain finite at all temperatures and the power-law divergences of the infinite system are replaced by rounded peaks. One of the roles of finite-size scaling theory is to describe the dependence of the peak-height and the range in temperature over which power-law behavior is observed on the system size L. In the second case, *e.g.* for an Ising model on an $L \times \infty \times \infty$ lattice, a true thermodynamic transition occurs even for finite L. However, the critical exponents observed will be those of the two-dimensional Ising model, at least asymptotically close to the critical point. On the other hand, if L is not too small there will also be a region of temperature in which *three*-dimensional critical exponents are observed and an intermediate *crossover* régime in which the two power laws join. We will also discuss this situation.

Consider first the finite system characterized by a length L in units of the lattice spacing. To be definite, we assume that the system is a magnet and focus on the behavior of the zero-field susceptibility $\chi(0,T)$ as function of temperature. We have already pointed out several times that the susceptibility per site can be expressed in terms of the spin-spin correlation function

$$k_B T\chi(0,T) = \frac{1}{V}\sum_{\mathbf{r},\mathbf{r}'}\Gamma(|\mathbf{r}-\mathbf{r}'|) = \frac{1}{L^d}\sum_{\mathbf{r},\mathbf{r}'}\left\{\langle S_{\mathbf{r}}S_{\mathbf{r}'}\rangle - \langle S_{\mathbf{r}}\rangle\langle S_{\mathbf{r}'}\rangle\right\} \ . \tag{6.104}$$

The asymptotic form of the correlation function of the infinite system is

$$\Gamma(|\mathbf{r}-\mathbf{r}'|) \sim \frac{\exp\{-|\mathbf{r}-\mathbf{r}'|/\xi(T)\}}{|\mathbf{r}-\mathbf{r}'|^{d-2+\eta}}$$

and it is clear that the susceptibility will saturate when the correlation length ξ becomes comparable to the system size L. One way of incorporating this effect into the scaling theory is to postulate the following form:

$$\chi(L,T) = |t|^{-\gamma} g\left(\frac{L}{\xi(t)}\right) \tag{6.105}$$

where $t = (T - T_c)/T_c$, T_c is the critical temperature of the corresponding infinite system, and where the scaling function g has the limiting behavior $g(x) \to$ const. as $x \to \infty$ and $g \sim x^{\gamma/\nu}$ as $x \to 0$. The first of these limits ensures that the correct power law behavior of the infinite system is recovered when $L \to \infty$. Since $\xi(t) \sim |t|^{-\nu}$, the second limit then produces a temperature-independent susceptibility as the correlation length (of the *infinite* system) becomes much larger than L. This form therefore also predicts that the maximum in the susceptibility will grow with system size as[3]

$$\chi_{max} \sim L^{\gamma/\nu} . \tag{6.106}$$

Clearly, there is nothing special about the susceptibility. The specific heat, for example will obey a similar scaling relation

$$C(L,t) = |t|^{-\alpha} f\left(\frac{L}{\xi(t)}\right) \tag{6.107}$$

and its maximum will scale with L as $C_{max} \sim L^{\alpha/\nu}$. The only essential assumption is that there is a single length ξ that determines the range of correlations close to the critical point.

In computer simulations of finite systems it is found that the temperature at which the maximum of thermodynamic functions such as C and χ occur is also a function of L, approaching the T_c of the infinite system as $L \to \infty$. If we denote the temperature of the peak in χ by $T_c(L)$, a natural assumption is

$$\xi\left(T_c(L) - T_c\right) = aL \tag{6.108}$$

where a is some constant. This yields

$$T_c(L) = T_c + bL^{-1/\nu} . \tag{6.109}$$

This result, as well as the scaling forms (6.105–6.107) have been verified in numerous finite-system calculations. As an example of this, we note that if we multiply a function $Q(|t|, L)$ that, in the thermodynamic limit, scales as $|t|^y$ by $L^{y/\nu}$ we will obtain the scaling form

$$L^{y/\nu} Q(|t|, L) = \left(L^{1/\nu}|t|\right)^y \phi\left(L^{1/\nu}|t|\right) .$$

If the finite-size scaling hypothesis holds the right-hand side depends only on the variable $L^{1/\nu}|t|$. Figure 6.7 shows such a plot for the magnetization of $m(|t|, L)$ of the two-dimensional Ising models for a number of different sizes and temperatures.

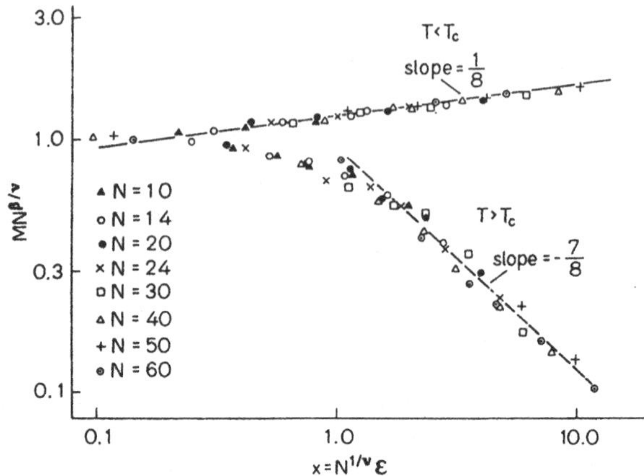

Figure 6.7: Plot of the scaled magnetization of two-dimensional Ising models for lattices of linear dimension N for several values of N both for $T > T_c$ and $T < T_c$ (taken from [38]).

We turn now to the case of finite-size effects in the slab geometry. To be specific, we assume that we have a system such as the Ising model that has a continuous phase transition in two dimensions as well as in three dimensions. We denote the critical exponents in the two-dimensional case by α_2, β_2, γ_2, *etc.*, and leave the three-dimensional exponents unsubscripted. Consider an $L \times \infty \times \infty$ slab and imagine decreasing the temperature from an initial temperature T_i that is much higher than the critical temperature of either the three- or two-dimensional system. Specifically, this means that $\xi(T_i) \ll L$. As the temperature is lowered, correlations will initially grow isotropically, at least in the regions of the sample that are far from either surface. Thus, if L is large enough that $\xi(T) \sim L$ occurs at a temperature that is sufficiently close to the critical temperature of the infinite three-dimensional system there will be a range of temperatures over which power-law behavior characterized by three-dimensional exponents is seen. Conversely, for $\xi(T) > L$ the correlations grow only in the two infinite directions and two-dimensional critical behavior is observed. The situation is shown schematically in Figure 6.8. The change from one form of critical behavior to another is called *crossover* and occurs as well in many other situations where the effect of a symmetry-breaking term is only seen very close to the critical point.

We construct our finite-size scaling theory in exactly the same way as above.

[3]For a finite system below $T_c(\infty)$ the susceptibility must be calculated using $\chi = \langle m^2 \rangle - \langle |m| \rangle^2$, since $\langle m \rangle$ is always zero.

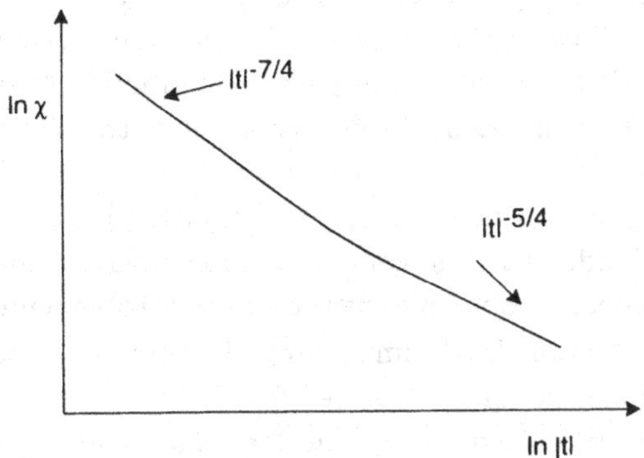

Figure 6.8: Illustration of crossover behavior in an Ising model in a slab geometry.

Taking the susceptibility as a specific example, we postulate the form

$$\chi(L,t) = |t|^{-\gamma} g\left(\frac{L}{\xi(t)}\right)$$

where $t = (T - T_c(L))/T_c(L)$. The correct three-dimensional scaling behavior is recovered if $g(x) \to$ const. as $x \to \infty$. On the other hand, for finite L, we have $\chi \sim |t|^{-\gamma_2}$. We obtain this form if we take $g(x) \to x^{[\gamma-\gamma_2]/\nu}$. Therefore, the "amplitude" of the two-dimensional piece of the susceptibility scales, in the large L limit as $L^{[\gamma-\gamma_2]/\nu}$. We can also estimate the crossover temperature as function of L. As in (6.109) we find $[T^* - T_c(L)]^{-\nu} \sim L$ where T^* denotes the temperature at which two-dimensional effects begin to dominate.

Finally, we note that we may construct a finite-size scaling form for the correlation length ξ in complete analogy with the discussion given above. We will return to this specific case in Section 7.6 because this particular scaling relation leads to a very powerful position-space renormalization group scheme.

6.5 Universality

We have seen in Section 6.2 that the critical exponents of the three-dimensional Ising model are the same no matter what the underlying lattice structure is. On the other hand, they differ from the critical exponents of the two-dimensional Ising model and from those of other three-dimensional spin systems such as the Heisenberg and XY models. It is natural to ask which features of system are important in determining the nature of the phase transition. Many possibilities spring to mind. The lattice structure has already been ruled out. Other possibilities include the range of interparticle interaction, size of spin, quantum-mechanical rather than classical spins, continuous translational freedom as in real liquids rather than lattice gases, spin space dimensionality ($n = 1, 2, 3, \ldots$, see Section 6.2), crystal field effects, and many others. A large number of investigations of different spin systems were carried out and it became clear that critical behavior was independent of a remarkable number of details.

Two parameters are clearly important. The spatial dimensionality, d, and the spin space dimensionality, n, have already been mentioned as parameters that do affect the critical exponents. On the other hand, one could ask whether a system with Hamiltonian

$$H = -J\sum_{\langle ij\rangle} \mathbf{S}_i{\cdot}\mathbf{S}_j - D\sum_i (S_i^z)^2 \tag{6.110}$$

which has spin space dimensionality $n = 3$ has the same critical exponents as the isotropic Heisenberg model ($D = 0$). It turns out that this model has the same critical behavior as the Ising model ($n = 1$) and some refinement of our concepts is required.

Empirically (*i.e.*, from series expansions) it was found that the symmetry of the ordered phase plays a crucial role. The Hamiltonian (6.110) has, for $D > 0$, as its ground state the state with all spins fully aligned along the z direction. The only transformation that leaves the ground-state energy, and at $T > 0$ the free energy, invariant is the transformation $S_i^z \to -S_i^z$, all i. On the other hand, the Heisenberg Hamiltonian has the full three-dimensional rotational symmetry—the vector $\mathbf{m} = 1/N\sum_i \mathbf{S}_i$ can be anywhere on the surface of a three-dimensional sphere. The parameter n which we have to use here to characterize the spin space dimensionality is now modified to denote the index n of the rotational group, $O(n)$, under which the free energy is invariant.

The fact that this index, n, is an important parameter was demonstrated by Jasnow and Wortis [141], who studied the classical spin-system with Hamiltonian

$$H = -J\sum_{\langle ij\rangle} \left(\mathbf{S}_i\cdot\mathbf{S}_j + \eta S_i^z S_j^z\right) . \tag{6.111}$$

The ground state of this Hamiltonian has $n = 3$ for $\eta = 0$, $n = 1$ for $\eta > 0$ and $n = 2$ for $-2 < \eta < 0$. The ground state thus changes from an XY ground state to a Heisenberg ground state and finally to an Ising ground state as the parameter η is varied. Jasnow and Wortis derived and analyzed susceptibility series for different values of η. Any finite series cannot yield completely unambiguous results, but their analysis strongly suggested that there are discontinuous changes in the exponent γ at the values of η at which the ground-state symmetry changes.

Most of the other parameters that we mentioned above were also eliminated. Some of these conclusions can be understood qualitatively. Since the correlation length diverges at T_c, one might expect that quantum mechanics would not play a role in critical phenomena: a cluster of ξ^d spins is effectively a classical object.[4] Similarly, the possibility of continuous rather than discrete translation should not be important. Localizing a fluid particle in a cell of size a and simply allowing it to hop to nearest-neighbor cells should not be important on a length scale of ξ. Indeed, classical fluids have, to within experimental error, the same critical exponents as the three-dimensional Ising model.

The range of interaction may, of course, be important. We now believe (Aharony, [7]) that as long as the range of the interaction is finite, or as long as the interaction decreases sufficiently rapidly as a function of separation, the critical behavior will be the same as that of a system with only nearest-neighbor interactions.

A wealth of empirical evidence led to the formulation of the *universality hypothesis* [145] or smoothness hypothesis [119]. This hypothesis simply states that the critical behavior of a system depends only on the spatial dimensionality, d, and the symmetry, n, of the ordered phase. The renormalization group has allowed us to understand in some detail this very general conclusion.

In the foregoing discussion we have considered only the symmetry group $O(n)$. Other discrete and continuous symmetries exist, of course. For example, a real spin system on a cubic lattice is in general subject to a crystal field that will align the ground-state magnetization with one of the symmetry directions of the crystal. This symmetry is neither Ising-like nor $O(n)$ with $n = 3$. Can this type of field have an effect? The answer to this question is "yes" and we will return to this point in Chapter 6. At this point we simply conclude by stating again that the concept of universality and the notion that seemingly very different systems on the microscopic level can be grouped into a small number of universality classes is an extremely powerful simplifying idea.

---

[4] An exception to this is the case of a phase transition that occurs at $T_c = 0$ as a function of some parameter in the Hamiltonian. Quantum mechanical phase transitions are discussed in the book by Sachdev [262].

6.6 Kosterlitz–Thouless Transition

To conclude this first chapter on critical phenomena we briefly discuss planar systems in which the ground state has a continuous symmetry as distinguished from, for example, the Ising model, which has only the discrete symmetry $m \to -m$. We single out these systems because in some of them there exists the possibility of an unusual finite-temperature transition to a low-temperature phase without long-range order. Such transitions are known as Kosterlitz–Thouless [158] transitions. Physical systems that are thought to display this transition are certain planar magnets, films of liquid helium (the two-dimensional version of the lambda transition), thin superconducting films, liquid-crystal monolayers, crystal surfaces which roughen as the temperature is increased and in some cases gases adsorbed on crystal surfaces. We will return to discuss some of these situations in more detail but first begin by showing, heuristically, that spatial dimensionality 2 separates simple critical behavior (as discussed in the rest of this chapter and in Chapter 6) from a nonordering situation.

We use, as an example, a two-dimensional XY model (Section 6.2) with spins modeled by classical vectors. In Chapter 9 we show that Bose condensation in an ideal Bose gas cannot occur for $d \leq 2$ and in Chapter 10 the reader is asked to demonstrate (Problem 10.6) the same result for the Heisenberg model in the spin wave approximation.

Consider the Hamiltonian

$$H = -J \sum_{\langle ij \rangle} (S_{ix}S_{jx} + S_{iy}S_{jy}) = -JS^2 \sum_{\langle ij \rangle} \cos(\phi_i - \phi_j) \qquad (6.112)$$

where the spins $\mathbf{S}_j$ are classical vectors of magnitude S constrained to lie in the $S_x - S_y$ plane. These spins can therefore be specified by their orientation ϕ_i $(0 \leq \phi_i < 2\pi)$ with respect to the S_x axis. For the present we assume that the sites i lie on a d-dimensional hypercubic lattice. We note that the Hamiltonian (1) has a continuous symmetry: the transformation $\phi_i \to \phi_i + \phi_0$, for all i, leaves the Hamiltonian invariant. In Chapter 11 we show that the same symmetry exists in the BCS theory of superconductivity and in the interacting Bose gas.

The ground state of (6.112) is the fully aligned state $\phi_i = \phi$ for all i, where ϕ can be any angle in the range 0 to 2π. We now assume that at low enough temperature $|\phi_i - \phi_j| \ll 2\pi$ for i, j nearest neighbors and approximate (6.112) by the expression

$$\begin{aligned} H &= -\frac{qNJS^2}{2} + \frac{1}{2}JS^2 \sum_{\langle ij \rangle} (\phi_i - \phi_j)^2 \\ &= E_0 + \frac{JS^2}{4} \sum_{\mathbf{r},\mathbf{a}} [\phi(\mathbf{r}+\mathbf{a}) - \phi(\mathbf{r})]^2 \end{aligned} \tag{6.113}$$

where E_0 is the ground-state energy and the sum over $\mathbf{a}$ runs over all nearest neighbors of site $\mathbf{r}$. If $\phi(\mathbf{r})$ is a slowly varying function of $\mathbf{r}$, we may further approximate (6.113) by a continuum model. Replacing the finite differences in (6.113) by derivatives and the sum over lattice sites by an integral, we obtain the expression

$$H = E_0 + \frac{JS^2}{2a^{d-2}} \int d^d r \, [\nabla\phi(\mathbf{r}) \cdot \nabla\phi(\mathbf{r})] \tag{6.114}$$

where a is the nearest-neighbor distance. The second term in (6.114) is the classical form of the spin wave energy (see Chapter 12).

The constraint that ϕ must be in the range 0 to 2π is inconvenient and we relax this condition and allow ϕ to range from $-\infty$ to ∞. It is then a trivial matter to calculate the partition function and other thermodynamic properties of the system. In particular, we wish to calculate the correlation function

$$g(r) = \langle \exp\{i[\phi(\mathbf{r}) - \phi(0)]\}\rangle \tag{6.115}$$

which, in a phase with conventional long-range order, will approach a constant as $r \to \infty$. In the ground state, of course, $g(r) = 1$. Using periodic boundary conditions and writing

$$\phi(\mathbf{r}) = \frac{1}{\sqrt{N}} \sum_{\mathbf{k}} \phi_{\mathbf{k}} e^{i\mathbf{k}\cdot\mathbf{r}} \tag{6.116}$$

we obtain

$$\begin{aligned} H &= E_0 + \frac{JS^2a^2}{2} \sum_{\mathbf{k}} k^2 \phi_{\mathbf{k}} \phi_{-\mathbf{k}} \\ &= E_0 + JS^2a^2 {\sum_{\mathbf{k}}}' k^2 (\alpha_{\mathbf{k}}^2 + \gamma_{\mathbf{k}}^2) \end{aligned} \tag{6.117}$$

where $\phi_{\mathbf{k}} = \alpha_{\mathbf{k}} + i\gamma_{\mathbf{k}} = (\phi_{-\mathbf{k}})^*$ and where $\sum'$ in the second expression indicates that we have combined the two terms for $\mathbf{k}$ and $-\mathbf{k}$ and are summing over half the Brillouin zone. The expectation value (6.115) can now be easily evaluated (Problem 6.6). The result is

$$g(r) = \exp\left\{ -\frac{k_B T}{NJS^2a^2} \sum_{\mathbf{k}} \frac{1 - \cos(\mathbf{k}\cdot\mathbf{r})}{k^2} \right\} . \tag{6.118}$$

We transform the sum over $\mathbf{k}$ in (6.118) to an integral in the usual way:

$$\frac{1}{N}\sum_{\mathbf{k}} \rightarrow \left(\frac{a}{2\pi}\right)^d \int d^dk$$

and arrive at the expression

$$g(r) = \exp\left\{-\frac{k_BTa^{d-2}}{(2\pi)^d JS^2}\int d^dk \frac{1-\cos(\mathbf{k}\cdot\mathbf{r})}{k^2}\right\} . \qquad (6.119)$$

If we now take $d = 2$, ignore the geometry of the Brillouin zone, and carry out the integration in polar coordinates, we have, using

$$\int_0^{2\pi} d\theta \cos(kr\cos\theta) = 2\pi J_0(kr)$$

where J_0 is the zeroth-order Bessel function,

$$g(r) = \exp\left\{-\frac{k_BT}{2\pi JS^2}\int_0^{\pi/a} dk \frac{1-J_0(kr)}{k}\right\} \qquad (6.120)$$

where the upper limit, π/a, is roughly the distance to the zone boundary. Substituting $x = kr$, we obtain for the integral in (6.120),

$$\int_0^{\pi r/a} dx \frac{1-J_0(x)}{x} .$$

For large r/a the dominant contribution comes from the region $x \gg 1$ in which we can ignore the Bessel function and we finally obtain

$$g(r) \approx \exp\left\{-\frac{k_BT}{2\pi JS^2}\ln\frac{\pi r}{a}\right\} = \left(\frac{\pi r}{a}\right)^{-k_BT/2\pi JS^2} = \left(\frac{\pi r}{a}\right)^{-\eta(T)} . \qquad (6.121)$$

We see, therefore, that for $d = 2$ the correlation function falls off algebraically at all finite temperatures and that there is no long-range order in the system. If, in (6.119), we take $d = 3$, we easily see that $g(r)$ approaches a constant as r becomes large (Problem 6.6). In physical terms, the absence of long-range order at finite temperatures is due to the excitation of long-wavelength low-energy spin waves which are weighted, in (6.119), by a phase space factor k^{d-1} that becomes more and more important as d is decreased.

The argument above can be made rigorous [203] and holds as well for the Heisenberg model, for superfluid films and superconductors [133] and, indeed, for any two-dimensional system with short-range interactions in which the ordered phase has a continuous symmetry. The continuous symmetry then implies that at least one branch of the spectrum of elementary excitations has the property that the energy approaches zero continuously as the wavelength becomes large (Goldstone boson; see *e.g.* Anderson [17] for a discussion).

The algebraic decay of the correlation functions in (6.121) is reminiscent of a system with a finite-temperature phase transition precisely at its critical point. Recall that we generally write, for the correlation function

$$\begin{aligned} g(r) &\approx \frac{\exp\{-r/\xi(T)\}}{r^{d-2+\eta}} \\ &\approx \frac{1}{r^{d-2+\eta}} \, . \end{aligned} \tag{6.122}$$

In these planar magnets we therefore find a temperature-dependent "critical" exponent $\eta(T)$. In our spin wave approximation the system seems to be at a critical point at all temperatures. This result is clearly unphysical. At high enough temperature we expect the correlation function to fall off exponentially. We now argue, following Kosterlitz and Thouless [158], that there exists another set of excitations which take the system from its low-temperature "critical" phase described by the spin wave approximation to a simple high-temperature disordered phase.

These excitations were identified by Kosterlitz and Thouless to be vortices which at low temperatures occur in tightly bound pairs that unbind at a critical temperature. To see that such a mechanism could give rise to a transition, consider first an isolated vortex which is displayed in Figure 6.9. We label the orientation of a spin at position r, θ by $\phi(r,\theta)$. In the continuum approximation $\phi(r,\theta) = n\theta$, where n is the strength of the vortex. Thus

$$\oint d\mathbf{l}\cdot\nabla\phi = 2\pi n \quad \text{and} \quad \nabla\phi = \frac{n}{r}\hat{\theta}$$

with $\hat{\theta}$ a unit vector in the direction of increasing θ. The energy of an isolated vortex is easily calculated:

$$E = \frac{JS^2}{2}\int d^2\mathbf{r}\nabla\phi(\mathbf{r})\cdot\nabla\phi(\mathbf{r}) = \pi JS^2n^2\int_a^L dr\frac{1}{r} = \pi JS^2n^2\ln\frac{L}{a} \tag{6.123}$$

where L is the linear dimension of the system and a the lattice constant. Thus the energy of an isolated vortex is infinite in the thermodynamic limit. The entropy associated with a single vortex is given by $S = k_B\ln(L/a)^2$ and the change in free energy due to formation of a vortex is

$$\Delta G = \left(\pi JS^2n^2 - 2k_BT\right)\ln\frac{L}{a} \, . \tag{6.124}$$

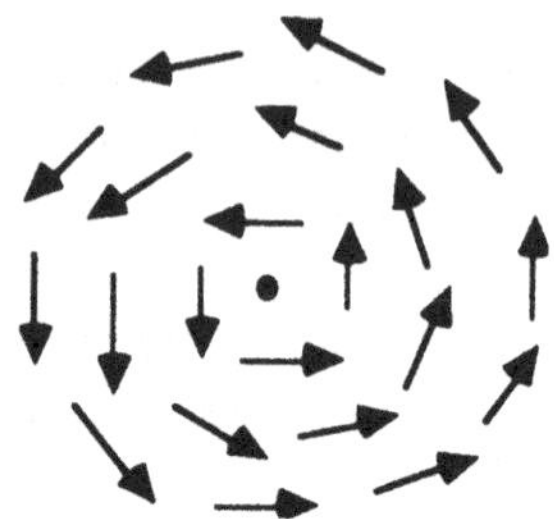

Figure 6.9: Schematic picture of vortex of unit strength in planar magnet.

This quantity is positive for $k_BT < \pi JS^2/2$ and isolated vortices will therefore not occur for temperatures lower than this value.

Consider next a pair of vortices separated by a distance r. We note that the ground-state configuration of the spin system is given by

$$\delta E(\{\phi\}) = \delta \int d^2r \frac{JS^2}{2} \left[\nabla\phi(\mathbf{r})\right]^2 = 0 \tag{6.125}$$

which yields

$$\nabla^2\phi(\mathbf{r}) = 0 \; . \tag{6.126}$$

This Laplace equation is supplemented by the condition

$$\oint_C \nabla\phi \cdot d\mathbf{l} = 2\pi n_1 \tag{6.127}$$

for a contour C which only encloses vortex 1, of strength n_1, and a similar condition for contours enclosing vortex 2 only.

Equation (6.127) is reminiscent of Ampère's law for the magnetic induction due to a current distribution:

$$\oint_C \mathbf{B} \cdot d\mathbf{l} = \mu I$$

where I is the current enclosed by the contour C. In this analogy we can identify the equivalent magnetic field with $\nabla\phi$. In SI units the correspondence is $I_1 = 2\pi JS^2 n_1$ and $\mu = 1/JS^2$, where μ is the "permittivity" and I_1 the equivalent current associated with a vortex of strength n_1. Using this analogy, one then obtains, for the interaction energy of a pair of vortices of strength n_1 and n_2,

$$E_{pair}(\mathbf{r_1}, \mathbf{r_2}) = -2\pi JS^2 n_1 n_2 \ln\left|\frac{\mathbf{r_1} - \mathbf{r_2}}{a}\right| \tag{6.128}$$

where we have set the energy of nearest-neighbor vortices equal to zero. This logarithmic dependence on separation of the interaction energy also occurs in the case of "two-dimensional" charged particles, or more accurately, lines of charge. Thus we also have an analogy between excitations of the planar magnet and the two-dimensional Coulomb gas. We note that (6.128) implies that the state of minimum energy for oppositely "charged" vortices is the tightly bound configuration in which they are nearest neighbors.

Since the size of the system does not appear in the expression (6.128), but does appear in the expression for the entropy, we see that the low-temperature state of the system will consist of an equilibrium density of bound vortex pairs. This equilibrium density is determined by pair-pair interactions which we have not considered. At higher temperatures, the vortex unbinding mechanism (6.124) will then destroy this condensed phase.

We also note that the arguments above yield no information on the nature of the vortex unbinding transition. They merely show that two qualitatively different states of the system can exist in different temperature ranges. It is possible to determine further properties of the transition using renormalization group methods. We refer the reader to the article by Jose *et al.* [144].

The foregoing discussion applies directly to the case of superfluid films. In a superfluid (see also Section 11.2) the quantity $\nabla\phi$ is proportional to the velocity of the film relative to the substrate on which it is adsorbed and the vortices quite literally represent circulation of material.

The case of melting of two-dimensional crystals is considerably more complicated. We first note that it is important, in physisorbed materials, to distinguish between lattice gases and floating monolayers. An example of a lattice gas (helium adsorbed onto the basal plane of graphite) is discussed in Section 7.5.2. In such a system the adsorbed layer does not have a continuous translational symmetry. To a first approximation, the atoms occupy discrete sites on the substrate and thermal excitation results in hopping of atoms between eligible sites. Such lattice gases have conventional long-range order below the critical point.

A floating monolayer, on the other hand, is not strongly perturbed by the periodic component of the interaction between adsorbate and substrate. The ground-state configuration is, in an ideal case, determined entirely by the interparticle interaction of the adsorbate and the entire layer can be displaced uniformly by any amount parallel to the substrate surface without cost in energy. For such floating monolayers one can show [202] that the two-dimensional crystal does not have long-range positional order at any nonzero temperature. The experimental manifestation of this result is that in a diffraction experiment one would not observe true Bragg peaks (δ-function peaks centered on the reciprocal lattice vectors of the two-dimensional crystal) but rather peaks with an intensity that falls off as a power law in the vicinity of the reciprocal lattice vectors. In terms of real space correlations, the analog of the spin-spin correlation function (6.115) is the function

$$g_G(\mathbf{R}) = \left\langle e^{i\mathbf{G}\cdot[\mathbf{u}(\mathbf{R})-\mathbf{u}(\mathbf{0})]}\right\rangle \tag{6.129}$$

where the positions of the atoms are given by $\mathbf{r} = \mathbf{R} + \mathbf{u}(\mathbf{R})$ and where $\mathbf{G}$ is a reciprocal lattice vector of the ground-state crystal. One can show (see *e.g.*, Nelson, [213]) that, in the harmonic approximation,

$$g_G(\mathbf{R}) \sim |\mathbf{R}|^{-\eta_G(T)} \tag{6.130}$$

where the "critical" exponent $\eta_G(T)$ depends linearly on the temperature and quadratically on the magnitude of the reciprocal lattice vector $\mathbf{G}$. Thus we have a similar situation as in the spin wave theory of the planar magnet (6.121).

There is, however, an added feature to the crystallization problem in two dimensions. It is possible for long-range *orientational* order to exist at finite temperature. We consider the case of a triangular lattice ground state and let

$$g_\theta(\mathbf{r}) = \left\langle e^{6i[\theta(\mathbf{r})-\theta(0)]}\right\rangle \tag{6.131}$$

where $\theta(\mathbf{r})$ is the angle of a nearest-neighbor bond between two atoms, one of which is at position $\mathbf{r}$. The function $g_\theta(\mathbf{r})$ is then a measure of orientational order. One can show that the long-wavelength phonons which destroy long-range positional order do not destroy orientational long-range order and that

$$\lim_{r\to\infty} g_\theta(\mathbf{r}) = \text{const.}$$

at low enough temperatures.

Although the case of melting of a two-dimensional crystal is more complicated than the disordering of a planar magnet, an analogous picture of the process can be constructed. The topological defects analogous to the vortices in the magnet are dislocations. These interact via a logarithmic potential (6.128) as do the vortices but the corresponding "charges" are the Burger's vectors of the dislocations and thus vector rather than scalar quantities. A dislocation-mediated theory of melting has been constructed by Halperin and Nelson and by Young (see [213] for a review and for the original references).

We now briefly discuss the experimental (computer and laboratory) situation. In the case of ^{4}He films the Kosterlitz–Thouless theory (with subsequent elaborations) predicts a universal discontinuity in the superfluid density

$\rho_s(T_c)/T_c$. The critical temperature can be varied by changing the thickness of the film and such experiments have been carried out, for example, by Bishop and Reppy [43] and Rudnick [260] and the results are consistent with the Kosterlitz–Thouless predictions. Computer experiments on planar magnets (Tobochnik and Chester, [303]; see also Saito and Müller–Krumbhaar [264] for an extensive review) are also consistent with both the low-temperature predictions of spin wave theory and the vortex unbinding mechanism.

The situation as far as two-dimensional melting is concerned is more controversial. Melting behavior consistent with Kosterlitz–Thouless theory has been observed in colloidal suspensions (Murray and Van Winkle [211]) while in other cases the melting transition seems to be a conventional first-order transition. Laboratory experiments are complicated by substrate effects and by long relaxation times; the latter also plague computer experiments. It seems clear, however, that the nature of the transition depends on microscopic parameters of the system in question and is therefore not a universal feature of two-dimensional melting. Rather than report on the results of specific calculations or experiments on these fascinating systems, we refer the reader to the reviews by Abraham [1], Saito and Müller–Krumbhaar [264], and Nelson [213].

6.7 Problems

6.1. *Approximate Solution of the Ising Model on the Square Lattice.*
Consider the modified transfer matrix

$$\mathbf{V} = (2\sinh 2K)^{M/2} \exp\left\{K^* \sum_{j=1}^{M} \sigma_{jX} + K \sum_{j=1}^{M} \sigma_{jZ}\sigma_{j+1,Z}\right\}$$

obtained from (6.15) if one ignores the fact that the operators $\mathbf{V}_1$ and $\mathbf{V}_2$ do not commute. The largest eigenvalue of this transfer matrix can be found by the methods of Sections 6.1.2 and 6.1.3 with some reduction in complexity. Calculate the free energy and show that the specific heat diverges logarithmically at the critical point.

6.2. *Internal Energy and Specific Heat of the Two-Dimensional Ising Model.*

(a) Supply the missing steps between (6.52) and (6.53). Hint: The incomplete elliptic integrals $F(\phi, q)$ and $E(\phi, q)$ are defined as follows:

$$F(\phi, q) = \int_0^\phi dx \frac{1}{\sqrt{1 - q^2 \sin^2 x}}$$
$$E(\phi, q) = \int_0^\phi dx \sqrt{1 - q^2 \sin^2 x} \ .$$

First show that

$$\frac{\partial F(\phi, q)}{\partial q} = \frac{1}{1 - q^2} \left[\frac{E(\phi, q) - (1 - q^2) F(\phi, q)}{q} - \frac{q \sin \phi \cos \phi}{\sqrt{1 - q^2 \sin^2 \phi}} \right]$$

and hence

$$\frac{dK_1(q)}{dq} = \frac{E_1(q)}{q(1 - q^2)} - \frac{K_1(q)}{q} \ .$$

(**b**) Show that (6.53) implies the logarithmic singularity of the specific heat (6.54).

6.3 *High-Temperature Series for the Susceptibility of the Heisenberg Model.* Consider the spin-$\frac{1}{2}$ Heisenberg model on the simple cubic lattice:

$$H = -J \sum_{\langle ij \rangle} \mathbf{S}_i \cdot \mathbf{S}_j - h \sum_i S_{iz} \ .$$

(**a**) Construct the high-temperature series for the susceptibility per spin

$$\chi(0, T) = \frac{\partial}{\partial h} \langle S_{iz} \rangle = \beta \frac{\mathrm{Tr}\, S_{iz} S_{jz} \exp\left\{\beta J \sum_{\langle nm \rangle} \mathbf{S}_m \cdot \mathbf{S}_n\right\}}{\mathrm{Tr} \exp\left\{\beta J \sum_{\langle nm \rangle} \mathbf{S}_m \cdot \mathbf{S}_n\right\}}$$

up to, and including, the term of order J^2.

(**b**) Analyze this short series by writing

$$\chi(0, T) = \frac{1}{T} \left(\frac{a_1 - a_2/T}{1 - a_3/T} \right)$$

which is a simple Padé approximant (see [106]). Find T_c.

(**c**) Compare with the critical temperature obtained from a two-term expansion and with the best estimate $k_B T_c / J\hbar^2 = 0.84$ obtained from longer series.

6.4. *Analysis of High-Temperature Series.*

Analyze the first 10 terms of the high-temperature series for the zero-field susceptibility of the Ising ferromagnet on the triangular lattice and obtain estimates of T_c and γ. The coefficients of the series can be found in [78].

6.5. *Scaling.*

For $T = T_c$ and h small we expect that the correlation length ξ will have the scaling form

$$\xi(h,0) \sim |h|^{-\nu_H}$$

and that the pair correlation function will have the approximate form

$$g(r,h,0) \sim \frac{e^{-r/\xi}}{r^{d-2+\eta_H}} \; .$$

(a) Use Landau–Ginzburg theory (Chapter 3) to derive the classical values of the critical exponents ν_H and η_H.

(b) We also expect that the susceptibility will diverge as $|h| \to 0$ on the critical isotherm with an exponent γ_H. Express γ_H in terms of ν_H and η_H.

(c) Using the scaling form (6.86), show that $\gamma_H \delta = \gamma/\beta$.

6.6. *Correlation Function in the Spin Wave Approximation.*

(a) Complete the calculation of the correlation function $g(r)$ defined in (6.115) to obtain equation (6.118).

(b) Show that in three dimensions the function $g(r)$ approaches a constant as r approaches infinity.

Chapter 7

Critical Phenomena II: The Renormalization Group

In this chapter we introduce the renormalization group approach to critical phenomena. In contrast to Chapter 6, we do not proceed historically, but begin in Section 7.1 with a renormalization treatment of a simple exactly solvable model, the familiar Ising chain (Section 3.6). We proceed in Section 7.2 to discuss properties of fixed points, the relation to scaling, and the notion of universality. Next, we show how these concepts work on an exactly solvable model that does exhibit a phase transition: the Ising model on a diamond fractal. Most applications of the renormalization group methods are of necessity appproximate and Section 7.4 continues with the cumulant approximation in the context of the position space renormalization group, and the critical exponents are calculated approximately, but in a nontrivial fashion, for a two-dimensional model. In Section 7.5 we describe the application of other position space renormalization methods to phase transitions. The "phenomenological renormalization group", based on the finite-size scaling ideas introduced in Section 6.4, is the subject of Section 7.6. We conclude this chapter with Section 7.7 in which the momentum space approach and the ϵ expansion are developed.

7.1 The Ising Chain Revisited

Consider again the Ising model for a one-dimensional chain with periodic boundary conditions (Section 3.6). The Hamiltonian is

$$\tilde{H} = -J\sum_{i=1}^{N}\sigma_i\sigma_{i+1} - \tilde{h}\sum_{i=1}^{N}\sigma_i \tag{7.1}$$

with $\sigma_i = \pm 1$ and $\sigma_{N+1} = \sigma_1$. We define the dimensionless Hamiltonian

$$H = -\beta\tilde{H} = K\sum_{i=1}^{N}\sigma_i\sigma_{i+1} + h\sum_{i=1}^{N}\sigma_i \tag{7.2}$$

with $K = \beta J$, $h = \beta\tilde{h}$. The partition function is given by

$$Z_c = \operatorname*{Tr}_{\{\sigma_i\}} e^H = \sum_{\{\sigma_i\}=\pm 1} \exp\left\{\sum_{i=1}^{N}[K\sigma_i\sigma_{i+1} + h\sigma_i)]\right\} . \tag{7.3}$$

We now carry out the sum over the degrees of freedom in two steps,

$$\sum_{\{\sigma_i\}=\pm 1} e^H = \sum_{\sigma_2=\pm 1}\sum_{\sigma_4=\pm 1}\cdots\sum_{\sigma_N=\pm 1}\left[\sum_{\sigma_1=\pm 1}\sum_{\sigma_3=\pm 1}\cdots\sum_{\sigma_{N-1}=\pm 1} e^H\right] . \tag{7.4}$$

The sums inside the brackets are easy to carry out. Each spin with an odd index is connected by the nearest-neighbor interaction only to spins with an even index. Thus the terms in H which involve σ_1, are simply

$$K\sigma_1(\sigma_N + \sigma_2) + h\sigma_1$$

and carrying out the trace over σ_1, we find that

$$\sum_{\sigma_1=\pm 1} e^{K\sigma_1(\sigma_N+\sigma_2)+h\sigma_1} = 2\cosh\left[K(\sigma_N + \sigma_2) + h\right] .$$

Using the property $\sigma^{2n} = 1$, $\sigma^{2n+1} = \sigma$ of Ising spins we write

$$\begin{aligned} &2e^{h(\sigma_N+\sigma_2)/2}\cosh\left[K(\sigma_N + \sigma_2) + h\right] \\ &\quad = \exp\left\{2g + K'\sigma_N\sigma_2 + \frac{1}{2}h'(\sigma_N + \sigma_2)\right\} \end{aligned} \tag{7.5}$$

where

$$K' = \frac{1}{4}\ln\frac{\cosh(2K+h)\cosh(2K-h)}{\cosh^2 h} \tag{7.6}$$

$$h' = h + \frac{1}{2}\ln\frac{\cosh(2K+h)}{\cosh(2K-h)} \tag{7.7}$$

and

$$g = \frac{1}{8}\ln\left[16\cosh(2K+h)\cosh(2K-h)\cosh^2 h\right] . \tag{7.8}$$

All other sums inside the brackets in (7.4) yield identical results and we have

$$\sum_{\sigma_1,\sigma_3\ldots\sigma_{N-1}} e^H = \exp\left\{Ng(K,h) + K'\sum_i \sigma_{2i}\sigma_{2i+2} + h'\sum_i \sigma_{2i}\right\} \tag{7.9}$$

where the sum over spins in the exponential is over the remaining even-numbered sites. We notice that the sum over the even spins constitutes a problem of exactly the same type as the calculation of the original partition function. The remaining spins of the thinned-out chain interact with their nearest neighbors through a "renormalized" coupling constant K' and are subject to a renormalized magnetic field h'. We therefore have the equation

$$Z_c(N,K,h) = e^{Ng(K,h)} Z_c\left(\frac{N}{2}, K', h'\right) . \tag{7.10}$$

The situation is shown schematically as

□ K □ K □ K □ K □ K □ K □ K □ K □ K □

K' □ K' □ K' □ K' □ K' □

K'' □ K'' □ K''

...

It is clear that the process may be continued indefinitely. Notice that from (7.10) we may obtain a formula for the free energy

$$\begin{aligned} -\beta G(N,K,h) &= \ln Z_c(N,K,h) \\ &= Ng(K,h) + \ln Z_c(\tfrac{1}{2}N, K', h') \end{aligned} \tag{7.11}$$

or

$$\begin{aligned} -\frac{\beta G}{N} &\equiv f(K,h) = g(K,h) + \frac{1}{2}g(K',h') + \frac{1}{4}g(K'',h'') + \cdots \\ &= \sum_{j=0}^{\infty}\left(\frac{1}{2}\right)^j g(K_j,h_j) . \end{aligned} \tag{7.12}$$

The important feature of this equation is that the same function g appears at each stage of the iteration since the renormalized Hamiltonian always has the same form. To discuss the convergence of the sum (7.12), we must understand the "flow" of the coupling constants K, h. Let us first consider the case $h = 0$. Then $h_j = 0$ for all j. From (7.6) we obtain

$$K' = \frac{1}{2} \ln \cosh 2K \leq K . \tag{7.13}$$

The equality $K' = K$ holds at the two special points $K = 0$ and $K = \infty$. These are called the *fixed point* of the renormalization transformation. For any finite K the successive thinning out of degrees of freedom produces a Hamiltonian in which the remaining spins are more weakly coupled. The flow in *coupling constant space* is thus toward a Hamiltonian which consists of noninteracting degrees of freedom. This fixed point, which can also be thought of as an infinite temperature fixed point, is *stable*, as long as h is kept equal to zero. Conversely, the other fixed point at $K = \infty$, or $T = 0$, is unstable — a dimensionless Hamiltonian that deviates from $K = \infty$ flows toward $K = 0$ under renormalization. It is now clear that the sum (7.12) will converge for any finite coupling constant K. From (7.8) we have

$$g(K, 0) = \frac{1}{2} \ln 2 + \frac{1}{4} \ln(\cosh 2K) . \tag{7.14}$$

As K becomes smaller, the second term in (7.14) approaches zero, and if we neglect the contribution of this term to the sum in (7.12) for $j > n$, we obtain

$$f(K, 0) = \sum_{j=0}^{n} g(K_j, 0) \left(\frac{1}{2}\right)^j + 2^{-(n+1)} \ln 2 . \tag{7.15}$$

The last term is simply the entropy/k_B per particle of the remaining $N/2^{n+1}$ spins, which are effectively noninteracting.

It is interesting to examine the flow of coupling constants for nonzero h. From (7.7) we see that

$$\frac{\partial h'}{\partial h} > 1$$

for all finite K. Thus a small magnetic field becomes larger under iteration, and since K becomes smaller, the flow is toward the line $K = 0$. This flow is shown schematically in Figure 7.1 for a number of different starting points in the $K - h$ plane.

In Section 3.6 we calculated the correlation length and found that $\xi = 0$ at $K = 0$ and $\xi = \infty$ at $K = \infty$. The flow of the coupling constants can be understood in terms of Kadanoff's scaling picture (Section 6.3.3) and the

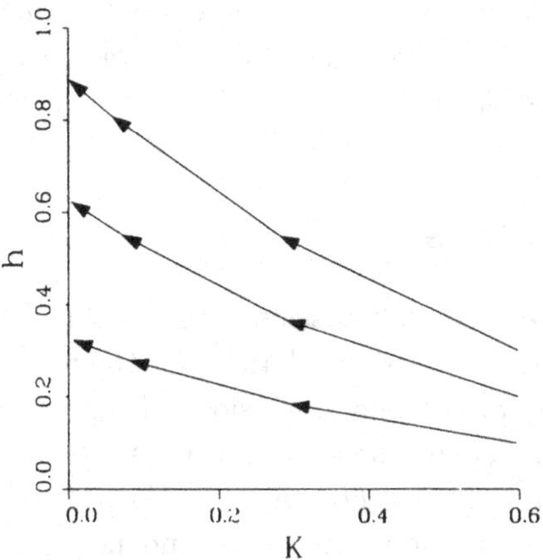

Figure 7.1: Renormalization flow for the one-dimensional Ising model.

behavior of the correlation length. Crudely speaking, we have replaced a pair of spins by a block spin. The block spins are separated by twice the site spin separation. In Kadanoff's scaling picture, one expects that the correlation length for block spins will be smaller than the correlation length for site spins, unless the system is critical ($\xi = \infty$), or noninteracting ($\xi = 0$). The two fixed points can therefore be understood as Hamiltonians for which the correlation lengths are invariant under rescaling. One Hamiltonian ($K = 0$) is trivial, one ($K = \infty$) is critical in this picture. In a higher-dimensional system, with a finite-temperature phase transition, we therefore expect a minimum of three fixed points. Two of these will be trivial (infinite and zero temperature) and one will be the critical fixed point.

Before we go on to discuss the properties of fixed points in more detail, we summarize the important results of this section. First, we have developed a new way of evaluating the partition function by successively thinning out degrees of freedom. Because the form of the Hamiltonian remained invariant under this "scale transformation", we were able to evaluate the free energy by means of a simple iterative scheme. The Kadanoff picture of decreasing correlation lengths emerges in this scheme as a flow toward Hamiltonians with successively smaller coupling constants. This procedure of thinning out degrees of freedom has been termed renormalization group by Wilson because two such operations performed sequentially have the property

$$R_b(R_b(\{K\})) = R_{b^2}(K)$$

where $R_b(K) = K'$ describes the effect on coupling constants of replacing b^d spins by one block spin. There is no inverse operation R_b^{-1} and therefore the word "group" is a misnomer.

7.2 Fixed Points

We now turn to a more general discussion of the renormalization group method. In Section 7.1 we derived recursion relations for the pair of dimensionless coupling constants (K, h) of the one-dimensional Ising model. We now wish to consider a system that is specified, on a d-dimensional lattice, in terms of a set of coupling constants $\{K\} = (K_1, K_2, \ldots, K_n)$. Here K_1 might correspond to nearest neighbor interactions, K_2 to second neighbor, K_3 to a magnetic field, and so forth. We suppose that this set of coupling constants is complete, in the sense that a renormalization transformation which replaces b^d degrees of freedom by one degree of freedom, results in a Hamiltonian with exactly the same type of interactions between the remaining dynamical variables. We describe the system in terms of a dimensionless Hamiltonian

$$H = -\beta\tilde{H} = \sum_{\alpha=1}^{n} K_\alpha \psi_\alpha(\sigma_i) \tag{7.16}$$

where, for example,

$$\psi_1 = \sum_{\langle ij \rangle} \sigma_i \sigma_j \qquad \psi_2 = \sum_{\{ij\}} \sigma_i \sigma_j \qquad \psi_3 = \sum_i \sigma_i \cdots$$

and where the notation $\langle ij \rangle$ indicates nearest neighbor pairs and $\{ij\}$ denotes second neighbors. A renormalization transformation will then produce a new Hamiltonian

$$H' = \sum_{\alpha=1}^{n} K'_\alpha \psi_\alpha(\sigma_I) + Ng(\{K\}) \tag{7.17}$$

where the remaining degrees of freedom $\{\sigma_I\}$ have the same algebraic properties as the original ones and the functional form of the ψ_α's is unchanged by the transformation. In (7.17) a term $g(\{K\})$ has been included because, as we saw in (7.8), there will in general be a spin-independent term as a result of the partial trace. Assuming that the thinning-out operation can be carried out, we have the relations

$$K'_\alpha = R_\alpha(K_1, K_2, \ldots, K_n) \tag{7.18}$$

and

$$\operatorname*{Tr}_{\{\sigma_i\}} e^{H(\{K\})} = \operatorname*{Tr}_{\{\sigma_I\}} e^{H'(\{K'\})} \tag{7.19}$$

or with

$$\begin{aligned} \operatorname{Tr} e^{H} &= e^{Nf(\{K\})} \\ \operatorname{Tr} e^{H'} &= e^{Nf(\{K'\})/b^d + Ng(\{K\})} \end{aligned} \tag{7.20}$$

we obtain

$$f(\{K\}) = g(\{K\}) + b^{-d} f(\{K'\}) \ . \tag{7.21}$$

Equation (7.11) is a special case of this formula. We leave the recursion relation for the free energy aside for the moment and concentrate on equation (7.18) for the coupling constants. We have already seen in Section 7.1 that the fixed points of this recursion relation correspond to either noninteracting or critical Hamiltonians. However, as we shall presently see, there will in general be critical points which are *not* fixed points. Let us imagine a two-dimensional space of coupling constants K_1, K_2 with a critical point at K_{1c}, K_{2c}. We note that this point is in general a point on a line of critical points. To see this, imagine a number of different systems with different ratios J_2/J_1 of second-neighbor to nearest-neighbor interactions. The critical temperature T_c will depend on this ratio. Thus as J_2/J_1, varies, the point

$$(K_{1c}, K_{2c}) = \left(\frac{J_1}{k_B T_c}, \frac{J_2}{k_B T_c} \right)$$

describes a curve in the K_1, K_2 plane. Each point on the curve corresponds to the critical point of a particular model in the family of Hamiltonians. This situation is depicted schematically in Figure 7.2.

The dotted line in Figure 7.2 is the path that a particular system follows in coupling constant space as the temperature is lowered from $T = \infty$ ($K_1 = K_2 = 0$) to $T = 0$ with $K_2/K_1 = J_2/J_1$ held fixed. We now attempt to relate the properties of the system, as it describes this path, to the flow of the coupling constants under a renormalization transformation. This flow has a number of simple properties. First, it is clear that the flow cannot approach the line of critical points. This is because the correlation length ξ is infinite on this curve and is finite everywhere else. We have argued that as the degrees of freedom are thinned out, the correlation length relative to the new spacing can only decrease. The states on the right of the line of critical points correspond to a low-temperature phase and are ordered. The states on the high-temperature

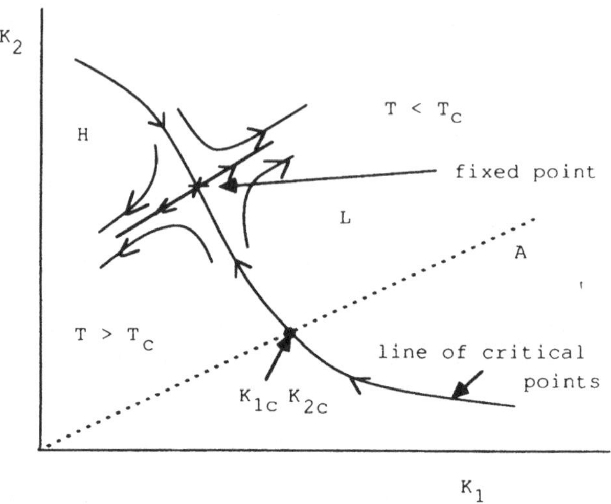

Figure 7.2: Lines with arrows indicate the direction of flow given by the recursion relation. The dashed line indicates possible states of a system with a given value of $K_2/K_1 = J_2/J_1$.

side are disordered. This aspect cannot be changed by the thinning out of the degrees of freedom, and the flow therefore cannot cross the critical line. We conclude that the flow for $T > T_c$ and $T < T_c$ must be as shown in Figure 7.2. In region H $(T > T_c)$ the flow will be toward a $T = \infty$ noninteracting fixed point $K_1 = K_2 = 0$. In region L $(T < T_c)$ the flow will be toward a zero temperature (ground state) fixed point. Conversely, the flow from points K_{1c}, K_{2c} on the critical line must remain on this line since $\xi = \infty$. It is possible that all points on the critical line are stationary (*i.e.*, are fixed points of the renormalization group transformation), but this turns out to be an exceptional case. Generically, one finds a finite number of isolated fixed points. Let us therefore assume that K_1^*, K_2^* is a fixed point on the critical line,

$$\begin{aligned} K_1^* &= R_1(K_1^*, K_2^*) \\ K_2^* &= R_2(K_1^*, K_2^*) \end{aligned} \tag{7.22}$$

and that the flow along the critical line approaches this point. Consider now the flow near K_1^*, K_2^*. Let $\delta K_1 = K_1 - K_1^*$, $\delta K_2 = K_2 - K_2^*$. To first order,

$$K_1' = R_1(K_1^* + \delta K_1, K_2^* + \delta K_2)$$

$$= \quad K_1^* + \delta K_1 \left.\frac{\partial R_1}{\partial K_1}\right|_{K_1^*,K_2^*} + \delta K_2 \left.\frac{\partial R_1}{\partial K_2}\right|_{K_1^*,K_2^*} \tag{7.23}$$

with a similar expression for K_2'. We write these expressions in matrix form,

$$\begin{aligned} \delta K_1' &= M_{11}\delta K_1 + M_{12}\delta K_2 \\ \delta K_2' &= M_{21}\delta K_1 + M_{22}\delta K_2 \end{aligned} \tag{7.24}$$

and find an appropriate coordinate system for describing the flow by solving the *left* eigenvalue problem

$$\sum_{i=1,2} \phi_{\alpha i} M_{ij} = \lambda_\alpha \phi_{\alpha j} = b^{y_\alpha} \phi_{\alpha j} \; . \tag{7.25}$$

In the last step of (7.25) we have replaced λ_α by b^{y_α} in view of the group property $\lambda_\alpha(b)\lambda_\alpha(b) = \lambda_\alpha(b^2)$. Therefore,

$$\lambda_\alpha = b^{y_\alpha} \tag{7.26}$$

defines the quantity y_α which in turn is independent of b. Consider now the new variables ($\alpha = 1, 2$)

$$U_\alpha = \delta K_1 \phi_{\alpha 1} + \delta K_2 \phi_{\alpha 2} \; . \tag{7.27}$$

We apply the linearized renormalization transformation (7.24) and use the fact that ϕ_α is a *left* eigenvector to obtain

$$\begin{aligned} U_\alpha' &= \delta K_1' \phi_{\alpha 1} + \delta K_2' \phi_{\alpha 2} \\ &= \lambda_\alpha U_\alpha = b^{y_\alpha} U_\alpha \; . \end{aligned} \tag{7.28}$$

Geometrically, U_α corresponds to a projection of the deviation (δK_1, δK_2) from the fixed point on the basis vector ϕ_α. The U_α's are called scaling fields for reasons that will become clear.

We can make one further statement regarding the flow near the fixed point. One of the exponents, say y_2, must be negative, the other, y_1 positive. The reason for this is the assumption that flows that originate on the critical line must tend toward the fixed point. Thus one of the basis vectors, ϕ_2, must be tangential to the critical surface at the fixed point. The other vector must point out of the critical surface. Since the matrix $\mathbf{M}$ is generally not symmetric, ϕ_1, and ϕ_2 are not necessarily orthogonal to each other, but this has no bearing on the discussion that follows. The exceptional case, $y_2 = 0$, corresponds to a line of fixed points rather than the case of an isolated fixed point.

Let us now return to the recursion relation (7.21) for the free energy. Since the first term $g(\{K\})$ arises from the removal of short distance fluctuations, which play no role in the phase transition, $g(\{K\})$ is expected to be an analytic function of the coupling constants. Therefore, the singular part of the free energy obeys the relation

$$f_s(\{K\}) = b^{-d} f_s(\{K'\}) \ . \tag{7.29}$$

We now suppose that the point $\{K\}$ is close enough to $\{K^*\}$ that we may use the linearized recursion relations (7.26). Re-expressing K, K' in terms of U_1, U_2 we have

$$f_s(U_1, U_2) = b^{-d} f_s(b^{y_1} U_1, b^{y_2} U_2) \tag{7.30}$$

that is, a scaling form of the free energy. We now assume that a change in temperature at constant field corresponds to a change in U_1 at constant U_2. The connection with the critical exponents defined in Chapter 6 can now be made by realizing that if $T \neq T_c$, $U_1 \neq 0$. Conversely, if $T = T_c$, $U_1 = 0$ as the system point must lie on the critical line. Defining

$$t = U_1 = \frac{T - T_c}{T_c}$$

we obtain

$$f_s(t, U_2) = b^{-d} f_s(b^{y_1} t, b^{y_2} U_2) \ . \tag{7.31}$$

This equation must be true for arbitrary b and we may therefore let $b = |t|^{-1/y_1}$ to obtain

$$f_s(t, U_2) = |t|^{d/y_1} \, f_s(t/|t|, |t|^{-y_2/y_1} U_2) \ . \tag{7.32}$$

This equation demonstrates two important features of critical points. First, the role of the fixed point is clarified: *the critical exponents are determined by the eigenvalues of the linearized recursion relations at the fixed point.* Since the specific heat is proportional to the second derivative of f with respect to t, we have $f_s \propto |t|^{2-\alpha}$, where α is the specific heat critical exponent. We thus have $d/y_1 = 2 - \alpha$ and $y_1 = \ln \lambda_1 / \ln b$. Next we see the concept of universality emerging from the theory. We argued above that $y_2 < 0$. As $t \to 0$ the term

$$|t|^{-y_2/y_1} \, U_2 \to 0$$

and the asymptotic behavior of the free energy is independent of U_2. In other words, *all systems whose Hamiltonians flow under renormalization to the same critical fixed point have the same critical exponents.* These are the most important qualitative results that follow from the renormalization group approach.

To obtain a complete description, we now generalize our analysis to higher-dimensional spaces. The fixed point in the n-dimensional space of coupling constants is given by

$$K_j^* = R_j(K_1^*, K_2^*, \ldots, K_n^*) \ .$$

The matrix $\mathbf{M}$ becomes an $n \times n$ matrix with

$$\delta K_j' = \sum_l M_{jl} \delta K_l \ .$$

There are now n eigenvalues corresponding to the solution to the eigenvalue problem

$$\phi_\alpha \mathbf{M} = b^{y_\alpha} \phi_\alpha \tag{7.33}$$

and the generalization of (7.30) is

$$f_s(U_1, U_2, \ldots, U_n) = b^{-d} f_s(b^{y_1} U_1, b^{y_2} U_2, \ldots, b^{y_n} U_n) \ . \tag{7.34}$$

Ordinary critical points are characterized (Section 6.3) by two independent exponents and we therefore expect that two of the y's, say y_1 and y_2, are positive, the rest negative. In the generalized Ising model, we expect that $U_1 \propto (T - T_c)/T_c$, $U_2 \propto h$ and that all other scaling fields (corresponding, for example, to $J_2/J_1 - J_2^*/J_1^*$ with J_2 the second-neighbor interaction) will play no role in the asymptotic critical behavior.

It is, of course, possible that the critical surface contains several fixed points with different domains of attraction. The critical surface of the anisotropic Heisenberg model (see Section 6.5) is presumably such a critical surface. In this case, there is one fixed point, the Heisenberg fixed point, that is unstable with respect to a third scaling field proportional to the anisotropy parameter η (6.111). For any nonzero η the flow in the critical surface is either toward the Ising or XY fixed point. On the basis of the analysis above, we see that it is possible to observe the Heisenberg critical exponents only if this scaling field or anisotropy parameter η is exactly zero. In any other situation we would observe either XY or Ising exponents as long as we were sufficiently close to the critical point. This statement, which follows very simply from (7.34), is consistent with the series expansion results of Jasnow and Wortis [141].

We mention also that there is a conventional terminology for the different types of scaling fields. A scaling field U_α with exponent $y_\alpha > 0$ is called *relevant*; if $y_\alpha < 0$ it is called *irrelevant* and in the special case $y_\alpha = 0$ it is *marginal*.

Before closing this section we note that we have made certain tacit assumptions. First, we have implicitly assumed that the renormalization transformation is analytic everywhere in coupling constant space and in particular at the fixed point. Moreover, we have assumed that a system can be characterized in terms of a finite number of coupling constants. That this is the case is by no means obvious and in practice it is usually a specific approximation that guarantees that the number of coupling constants remains finite and that the recursion relations are analytic. We shall have more to say about these points later.

We have also assumed, in a rather cavalier manner, that the free energy $f(K)$ which is given by (7.21) has a singular piece that obeys the simple relation

$$f_s(\{K\}) = b^{-d} f_s(\{K'\}) \ .$$

We could use (7.21) to obtain an infinite series for the free energy in terms of the analytic function $g(\{K\})$

$$f(\{K\}) = \sum_{j=1}^{\infty} b^{-jd} g(\{K^{(j)}\})$$

and attempt to show how the singular piece emerges from this sum. This has been done by Niemeijer and van Leeuwen [217] and the interested reader is referred to this article.

Finally, we have assumed that the linear approximation is valid even far from the fixed point as long as the system is close to the critical surface. This assumption may also be removed by systematically including higher-order terms in (7.23) [217]. In the next two sections, we turn toward some concrete examples of renormalization group calculations for systems that, unlike the one-dimensional Ising model, do display a phase transition.

7.3 An Exactly Solvable Model: Ising Spins on a Diamond Fractal

We next describe one of the simplest models that exhibits phase change with non mean-field critical behavior. Our discussion is inspired by Saleur *et al.* [265], but our emphasis will be rather different. Consider a system of Ising spins that are located at the vertices of the *diamond fractal*. This geometric object is obtained by starting at magnification 0 with a single bond connecting two spins. At magnification 1 this bond is replaced by four bonds and two additional spins as shown below:

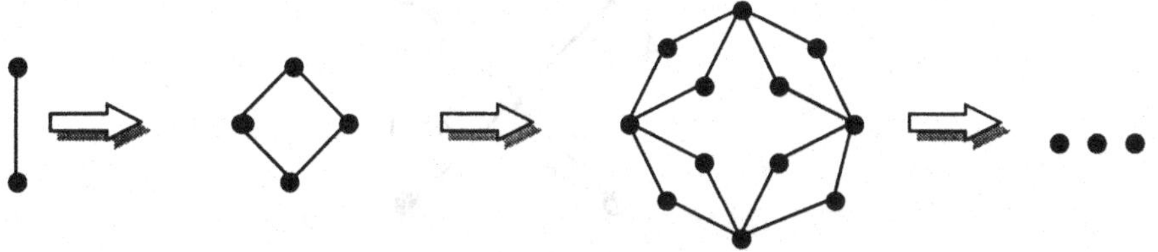

Figure 7.3: The diomond fractal.

The process is continued p times. A bit of reflection may convince the reader that at level p there will be 4^p bonds connecting $\frac{2}{3}(2+4^p)$ spins. Each of these spins may take on the value $\sigma = \pm 1$.

We label the bonds at the highest magnification $l = 1, 2, \cdots, N$ and let $i(l), j(l)$ be respectively the site at the top and bottom end of the bond. The Hamiltonian for the Ising model on this fractal is thus

$$H = -\sum_{l=1}^{N} \hat{J}\sigma_{i(l)}\sigma_{j(l)} - \hat{h}\sum_i \sigma_i \ .$$

To simplify our notation let

$$J = \beta\hat{J}; \ \ h = \beta\hat{h}; \ \ \Theta = e^J; \ \ \Gamma = e^h$$

$$b(l) = \sigma_{i(l)}\sigma_{j(l)}$$

where b_l can take on the values ± 1. The canonical partition function for the system is

$$Z = \sum_{\sigma_i = \pm 1} (\prod_{l=1}^{N} \Theta^{b(l)})(\prod_i \Gamma^{\sigma_i}) \ .$$

We now sum over the middle two spins of the diamonds at the highest magnification to get (see Figure 7.4)

$$Z = \sum_{\sigma_{i(4n)} = \pm 1} \left(\prod_{n=1}^{N/4} A_n(\sigma_{i(4n)}, \sigma_{i(4n+4)}) \right) \left(\prod_{i'} \Gamma^{\sigma_{i'}} \right)$$

where i' labels the remaining spins and

$$A_n = \sum_{\sigma_{i(4n+1)} = \pm 1, \sigma_{i(4n+2)} = \pm 1} \Theta^{b(4n+1)}\Theta^{b(4n+2)}\Theta^{b(4n+3)}\Theta^{b(4n+4)}\Gamma^{\sigma_{i(4n+1)}}\Gamma^{\sigma_{i(4n+2)}} \ .$$

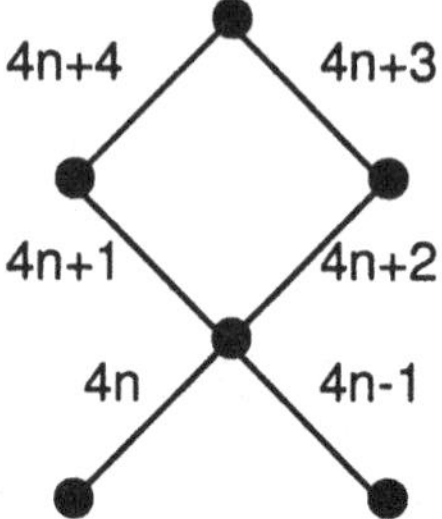

Figure 7.4: Bond labels in recursion relation.

We have

$$A(1,1) = (\Theta^2\Gamma + \frac{1}{\Theta^2\Gamma})^2$$
$$A(-1,1) = A(1,-1) = (\Gamma + \frac{1}{\Gamma})^2$$
$$A(-1,-1) = (\frac{\theta^2}{\Gamma} + \frac{\Gamma}{\theta^2})^2 \; .$$

We wish to rewrite the partition function as a sum over contributions from the remaining spins

$$Z = \sum_{\sigma_{i'}=\pm 1} \left(\prod_{n=1}^{N/4} C\Theta_{new}^{b(n)} \right) \left(\prod_{i'} \Gamma_{new}^{\sigma_{i'}} \right)$$

where C is a constant. We find

$$C\Theta_{new}\Gamma_{new} = A(1,1)\Gamma$$
$$\frac{C}{\Theta_{new}} = A(1,-1)$$
$$\frac{C\Theta_{new}}{\Gamma_{new}} = \frac{A(-1,-1)}{\Gamma}$$

with solution

$$C = [A(1,1)A(1,-1)A(-1,1)A(-1,-1)]^{1/4}$$

$$\Theta_{new} = \left[\frac{A(1,1)A(-1,-1)}{A(1,-1)^2} \right]^{1/4}$$

$$\Gamma_{new} = \left[\frac{A(1,1)}{A(-1,-1)} \right]^{1/2} \Gamma \; .$$

We define a new quantity g from $C = e^{4g}$ and let $J' = \ln \Theta_{new}$, and $h' = \ln \Gamma_{new}$, the renormalized coupling constants. Using

$$\Theta^2\Gamma + \frac{1}{\Theta^2\Gamma} = 2\cosh(2J+h)$$

$$(\frac{\theta^2}{\Gamma} + \frac{\Gamma}{\theta^2}) = 2\cosh(2J-h)$$

$$\Gamma + \frac{1}{\Gamma} = 2\cosh h$$

we find for the partition function the recursion relation

$$Z(N,J,H) = e^{Ng(J,h)} Z(N/4, J', h')$$

with

$$g = \frac{1}{8}\ln[16\cosh(2J+h)\cosh(2J-h)\cosh^2 h]$$

$$J' = \frac{1}{2}\ln\left[\frac{\cosh(2J+h)\cosh(2J-h)}{\cosh^2 h}\right] \equiv R_1(J,h) \tag{7.35}$$

$$h' = h + \ln\frac{\cosh(2J+h)}{\cosh(2J-h)} \equiv R_2(J,h) \ . \tag{7.36}$$

We can use these results to calculate the free energy of the system to arbitrary accuracy. Let f/β be the free energy per bond

$$-f(J,h) = \frac{-\beta G}{N} = g(J,h) - \frac{1}{4}f(J',h')$$

or

$$-f(J,h) = g(J,h) + \frac{1}{4}g(J',h') + \frac{1}{16}g(J'',h'') + \cdots$$

where

$$J'' = R_1(J',h')$$

$$h'' = R_2(J',h') \ .$$

Some renormalization flows using (7.35) and (7.36) are shown in Figure 7.5 for different starting values of J and h. Note that all recursion flows, not starting on the $h=0$ axis, end up on the $J=0$ axis.

To better understand the nature of the solution let us first consider the *fixed points* J^*, h^* of the renormalization transformation. These are the points for which the coupling constants are unchanged under the scale transformation

$$J^* = R_1(J^*, h^*)$$

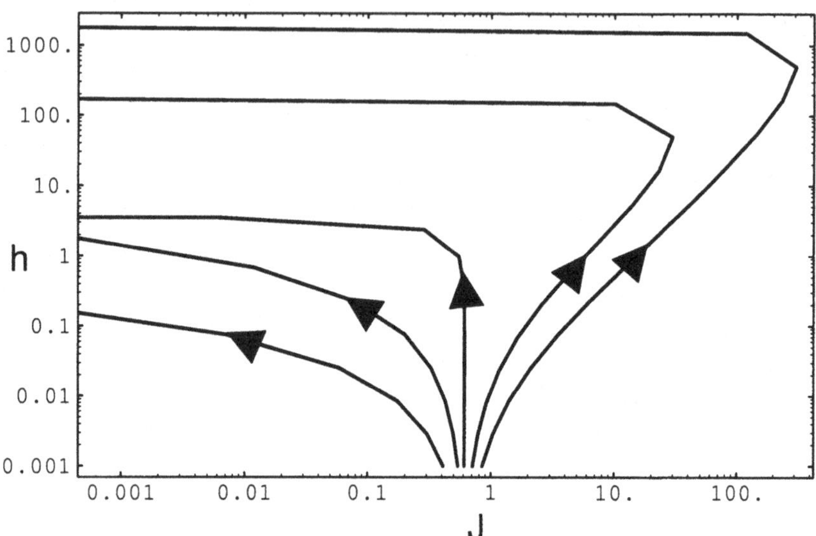

Figure 7.5: Renormalization flows for the Ising model on the diamond fractal. The starting values of the couplings constants are, respectively $h = 0.001$, $J = J^* - .2$, $J = J^* - 0.07$, $J = J^*$, $J = J^* + 0.1$, $J = J^* + 0.24$. Note the logarithmic scale.

$$h^* = R_2(J^*, h^*) \ .$$

First we note that if $J > 0$, the recursion relation for the field h will always make h grow in magnitude, while the magnitude remains unchanged if $J = 0$. If $J < 0$ initially (antiferromagnetic coupling) J will become positive and will not change sign in rescaling. The fixed points therefore have to lie on the positive $h = 0$ axis or on the $J = 0$ axis of the $J - h$ plane. If $h = 0$ the recursion relation simplifies to

$$J' = \ln[\cosh(2J)] \ .$$

By plotting the right hand side of this equation it is easy to see that there are the following fixed points:

- All states $J = 0$, h arbitrary.
- $J = \infty$, $h = 0$ corresponding to the $T = 0$ ordered state.
- $J = J^* = 0.60938 \cdots$ corresponding to $k_B T^* = 1.64102\hat{J}$.

As we shall see the last fixed point describes the phase transition. The $J = 0$ fixed points are marginally stable while the others are unstable. If $J = 0$, $T = \infty$, we have $g(0,0) = \frac{1}{2}\ln 2$. We find

$$f(0,0) = -\frac{1}{2}\ln 2[1 + \frac{1}{4} + \frac{1}{16}\cdots] = -\frac{2}{3}\ln 2 \ .$$

For large magnification p there will be $\frac{2}{3}$ spins/bond. At high temperatures we expect the average energy to be zero while the entropy per spin will be $k_B \ln 2$. We thus expect the free energy/spin to be $-k_B T \ln 2$ in agreement with the above formula.

Next let J be very large, *i.e.*, the temperature is near absolute zero. The recursion relation is now

$$J' \approx 2J - \ln 2 \approx 2J$$

and

$$g(J,h) \approx \frac{1}{8}\ln(4e^{4J}) \approx \frac{J}{2}$$

giving

$$f \approx -\frac{J}{2}[1 + \frac{2}{4} + \frac{4}{16}\cdots] = J$$

in agreement with the fact that at low temperatures the energy/bond will be approximately $\hat{J}$ and the entropy will be very small.

Near the $(J = J^*, h = 0)$ fixed point we assume that

$$t = J^* - J \propto T - T^*$$

and h are very small. We can therefore linearize the recursions near the critical point

$$t' = \left.\frac{\partial R_1}{\partial J}\right|_{J=J^*,h=0} t + \left.\frac{\partial R_1}{\partial h}\right|_{J=J^*,h=0} h$$

$$h' = -\left.\frac{\partial R_2}{\partial J}\right|_{J=J^*,h=0} t + \left.\frac{\partial R_2}{\partial h}\right|_{J=J^*,h=0} h \ .$$

If we differentiate R_1 and R_2 we find that

$$\left.\frac{\partial R_1}{\partial h}\right|_{J=J^*,h=0} = \left.\frac{\partial R_2}{\partial J}\right|_{J=J^*,h=0} = 0$$

and substituting the value of J^* at the critical point we find

$$\left.\frac{\partial R_1}{\partial J}\right|_{J=J^*,h=0} = 1.67857\cdots$$

$$\left.\frac{\partial R_2}{\partial h}\right|_{J=J^*,h=0} = 2.67857\cdots \ .$$

The function $g(J,h)$ is perfectly well behaved near the critical point. The singular part of the free energy must thus be of the scaling form

$$f_s(t,h) = \frac{1}{4} f_s(\frac{\partial R_1}{\partial J} t, \frac{\partial R_2}{\partial h} h) = \frac{1}{\lambda} f_s(\lambda^y t, \lambda^x h)$$

with $\lambda = 4$ and

$$y = \frac{\ln 1.67857}{\ln 4} = 0.373618$$
$$x = \frac{\ln 2.67857}{\ln 4} = 0.710732$$

from which we find the critical exponents

$$\alpha = \frac{2y-1}{y} = -0.676532$$
$$\beta = \frac{1-x}{y} = 0.774234$$
$$\gamma = \frac{2x-1}{y} = 1.12806$$
$$\delta = \frac{x}{1-x} = 2.45701 \ . \tag{7.37}$$

Let us now compute some of the thermodynamic quantities. As a check of the formalism consider first the case $J=0$, $h \neq 0$. In this case $J' = J = 0$ and $h' = h$. We find

$$g(0,h) = \frac{1}{8}\ln[16\cosh^4 h] = \frac{1}{2}\ln(2\cosh h)$$

and

$$f = -\frac{1}{2}\ln(2\cosh h)[1 + \frac{1}{4} + \frac{1}{16} + \cdots] = -\frac{2}{3}\ln(2\cosh h) \ .$$

The magnetization per bond is

$$m_b = -\frac{\partial G}{\partial \hat{h}} = -\frac{\partial f}{\partial h} = \frac{2}{3}\tanh h \ .$$

For large N there are 2/3 sites per bond giving for the magnetization/spin

$$m_s = \tanh h = \frac{e^{\beta\hat{h}} - e^{-\beta\hat{h}}}{e^{\beta\hat{h}} + e^{-\beta\hat{h}}}$$

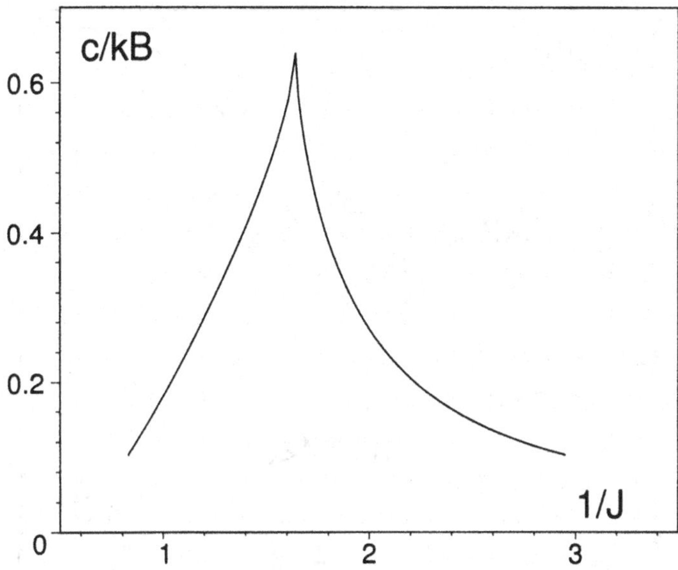

Figure 7.6: Specific heat obtained evaluating $f(J,h)$ by iterating the recursion formula until convergence, and then differentiating the result numerically using (7.38).

which is the expected result for a system of independent spins.

We next discuss the specific heat in zero magnetic field. For simplicity we let $\hat{J} = 1$ so that $J = \frac{1}{k_B T}$. We have for the entropy per bond.

$$s = -\frac{\partial G}{N\partial T} = -k_B f(J,0) + k_B J \frac{\partial f}{\partial J} \ .$$

The specific heat per bond then becomes

$$c = T\frac{\partial s}{\partial T} = T\frac{\partial J}{\partial T}\frac{\partial s}{\partial J} = -k_B J^2 \frac{\partial^2 f(J,0)}{\partial J^2}$$

$$\frac{\partial^2 f(J,0)}{\partial J^2} \approx \frac{f(J+\delta,0) + f(J-\delta,0) - 2f(J,0)}{\delta^2} \ . \tag{7.38}$$

We see that the specific heat is continuous, with a cusp at the critical point. The singular behavior of the derivative of the specific heat is in agreement with the fact that the specific heat exponent $-1 < \alpha < 0$.

The magnetic properties are more complicated. Above the critical temperature we can calculate the susceptibility by evaluating numerically the second

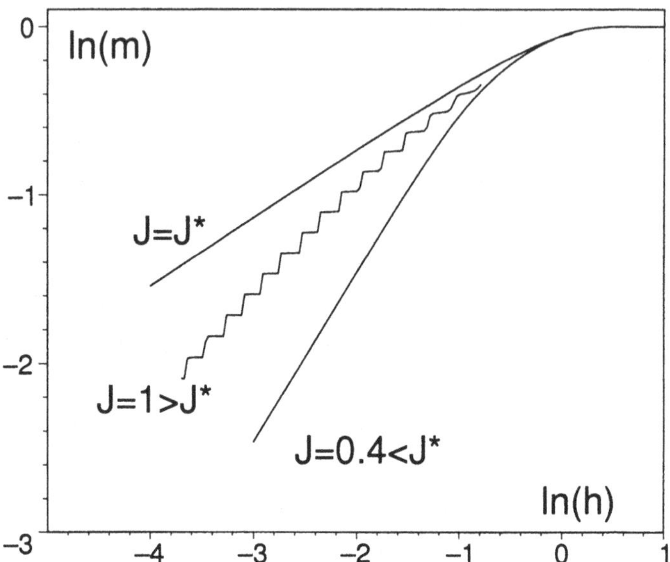

Figure 7.7: Log-log plot of the magnetization for the Ising model on the diamond fractal for three values of the coupling constant J.

derivative of the free energy with respect to the field, much as we did for the specific heat

$$\left.\frac{\partial^2 f(J,h)}{\partial h^2}\right|_{h=0} = \lim_{\delta\to 0} \frac{f(J,\delta) + f(J,-\delta) - 2f(J,0)}{\delta^2} \quad .$$

Since the susceptibility exponent γ is positive the result will diverge at the critical point as more terms in series for f are included. Below the critical temperature the susceptibility will diverge.

The magnetization as a function of the field is plotted in Figure 7.7 for values of J below, above and at the critical value. The curves are computed by iterating the recursion relation for the free energy and differentiating the result term by term numerically until convergence using

$$m = -\frac{3\partial G}{2N\partial h} \quad . \tag{7.39}$$

The factor 3/2 in (7.39) reflects the fact that there are 2/3 as many spins as there are bonds in the thermodynamic limit.

For values of J below the transition the magnetization rises linearly with the field for a weak field and eventually saturates. At the critical value of J

the magnetization increases with field as $m = h^{1/\delta}$ for weak field and the slope of the log-log plot is compatible with the critical exponent $\delta = 2.45701$.

When J is above its critical value (the temperature is below the critical temperature), one would naively expect a spontaneous magnetization, since the exponent $\beta < 0$. The question of spontaneous magnetization is a bit tricky since

$$\left.\frac{\partial g(J,h)}{\partial h}\right|_{h=0} = 0$$

and all the terms in the recursion series for m will be zero. The correct formula for the magnetization is thus

$$m = \lim_{\delta \to 0} \lim_{N \to \infty} \left.\frac{\partial f(J,h)}{\partial h}\right|_{h=\delta}$$

i.e., we should take the thermodynamic limit $N \to \infty$ before we let $h \to 0$. However, when we numerically sum the recursion series to convergence in a finite field and gradually reduce the field we find that the magnetization approaches zero, *i.e.*, there is no spontaneous magnetization. Instead we encounter a new phenomenon *log-periodic oscillations* (*i.e.* the magnetization follows a power law with a complex exponent). As the coupling approaches the critical value the log-periodic oscillations vanish for moderate fields and only exist for very weak fields. For larger values of the coupling constant the oscillations are ubiquitous and extend almost to saturation. We illustrate this behavior in Figure 7.8.

The log-periodic behavior can be understood from the hierarchical organization of the spins. Most spins are connected to two neighbors, while some are connected to four, fewer still to eight and so on. When the diamond fractal is polarized by an external field the highly connected spins exhibit a much stronger effective field than the less connected ones. For very weak fields only the most connected spins are polarized while the "ordinary" spins are disordered. As the field increases, layer after layer become polarized, and for very low temperatures the polarization of a layer is effectively saturated before the next layer feels the field. In the present case, the log-periodic oscillations is not a critical phenomenon, in fact at the critical point nothing much happens and only the very few superconnected spins experience a change.

Log-periodic oscillations have recently been given much attention (see *e.g.* the book and review by Sornette [279] [278]). In this literature the log-periodic events are associated with catastrophic events at the critical point, and evidence is presented for their importance in a number of instances such as rupture

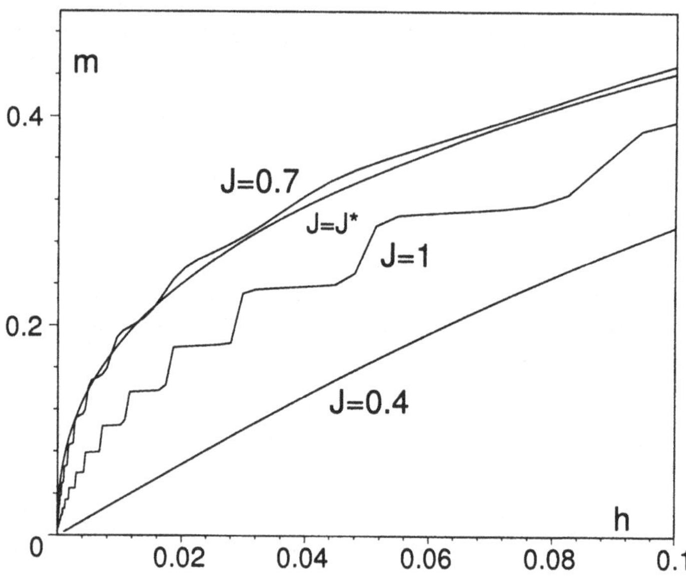

Figure 7.8: Linear plot of magnetization *vs.* field for some values of J.

of gas containers at high pressure [20], stock market crashes [279], and earthquakes [143]. Considering the importance of forecasting such events, it is not surprising that these findings are controversial (the difficulty lies in the necessity of analyzing noisy data with more parameters, and less data points, than could be desired). In the present case the log-periodic events do not signal a catastrophe at the critical point, rather they die out. However, if we interpret the highly connected spins as rulers and the less connected as their subjects, we note that near the critical point the rulers are quite out of touch with their subjects. This, of course, can prove quite catastrophic, for the rulers, come election time!

7.4 Position Space Renormalization: Cumulant Method

We next turn to systems where exact solutions cannot be found. While the solution in the previous section illustrated a number of the key features of the renormalization approach, the hierarchical structure of the model produced

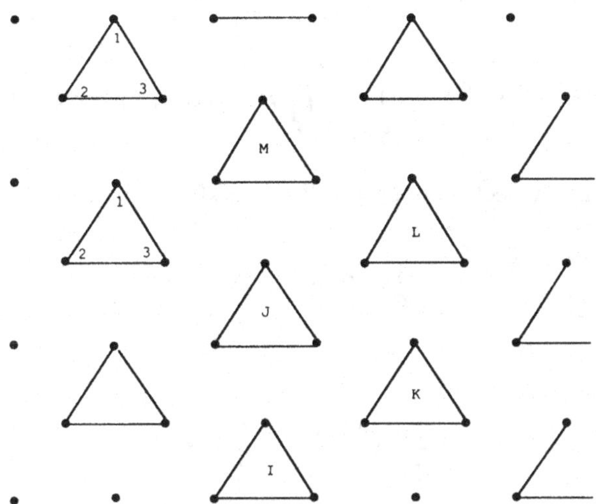

Figure 7.9: Partitioning of a triangular lattice into blocks of three spins. Numbers indicate labeling within each block; capital letters label blocks.

some atypical features. We now carry out the analogous calculation for the two-dimensional model on the triangular lattice [217]. These calculations are now only approximate, but illustrate some further features discussed in Section 7.2, and also highlights a number of difficulties that arise in such calculations for more realistic systems. The dimensionless Hamiltonian is

$$H = \frac{1}{2}\sum_{ij} K_{ij}\sigma_i\sigma_j + h\sum_{i}\sigma_i \tag{7.40}$$

where $\sigma_j = \pm 1$, with the spins occupying the sites of a triangular lattice, and the sum no longer restricted to nearest-neighbor sites.

We see from Figure 7.9 that the lattice can be divided into triangular blocks each containing three spins. These triangles in turn form a triangular lattice with separation $\sqrt{3}\,\times$ the original separation. Our procedure will be to map the original system onto a system of the same form as (7.40) but with block spins $\mu_J = \pm 1$ representing the state of the three spins σ_{1I}, σ_{2I}, σ_{3I} composing block I. We formulate this mapping process in terms of a projection operator

$$P\left(\mu_I, \sigma_{1I}, \sigma_{2I}, \sigma_{3I}\right) = P\left(\mu_I, \{\sigma_I\}\right) \tag{7.41}$$

with

$$e^{Ng(K,h)+H'(\{\mu\},K',h')} = \operatorname*{Tr}_{\{\sigma\}} \left(\prod_I P\left(\mu_I, \{\sigma\}\right) \right) e^{H(\{\sigma\},K,h)} \ . \tag{7.42}$$

If we require the projection operator P to satisfy the relation

$$\operatorname*{Tr}_{\{\mu\}} P\left(\mu_I, \{\sigma\}\right) = 1 \tag{7.43}$$

then

$$\operatorname*{Tr}_{\{\mu\}} e^{Ng+H'} = \operatorname*{Tr}_{\{\sigma\}} e^{H} \tag{7.44}$$

and the free energy is preserved exactly by the transformation, provided that we can carry out the operation (7.42). A possible candidate for the projection operator P is

$$P\left(\mu, \{\sigma\}\right) = \delta_{\mu,\phi(\{\sigma\})} \tag{7.45}$$

where $\phi(\{\sigma\}) = \frac{1}{2}\left(\sigma_1 + \sigma_2 + \sigma_3 - \sigma_1\sigma_2\sigma_3\right)$ and $\delta_{\mu,\phi}$ is the Kronecker symbol. Note that $\phi(\{\sigma\}) = 1$ whenever two or more spins on the triangle are $+1$ and $\phi(\{\sigma\}) = -1$ whenever two or more spins are -1. The projection operator thus assigns the block spin μ the value $+1$ or -1 according to a *majority rule.*

At this point our formulation is still exact, but the problem of evaluating (7.42) is intractable without some approximation. Before carrying out an approximate calculation, we note that the question of whether (7.42) and (7.45) define an analytic recursion relation (see Section 7.2) constitutes a difficult mathematical problem. This problem was addressed by Griffiths and Pearce [120], who showed that this renormalization transformation is in general not analytic. However, approximate versions of (7.42) with (7.45) do produce analytic recursion relations, as do more sophisticated renormalization transformations.

Many different approximate treatments of (7.42) are possible. We outline first the cumulant approach of Niemeijer and van Leeuwen [217], which is perhaps the simplest technically. We divide the Hamiltonian into two parts,

$$H(\{\sigma\}, K, h) = H_0(\{\sigma\}, K, h) + V(\{\sigma\}, K, h) \ . \tag{7.46}$$

H_0 represents the part of the Hamiltonian that does not involve couplings between spins in different blocks, and V contains the coupling between blocks.

Thus

$$H_0 = \sum_I K_1 \left(\sigma_{1I}\sigma_{2I} + \sigma_{1I}\sigma_{3I} + \sigma_{2I}\sigma_{3I}\right) + h \sum_I \left(\sigma_{1I} + \sigma_{2I} + \sigma_{3I}\right) \tag{7.47}$$

and

$$V = \sum_{I,J,n} K_n \sum_{\alpha,\beta} \sigma_{\alpha I}\sigma_{\beta J} \tag{7.48}$$

where K_n is the nth nearest-neighbor coupling constant and the labels α, β run over the appropriate labels on blocks I, J. From Figure 7.9 we see that in the case of nearest-neighbor interaction, adjacent blocks interact through two types of coupling terms.

$$\begin{aligned} V_{IJ} &= K_1(\sigma_{1I}\sigma_{2J} + \sigma_{1I}\sigma_{3J}) \\ V_{IK} &= K_1(\sigma_{1I}\sigma_{2K} + \sigma_{3I}\sigma_{2K}) \ . \end{aligned} \tag{7.49}$$

We may now write (7.42) in the form

$$\operatorname*{Tr}_{\{\sigma\}} \left(\prod_I P\left(\mu_I, \{\sigma_I\}\right) \right) e^H = \operatorname*{Tr}_{\{\sigma\}} \left(\prod_I^{N/3} P\left(\mu_I, \{\sigma_I\}\right) \right) e^{H_0} e^V = Z_0 \left\langle e^V \right\rangle \tag{7.50}$$

where

$$Z_0 = \operatorname*{Tr}_{\{\sigma\}} \left(\prod_I^{N/3} P\left(\mu_I, \{\sigma_I\}\right) \right) e^{H_0} \tag{7.51}$$

and

$$\langle A \rangle = \frac{1}{Z_0} \operatorname*{Tr}_{\{\sigma\}} \left(\prod_I^{N/3} P\left(\mu_I, \{\sigma_I\}\right) \right) A e^{H_0} \ . \tag{7.52}$$

This formula is still exact. We now approximate $\langle e^V \rangle$ by a truncated cumulant expansion

$$\begin{aligned} \left\langle e^V \right\rangle &= \left\langle 1 + V + \frac{V^2}{2!} + \cdots + \frac{V^n}{n!} + \cdots \right\rangle \\ &= \exp\left\{ \langle V \rangle + \frac{1}{2!}(\langle V^2 \rangle - \langle V \rangle^2) + \frac{1}{3!}(\langle V^3 \rangle - 3\langle V \rangle \langle V^2 \rangle + 2\langle V \rangle^3) + \cdots \right\} \\ &= \exp\left\{ C_1 + C_2 + C_3 + \cdots \right\} \ . \end{aligned} \tag{7.53}$$

The jth cumulant approximation then corresponds to retaining the first j cumulants $C_1, \ldots, C_j$.

7.4.1 First-order approximation

Let us assume that only nearest-neighbor interactions are present and set $K_1 = K$, $K_2 = K_3 = \cdots = 0$. The trace (7.51) over the decoupled blocks is easy to carry out. Writing

$$Z_0 = \prod_I e^{A+B\mu_I} \tag{7.54}$$

we obtain

$$e^{A+B} = e^{3K+3h} + 3e^{-K+h}$$
$$e^{A-B} = e^{3K-3h} + 3e^{-K-h}$$

or

$$\begin{aligned} A &= \frac{1}{2}\ln\left[(e^{3K+3h} + 3e^{-K+h})(e^{3K-3h} + 3e^{-K-h}\right] \\ B &= \frac{1}{2}\ln\frac{e^{3K+3h} + 3e^{-K+h}}{e^{3K-3h} + 3e^{-K-h}} \ . \end{aligned} \tag{7.55}$$

Expectation values of the type $\langle\sigma_{\alpha I}\sigma_{\beta J}\rangle$ can be factored since they are to be evaluated with respect to the Hamiltonian H_0 which contains no coupling between blocks I and J. For this reason the expectation value $\langle V\rangle$ is also easy to obtain:

$$\langle V\rangle = K\sum_{\alpha I,\beta J}\langle\sigma_{\alpha I}\rangle\langle\sigma_{\beta J}\rangle \ . \tag{7.56}$$

The expectation value $\langle\sigma_{\alpha I}\rangle$ is independent of α because of translational invariance and can be written as

$$\langle\sigma_{\alpha I}\rangle = C + D\mu_I \tag{7.57}$$

with

$$\begin{aligned} C &= \frac{1}{2}\left(\frac{e^{3K+3h} + e^{-K+h}}{e^{3K+3h} + 3e^{-K+h}} - \frac{e^{3K-3h} + e^{-K-h}}{e^{3K-3h} + 3e^{-K-h}}\right) \\ D &= \frac{1}{2}\left(\frac{e^{3K+3h} + e^{-K+h}}{e^{3K+3h} + 3e^{-K+h}} + \frac{e^{3K-3h} + e^{-K-h}}{e^{3K-3h} + 3e^{-K-h}}\right) \ . \end{aligned} \tag{7.58}$$

Combining these results, we have

$$\begin{aligned} &Ng(K,h) + H'(\{\mu_I\}, K', h') \\ &\quad = \frac{1}{3}NA(K,h) + B\sum_I \mu_I + 2K\sum_{\langle IJ\rangle}(C + D\mu_I)(C + D\mu_J) \end{aligned} \tag{7.59}$$

where the functions A, B, C, D are given by (7.55–7.58). Rewriting the renormalized Hamiltonian H' in the original form (7.40),

$$H' = K' \sum_{\langle IJ \rangle} \mu_I \mu_J + h' \sum_I \mu_I \tag{7.60}$$

we obtain the recursion relations

$$\begin{aligned} K' &= 2KD^2(K,h) \\ h' &= B(K,h) + 12KC(K,h)D(K,h) \\ g(K,h) &= \frac{1}{3}A(K,h) + 2KC^2(K,h) \ . \end{aligned} \tag{7.61}$$

The flow in the $K - h$ plane from the recursion relations (7.60) is indicated in Figure 7.10. Since the phase transition to the ferromagnetic state takes place at $h = 0$, we look for a fixed point $(K^*, h^*) = (K_c, 0)$ where $K_c = J/k_B T_c$. For $h = 0$, $B = C = 0$ and our recursion relations (7.60) reduce to

$$K' = 2K \left(\frac{e^{3K} + e^{-K}}{e^{3K} + 3e^{-K}} \right)^2 \tag{7.62}$$

$$h' = h = 0 \ . \tag{7.63}$$

The recursion relation (7.62) has very simple limiting behavior. For $K \ll 1$, $K' \approx (2K) \left(\frac{1}{2}\right)^2 = \frac{1}{2}K$, while for $K \gg 1$, $K' \approx 2K$ and we see that the flow reverses itself at some finite value of K. For small K the flow is toward the noninteracting high-temperature fixed point; for large K toward the $K = \infty$ ground-state fixed point. The critical point is given by

$$\frac{e^{3K^*} + e^{-K^*}}{e^{3K^*} + 3e^{-K^*}} = \frac{1}{\sqrt{2}} \ . \tag{7.64}$$

This equation is easily solved analytically, by making the substitution $x = \exp\{4K^*\}$ and solving for x, to yield $K^* = J/k_B T_c \approx 0.3356$. The exact result for this model is $K^* = 0.27465\ldots$ and the mean field result is $K^* = \frac{1}{6}$. We see that the critical temperature in the first-order approximation does not agree particularly well with the exact result. Nevertheless, this treatment of the model, because of the very structure of the renormalization theory, does produce nontrivial (*i.e.*, non-mean-field-like) critical exponents. By symmetry,

$$\left.\frac{\partial h'}{\partial K}\right|_{K^*,h^*} = \left.\frac{\partial K'}{\partial h}\right|_{K^*,h^*} = 0$$

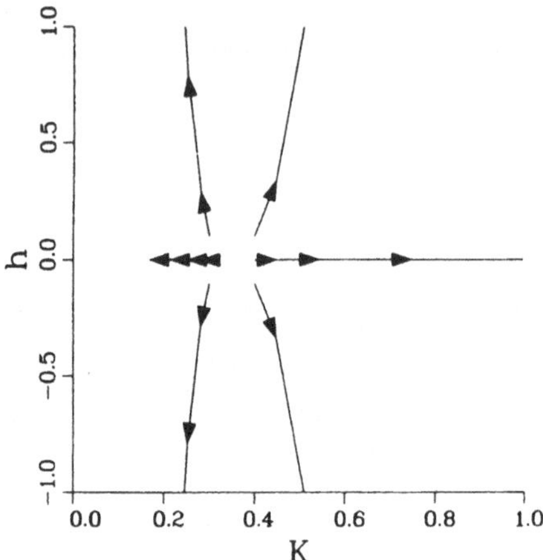

Figure 7.10: Flow of coupling constants for the two-dimensional Ising model in the first-order cumulant approximation.

so that we quickly obtain

$$b^{y_t} = \left.\frac{\partial K'}{\partial K}\right|_{K^*,h^*} \qquad\qquad b^{y_h} = \left.\frac{\partial h'}{\partial h}\right|_{K^*,h^*} \tag{7.65}$$

with $b = \sqrt{3}$. Evaluating the derivatives at the fixed point we find $y_t = 0.882$, $y_h = 2.034$. The specific heat exponent $\alpha = 2 - d/y_t = -0.267$ and the magnetization exponent is $\beta = (d - y_h)/y_t = -0.039$ in the first-order approximation. Again these results are not particularly accurate ($\alpha = 0$, $\beta = 1/8$ in the exact theory), but we have developed a theory that captures the essence of critical behavior and can be improved systematically.

7.4.2 Second-order approximation

The next order approximation in the hierarchy (7.53) consists of retaining the terms

$$\langle V\rangle + \frac{1}{2}\left(\langle V^2\rangle - \langle V\rangle^2\right) \ .$$

The calculation is then more cumbersome, but we outline it here since it will illustrate some features of the qualitative discussion of Section 7.2 and also highlight some of the technical problems in renormalization calculations.

Consider again Figure 7.9, in which some typical cells have been drawn. Cells I, J are nearest neighbors, while cells I, L are second neighbors and I,

M third neighbors in the triangular lattice of cells. It is clear that the second-order cumulant $\langle V^2\rangle - \langle V\rangle^2$ contains interactions of longer range than nearest neighbor. For example, the third nearest-neighbor cells I and M are coupled by terms of the form

$$\begin{aligned}
&K^2\langle\sigma_{1I}(\sigma_{2J}+\sigma_{3J})\sigma_{1J}(\sigma_{2M}+\sigma_{3M})\rangle \\
&-K^2\langle\sigma_{1I}(\sigma_{2J}+\sigma_{3J})\rangle\langle\sigma_{1J}(\sigma_{2M}+\sigma_{3M})\rangle \\
&\quad = 4K^2\left\{\langle\sigma_{1I}\rangle\langle\sigma_{1J}\sigma_{2J}\rangle\langle\sigma_{1M}\rangle - \langle\sigma_{1I}\rangle\langle\sigma_{2J}\rangle^2\langle\sigma_{2M}\rangle\right\} . \qquad (7.66)
\end{aligned}$$

Let us consider the simple case $h = 0$ for which (7.66) can be easily evaluated. We need one new expectation value

$$\langle\sigma_{1J}\sigma_{2J}\rangle = \frac{e^{3K} - e^{-K}}{e^{3K} + 3e^{-K}} = E(K) \qquad (7.67)$$

which is independent of μ_J. The term (7.66) appears twice in the expansion of $\langle V^2\rangle$ canceling the factor 1/2 in the cumulant expansion (7.53) and we see that the renormalized Hamiltonian contains a third nearest-neighbor coupling term $K_3'\mu_I\mu_M$. Noting that for $h = 0$, $B = C = 0$, we find that

$$K_3' = 4K^2\left[D^2(K)E(K) - D^4(K)\right] \qquad (7.68)$$

where $D(K)$ and $E(K)$ are given by (7.58) and (7.67), respectively. In the next iteration the second cumulant will generate still longer-range interactions through terms of order $(K_3')^2$. Therefore, we must find some way of truncating the system of recursion relations.

The first approach of Niemeijer and van Leeuwen [217] ordered the coupling constants into a hierarchy according to the power of the nearest-neighbor coupling constant at which they are first generated. Thus the unique term of order 1 is the nearest-neighbor coupling constant; the terms of order 2 are the second and third neighbor constants K_2 and K_3. In the second cumulant approximation the second and third neighbor interactions are included only in $\langle V\rangle$, while the first neighbor interaction is retained in $\langle V\rangle$ and $\langle V^2\rangle - \langle V\rangle^2$. With this choice a fixed number of coupling constants appears at every iteration. This classification of interactions is rather arbitrary, and we will describe a better method, the finite cluster approximation, in Section 7.5.1. This method was also developed by Niemeijer and van Leeuwen [217]. However, we note that all position space calculations on models that are not exactly solvable do require some *ad hoc* approximation procedure, to avoid dealing with the infinite number of coupling constants implied by (7.42). The rationalization for such a truncation is that at the fixed point there appear to be only a small number of relevant scaling fields (Section 7.2). One hopes that by a proper choice of a finite set of coupling constants, one can obtain an accurate representation of the relevant scaling fields.

The second-order recursion relations are

$$\begin{aligned} K_1' &= 2K_1D^2 + 4(D^2 + D^2E - 2D^4)K_1^2 + 3D^2K_2 + 2D^2K_3 \\ K_2' &= K_1^2(7D^2E + D^2 - 8D^4) + K_3D^2 \\ K_3' &= 4K_1^2(D^2E - D^4) \end{aligned} \tag{7.69}$$

where D and E are given by (7.58) and (7.67). The fixed point is located at

$$K_1^* = 0.27887 \qquad K_2^* = -0.01425 \qquad K_2^* = -0.01523 \ .$$

The critical point in the nearest-neighbor model can be located by finding the intersection of the critical surface and the K_1 axis. One finds $K_{1c} = 0.2575$ in better agreement with the exact result ($K_{1c} = 0.27465$) than the first-order approximation.

The linearized recursion relations at $\{K^*\}$ yield the matrix $\mathbf{M}$ (see Section 7.2). Numerically,

$$\mathbf{M} = \begin{bmatrix} 1.8313 & 1.3446 & 0.8964 \\ -0.0052 & 0 & 0.4482 \\ -0.0781 & 0 & 0 \end{bmatrix}$$

with eigenvalues $\lambda_1 = 1.7728$, $\lambda_2 = 0.1948$, and $\lambda_3 = -0.1364$. The relevant scaling field corresponds to λ_1 and has exponent $y_t = y_1 = 1.042$, which is close to the exact value $y_t = 1.0$. The specific heat exponent is $\alpha = 0.081$ (*i.e.*, quite small). In the exact solution there is only a logarithmic singularity in the specific heat ($\alpha = 0$).

One might ask if still better results can be obtained with higher order cumulants, larger cells, or weighting functions other than the majority rule. The experience so far has been rather discouraging. However, there are ways other than the cumulant method of implementing the position space renormalization method that yield much better convergence. We describe some of these methods in Section 7.5.

7.5 Other Position Space Renormalization Group Methods

7.5.1 Finite lattice methods

The finite lattice approach or cluster approximation represents one of the most useful renormalization group techniques. The basic notion is that the recursion relations for an infinite system can be modeled by exact recursion relations for a small system. We will not attempt any comprehensive review of the method here, and we refer the interested reader to the review article by Niemeijer and van Leeuwen [217]. We illustrate the procedure by considering the smallest cluster that can model the Ising ferromagnet on the triangular lattice. A further example involving a system of more direct physical interest is discussed in the next subsection.

Consider the system made up of a pair of nearest-neighbor cells of three spins (see Figure 7.11). We will use the "majority rule" projection (7.45), just as we did in our discussion of the cumulant approximation. It is now no longer necessary to approximate $\langle e^V\rangle$; we can evaluate this expression simply by carrying out the trace over the 2^6 configurations of the six spins. For simplicity we only consider $h=0$. In this case the renormalized Hamiltonian takes the form

$$\begin{aligned} e^{g(K)+H'(K',\mu)} &= e^{g+K'\mu_I\mu_J} \\ &= \operatorname*{Tr}_{\sigma} P\left(\mu,\{\sigma\}\right) \exp\left\{K\sum_{\langle ij\rangle}\sigma_i\sigma_j\right\} . \end{aligned} \tag{7.70}$$

The functions g and K' can be determined by noting that

$$\begin{aligned} e^{g+K'} &= \operatorname*{Tr}_{\sigma} P\left(\mu_i=+1,\{\sigma\}\right) P\left(\mu_J=+1,\{\sigma\}\right) e^{H} \\ e^{g-K'} &= \operatorname*{Tr}_{\sigma} P\left(\mu_i=+1,\{\sigma\}\right) P\left(\mu_J=-1,\{\sigma\}\right) e^{H} . \end{aligned} \tag{7.71}$$

Because of the restrictions placed on the trace by the projection operator, sixteen terms contribute in each case. We obtain

$$\begin{aligned} e^{g+K'} &= e^{8K}+3e^{4K}+2e^{2K}+3+6e^{-2K}+e^{-4K} \\ e^{g-K'} &= 2e^{4K}+2e^{2K}+4+6e^{-2K}+2e^{-4K} \end{aligned} \tag{7.72}$$

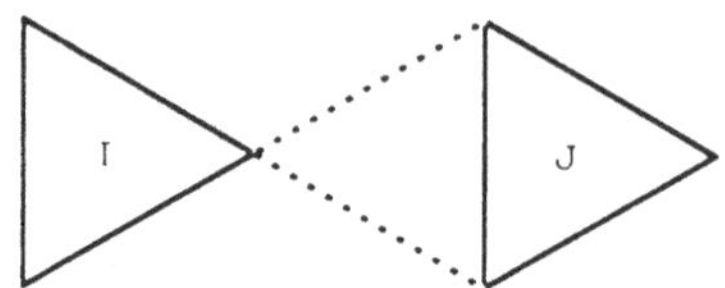

Figure 7.11: Six-spin cluster.

or

$$K' = \frac{1}{2} \ln \frac{e^{8K} + 3e^{4K} + 2e^{2K} + 3 + 6e^{-2K} + e^{-4K}}{2e^{4K} + 2e^{2K} + 4 + 6e^{-2K} + 2e^{-4K}} . \tag{7.73}$$

This recursion relation yields a fixed point $K^* = K_c = 0.3653$ and an exponent $y_t = 0.7922$. The results of this simple calculation are not impressive, but Niemeijer and van Leeuwen [217] have used larger and more symmetric clusters to obtain very good convergence for both the critical temperature and the critical exponents for the Ising ferromagnet on the triangular lattice. It is not apparent from the simple example above how longer-range coupling constants enter into the calculation. A little reflection will convince the reader that all couplings consistent with the symmetry of the Hamiltonian and the cluster will eventually be generated under iteration. Thus, for the Ising model on the square lattice in zero magnetic field a 16-spin cluster divided into four-spin cells allows the first and second nearest-neighbor interactions K_1 and K_2 and a four-spin interaction of the form $K'_4\mu_1\mu_2\mu_3\mu_4$. The calculation $K_1, K_2, K_4 \to K'_1, K'_2, K'_4$ has been carried out by Nauenberg and Nienhuis [212], who obtained excellent agreement with the exact results of Onsager. Since there is an even number of spins in each block, it is necessary to modify the majority rule to handle tie votes.

At this point, the reader may wonder whether the renormalization group approach, aside from the fact that it automatically produces a scaling free energy and universality, is limited as a calculational tool to models that are already well understood. In the next section we discuss a model that is relevant to experimental situations, and where mean field theory gives misleading results.

7.5.2 Adsorbed monolayers: Ising antiferromagnet

A number of physisorbed systems have been studied intensely in recent years, since they provide realizations of a rich variety of phase transitions, some of

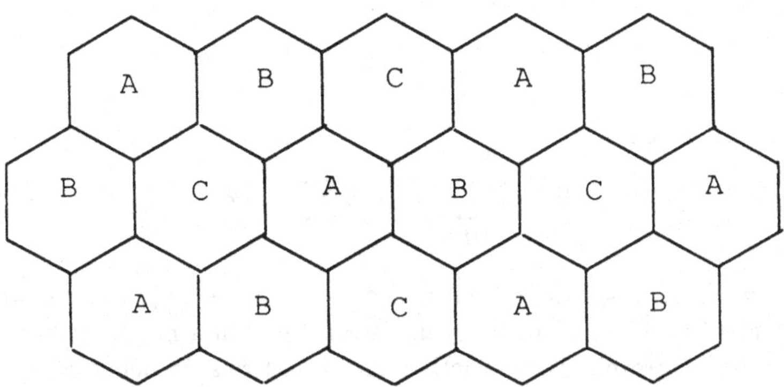

Figure 7.12: Honeycomb structure of a graphite monolayer.

which are peculiar to two dimensions. We discuss one such system here, namely helium on the surface of graphite.

The graphite surface has a honeycomb structure, as indicated in Figure 7.12 and the graphite-helium interaction gives rise to preferred adsorption sites directly above the honeycomb centers. In order to move from one preferred adsorption site to the next, a helium atom has to pass over a potential barrier. We will assume that these barriers are large enough that we can treat the system as a two-dimensional lattice gas with the adsorption sites being either filled ($n_j = 1$) or empty ($n_j = 0$). If there is a pairwise He-He interaction, $V(r_{ij})$, we obtain a Hamiltonian of the form

$$H = \sum_{i<j} V(r_{ij}) n_i n_j \ . \tag{7.74}$$

The He-He interaction can be approximated by a Lennard–Jones potential (5.7), with a minimum somewhere between the nearest-neighbor distance (0.246 nm) and the second nearest honeycomb to honeycomb distance ($\sqrt{3} \times 0.246$ nm $= .426$ nm). We neglect the second- and higher-neighbor interactions and work with the idealized Hamiltonian

$$H = V_0 \sum_{\langle ij \rangle} n_i n_j \tag{7.75}$$

with $V_0 > 0$ and the sum extending over nearest-neighbor pairs of hexagon centers. It is convenient to work in the grand canonical ensemble (in an experiment the adsorbed He atoms are in equilibrium with He vapor). We therefore add a term $-\mu \sum_i n_i$ to the Hamiltonian. To express the Hamiltonian in the Ising form, we make the transformation

$$n_i = \frac{1}{2}(1 + \sigma_i) \tag{7.76}$$

with $\sigma_i = \pm 1$. Then

$$H = J \sum_{\langle ij \rangle} \sigma_i \sigma_j - h \sum_i \sigma_i + c \tag{7.77}$$

where $J = V_0/4$, $h = \frac{1}{2}(\mu - 3V_0)$, and $c = \frac{1}{4}N(3V_0 - 2\mu)$, that is, we have the equivalent problem of an Ising antiferromagnet in a magnetic field. Zero magnetization corresponds to exactly half the lattice sites being occupied. In the special case $h = 0$ the model has been solved by Houtappel [135] and Husimi and Syozi [137]. They found that the system remains disordered for all nonzero temperatures and that there is no phase transition. The physical reason for this absence of a phase transition is the very high degeneracy of the ground state. To see this we note that the triangular lattice may be divided into three sublattices (labeled A, B, and C in Figure 7.12). All sites on one sublattice have nearest neighbors on the two other sublattices but none on its own kind. Some of the degenerate ground-state configurations have $\sigma_i = +1$ on one of the sublattices (say, A) and $\sigma_i = -1$ on another (*e.g.*, B). Because of the antiferromagnetic coupling this lowers the energy relative to the completely disordered configuration. However, once the assignment of the A and B spins have been made the C spins are completely "frustrated" (*i.e.*, it does not matter what their orientation is). The degeneracy of the ground state is therefore greater than $2^{N/3}$ and there will be a residual entropy even at zero temperature. On the other hand, if $h \neq 0$ in (7.77), the degeneracy of the ground state is broken and we expect a phase diagram of the type sketched in Figure 7.13.

We now briefly outline a renormalization group treatment of the Hamiltonian (7.77) due to Schick *et al.* [267]. In dealing with models in which there is an underlying symmetry, it is important to preserve this symmetry under renormalization. The reason for this is that although formulas such as (7.42) and (7.50) are exact and hold for any choice of blocking, they cannot be evaluated without approximation. In an approximate calculation a disregard of symmetry may result in a renormalized Hamiltonian which belongs to a different universality class than the original Hamiltonian.

In the present problem we wish to retain the equivalence of the three sublattices under renormalization. The $\sqrt{3} \times \sqrt{3}$ ordered state, corresponding to

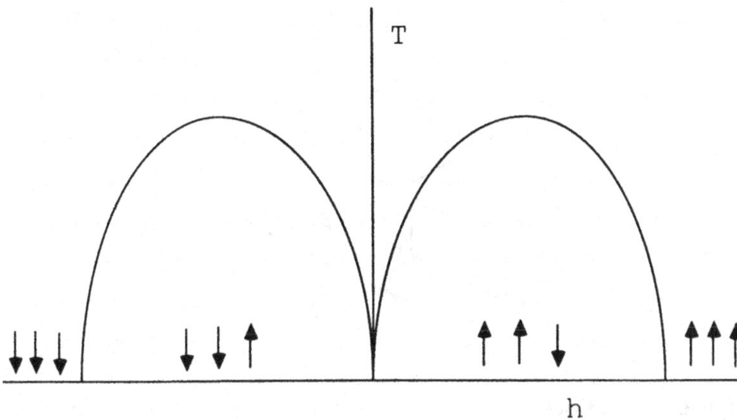

Figure 7.13: Schematic phase diagram for the two-dimensional Ising antiferromagnet on the triangular lattice.

a coverage of $\frac{1}{3}$, can have either the A, B, or C site predominantly occupied and we wish to preserve this feature under renormalization. We note that the ordered state of this system, the case where the coverage is exactly $\frac{1}{3}$, can be described by a two-component order parameter. In the disordered phase the density on each of the sublattices is $\frac{1}{3}$ and the degree to which one sublattice is preferred is given by the three numbers $\langle n_A\rangle - \frac{1}{3}$, $\langle n_B\rangle - \frac{1}{3}$, $\langle n_C\rangle - \frac{1}{3}$. At fixed density these three numbers must add to zero, and there are only two independent variables which describe the system. The discrete threefold symmetry of the system (corresponding to rotation of the order parameter by $2\pi/3$) is often referred to as the three-state Potts symmetry, and we expect the phase transitions of this system to be the same as for the three-state Potts [248] model previously introduced in Section 3.8.1. We may use the parametrization of (3.44) to describe the two-component order parameter,

$$\begin{aligned}\langle n_A\rangle &= \frac{1}{3} + \frac{2}{3}y \\ \langle n_B\rangle &= \frac{1}{3} + \frac{1}{\sqrt{3}}x - \frac{1}{3}y \\ \langle n_C\rangle &= \frac{1}{3} - \frac{1}{\sqrt{3}}x - \frac{1}{3}y\end{aligned} \tag{7.78}$$

and the allowed values of x and y are shown in Figure 3.11. The Landau free energy of the Potts model (or our antiferromagnet) contains cubic terms signifying a first order transition. In this case the Landau theory is qualitatively

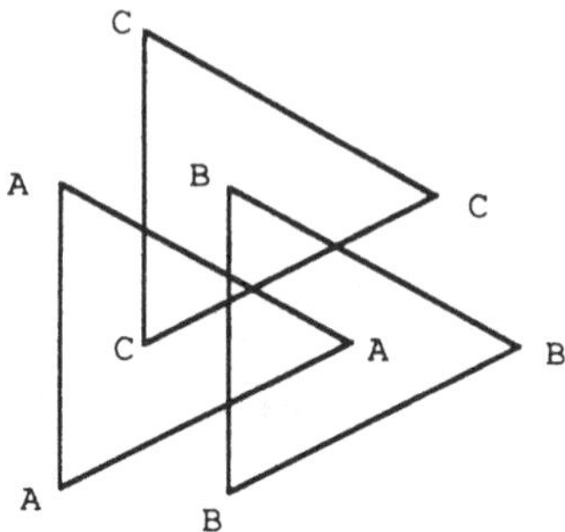

Figure 7.14: Blocking scheme preserving the threefold symmetry.

incorrect—it can be proven rigorously [29] that the transition is continuous in two dimensions. In three dimensions the Landau result seems to be correct.

In view of the discussion above it is clear that a renormalization transformation should preserve the identity of the three sublattices. The simplest blocking scheme consistent with this criterion is shown in Figure 7.14. The three interpenetrating triangles are three blocks of equivalent sites, and we can associate a block spin (e.g., σ_A) with the three site spins on the original lattice. At this point we may choose an approximation scheme (cluster method, cumulant method, etc.). The resulting renormalized Hamiltonian should have the proper threefold symmetry. We shall not carry out this calculation here. The details of a finite cluster calculation may be found in the article by Schick *et al.* [267].

7.5.3 Monte Carlo renormalization

In the preceding sections we have developed the renormalization group formalism and described a few methods of carrying out renormalization group calculations. None of these methods were extremely successful in producing accurate values of the critical coupling constants or exponents. Their value today lies more in the insights they provide (*e.g.,* when there are a number of competing fixed points) than in the accuracy of the specific results. We now turn to a renormalization group scheme which is capable of producing, at least in the case of classical spins, highly accurate results for the critical exponents.

The Monte Carlo renormalization group method was invented by Ma [183] and further developed by Swendsen and co-workers [292, 293, 294, 295, 44, 236].

Consider an Ising system on a d-dimensional lattice. We write the Hamiltonian in the compact form

$$H = \sum_{\alpha} K_{\alpha}\psi_{\alpha}(\sigma_i) \tag{7.79}$$

where the index α refers to a generic type of coupling (7.16). At this point we make no specific restrictions on the type of couplings that we will include in H. We note that the expectation values and correlation functions of the spin operators are given by

$$\langle\psi_{\alpha}\rangle = \frac{\partial \ln Z}{\partial K_{\alpha}} \tag{7.80}$$

where

$$Z = \mathrm{Tr} e^{H}$$

and

$$\langle\psi_{\alpha}\psi_{\beta}\rangle - \langle\psi_{\alpha}\rangle\langle\psi_{\beta}\rangle = \frac{\partial\langle\psi_{\alpha}\rangle}{\partial K_{\beta}} \equiv S_{\alpha\beta} \ . \tag{7.81}$$

Consider now a large finite system with the Hamiltonian (7.79) and (usually) periodic boundary conditions. We will describe how such a system can be simulated by the Monte Carlo method in Chapter 9. The expectation values (7.80) and (7.81) can be obtained to arbitrary accuracy (for the finite system) by sampling for a sufficiently long time.

We now divide the finite system into blocks. For example, on a d-dimensional cubic lattice one may choose blocks of 2^d spins and the $N/2^d$ blocks will once again form a cubic lattice. We also define a rule (perhaps the majority rule) which assigns values to the block spins on the basis of the configuration of the original spins. As we progress through a Monte Carlo run we can thus keep track of the configurations of both the block and site spins. Indeed, if the original lattice is large enough, we may accumulate information on several generations of spins (N_0 site spins, $N_1 = N_0/2^d$ block spins, $N_2 = N_1/2^d$ second-generation block spins, etc.).

We suppose that the Hamiltonian for the nth generation of the block spins can be written in the form

$$H^n = \sum_{\alpha} K_{\alpha}^n \psi_{\alpha}(\sigma_i^n) \tag{7.82}$$

where the functions ψ_{α} are the same at all levels. Our goal is to obtain a recursion relation $K_{\alpha}^n = R_{\alpha}(K_1^{n-1}, K_2^{n-1}, \ldots)$. We write

$$\exp\{H^n\} = \operatorname*{Tr}_{\sigma^{n-1}} P\left(\sigma^n, \sigma^{n-1}\right) \exp\left\{H^{n-1}\right\} \ . \tag{7.83}$$

We define $S^n_{\alpha\beta}$ to be the correlation function (7.81) computed at the nth level of blocking, and denote by $S^{n,n-1}_{\alpha\beta}$ the correlation function connecting the two blocking levels n and $n-1$:

$$S^{n,n-1}_{\beta\alpha} \equiv \frac{\partial \langle \psi^n_\beta \rangle}{\partial K^{n-1}_\alpha} = \frac{\partial}{\partial K^{n-1}_\alpha} \frac{\operatorname*{Tr}_{\sigma^n} \psi^n_\beta \operatorname*{Tr}_{\sigma^{n-1}} P(\sigma^n, \sigma^{n-1}) \exp\left\{H^{n-1}\right\}}{\operatorname*{Tr}_{\sigma^n} \operatorname*{Tr}_{\sigma^{n-1}} P(\sigma^n, \sigma^{n-1}) \exp\left\{H^{n-1}\right\}} \,. \tag{7.84}$$

Differentiating the right-hand side of (7.84) and using

$$\operatorname*{Tr}_{\sigma^n} P(\sigma^n, \sigma^{n-1}) = 1$$

we obtain

$$S^{n,n-1}_{\beta\alpha} = \frac{\operatorname*{Tr}_{\sigma^n} P(\sigma^n, \sigma^{n-1}) \psi^n_\beta \psi^{n-1}_\alpha \exp\left\{H^{n-1}\right\}}{\operatorname{Tr} \exp\left\{H^{n-1}\right\}} - \frac{\operatorname*{Tr}_{\sigma^n} \psi^n_\beta \exp\left\{H^n\right\}}{\operatorname*{Tr}_{\sigma^n} \exp\left\{H^n\right\}} \frac{\operatorname*{Tr}_{\sigma^{n-1}} \psi^{n-1}_\alpha \exp\left\{H^{n-1}\right\}}{\operatorname*{Tr}_{\sigma^{n-1}} \exp\left\{H^{n-1}\right\}} \,. \tag{7.85}$$

Using the chain rule of differentiation, we find that

$$S^{n,n-1}_{\beta\alpha} \equiv \frac{\partial \langle \psi^n_\beta \rangle}{\partial K^{n-1}_\alpha} = \sum_\gamma \frac{\partial \langle \psi^n_\beta \rangle}{\partial K^n_\gamma} \frac{\partial K^n_\gamma}{\partial K^{n-1}_\alpha} \tag{7.86}$$

or

$$S^{n,n-1}_{\beta\alpha} = \sum_\gamma S^n_{\beta\gamma} \frac{\partial K^n_\gamma}{\partial K^{n-1}_\alpha} \,. \tag{7.87}$$

Equation (7.87) is the basic equation of the Monte Carlo renormalization group method. The matrix elements $S^n_{\beta\gamma}$ and $S^{n,n-1}_{\beta\alpha}$ can be calculated by averaging over the configurations generated in a Monte Carlo run. They depend only on the coupling constants of the site spin Hamiltonian and on the projection operator P that defines the blocking. If the site spin Hamiltonian is on a critical surface, the recursion relations $\partial K^n_\gamma / \partial K^{n-1}_\alpha$ will, as n becomes large, approach the linearized recursion relations at the stable fixed point, that is,

$$T^{n,n-1}_{\gamma\alpha} \equiv \frac{\partial K^n_\gamma}{\partial K^{n-1}_\alpha} \to M_{\gamma\alpha} \ . \tag{7.88}$$

In matrix notation,

$$\mathbf{M} = [\mathbf{S}^n]^{-1} \left[\mathbf{S}^{n,n-1}\right] \ . \tag{7.89}$$

The eigenvalues of this matrix yield the critical exponents.

To this point we have been rather vague about a number of important technical points. As we have noted previously, the space of coupling constants is infinite dimensional. In a finite-size spin system we can accommodate only a finite number of independent spin operators and coupling constants. In the Monte Carlo approach outlined here, the number of coupling constants that can be used is determined by the number of block spins in the final generation. It is straightforward, however, to include more coupling constants by going to larger site spin systems while using the same number of generations or by blocking fewer times.

In (7.89) we have implicitly assumed that we will, with a finite number of iterations, come sufficiently close to the fixed point that the matrix $\mathbf{T}$ has effectively become the same as the matrix $\mathbf{M}$. In practice the relevant eigenvalues of the matrix $\mathbf{T}^{n,n-1}$ rather rapidly approach stationary values as functions of n for a small range of coupling constants near the critical surface, and there is therefore good reason to believe that the procedure is sensible. For further technical details, and the results of some of the more important applications of the Monte Carlo renormalization group method, we refer to the references quoted earlier in this section.

7.6 Phenomenological Renormalization Group

The term 'phenomenological renormalization group' is sometimes used to denote a technique due to Nightingale [218] that is particularly powerful in the case of two-dimensional systems for which one can construct a transfer matrix. Such systems include the Ising model, the Potts model and many other discrete-spin models with short-range interactions. This technique is also closely related to finite-size scaling theory discussed in Section 6.4. We illustrate the method by applying it to the two-dimensional Ising model but begin with a somewhat more general discussion. We consider a two-dimensional spin system with periodic boundary conditions. Assume that the partition function can be calculated by means of a transfer matrix (Section 6.1) $\mathbf{V}$ which we leave unspecified for the moment:

$$Z = \operatorname*{Tr}_{\{\sigma_1\}} \operatorname*{Tr}_{\{\sigma_2\}} \cdots \operatorname*{Tr}_{\{\sigma_M\}} \mathbf{V}_{\sigma_1,\sigma_2}\mathbf{V}_{\sigma_2,\sigma_3}\cdots\mathbf{V}_{\sigma_M,\sigma_1} \ . \tag{7.90}$$

In (7.90) the operation $\mathrm{Tr}_{\{\sigma_j\}}$ refers to the trace over all the spin variables in row j. As we have shown previously, the partition function is simply the Mth power of the largest eigenvalue λ_0 of the matrix $\mathbf{V}$. We show now that the correlation length ξ can be expressed in terms of the two largest eigenvalues of this matrix. Consider the simplest correlation function of two spins in the same row, j columns apart:

$$g(j) = \langle \sigma_{i,1}\sigma_{i,j+1}\rangle \ . \tag{7.91}$$

Let λ_n denote the nth largest eigenvalue of $\mathbf{V}$ and $|n\rangle$ the corresponding normalized eigenvector. We implicitly define a matrix $\mathbf{W}$ by requiring that (7.91) can be expressed as

$$g(j) = \frac{\operatorname*{Tr}_{\{\sigma_1\}} \operatorname*{Tr}_{\{\sigma_2\}} \cdots \operatorname*{Tr}_{\{\sigma_M\}} \mathbf{W}\mathbf{V}^{j-1}\mathbf{W}\mathbf{V}^{M-j-1}}{\lambda_0^M} \tag{7.92}$$

where the matrix $\mathbf{W}$ differs from $\mathbf{V}$ because of the factors $\sigma_{i,1}$, $\sigma_{i,j+1}$ multiplying the Boltzmann weight for columns 1 and $j+1$. Because of the factor $\mathbf{V}^{M-j-1}$ in the numerator, the expectation value (7.92) reduces to

$$g(j) = \frac{\langle 0|\mathbf{W}\mathbf{V}^{j-1}\mathbf{W}|0\rangle}{\lambda_0^{j+1}} \tag{7.93}$$

in the thermodynamic limit $M \to \infty$. Using $\sum_n |n\rangle\langle n| = \mathbf{1}$ we may further simplify this equation to obtain

$$g(j) = \frac{\langle 0|\mathbf{W}|0\rangle^2}{\lambda_0^2} + \sum_{n>0} \frac{\langle n|\mathbf{W}|0\rangle^2}{\lambda_0\lambda_n}\left(\frac{\lambda_n}{\lambda_0}\right)^j \ . \tag{7.94}$$

The first term in (7.94) is clearly the square of the order parameter $\langle\sigma_{ij}\rangle^2$ and the second term is the spin-spin correlation function $\Gamma(j)$. When we let

j become large, only the contribution from the second-largest eigenstate is significant:

$$\Gamma(j) \sim \frac{\langle 1|\mathbf{W}|0\rangle^2}{\lambda_0\lambda_1}\left(\frac{\lambda_1}{\lambda_0}\right)^j = ae^{-j/\xi} \tag{7.95}$$

with

$$\xi^{-1} = -\ln\frac{\lambda_1}{\lambda_0} . \tag{7.96}$$

We have already seen an example of this expression in Section 3.6 where the pair correlation function of the one-dimensional Ising model is derived.

We have made a number of assumptions in the foregoing derivation. The most important is that the matrix element $\langle 1|\mathbf{W}|0\rangle$ is nonzero. While we are not aware of a general proof, this certainly holds for the two-dimensional Ising model and should be the generic case. We have also calculated a rather special correlation function *i.e.,* that of two spins in the same row. This allowed us to proceed without specifying the transfer matrix. We may now appeal to symmetry to argue that the result (7.96) is in fact quite general. The division of the lattice into rows and columns is a technical artifact: the correlation function must be the same along a row as along a column. Similarly, at large distances, only the separation between two points should enter into the correlation function, not the direction of the line joining the two spins.

We use these results to construct an approximate renormalization procedure. Recall first the finite-size scaling theory of Section 6.4. There it was postulated that in the vicinity of the critical point a quantity $Q(L,T)$ that, for the infinite system has a power law singularity at T_c can be written in the form[1]

$$Q(L,T) = |t|^{-\rho} f\left(\frac{L}{\xi(t)}\right)$$

with appropriate limiting forms for the scaling function f. For our present purposes it is more convenient to rewrite this in the equivalent form

$$Q(L,T) = L^{\rho/\nu} g(|t|L^{1/\nu}) \tag{7.97}$$

where for $L \to \infty$ we must have $g(x) \sim |x|^{-\rho}$ so as to recover the $|t|^{-\rho}$ behavior. We may now use this form for the correlation length ξ of a finite system and obtain

$$\xi(L,T) = Lg(|t|L^{1/\nu}) . \tag{7.98}$$

[1] For a derivation of this scaling form on the basis of position-space renormalization group methods, see [24].

Comparing two systems of linear dimension L and L' we have the relation

$$\frac{\xi(L,T)}{L} = \frac{\xi(L',T')}{L'} \tag{7.99}$$

where the equality defines the relation between the temperatures T and T' of the two systems. This equation therefore defines a renormalization transformation since we can think of the smaller of the two systems, say the one of size L, as having been obtained from that of size L' by a block-spin transformation. The following procedure is especially effective. Consider two infinite strips of width L and L'. The transfer matrix for such strips is of finite dimension ($2^L \times 2^L$ for the Ising model) and it is easy to calculate the largest two eigenvalues for quite large matrices. Using equations (7.96) and (7.99) then produces the recursion relation for the temperature-like coupling. The critical or fixed point $T^*(L,L')$ is then given by $\xi(L,T^*)/L = \xi(L',T^*)/L'$, and we expect that $T^*(L,L')$ will converge to the critical temperature of the infinite system as $L, L' \to \infty$. In the vicinity of this fixed point we have

$$\frac{\xi(L,T^*+dT)}{L} = \frac{\xi(L',T^*+dT')}{L'}$$

or

$$\left.\frac{dT}{dT'}\right|_{T^*} = b^{y_t} = \left.\frac{L\xi'(L',T')}{L'\xi'(L,T)}\right|_{T^*} \tag{7.100}$$

where $b = L'/L$ and the thermal exponent $y_t = 1/\nu$.

We illustrate this procedure by a simple calculation for the two-dimensional Ising model. Consider two strips of width $L' = 4$ and $L = 2$. We could easily diagonalize the transfer matrices, but because of the exact solution of Section 6.1, we have analytic expressions for the largest eigenvalues. Referring back to equations (6.37) and (6.38) and noting that the two largest eigenvalues of $\mathbf{V}$ are the two largest eigenvalues in the even and odd subspaces we have

$$\xi^{-1}(M,K) = -2K + \sum_{q_1\neq 0,\pi} \epsilon(q_1) - \sum_{q_2} \epsilon(q_2)$$

where $q_1 = (0, \pm 2, \pm 4, \ldots, M)\pi/M$ and $q_2 = (\pm 1, \pm 3, \ldots, \pm(M-1))\pi/M$ for a strip of width M and where $\cosh\epsilon(q) = \cosh 2K \coth 2K + \cos q$. With the eigenvalues in hand, it is easy to determine the fixed point for $L' = 4$, $L = 2$. The value is $K^* = J/k_BT_c = 0.4266$ which is in respectable agreement with the exact result $K_c = 0.4407$. Similarly, the thermal eigenvalue obtained from (7.100) is $y_t = 1.075$ which again compares well with the exact value $y_t = 1.0$.

It turns out that with a bit more work, the agreement between the phenomenological renormalization group and the exact results can be improved substantially. However, more important is the fact that it can be applied to a large number of other two-dimensional systems, generally with very good results. For a more extensive discussion the reader is referred to [24] and references therein.

7.7 The ϵ-Expansion

We next develop the renormalization group from a different point of view. Instead of dealing with very specific models, such as the Ising model on a particular lattice, we wish to exploit the notion of universality and retain in our Hamiltonian only what we believe to be the most essential features. Our development follows Wilson's original formulation of the renormalization group [325],[327]. To start with, consider an n-component spin system on a d-dimensional cubic lattice. The Hamiltonian is taken to be

$$\tilde{H} = -\sum_{\mathbf{r},\mathbf{a}} J(\mathbf{a})\mathbf{S}_{\mathbf{r}} \cdot \mathbf{S}_{\mathbf{r}+\mathbf{a}} - \tilde{h}\sum_{\mathbf{r}} S_{\mathbf{r}}^{\alpha} \tag{7.101}$$

where $J(\mathbf{a})$ is an interaction energy, $\tilde{h}$ a magnetic field pointing in the α direction, and $\mathbf{r}$ the points on the d-dimensional lattice. The spin variables are taken to be continuous (classical) n-component vectors $\mathbf{S}_{\mathbf{r}} = (S_{\mathbf{r}}^{x}, S_{\mathbf{r}}^{y}, \ldots)$ with

$$\sum_{\gamma=1}^{n} (S_{\mathbf{r}}^{\gamma})^2 = 1 \,.$$

It is convenient to remove the restriction that the spins have a fixed magnitude and instead to use a weighting function that allows the magnitude of the spins to fluctuate:

$$Z = \left(\prod_{\mathbf{r},\gamma}\int_{|S_{\mathbf{r}}|=1} dS_{\mathbf{r}}^{\gamma}\right) e^{-\beta\tilde{H}} \to \left(\prod_{\mathbf{r},\gamma}\int_{-\infty}^{\infty} dS_{\mathbf{r}}^{\gamma} W(|S_{\mathbf{r}}|)\right) e^{-\beta\tilde{H}} \,. \tag{7.102}$$

If we choose

$$W(|S_{\mathbf{r}}|) = \delta\left(\sum_{\gamma}(S_{\mathbf{r}}^{\gamma})^2 - 1\right)$$

the replacement (7.102) is an identity. It is, however, possible to simplify the calculation by replacing the δ-function by a continuous function such as

$$W(\mathbf{S}_{\mathbf{r}}) = e^{-(1/2)b|\mathbf{S}_{\mathbf{r}}|^2 - c(\mathbf{S}_{\mathbf{r}}\cdot\mathbf{S}_{\mathbf{r}})^2} \,. \tag{7.103}$$

With $b < 0$, $c = -b/4$, this function has a maximum at $|\mathbf{S}| = 1$ and, if $|b|$ is large, W decreases rapidly as $|\mathbf{S}|$ deviates from unity. We incorporate the logarithm of the weighting function in the effective Hamiltonian and also re-express it in terms of the momentum-space representation of the spin variables using

$$\mathbf{S}_{\mathbf{r}} = \frac{1}{\sqrt{N}} \sum_{\mathbf{q}} \mathbf{S}_{\mathbf{q}} e^{i\mathbf{q}\cdot\mathbf{r}} . \tag{7.104}$$

Here the vectors $\mathbf{q}$ are restricted to the first Brillouin zone of the simple cubic d-dimensional lattice $(-\pi/a \le q_i < \pi/a)$, $i = 1, 2, \ldots, d$. Substituting, we obtain

$$\begin{aligned} H &= -\beta\tilde{H} + \ln W(\{\mathbf{S}_{\mathbf{r}}\}) \\ &= -\sum_{\mathbf{q}} \left[\frac{b}{2} - K(\mathbf{q})\right] \mathbf{S}_{\mathbf{q}} \cdot \mathbf{S}_{-\mathbf{q}} \\ &\quad -\frac{c}{N} \sum_{\mathbf{q}_1,\mathbf{q}_2,\mathbf{q}_3} \mathbf{S}_{\mathbf{q}_1} \cdot \mathbf{S}_{\mathbf{q}_2} \mathbf{S}_{\mathbf{q}_3} \cdot \mathbf{S}_{-\mathbf{q}_1-\mathbf{q}_2-\mathbf{q}_3} + \sqrt{N} h_0 S_{\mathbf{0}}^{\alpha} \end{aligned} \tag{7.105}$$

where

$$K(\mathbf{q}) = \beta \sum_{\mathbf{a}} J(\mathbf{a}) e^{-i\mathbf{q}\cdot\mathbf{a}} \tag{7.106}$$

and $h_0 = \beta\tilde{h}$. Consider a lattice with nearest-neighbor spacing a and nearest-neighbor interactions only. With $\beta J(\mathbf{a}) = \frac{1}{2}K_0$, we have, in the limit of long wavelength (small q),

$$K(\mathbf{q}) = K_0 \sum_{j=1}^{d} \cos q_j a \approx dK_0 - \frac{1}{2} K_0 a^2 q^2 . \tag{7.107}$$

In (7.107) we have only retained terms to order q^2, since the important fluctuations near the critical point are the long-wavelength ones, and the essential physics is contained in the leading term[2]. We next rescale the spin variables so as to make the coefficient of q^2 in the effective Hamiltonian equal to 1/2 and finally obtain the form

$$\begin{aligned} H(\{\mathbf{S}_{\mathbf{q}}\}) &= -\frac{1}{2} \sum_{\mathbf{q}} (r + q^2)\, \mathbf{S}_{\mathbf{q}} \cdot \mathbf{S}_{-\mathbf{q}} + \sqrt{N} h_0 S_{\mathbf{0}}^{\alpha} \\ &\quad -\frac{u}{N} \sum_{\mathbf{q}_1,\mathbf{q}_2,\mathbf{q}_3} (\mathbf{S}_{\mathbf{q}_1} \cdot \mathbf{S}_{\mathbf{q}_2})(\mathbf{S}_{\mathbf{q}_3} \cdot \mathbf{S}_{-\mathbf{q}_1-\mathbf{q}_2-\mathbf{q}_3}) \end{aligned} \tag{7.108}$$

[2] See Chaikin and Lubensky, Section 5.8.4, for a more detailed discussion of this point.

where

$$r = \frac{b - 2dK_0}{K_0 a^2}\,, \qquad h = \frac{h_0}{(K_0 a^2)^{1/2}}\,, \qquad u = \frac{c}{K_0^2 a^4}\,.$$

In what follows we shall, invoking universality again, ignore the geometry of the simple cubic Brillouin zone and rather restrict the now dimensionless wave vectors $\mathbf{q}$ to the interior of a d-dimensional unit sphere. The desired partition function is then

$$Z = \left(\prod_{q\le 1} \int d\mathbf{S_q}\right) e^{H(\{\mathbf{S_q}\})}\,. \tag{7.109}$$

To illustrate the momentum space renormalization group, we first consider the simple case $u = 0$ which is known as the Gaussian model.

7.7.1 The Gaussian model

The partition function of the Gaussian model

$$Z = \left(\prod_{\mathbf{q},\alpha} \int dS_{\mathbf{q}}^{\alpha}\right) \exp\left\{-\frac{1}{2}\sum_{\mathbf{q}} \left(r + q^2\right) \mathbf{S_q}\cdot\mathbf{S_{-q}} + \sqrt{N} h S_0^{\alpha}\right\} \tag{7.110}$$

can obviously be calculated directly. However, we will carry out a renormalization group calculation instead for illustrative purposes. The steps are as follows:

1. Carry out the functional integration over all $\mathbf{S_q}$ with $q > q_l = 1/l$, where $l > 1$ is a parameter. In the position space approach, this corresponds to the operation of coarse graining or choosing a block spin. In carrying out the integration over $\mathbf{S_q}$ with $q > 1/l$, the minimum-length scale is changed from 1 to l; correlations at distances shorter than l can no longer be resolved. The integration produces the result (with A a constant)

$$Z = A\left(\prod_{q<q_l,\alpha} \int dS_{\mathbf{q}}^{\alpha}\right) \exp\left\{-\frac{1}{2}\sum_{q<q_l} \mathbf{S_q}\cdot\mathbf{S_{-q}}(r + q^2) + \sqrt{N} h S_0^{\alpha}\right\}\,. \tag{7.111}$$

2. Rescale lengths to the original scale (*i.e.*, let $\mathbf{q} = \mathbf{q}'/l$ with $q' \le 1$). With this transformation the exponent in (7.111) becomes

$$H' = -\frac{1}{2}\sum_{\mathbf{q}'} \left[r + \left(\frac{q'}{l}\right)^2\right] l^{-d} \mathbf{S}_{\mathbf{q}'/l}\cdot\mathbf{S}_{-\mathbf{q}'/l} + \sqrt{N} h S_0^{\alpha}\,. \tag{7.112}$$

The factor l^{-d} compensates for the extra degrees of freedom that have been introduced by the expansion (at constant density of points in $\mathbf{q}$ space) of the remaining part of the spherical 'Brillouin zone'.

3. We require that, aside from additive constants, H' must have precisely the same form as H. We have stipulated that the coefficient of the term $\frac{1}{2}q^2\mathbf{S}_\mathbf{q}\cdot\mathbf{S}_{-\mathbf{q}}$ in H should be unity. The corresponding term in (7.112) is $\frac{1}{2}q'^2 l^{-(d+2)}\mathbf{S}_{\mathbf{q}'/l}\cdot\mathbf{S}_{-\mathbf{q}'/l}$ and in order to satisfy the condition that this term has a fixed coefficient, independent of l, we make the spin rescaling transformation

$$\begin{aligned}\mathbf{S}_{\mathbf{q}'/l} &= \zeta(l)\mathbf{S}_{\mathbf{q}'}\\ \zeta(l) &= l^{1+d/2}\ .\end{aligned} \tag{7.113}$$

This spin rescaling operation is similar in spirit to the requirement that the block spin in the position space renormalization group (Sections 7.2 and 7.4) be a variable of the same type as the site spin (see Pfeuty and Toulouse [242], p. 66, for further discussion of this point).

The final form of H' is therefore

$$\begin{aligned} H' &= -\frac{1}{2}\sum_{\mathbf{q}'}\left(rl^2+q'^2\right)\mathbf{S}_{\mathbf{q}'}\cdot\mathbf{S}_{-\mathbf{q}'}+\sqrt{N}hl^{1+d/2}S_0^\alpha \\ &= -\frac{1}{2}\sum_{\mathbf{q}}\left(r'+q^2\right)\mathbf{S}_{\mathbf{q}}\cdot\mathbf{S}_{-\mathbf{q}}+\sqrt{N}h'S_0^\alpha \end{aligned} \tag{7.114}$$

with

$$r' = rl^2 \qquad h' = hl^{1+d/2}\ . \tag{7.115}$$

Equations (7.115) are the recursion relations for the Gaussian model. We see that there are three fixed points for $h=0$: $r=+\infty$ corresponding to $T=\infty$; $r=-\infty$ corresponding to $T=0$, and $r=0$, which we shall see is the critical fixed point. Assuming that r is a temperature-like variable we have the scaling form of the singular part of the free energy (6.94)

$$g(t,h) = l^{-d}g(tl^2, hl^{1+d/2})\ . \tag{7.116}$$

The critical exponents are easily found to be

$$\alpha = 2-\frac{1}{2}d \qquad \beta = \frac{d}{4}-\frac{1}{2} \qquad \gamma = 1$$

and we see that these exponents become identical with the Landau exponents when $d=4$.

Consider now the correlation function $\Gamma(\mathbf{r}) = \langle S_0^\alpha S_\mathbf{r}^\alpha \rangle$:

$$\begin{aligned}\Gamma(\mathbf{r}) &= \frac{1}{N}\sum_{\mathbf{x}} \langle S_\mathbf{x}^\alpha S_{\mathbf{x}+\mathbf{r}}^\alpha \rangle = \frac{1}{N^2}\sum_{\mathbf{q}\mathbf{q}'\mathbf{x}} \langle S_\mathbf{q}^\alpha S_{\mathbf{q}'}^\alpha \rangle e^{i\mathbf{q}\cdot\mathbf{x}} e^{i\mathbf{q}'\cdot(\mathbf{x}+\mathbf{r})} \\ &= \frac{1}{N}\sum_{\mathbf{q}} \langle S_\mathbf{q}^\alpha S_{-\mathbf{q}}^\alpha \rangle e^{-i\mathbf{q}\cdot\mathbf{r}} = \frac{1}{N}\sum_{\mathbf{q}} \Gamma(\mathbf{q}) e^{-i\mathbf{q}\cdot\mathbf{r}} \end{aligned} \tag{7.117}$$

where

$$\Gamma(\mathbf{q}) = \frac{\int dS_\mathbf{q}^\alpha dS_{-\mathbf{q}}^\alpha \, S_\mathbf{q}^\alpha S_{-\mathbf{q}}^\alpha \exp\left\{-\left(r+q^2\right) S_\mathbf{q}^\alpha S_{-\mathbf{q}}^\alpha\right\}}{\int dS_\mathbf{q}^\alpha dS_{-\mathbf{q}}^\alpha \exp\left\{-\left(r+q^2\right) S_\mathbf{q}^\alpha S_{-\mathbf{q}}^\alpha\right\}} . \tag{7.118}$$

To obtain (7.118) we have taken $h = 0$ and carried out the integral over all spin variables except $S_\mathbf{q}^\alpha$, $S_{-\mathbf{q}}^\alpha$. The factors $\frac{1}{2}$ in the exponentials have disappeared because the term $S_\mathbf{q}^\alpha S_{-\mathbf{q}}^\alpha$ appears twice in the Hamiltonian. Separating the spin variables into real and imaginary parts,

$$S_\mathbf{q}^\alpha = x + iy \qquad S_{-\mathbf{q}}^\alpha = x - iy$$

we carry out the integration and find

$$\begin{aligned}\Gamma(\mathbf{q}) &= \frac{\int_{-\infty}^{\infty} dx \int_{-\infty}^{\infty} dy (x^2+y^2) \exp\left\{-(r+q^2)(x^2+y^2)\right\}}{\int_{-\infty}^{\infty} dx \int_{-\infty}^{\infty} dy \exp\left\{-(r+q^2)(x^2+y^2)\right\}} \\ &= \frac{1}{r+q^2} . \end{aligned} \tag{7.119}$$

The spatial correlation function is given by

$$\Gamma(\mathbf{x}) = \left(\frac{1}{2\pi}\right)^d \int d^d q \, \frac{1}{r+q^2} e^{i\mathbf{q}\cdot\mathbf{x}} . \tag{7.120}$$

The form of this integral depends on the dimensionality d, but for $r \to 0$ is dominated by the contribution from the pole at $q = ir^{1/2}$ and we therefore obtain

$$\Gamma(x) \sim e^{-|x| r^{1/2}} \equiv e^{-|x|/\xi} \tag{7.121}$$

with $\xi = r^{-1/2}$. Since $\xi \sim |t|^{-\nu}$ we have, in any dimension, $\nu = \frac{1}{2}$ which is the Landau–Ginzburg result.

It is interesting to carry out the renormalization procedure described above for the correlation function $\Gamma(q)$ for $q < 1/l$. We modify the Hamiltonian slightly by adding a **q**-dependent magnetic field

$$H = -\frac{1}{2}\sum_{\mathbf{q}} (r+q^2) \mathbf{S}_\mathbf{q} \cdot \mathbf{S}_{-\mathbf{q}} + \sum_{\mathbf{q}} \mathbf{h}_\mathbf{q} \cdot \mathbf{S}_{-\mathbf{q}} . \tag{7.122}$$

Then

$$\langle \mathbf{S_q} \rangle = \frac{\partial \ln Z}{\partial \mathbf{h_{-q}}}$$

and

$$\Gamma(\mathbf{q}, r, \mathbf{h}) = \langle \mathbf{S_q} \cdot \mathbf{S_{-q}} \rangle - \langle \mathbf{S_q} \rangle \cdot \langle \mathbf{S_{-q}} \rangle = \frac{\partial^2 \ln Z}{\partial \mathbf{h_q} \partial \mathbf{h_{-q}}} . \tag{7.123}$$

We now integrate over the spin variables with $q > 1/l$ and use (7.113) with $\mathbf{q} = \mathbf{q}'/l$ and obtain

$$\begin{aligned} H' &= -\frac{1}{2} \sum_{\mathbf{q}'} \left(rl^2 + q'^2 \right) \mathbf{S_{q'}} \cdot \mathbf{S_{-q'}} + \sum_{\mathbf{q}'} \mathbf{h}_{\mathbf{q}'/l} \zeta l^{-d} \mathbf{S_{q'}} \\ &= -\frac{1}{2} \sum_{\mathbf{q}} \left(r' + q^2 \right) \mathbf{S_q} \cdot \mathbf{S_{-q}} + \sum_{\mathbf{q}} \mathbf{h}'_{\mathbf{q}} \cdot \mathbf{S_{-q}} \end{aligned} \tag{7.124}$$

with $r' = rl^2$, $\mathbf{h}'_{\mathbf{q}} = \mathbf{h}_{\mathbf{q}/l} \zeta l^{-d}$. The correlation function $\Gamma(\mathbf{q}, r', \mathbf{h}')$ for the block spin system is given by

$$\Gamma(\mathbf{q}, r', \mathbf{h}') = \frac{\partial^2 \ln Z'(r', \mathbf{h}')}{\partial \mathbf{h}'_{\mathbf{q}} \partial \mathbf{h}'_{-\mathbf{q}}} = \zeta^{-2} l^{2d} \frac{\partial^2 \ln Z'(r', \mathbf{h}')}{\partial \mathbf{h}_{\mathbf{q}/l} \partial \mathbf{h}_{-\mathbf{q}/l}} . \tag{7.125}$$

Finally, we let $\mathbf{q} \to l\mathbf{q}$, to obtain

$$\Gamma(l\mathbf{q}, r', \mathbf{h}') = \zeta^{-2} l^{2d} \frac{\partial^2 \ln Z'(r', \mathbf{h}')}{\partial \mathbf{h_q} \partial \mathbf{h_{-q}}} = \zeta^{-2} l^d \Gamma(\mathbf{q}, r, \mathbf{h}) \tag{7.126}$$

where we have used the fact (7.116) that $\ln Z' = l^{-d} \ln Z$. We next let $\mathbf{h} \to 0$ to find the scaling relation for the correlation function

$$\Gamma(l\mathbf{q}, r') = l^{-2} \Gamma(\mathbf{q}, r) . \tag{7.127}$$

At the fixed point $r = 0$, and letting $l = q^{-1}$ we find that

$$\Gamma(\mathbf{q}, 0) = q^{-2} \Gamma(\hat{q}, 0) .$$

Conventionally, one writes $\Gamma(\mathbf{q}, 0) \sim q^{-2+\eta}$ and we therefore find as in (7.119), that in the Gaussian model $\eta = 0$.

7.7.2 The S^4 model

We now return to the more general case and consider the effective Hamiltonian

$$H = -\frac{1}{2} \sum_{\mathbf{q}} (r + q^2) \mathbf{S_q} \cdot \mathbf{S_{-q}}$$

$$-\frac{u}{N}\sum_{\mathbf{q}_1,\mathbf{q}_2,\mathbf{q}_3}(\mathbf{S}_{\mathbf{q}_1}\cdot\mathbf{S}_{\mathbf{q}_2})(\mathbf{S}_{\mathbf{q}_3}\cdot\mathbf{S}_{-\mathbf{q}_1-\mathbf{q}_2-\mathbf{q}_3})$$
$$-\frac{w}{N^2}\sum_{\mathbf{q}_1\ldots\mathbf{q}_5}(\mathbf{S}_{\mathbf{q}_1}\cdot\mathbf{S}_{\mathbf{q}_2})(\mathbf{S}_{\mathbf{q}_3}\cdot\mathbf{S}_{\mathbf{q}_4})$$
$$\times(\mathbf{S}_{\mathbf{q}_5}\cdot\mathbf{S}_{-\mathbf{q}_1-\mathbf{q}_2-\mathbf{q}_3-\mathbf{q}_4-\mathbf{q}_5})+\cdots+h\sqrt{N}S_0^\alpha \tag{7.128}$$

where sixth- and higher-order terms have been added to (7.108) because we expect them to be generated by the renormalization transformation. The Hamiltonian (7.128) is commonly referred to as the Landau–Ginzburg–Wilson Hamiltonian because of its similarity to the Landau–Ginzburg free-energy functional (Section 3.10). Our renormalization procedures will be the same as in the preceding section, but since the integrals over the fourth- and higher-order spin terms are difficult, we construct a cumulant expansion for the partition function. We write

$$H = H_0 + H_1 \tag{7.129}$$

where

$$H_0 = -\frac{1}{2}\sum_{\mathbf{q}}(r+q^2)\mathbf{S}_{\mathbf{q}}\cdot\mathbf{S}_{-\mathbf{q}}+h\sqrt{N}S_0^\alpha \tag{7.130}$$

and

$$H_1 = -\frac{u}{N}\sum_{\mathbf{q}_1\ldots\mathbf{q}_4}(\mathbf{S}_{\mathbf{q}_1}\cdot\mathbf{S}_{\mathbf{q}_2})(\mathbf{S}_{\mathbf{q}_3}\cdot\mathbf{S}_{\mathbf{q}_4})\Delta(\mathbf{q}_1+\mathbf{q}_2+\mathbf{q}_3+\mathbf{q}_4)+\cdots\,. \tag{7.131}$$

In (7.131), $\Delta(\mathbf{q}_1+\mathbf{q}_2+\mathbf{q}_3+\mathbf{q}_4)$ is the d-dimensional Kronecker delta. Proceeding as in Section 7.4 (7.53), we have

$$\begin{aligned} e^{H'} &\propto \left(\prod_{\alpha,q>q_l}\int dS_{\mathbf{q}}^\alpha\right)e^{H_0+H_1} \\ &= \left(\prod_{\alpha,q>q_l}\int dS_{\mathbf{q}}^\alpha\right)\left(1+H_1+\frac{1}{2}H_1^2+\cdots\right)e^{H_0} \\ &= e^{H_0}\exp\left\{\langle H_1\rangle+\frac{1}{2}(\langle H_1^2\rangle-\langle H_1\rangle^2)+\cdots\right\} \end{aligned} \tag{7.132}$$

where

$$\langle A\rangle = \frac{\prod_{\alpha,q>q_l}\int dS_{\mathbf{q}}^\alpha\,A\,e^{H_0}}{\prod_{\alpha,q>q_l}\int dS_{\mathbf{q}}^\alpha\,e^{H_0}}\,. \tag{7.133}$$

As in the cumulant approximation in the position space renormalization group, a systematic approximation scheme is to retain more and more cumulants. We will carry out this calculation to second order since this produces the first non-trivial result for the critical exponents. Even this low level calculation involves considerable algebra and we have left most of this to an Appendix to the present Chapter. However, we describe the procedure here, carry out the first order calculation and then discuss the fixed point structure and critical exponents in the second order approximation. Consider the first-order truncation of the cumulant series:

$$H' = H_0 + \langle H_1 \rangle \quad . \tag{7.134}$$

In H_1 we have a piece for which all wavevectors $\mathbf{q}$ are in the inner shell $q < 1/l$. These are unaffected by the partial integration over spins in the outer shell and result in a contribution to H' of the form

$$\begin{aligned} H_1' &= -\frac{1}{2} \sum_{q<1/l} (r + q^2)\mathbf{S}_{\mathbf{q}} \cdot \mathbf{S}_{-\mathbf{q}} - \frac{u}{N} \sum_{q_j<1/l} \mathbf{S}_{\mathbf{q}_1} \cdot \mathbf{S}_{\mathbf{q}_2} \mathbf{S}_{\mathbf{q}_3} \cdot \mathbf{S}_{\mathbf{q}_4} \\ &\quad - \frac{w}{N^2} \sum_{q_j<1/l} \mathbf{S}_{\mathbf{q}_1} \cdot \mathbf{S}_{\mathbf{q}_2} \mathbf{S}_{\mathbf{q}_3} \cdot \mathbf{S}_{\mathbf{q}_4} \mathbf{S}_{\mathbf{q}_5} \cdot \mathbf{S}_{\mathbf{q}_6} + \sqrt{N} h S_0^\alpha \end{aligned} \tag{7.135}$$

where the wave vectors in the middle two terms must add to zero. The pieces of H_1 with some or all wave vectors in the outer shell fall into the following categories (for the fourth order term):

1. $\mathbf{q}_1$, $\mathbf{q}_2$, $\mathbf{q}_3$, $\mathbf{q}_4$ are all in the region $q > q_l$. The contribution from such terms to the numerator is simply a constant, since all variables are integrated out. These terms contribute to the free energy but not to the renormalized Hamiltonian.

2. One, or three, of the vectors $\mathbf{q}_i$ lie in the region $q > q_l$. The contribution from such terms is zero by symmetry.

3. Two of the vectors $\mathbf{q}_i$ are larger than q_l in magnitude. There are two distinct possibilities:

   (a) $\mathbf{q}_1 = -\mathbf{q}_2$, $\mathbf{q}_3 = -\mathbf{q}_4$. Let the latter two vectors be the ones which are larger than q_l. The expectation value of such a term is then

   $$n\Gamma(q_3) \sum_{\mathbf{q}_1,\alpha} S^\alpha_{\mathbf{q}_1} S^\alpha_{-\mathbf{q}_1}$$

   where $\Gamma(q)$ is the Gaussian correlation function calculated above and n is the number of components of the spin vectors.

(b) $\mathbf{q}_1 = -\mathbf{q}_3$, $\mathbf{q}_2 = -\mathbf{q}_4$ (or $\mathbf{q}_1 = -\mathbf{q}_4$, $\mathbf{q}_2 = -\mathbf{q}_3$). If we select the first possibility with the two first vectors smaller than q_l, we obtain contributions of the form

$$\sum_{\alpha} S^{\alpha}_{\mathbf{q}_1} S^{\alpha}_{\mathbf{q}_3} \langle S^{\alpha}_{\mathbf{q}_2} S^{\alpha}_{\mathbf{q}_4} \rangle \Delta(\mathbf{q}_1 + \mathbf{q}_2 + \mathbf{q}_3 + \mathbf{q}_4) = \mathbf{S}_{\mathbf{q}_1} \cdot \mathbf{S}_{-\mathbf{q}_1} \Gamma(q_2) \,.$$

Adding up terms, we find that

$$\begin{aligned} H' &= -\frac{1}{2} \sum_{q<q_l} \left[r + q^2 + \frac{u}{N} 4(n+2) \sum_{q'>q_l} \Gamma(q') \right] \mathbf{S}_{\mathbf{q}} \cdot \mathbf{S}_{-\mathbf{q}} \\ &\quad - \frac{u}{N} \sum_{q_1 \cdots q_4 < q_l} (\mathbf{S}_{\mathbf{q}_1} \cdot \mathbf{S}_{\mathbf{q}_2})(\mathbf{S}_{\mathbf{q}_3} \cdot \mathbf{S}_{\mathbf{q}_4}) \Delta(\mathbf{q}_1 + \mathbf{q}_2 + \mathbf{q}_3 + \mathbf{q}_4) \\ &\quad - \frac{w}{N^2} \sum_{q_j < q_l} \mathbf{S}_{\mathbf{q}_1} \cdot \mathbf{S}_{\mathbf{q}_2} \mathbf{S}_{\mathbf{q}_3} \cdot \mathbf{S}_{\mathbf{q}_4} \mathbf{S}_{\mathbf{q}_5} \cdot S_{\mathbf{q}_6} \\ &\quad \times \Delta(\mathbf{q}_1 + \mathbf{q}_2 + \mathbf{q}_3 + \mathbf{q}_4 + \mathbf{q}_5 + \mathbf{q}_6) \,. \end{aligned} \tag{7.136}$$

We follow the rescaling procedure (7.111)–(7.115) to obtain the recursion relations

$$r' = rl^2 + 4(n+2) l^2 \frac{u}{N} \sum_{q_1 > 1/l} \Gamma(q_1) \tag{7.137}$$

$$u' = ul^{-3d} \zeta^4 = ul^{4-d} \tag{7.138}$$

$$w' = wl^{-5d} \zeta^6 = wl^{6-2d} \tag{7.139}$$

$$h' = hl^{1+\frac{1}{2}d} \,. \tag{7.140}$$

We note that there are contributions of order w to u' but we shall show in the Appendix that these are irrelevant. Without evaluating the sum over q_1 in (7.137), we see from (7.138) that $u' < u$ if $d > 4$. In this case the Gaussian fixed point is stable with respect to the addition of a fourth-order term. Conversely, for $d < 4$, $u' > u$ and the fourth-order term grows, indicating that another fixed point with nonzero values of r^*, u^*, determines the critical exponents of the system. The structure of (7.138) is very suggestive. Suppose that we carried out the cumulant expansion to order u^2. We then expect (7.138) to become

$$u' = ul^{4-d} + \psi(l, r) u^2 \tag{7.141}$$

where $\psi(l, r)$ is some function. This equation can be used to determine u^*:

$$u^* = -\frac{l^{4-d} - 1}{\psi(l, r^*)} = -\frac{l^{\epsilon} - 1}{\psi(l, r^*)} \approx -\frac{\ln l}{\psi(l, r^*)} \epsilon$$

where $\epsilon = 4 - d$. The parameter ϵ is a natural expansion parameter for this problem and the celebrated ϵ-expansion of Wilson and Fisher [326] is based on this notion. One classifies the coupling constants $u, w, \ldots$ in (7.128) according to the leading power of ϵ with which they appear in the fixed-point equation and systematically constructs the fixed point to order ϵ^0, ϵ, ϵ^2, and so on. The critical exponents are then found to be given by a power series in ϵ which one can attempt to evaluate for $\epsilon = 1$ ($d = 3$). It turns out, as we shall see below, that the sixth-order coupling constant $w^* \propto \epsilon^3$, and higher-order coupling constants in turn are proportional to still-higher powers of ϵ at the fixed point.

We must now consider the second order term in the cumulant expansion (7.132). To simplify some of the technical aspects of the calculation, we take the rescaling parameter $l = 1 + \delta$ where δ is an infinitesimal. The combinatorical work to identify the nonvanishing contributions to u' is rather involved and can be found in the Appendix at the end of this chapter. The results for the recursion relation sufficient to yield exponents to order ϵ are

$$\begin{aligned} r' &= rl^2 + \frac{4(n+2)ul^2C\delta}{1+r} \\ u' &= l^{4-d}\left[u - \frac{4(n+8)C\delta}{(1+r)^2}u^2\right] \\ h' &= hl^{1+d/2}, \end{aligned} \tag{7.142}$$

where C is a constant. The first two of these equations can be written in the form

$$r' - r = \delta\frac{\partial r}{\partial \delta} = \left[(1+\delta)^2 - 1\right]r + \frac{4(n+2)C\delta}{1+r}u \tag{7.143}$$

$$u' - u = \delta\frac{\partial u}{\partial \delta} = \left[(1+\delta)^{4-d} - 1\right]u + \frac{4(n+8)C\delta}{(1+r)^2}u^2 . \tag{7.144}$$

At the fixed point $r = r' = r^*$, $u = u' = u^*$, and expanding to lowest order in δ we find that

$$u^* = \frac{\epsilon(1+r^*)^2}{4(n+8)C} \qquad r^* = -\frac{2(n+2)C}{1+r^*}u^* . \tag{7.145}$$

Retaining only the first-order term in ϵ, we obtain

$$u^* = \frac{\epsilon}{4(n+8)C} \qquad r^* = -\frac{(n+2)\epsilon}{2(n+8)} . \tag{7.146}$$

Let us now consider the linearized recursion relations at this fixed point. From (7.142) we have, again to first order in ϵ and δ,

$$\begin{aligned}
\frac{\partial r'}{\partial r} &= l^2 - \frac{4(n+2)l^2 C\delta u^*}{(1+r^*)^2} = l^2 - \frac{(n+2)\delta\epsilon}{n+8} \\
&= 1 + 2\delta - \delta\epsilon\frac{n+2}{n+8} \approx (1+\delta)^{2-\epsilon(n+2)/(n+8)} && (7.147) \\
\frac{\partial r'}{\partial u} &= \frac{4(n+2)C\delta}{1+r^*} && (7.148) \\
\frac{\partial u'}{\partial r} &= O(\epsilon^2) && (7.149) \\
\frac{\partial u'}{\partial u} &= (1+\delta)^\epsilon - 8(n+8)U^*C\delta = (1+\delta)^\epsilon - 2\epsilon\delta \approx (1+\delta)^{-\epsilon} . && (7.150)
\end{aligned}$$

The matrix **M** which determines the eigenvalues, and thus the critical exponents, takes the form

$$\mathbf{M} = \begin{bmatrix} l^{2-\epsilon(n+2)/(n+8)} & 4(n+2)C\delta \\ 0 & l^{-\epsilon} \end{bmatrix} \tag{7.151}$$

and we obtain

$$y_1 = 2 - \epsilon\frac{n+2}{n+8} \qquad y_2 = -\epsilon . \tag{7.152}$$

To this point we have ignored the magnetic field. The field-dependent term in the Hamiltonian is simply rescaled under renormalization as in (7.115):

$$h'_\alpha = l^{1+d/2}h_\alpha .$$

Thus

$$y_h = 1 + \frac{1}{2}d = 3 - \frac{1}{2}\epsilon . \tag{7.153}$$

The usual critical exponents are determined from the scaling form of the free energy. In terms of the scaling fields u_1 and u_2 which are the eigenvectors of **M** (7.151), we have for the free energy per spin,

$$g(u_1, u_2, h) = l^{-d}g(u_1 l^{y_1}, u_2 l^{y_2}, h l^{y_h}) \tag{7.154}$$

which leads to the critical exponents (Section 6.3)

$$\begin{aligned}
\alpha &= 2 - \frac{d}{y_1} = \epsilon\left[\frac{1}{2} - \frac{n+2}{n+8}\right] + O(\epsilon^2) \\
\beta &= \frac{d-y_h}{y_1} = \frac{1}{2} - \frac{3\epsilon}{2(n+8)} + O(\epsilon^2)
\end{aligned}$$

$$\begin{aligned}\gamma &= \frac{2y_h - d}{y_1} = 1 + \frac{(n+2)\epsilon}{2(n+8)} + O(\epsilon^2)\\ \delta &= \frac{y_h}{d - y_h} = 3 + \epsilon + O(\epsilon^2)\ .\end{aligned} \tag{7.155}$$

The susceptibility exponent γ for the Ising model ($n = 1$) in three dimensions to order ϵ is therefore $\frac{7}{6}$. For the XY model ($n = 2$) $\gamma = \frac{6}{5}$ and for the Heisenberg model ($n = 3$) we obtain $\gamma = \frac{27}{22}$. These estimates are not very accurate, but do display the correct qualitative trend to larger γ as n increases. The best series expansion estimates for γ are 1.24, 1.33, and 1.43 for the Ising, XY, and Heisenberg models.

We also recover, at least to order ϵ, the property of universality of critical exponents. Since the exponent y_2 is negative, the scaling field u_2 is irrelevant and thus does not affect the critical exponents. We note, however, that the four-spin interaction in (7.128) does not break the n-dimensional rotational symmetry of the Hamiltonian. To determine which of the fixed points (Ising, XY, or Heisenberg) governs the critical behavior of the anisotropic Heisenberg model, one has to introduce a symmetry breaking term into (7.128) and examine the stability of the Heisenberg fixed point with respect to such fields (Problem 7.5).

Finally, we briefly discuss the flows in coupling constant space. An examination of the linearized recursion relations near the nontrivial fixed point shows that the coupling constants approach the fixed point along the line $(r - r^*)/(u - u^*) = -2(n+2)/C$. This line connects the unphysical (for $\epsilon > 0$) Gaussian fixed point with the nontrivial fixed point. The coupling constants move away from the Gaussian fixed point along the line $u^* = u$. Thus the flows behave qualitatively as sketched in Figure 7.15. As ϵ decreases, the physical fixed point approaches the Gaussian fixed point along the critical surface, and at $\epsilon = 0$ the two fixed points merge. For $\epsilon < 0$ ($d > 4$), $u^*(\epsilon) < 0$ and the nontrivial fixed point becomes unstable, and thus unphysical. For all $d > 4$ the critical properties of these spin systems are determined by the Gaussian fixed point. The magnetic field direction is not depicted in the diagram. It may be thought of as a third axis orthogonal to both the r and u axes.

7.7.3 Conclusion

The preceding derivation and that in the Appendix are rather lengthy. This is in part due to our assumption that the reader may not be familiar with Feynman diagram techniques. More elegant derivations of the foregoing results can be found in the books of Ma [182] or Pfeuty and Toulouse [242] and the

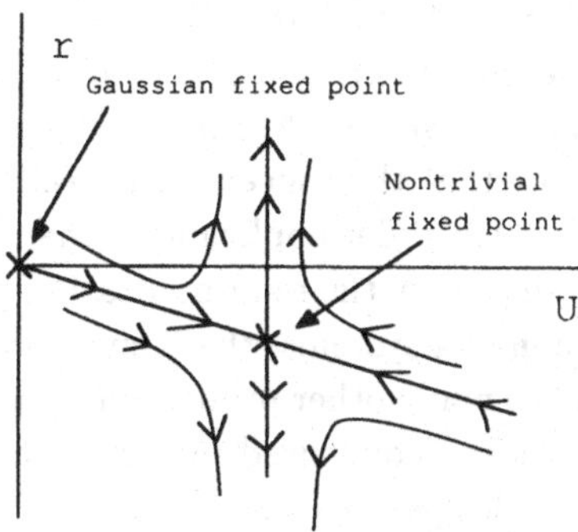

Figure 7.15: Coupling-constant flow for the S^4 model when $d < 4$ ($\epsilon > 0$).

reviews of Wilson and Kogut [327] and Fisher [96]. With considerable effort the epsilon expansion can be extended to higher orders—presently the state of the art is ϵ^5 (see [114] and [173]). Although the epsilon expansion is an asymptotic rather than a convergent series, powerful summation techniques can be used to evaluate the critical exponents in three and even two dimensions. These estimates are found to be in excellent agreement with high-temperature series results in $d = 3$ and in remarkably good agreement with the exactly known Ising model exponents in $d = 2$ [173].

Other methods of calculating the critical exponents of the Landau–Ginzburg–Wilson Hamiltonian (7.128) have also been developed. In particular, field-theoretic methods [52] have proven to be very effective and the results of Baker *et al.* [22] for the Ising model in three dimensions were in fact the first convincing demonstration that the renormalization group could yield critical exponents of the same accuracy as high-temperature series.

The most important result, in our opinion, to come from the renormalization group method is the elaboration of the notion of universality. One now has a tool that can be used to determine which features of a microscopic Hamiltonian are important in determining critical behavior. For example, it is quite straightforward to incorporate symmetry breaking terms (cubic or uniaxial anisotropy), or dipolar and isotropic long-range interactions, into the Landau–Ginzburg–Wilson Hamiltonian and to determine the flow on the critical surface to low order in ϵ. It is generally believed that the stability of fixed points for $d < 4$ does not depend sensitively on ϵ and that reliable conclusions can be drawn from low-order calculations. Thus an almost global flow diagram for the critical surface can be constructed, and the asymptotic behavior of a system close to its critical point is then determined by the fixed point into whose basin of attraction its critical point falls. For a discussion of this mapping out of the critical surface, the reader is referred to [96] or [7]. Before leaving this subject we note that the concepts and techniques of the renormalization group have found much wider applicability than we have demonstrated in this chapter. Since its discovery the renormalization group has been applied to critical dynamics, dynamical systems, the Kondo problem, conductivity in disordered materials, and to many other problems (see also Chapter 13). It has now become a standard tool of condensed matter theory, and for this reason we have devoted considerable space to it.

Appendix: Second Order Cumulant Expansion

We wish to evaluate the contribution from the term

$$\frac{1}{2}\left(\langle H_1^2\rangle - \langle H_1\rangle^2\right)$$

in (7.132) to the recursion relations for r and u to the extent necessary to obtain the fixed point and exponents to order ϵ. We first revisit the sixth order term with coefficient w which we have stated is less relevant than the fourth order term. To simplify the formalism we will take the rescaling parameter l to be $1+\delta$, where δ is an infinitesimal. The recursion relation for the coupling constant w is schematically:

$$\begin{aligned} w' &= l^{-5d}\zeta^6 w + O(u^2) \\ &= l^{-2d+6} w + O(u^2)\ . \end{aligned} \tag{7.156}$$

Since the term of order u^2 is of order ϵ^2 at the fixed point, it would seem that $w^* \propto \epsilon^2$. We show, however, that the term of order u^2 in (7.156) vanishes and that the leading term is at least of order u^3. To see this we must consider that contribution of order u^2 to the renormalized Hamiltonian, which is of sixth order in the renormalized spin variables. In general,

$$\begin{aligned} &\frac{1}{2}\left(\langle H_1^2\rangle - \langle H_1\rangle^2\right) \\ &\quad = \frac{u^2}{2N^2}\sum_{\mathbf{q}_1\ldots\mathbf{q}_8} \langle(\mathbf{S}_{\mathbf{q}_1}\cdot\mathbf{S}_{\mathbf{q}_2})(\mathbf{S}_{\mathbf{q}_3}\cdot\mathbf{S}_{\mathbf{q}_4})(\mathbf{S}_{\mathbf{q}_5}\cdot\mathbf{S}_{\mathbf{q}_6})(\mathbf{S}_{\mathbf{q}_7}\cdot\mathbf{S}_{\mathbf{q}_8})\rangle \\ &\quad\quad \times\Delta(\mathbf{q}_1\ldots\mathbf{q}_4)\Delta(\mathbf{q}_5\ldots\mathbf{q}_8) - \frac{1}{2}\langle V\rangle^2\ . \end{aligned} \tag{7.157}$$

Because of the term $-\frac{1}{2}\langle H_1\rangle^2$ in (7.157) we require that some of the wave vectors $\mathbf{q}_1 \ldots \mathbf{q}_4$ be paired with vectors from the set $\mathbf{q}_5 \ldots \mathbf{q}_8$. Thus, in sixth order in spin variables we must have one $\mathbf{q}$, say $\mathbf{q}_4$, from $\mathbf{q}_1 \ldots \mathbf{q}_4$ in the outer shell $1/(1+\delta) < q_4 < 1$. Similarly, one vector, say $\mathbf{q}_8$, must be in the same shell. From condition 2 following (7.135) we have $\langle V\rangle = 0$ for this assignment of wave vectors and we obtain a contribution of the form

$$\frac{u^2}{2N^2}\sum_{\mathbf{q}_1\ldots\mathbf{q}_7}(\mathbf{S}_{\mathbf{q}_1}\cdot\mathbf{S}_{\mathbf{q}_2})(\mathbf{S}_{\mathbf{q}_3}\cdot\mathbf{S}_{\mathbf{q}_7})(\mathbf{S}_{\mathbf{q}_5}\cdot\mathbf{S}_{\mathbf{q}_6})\Gamma(q_4)$$
$$\times\Delta(\mathbf{q}_1+\mathbf{q}_2+\mathbf{q}_3+\mathbf{q}_4)\Delta(\mathbf{q}_5+\mathbf{q}_6+\mathbf{q}_7-\mathbf{q}_4)\ . \tag{7.158}$$

The product of Kronecker deltas can be put into the form

$$\Delta(\mathbf{q}_1+\mathbf{q}_2+\mathbf{q}_3+\mathbf{q}_5+\mathbf{q}_6+\mathbf{q}_7)\Delta(\mathbf{q}_5+\mathbf{q}_6+\mathbf{q}_7-\mathbf{q}_4)\ .$$

Since $\mathbf{q}_4$ is in a shell of width δ, the second Kronecker delta restricts the sum of $\mathbf{q}_5+\mathbf{q}_6+\mathbf{q}_7$ to a shell of width δ. As $\delta\to 0$ these terms vanish and we have no contribution of order u^2. Thus it is safe to ignore w as long as we are only interested in fixed points and exponents to order ϵ. We may also ignore the contribution of order u^2 to r' as we already have $r^* = O(u^*) = O(\epsilon)$ [see (7.137)]. Our task is therefore to evaluate the contribution to the renormalized four-spin term from (7.157).

As in the evaluation of the sixth-order terms, we may, because of the presence of the term $-\frac{1}{2}\langle H_1\rangle^2$, only consider contributions in which two of the momenta $\mathbf{q}_1 \ldots \mathbf{q}_4$ and of $\mathbf{q}_5 \ldots \mathbf{q}_8$ lie in the outer shell. There are two main cases which we illustrate below.

1. The two vectors in the outer shell from each set come from the same scalar product. A typical term has $\mathbf{q}_3 = -\mathbf{q}_7$ and $\mathbf{q}_4 = -\mathbf{q}_8$ in the outer shell. There are eight such terms, which add up to

$$\frac{4u^2}{N^2}\sum_{\substack{q_1,q_2,q_5,q_6<q_l\\ q_3,q_4>q_l}}(\mathbf{S}_{\mathbf{q}_1}\cdot\mathbf{S}_{\mathbf{q}_2})(\mathbf{S}_{\mathbf{q}_5}\cdot\mathbf{S}_{\mathbf{q}_6})\sum_{\alpha\beta}\langle S^{\alpha}_{\mathbf{q}_3}S^{\beta}_{-\mathbf{q}_3}S^{\alpha}_{\mathbf{q}_4}S^{\beta}_{-\mathbf{q}_4}\rangle$$
$$\times\Delta(\mathbf{q}_1+\mathbf{q}_2+\mathbf{q}_3+\mathbf{q}_4)\Delta(\mathbf{q}_5+\mathbf{q}_6-\mathbf{q}_3-\mathbf{q}_4)$$
$$=\frac{4nU^2}{N}\sum_{q_1,q_2,q_5,q_6<q_l}(\mathbf{S}_{\mathbf{q}_1}\cdot\mathbf{S}_{\mathbf{q}_2})(\mathbf{S}_{\mathbf{q}_5}\cdot\mathbf{S}_{\mathbf{q}_6})\Delta(\mathbf{q}_1+\mathbf{q}_2+\mathbf{q}_5+\mathbf{q}_6)$$
$$\times\frac{1}{N}\sum_{q_3,q_4>q_l}\Gamma(q_3)\Gamma(q_4)\Delta(\mathbf{q}_5+\mathbf{q}_6-\mathbf{q}_3-\mathbf{q}_4)\ . \tag{7.159}$$

2. The remaining 64 pairings of two vectors from the sets $\mathbf{q}_1 \dots \mathbf{q}_4$ and $\mathbf{q}_5 \dots \mathbf{q}_8$ all yield the same expectation values, given here for the case $\mathbf{q}_2 = -\mathbf{q}_6$ and $\mathbf{q}_4 = -\mathbf{q}_8$ by

$$
\begin{aligned}
&\frac{32u^2}{N^2} \sum_{\substack{q_1,q_3,q_5,q_7<q_l \\ q_2,q_4>q_l,\alpha,\beta,\gamma,\nu}} S^{\alpha}_{\mathbf{q}_1} S^{\beta}_{\mathbf{q}_3} S^{\gamma}_{\mathbf{q}_5} S^{\nu}_{\mathbf{q}_7} \langle S^{\alpha}_{\mathbf{q}_2} S^{\gamma}_{-\mathbf{q}_2} S^{\beta}_{\mathbf{q}_4} S^{\nu}_{-\mathbf{q}_4} \rangle \\
&\quad \times \Delta(\mathbf{q}_1 + \mathbf{q}_2 + \mathbf{q}_3 + \mathbf{q}_4) \Delta(\mathbf{q}_5 - \mathbf{q}_2 + \mathbf{q}_7 - \mathbf{q}_4) \\
&= \frac{32u^2}{N^2} \sum_{\substack{q_1,q_3,q_5,q_7<q_l \\ q_2,q_4>q_l,\alpha,\beta}} S^{\alpha}_{\mathbf{q}_1} S^{\alpha}_{\mathbf{q}_5} S^{\beta}_{\mathbf{q}_3} S^{\beta}_{\mathbf{q}_7} \Gamma(q_2)\Gamma(q_4) \\
&\quad \times \Delta(\mathbf{q}_1 + \mathbf{q}_2 + \mathbf{q}_3 + \mathbf{q}_4) \Delta(\mathbf{q}_5 - \mathbf{q}_2 + \mathbf{q}_7 - \mathbf{q}_4) \\
&= \frac{32u^2}{N} \sum_{q_1,q_3,q_5,q_7<q_l} (\mathbf{S}_{\mathbf{q}_1} \cdot \mathbf{S}_{\mathbf{q}_5})(\mathbf{S}_{\mathbf{q}_3} \cdot \mathbf{S}_{\mathbf{q}_7}) \Delta(\mathbf{q}_1 + \mathbf{q}_2 + \mathbf{q}_5 + \mathbf{q}_7) \\
&\quad \times \frac{1}{N} \sum_{q_2,q_4>q_l} \Gamma(q_2)\Gamma(q_4) \Delta(\mathbf{q}_2 + \mathbf{q}_4 - \mathbf{q}_5 - \mathbf{q}_7) \, .
\end{aligned}
\tag{7.160}
$$

Cases such as when $\mathbf{q}_3 = -\mathbf{q}_4 = \mathbf{q}_7 = -\mathbf{q}_8$ are all in the outer shell have vanishingly small phase space and can be ignored.

Thus the renormalized fourth-order term in the Hamiltonian becomes, after rescaling,

$$
\begin{aligned}
&\frac{u}{N} l^{4-d} \sum_{\mathbf{q}_1,\mathbf{q}_2,\mathbf{q}_3,\mathbf{q}_4} (\mathbf{S}_{\mathbf{q}_1} \cdot \mathbf{S}_{\mathbf{q}_2})(\mathbf{S}_{\mathbf{q}_3} \cdot \mathbf{S}_{\mathbf{q}_4}) \Delta(\mathbf{q}_1 + \mathbf{q}_2 + \mathbf{q}_3 + \mathbf{q}_4) \\
&\quad \times \left[1 - 4(n+8) \frac{u}{N} \sum_{q,q'>q_l} \Gamma(q)\Gamma(q') \Delta(\mathbf{q}_3 + \mathbf{q}_4 - \mathbf{q} - \mathbf{q}') \right] .
\end{aligned}
\tag{7.161}
$$

For the second-order term we have

$$
r' = rl^2 + 4(n+2)ul^2 \frac{1}{N} \sum_{q>q_l} \Gamma(q) \, . \tag{7.162}
$$

We now convert the sum appearing in (7.162) to an integral:

$$
\frac{1}{N} \sum_{q>q_l} \Gamma(q) = \left(\frac{1}{2\pi} \right)^d \int_{1>q>1/(1+\delta)} d^d q \frac{1}{r+q^2} = \frac{C}{1+r} , \tag{7.163}
$$

where $C = A_d/(2\pi)^d$ and A_d is the area of a d-dimensional unit sphere. Therefore, we have

$$
r' = rl^2 + 4(n+2)Ul^2 \frac{C\delta}{1+r} \, . \tag{7.164}
$$

The sums appearing in the four-spin term (7.161) are of the form

$$\frac{1}{N}\sum_{q,q'>q_l}\Gamma(q)\Gamma(q')\Delta(\mathbf{q_3}+\mathbf{q_4}-\mathbf{q}-\mathbf{q}')\,. \tag{7.165}$$

The corresponding integrals are q-dependent but, as we have stated earlier, we only need the lowest order in q, *i.e.*, $\mathbf{q}_3+\mathbf{q}_4=0$ and, therefore,

$$\frac{1}{N}\sum_{q>q_l}\Gamma^2(q)=\frac{C\delta}{(1+r)^2}\,. \tag{7.166}$$

Substituting, we finally obtain

$$u'=l^{4-d}\left[u-\frac{4(n+8)u^2C\delta}{(1+r)^2}\right]\,. \tag{7.167}$$

7.8 Problems

7.1. *Renormalization of the Ising Chain.*

a: Carry out the iteration procedure indicated by (7.13)–(7.15) to obtain K_j, $g(K_j,0)$ to fourth order starting with $K=1$. Use the results to obtain an approximate expression for the dimensionless free energy per spin f. Your result should be quite close to the exact value $f=1.127\ldots$.

b: Show that the zero field recursion relation (7.13) can also be written

$$\tanh K'=\tanh^2 K$$

and that this implies that the recursion relation for the correlation length is

$$\xi'(K',0)=\frac{1}{2}\xi(K,0)\ .$$

7.2. *First-Order Cumulant Approximation.*

Find the critical exponents of the two-dimensional spin-$\frac{1}{2}$ Ising model on the (a) square, (b) triangular, and (c) honeycomb lattices in the first-order cumulant approximation using blocks of spins as in Figure 7.16. Along the way you will also find the critical temperatures for these lattices. The values obtained from the exact solutions are $K_{c,sq}\approx 0.441$, $K_{c,triang}\approx 0.275$, $K_{c,hon}\approx 0.658$.

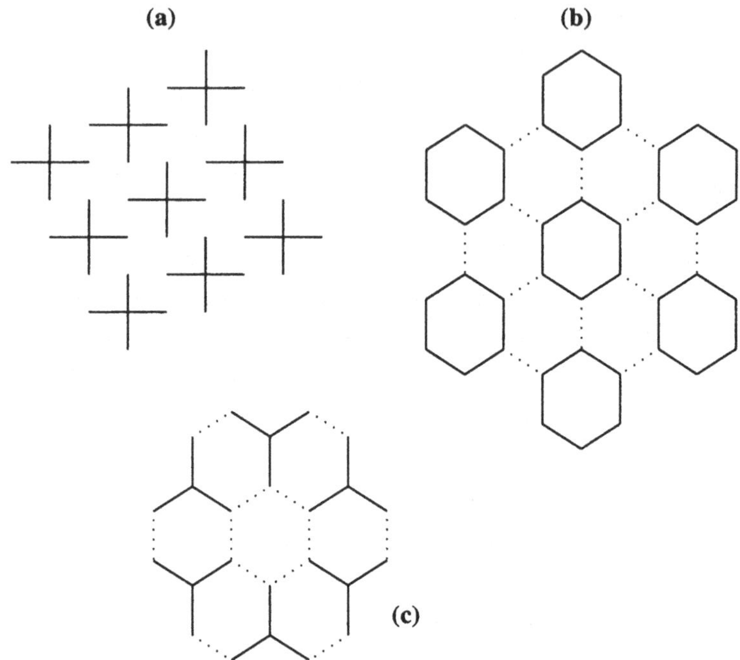

Figure 7.16: Selection of blocks for Problem 7.2.

7.3. *Second-Order Cumulant Approximation.*

Use the truncation procedure outlined below (7.68) to obtain the recursion relations (7.69) for the second-order cumulant approximation to the Ising model on a triangular lattice.

7.4. *Migdal-Kadanoff Transformation.*

Consider the Ising model on the square lattice and construct a renormalization transformation according to the following two-step process:

Step 1. Shift one-half of the horizontal bonds by one lattice spacing to obtain the modified interactions shown in Figure 7.17. It is now easy to perform the trace over the spins on the sites denoted by open circles to produce the anisotropic renormalized Hamiltonian shown in Figure 7.18.

Step 2. Shift one-half of the vertical bonds by one lattice spacing to obtain Figure 7.19. Once again carry out the trace over the variables at

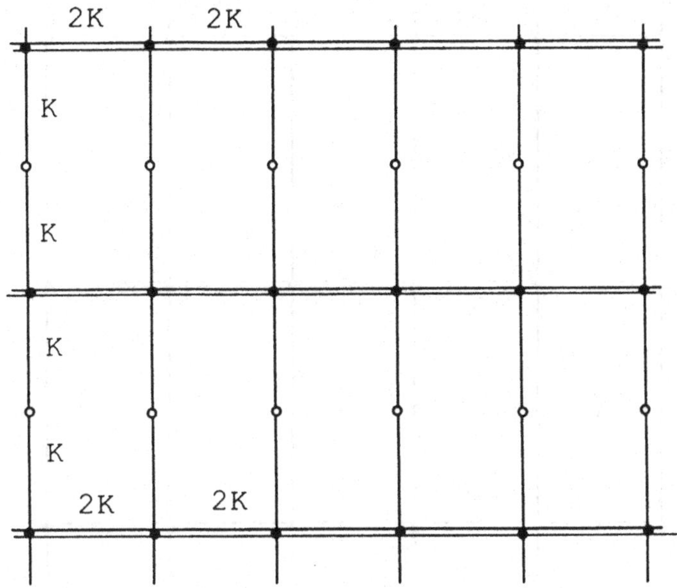

Figure 7.17: First step in Migdal–Kadanoff procedure.

the open circles to obtain the new renormalized Hamiltonian indicated in Figure 7.20. Symmetrize the Hamiltonian by defining

$$K' = \frac{1}{2}(2\gamma' + 2\gamma) = \gamma'(K) + \gamma(K) \ .$$

The rescaling parameter is $b = 2$ in this case. Solve for the fixed point of the transformation and obtain the leading thermal exponent y_t. Compare with the exact critical temperature and specific heat exponent.

7.5. *ϵ-Expansion for the Anisotropic Heisenberg Model.*

Apply the methods of Section 7.7 to the anisotropic n-vector model with Landau–Ginzburg–Wilson Hamiltonian

$$\begin{aligned} H &= -\frac{1}{2}\sum_{\mathbf{q}}\sum_{\alpha=1}^{n-1}(r_s + q^2)S^\alpha_{\mathbf{q}}S^\alpha_{-\mathbf{q}} - \frac{1}{2}\sum_{\mathbf{q}}(r_n + q^2)S^n_{\mathbf{q}}S^n_{-\mathbf{q}} \\ &\quad + \frac{V_1}{N}\sum_{\{\mathbf{q}_j\}}\sum_{\alpha,\beta<n-1} S^\alpha_{\mathbf{q}_1}S^\alpha_{\mathbf{q}_2}S^\beta_{\mathbf{q}_3}S^\beta_{-\mathbf{q}_1-\mathbf{q}_2-\mathbf{q}_3} \\ &\quad + \frac{2V_2}{N}\sum_{\{\mathbf{q}_j\}}\sum_{\alpha<n} S^\alpha_{\mathbf{q}_1}S^\alpha_{\mathbf{q}_2}S^n_{\mathbf{q}_3}S^n_{-\mathbf{q}_1-\mathbf{q}_2-\mathbf{q}_3} \end{aligned}$$

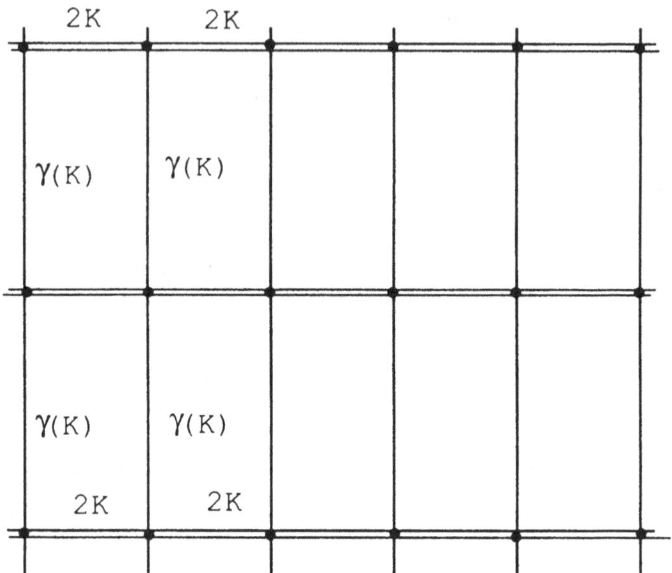

Figure 7.18: Anisotropic Ising lattice after first step.

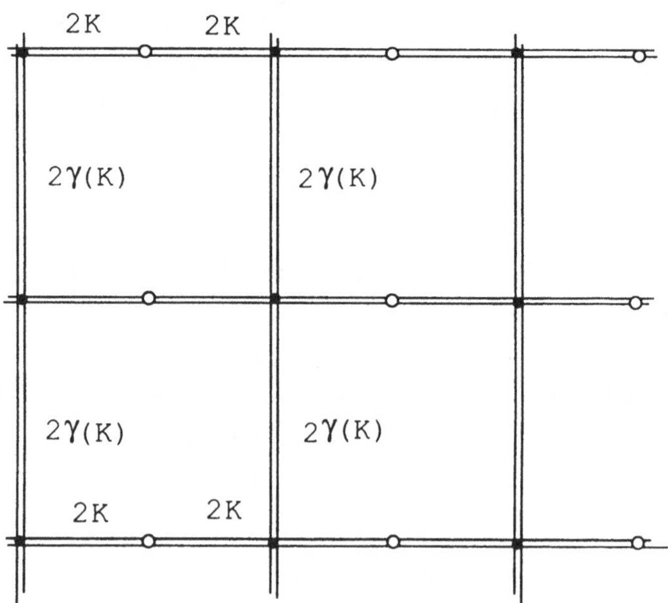

Figure 7.19: Second step in Migdal–Kadanoff transformation.

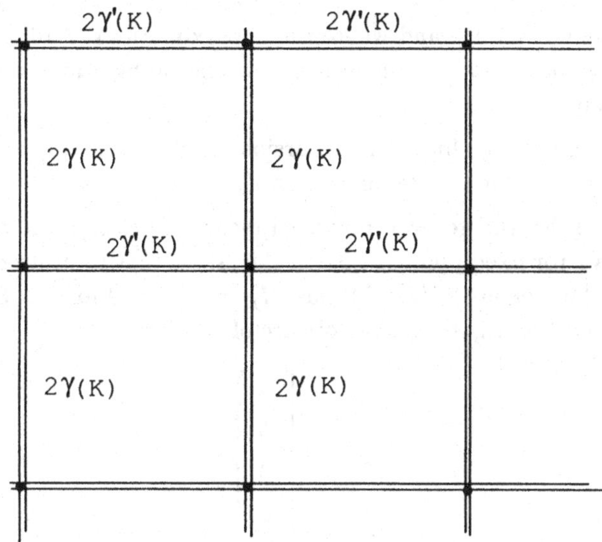

Figure 7.20: Renormalized Ising lattice after second step.

$$+\frac{V_3}{N}\sum_{\{\mathbf{q}_j\}} S^n_{\mathbf{q}_1} S^n_{\mathbf{q}_2} S^n_{\mathbf{q}_3} S^n_{-\mathbf{q}_1-\mathbf{q}_2-\mathbf{q}_3} \quad . \tag{7.168}$$

The isotropic n-vector model is clearly the special $r_s = r_n$, $V_1 = V_2 = V_3$. The Ising model corresponds to $r_s = \infty$, $V_1 = V_2 = 0$ and the $m = (n-1)$-vector model to $r_n = \infty$, $V_2 = V_3 = 0$.

(a) Show that the recursion relations to order ϵ are given by

$$\begin{aligned}
r'_s &= l^2\left[r_s - \frac{4(n+1)c\delta}{1+r_s}V_1 - \frac{4c\delta}{1+r_n}V_2\right] \\
r'_n &= l^2\left[r_n - \frac{4(n-1)c\delta}{1+r_s}V_2 - \frac{12c\delta}{1+r_n}V_3\right] \\
V'_1 &= l^\epsilon\left[V_1 + \frac{4(n+7)c\delta}{(1+r_s)^2}V_1^2 + \frac{4c\delta}{(1+r_n)^2}V_2^2\right] \\
V'_2 &= l^\epsilon\left[V_2 + \frac{16c\delta}{(1+r_s)(1+r_n)}V_2^2\right. \\
&\qquad \left. + \frac{4(n+1)c\delta}{(1+r_s)^2}V_1V_2 + \frac{12c\delta}{(1+r_n)^2}V_2V_3\right] \\
V'_3 &= l^\epsilon\left[V_3 + \frac{36c\delta}{(1+r_n)^2}V_3^2 + \frac{4(n-1)c\delta}{(1+r_s)^2}V_2^2\right]
\end{aligned} \tag{7.169}$$

where $l = 1 + \delta$ and where the approximations (7.163) and (7.166) have been made in integrating correlation functions over the outer shell.

(b) Show that the linearized recursion relations (7.147) and (7.151) are recovered for the Heisenberg, XY, and Ising cases.

(c) Consider the general linearized recursion relations at the *isotropic* n-vector fixed point. In particular, show that to order ϵ the matrix $\mathbf{M}$ [analogous to (7.151)] has $M_{j1} = M_{j2} = 0$ for $j = 3, 4, 5$. Thus two of the right eigenvectors are of the form

$$\begin{bmatrix} a \\ b \\ 0 \\ 0 \\ 0 \end{bmatrix} .$$

Show that the corresponding exponents are given by

$$y_1 = 2 - \frac{n+2}{n+8}\epsilon \qquad y_2 = 2 - \frac{2}{n+8}\epsilon \quad .$$

For small anisotropy

$$a = \frac{r_n - r_s}{r_s}$$

we may express the singular part of the zero-field free energy near the critical point in the scaling form

$$g_s(t, a, 0) = l^{-d} g_s(t l^{y_1}, a l^{y_2}, 0) = |t|^{d/y_1} \psi(a/|t|^{\phi})$$

where

$$\phi = \frac{y_2}{y_1} = 1 + \frac{n}{2(n+8)}\epsilon$$

is called the "crossover" exponent. Note that the stable fixed point in this case is the Ising fixed point, not the isotropic fixed point.

(d) Consider an anisotropic Heisenberg magnet with $a = 10^{-3}$. For what range of temperatures could one expect to measure the asymptotic Ising critical exponent?

(e) Carry out the analysis of part (c) (*i.e.*, for $n = 3$) at the Ising and XY fixed points and show that the exponent y_2 is negative (*i.e.*, both fixed points are stable).

7.6. *Ising model on a fractal lattice.* Consider a fractal object constructed by starting at magnification zero with a single bond and two Ising spins that can take on the values ± 1. At the next level each bond is replaced by six bonds and three additional Ising spins as shown:

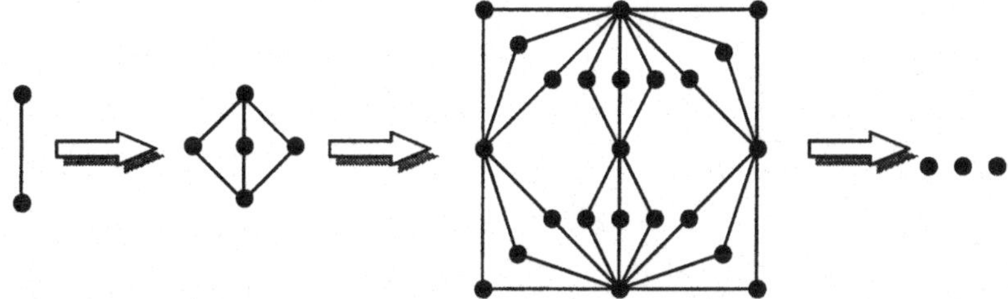

The spins interact along the bonds at the highest magnification with a ferromagnetic coupling $\hat{J}$, and are subject to a magnetic field $\hat{h}$. Let

$$J = \frac{\hat{J}}{k_B T}; \quad h = \frac{\hat{h}}{k_B T} \ .$$

(a) Use a renormalization transformation to find the ferromagnetic transition temperature T_c and the critical exponents α, β, γ, δ (for the specific heat, order parameter, susceptibility and critical isotherm, respectively). **(b)** Check your algebra by calculating the free energy in the limits

$h = 0, \ J \to 0$ and $J \to \infty$.

$J = 0, \ h \neq 0$.

Chapter 8

Stochastic Processes

To this point we have mainly studied very large systems where, in the thermodynamic limit, the behavior is governed by deterministic laws. The modern advances in biophysics, and electronics have to a large extent involved systems of smaller and smaller size. In the present chapter we will be concerned with *mesoscopic systems* that are not small or cold enough that quantum coherence is important, nor are they so large that fluctuation and noise can be neglected. We will be concerned with processes that are governed by random, unpredictable, events. The stochastic nature may arise from the probabilities of incoherent quantum processes, or derive from thermal noise that we do not care to study in detail. Alternatively, the randomness may arise because systems are open and subject to external influences that cannot be predicted. The latter source of randomness is particularly important when extending the techniques of statistical mechanics to the multiagent systems of ecology and economics.

In Section 8.1 we specialize to systems which undergo transitions from microstate to microstate at rates which depend only on the current state of the system, and not explicitly on the history. Such processes are called Markov processes, and the equation governing the transitions is referred to as the master equation. Birth and death processes represent an inportant special case, and we illustrate these in Section 8.2 with a stochastic version of the insect infestation model of Section 4.5 in which the system undergoes a first order transition in the large size limit. Birth and death processes can sometimes be described as branching processes which are the topic of Section 8.3. In this case we can solve for the dynamics of the model. When the microscopic processes are discrete, but take place on scales much smaller than those of interest, we show in Section 8.4 that the master equation can be converted into a partial differential equation, the Fokker–Planck equation. We apply this equation in Section 8.5 to the SIR (Susceptible-Infected-Removed) model of epidemiology. The master equation for continuous processes can also be converted to a Fokker–Planck equation by an expansion in *jump moments*, as shown in Section 8.6. We illustrate the method by considering Brownian motion, the Rayleigh equation for velocity decay, and the Kramers equation for Brownian motion with inertia. In Section 8.7 we discuss absorbing states and distinguish between natural and artificial boundary conditions (Section 8.7.2). The former are illusrated by the Kimura–Weiss genetic drift model in Section 8.7.1, while the latter are related to first passage time problems in Section 8.7.3. The Kramers formula for the escape rate of a particle from a shallow trap is discussed in subsection 8.7.4. The derivations of the Fokker–Planck equation generally give rise to non-hermitian differential equations. In Section 8.8 we show how to transform the Fokker–Planck by eliminating a heterogeneous diffusion term and to make it self adjoint by transforming it to a Schrödinger-like equation. The eigenvalues have an interpretation in terms of rates for different processes and the conversion allows one to separate drift and diffusive processes.

Our approach to probability theory and stochastic processes will be intuitive. For a more rigorous, but still readable, introduction we recommend the book by Durrett [81]. On the applied side Riley *et al.* [255] has a readable summary of the properties of the most common distributions. The text by van Kampen [309] is a good general reference for much of the material in this chapter. In a large part of this chapter we are concerned with solving the Fokker–Planck equation in various forms. A most valuable resource for methods to solve this equation is the book by Risken [256].

8.1 Markov Processes and the Master Equation

A stochastic variable q may take on different values depending on the *realization* (or the instance). The most complete description of such a variable is obtained by specifying its probability distribution $P(q)$. For continuous distributions $P(q)dq$ represents the probability that the stochastic variable takes on a value between q and $q + dq$. Often $P(q)$ is discrete, *i.e.*, q can only take on a countable number of values (*e.g.*, integer values for population sizes). Any acceptable distribution must be positive semidefinite and satisfy a normalization condition

$$P(q) \geq 0; \ \sum_q P(q) = 1 \text{ or } \int P(q)dq = 1 \,. \tag{8.1}$$

We wish to describe situations in which q represents the *microscopic* state of a dynamical system and assume that the system changes in time through *stochastic transitions* which occur at *rates* $W(q_{new}|q_{old})$ for $q_{old} \Rightarrow q_{new}$. These rates specify the model. There are many possible applications, *e.g.*,

- If we ignore entanglements and quantum coherence, the dynamics of *quantum mechanics* is commonly described by Fermi's golden rule. The states q could then be the ground and excited states of an atom or molecule, the phonon states of a solid, or the conformations of a polymer chain.

- In genetics q could represent the frequency of occurrence of different alleles in a population which changes through birth and death processes.

- Population biology also concerns itself with the size of populations due to birth and death, predator–prey interactions, and the immigration and emigration of populations.

- Epidemiology is concerned with transitions between different categories such as susceptible, infected, recovered, or more recently the spreading of viruses and worms between computers.

- Finance attempts to describe how prices and transaction volumes of stocks, commodities and currencies change in time in a noisy environment.

In general we will be interested in situations where microscopic fluctuations take place on timescales which are *short* compared to other times of interest. One may describe the evolution of the system by looking at individual *realizations* of the process through simulations (Chapter 9), and study the statistics of different possible outcomes. In the present chapter we attempt a more complete description by considering the *probability* $P(q,t)$ that the system is in state q at time t. This probability will satisfy the integro-differential equation

$$\frac{\partial P(q,t)}{\partial t} = \int dq'[W(q|q')P(q',t) - W(q'|q)P(q,t)] \tag{8.2}$$

while for discrete system the integral is replaced by a sum. This equation is called the *master* equation.

The physical picture is that dynamics consists of changes by transitions *in* and *out* of states. We stress that the master equation describes evolution of the probability distribution, it does not describe the evolution of the actual state of the system. In all cases we assume that the system is *intrinsically* stochastic and evolves through a *Markov process.* A Markov process is one in which the transition rates depend only on the state of the system – the process has *no memory.* Atoms and molecules retain no memory of how they got into the state they are in, but people do (sometimes less, sometimes more than they should). We can partially get around this problem by increasing the state space to include "the state of the brain". Some processes do not occur a constant rate, but be more likely after a certain latency period. We can in part get around that problem by introducing new, "latent" states.

The system may eventually approach a time independent solution $P_s(q)$. We call this the *steady state.* Sometimes the steady state is also an *equilibrium state.* If this is the case we require that the solution can be written in the form

$$P_s(q) = \frac{\exp(-\beta F(q))}{\sum_q \exp(-\beta F(q))}$$

where $F(q)$ is a free energy.

The master equation (8.2) describes *continuous time* evolution. In Section 9.2 we describe simulations employing the Monte Carlo method. Most commonly the simulations are carried out as a process in which the transitions takes place at discrete times. We show that the steady state frequencies of occurrence of the different microstates in any instance of the process follow the equilibrium distribution, if the transition probabilities satisfy *detailed balance.* If we are only interested in equilibrium properties it will then not matter if the transition probabilities describe the actual dynamics.

8.2 Birth and Death Processes

If the states of the system can be represented by an integer n, the master equation is discrete

$$\frac{\partial P(n,t)}{\partial t} = \sum_{n'} [W(n|n')P(n',t) - W(n'|n)P(n,t)] \ . \tag{8.3}$$

Often the transitions $n' \to n$ can be described as birth and death processes that take place one at a time. We then have a *one step process* and we write

$$W(n|n') = d(n')\delta_{n,n'-1} + b(n')\delta_{n,n'+1} \tag{8.4}$$

where $d =$ death rate, $b =$ birth rate. It is of interest to identify possible steady states. For one step processes we must have for the stationary probability distribution

$$0 = d(n+1)P_s(n+1) + b(n-1)P_s(n-1) - d(n)P_s(n) - b(n)P_s(n) \ . \quad (8.5)$$

We cannot have a negative number of individuals, so for birth and death processes we must have $d(0) = 0$ and $P_s(-1) = 0$. This gives if we substitute $n = 0$ in (8.5)

$$d(1)P_s(1) = b(0)P_s(0) \ .$$

We can repeat the procedure for $n = 1, 2 \cdots$ and find

$$d(n)P_s(n) = b(n-1)P_s(n-1) \quad (8.6)$$

allowing us to solve the master equation by first expressing $P_s(n)$ in terms of $P_s(0)$:

$$P_s(n) = \frac{\prod_{i=0}^{n-1} b(i)}{\prod_{i=1}^{n} d(i)} P_s(0) \quad (8.7)$$

and evaluate $P_s(0)$ from the normalization condition

$$\sum_n P_s(n) = 1 \ . \quad (8.8)$$

The above method for finding the steady state solution works only when there is only one independent variable n. When there are more variables we need a different approach (see Section 8.5).

As an example we now reformulate the insect infestation model of Section 4.5 as a one-step birth and death process, and we refer to that section for the definition of the different constants. One difference is that we now need to specify both the net birth rate β and the death rate $r_B - \beta$ not just the reproduction rate r_B. We take the rate $b(n)$ (4.45) to be

$$b(N) = \beta N + \Delta \quad (8.9)$$

and write for the removal rate

$$d(N) = (\beta - r_B)N + \frac{r_B N^2}{K_B} + \frac{BN^2}{A^2 + N^2} \ . \quad (8.10)$$

Equations (8.9) and (8.10) define our stochastic version of the model. For $\Delta = 0$ we have $b(0) = 0$ and $P_s(N) = 0$ for all $N > 0$. If Δ is nonzero, but very small we can use some other reference number and find

$$\ln \frac{P_s(N_2)}{P_s(N_1)} \approx \int_{N_1}^{N_2} dN \ln\left(\frac{b(N)}{d(N)}\right) \ . \quad (8.11)$$

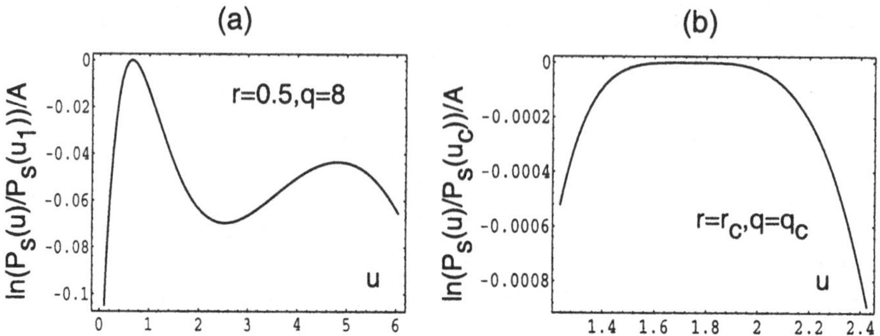

Figure 8.1: (**a**) Logplot of the probability distribution $p(u)$ for $r = 0.5$ and $q = 0.8$ in region where there are two maxima. Near the maximum $p(u)$ is approximately Gaussian. (**b**) Logplot of $p(u)$ for $r = r_c$, $q = q_c$ near $u = u_c$ $p(u) \propto \exp(-(u-u_c)^4/a)$.

We have approximated the sum over N by an integral, something which is allowed if N is large. If $b(N) > d(N)$, $P_s(N)$ will be an increasing function of N while the opposite is true when $b(N) < d(N)$. Comparing (4.45) and (8.9), (8.10) we see that the maxima of $P_s(N)$ will be the same as the stable steady states of the mean field model in Section 4.5. Let us introduce the same reduced variables as in (4.49)

$$u = \frac{N}{A};\ r = \frac{Ar_B}{B},\ q = \frac{K_B}{A},\ \delta = \frac{\Delta}{B}$$

and define $\rho = A\beta/B$ and $p_s(u) = AP_s(N)$. We find

$$I(u_1, u_2) \equiv \ln\left(\frac{p_s(u_2)}{p_s(u_1)}\right) = A\int_{u_1}^{u_2} du \ln\frac{\rho u + \delta}{(\rho - r)u + \frac{ru^2}{q} + \frac{u^2}{1+u^2}}\ . \tag{8.12}$$

Suppose u_1 and u_2 are the stable steady state solutions of the mean field model in Section 4.5 in the region of parameter space in which there are two stable steady states. If $I(u_1, u_2) > 0$ then u_2 represents the global maximum of $p_s(u)$ while u_1 will be the maximum if $I(u_1, u_2) < 0$. It is natural to interpret this maximum as a true equilibrium state for a large system. We leave it as an exercise to show that this result is qualitatively similar, but different in detail, from what we obtained in Section 4.5 where we selected the solution with the lowest "free energy" g given by (4.51). A new feature is that the result depends not only on the effective birth rate r_B but on both the natural birth rate β and the effective rate.

In Figure 8.1 we plot the logarithm of $p(u)$ for parameter values in the region where there are two maxima, and at the critical point. At the steady state value of $u = u_s$, the death and birth rate are the same and

$$I(u - u_s) \propto A(u - u_s)^2.$$

This means that fluctuations in N will be Gaussian, and typically of the order $\sqrt{N}$. We leave it as an exercise to show that near the critical point

$$I(u - u_c) \propto A(u - u_c)^4 \tag{8.13}$$

so that typical fluctations will be of the order $N^{3/4}$.

8.3 Branching Processes

We next consider a particularly simple birth and death process in which the individuals, independently of each other, either reproduce or die

$$A \to A + A \text{ or } A \to 0$$

with rates β and δ respectively. This process can be visualized as a branching process.

The state of zero individuals is an *absorbing state.* If the system reaches this state it stays there. The corresponding master equation is

$$\frac{\partial P_n(t)}{\partial t} = \beta(n-1)P_{n-1}(t) + \delta(n+1)P_{n+1}(t) - (\delta + \beta)nP_n(t) \ . \tag{8.14}$$

Suppose we start with one individual at $t = 0$. We would like to know something about the size and duration of the resulting process, or cascade. One quantity of interest is the *survival probability*

$$P_a(t) = 1 - P_0(t) \tag{8.15}$$

which is the fraction of the cascades that last longer than time t. We are particularly interested in its asymptotic value

$$P_\infty = \lim_{t \to \infty} P_a(t) \tag{8.16}$$

P_∞ could *e.g.* represent the probability that a new mutation will survive when introduced into a population.

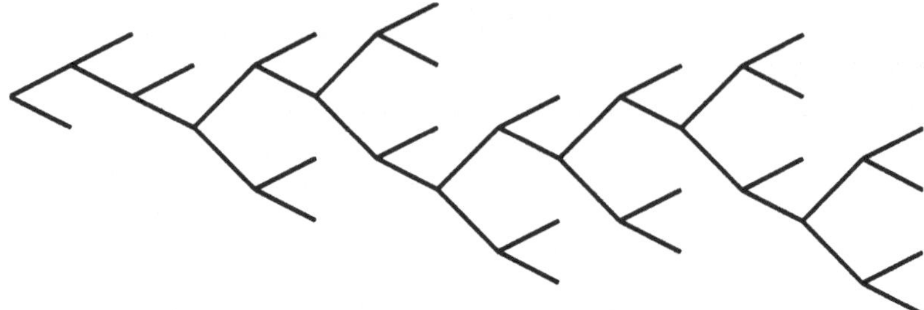

Figure 8.2: Branching process

When $\Delta \equiv \beta - \delta = 0$, we will find that several quantities exhibit power law (critical) behavior:

$$P_a(t) \sim t^{-\gamma} \tag{8.17}$$

$$P_\infty \sim \Delta^\theta \ . \tag{8.18}$$

We will derive formulas for these exponents.

Many probability problems can be solved using the *generating function* defined as

$$G(z,t) = \sum_{n=0}^{\infty} P_n(t) z^n \ . \tag{8.19}$$

From this function, we can find the *moments* by taking derivative with respect to z and then setting $z = 1$. For example:

$$\langle n \rangle = \left. \frac{\partial G(z)}{dz} \right|_{z=1} \tag{8.20}$$

and

$$\langle n(n-1) \rangle = \left. \frac{\partial^2 G(z)}{dz^2} \right|_{z=1} \ . \tag{8.21}$$

The moments

$$\phi_m = \langle n(n-1)(n-2)\cdots(n-m+1) \rangle = \left. \frac{d^m G(z)}{dz^m} \right|_{z=1} \tag{8.22}$$

are referred to as *factorial moments.* From the factorial moments, one can get back the probabilities

$$P_n = \frac{1}{n!} \sum_{m=n}^{\infty} \frac{(-1)^k}{k!} \phi_m, \tag{8.23}$$

with $k = m - n$. We define the *conditional probability*

$$P_{n/m}(t_2, t_1) = P_{n/m}(t_2 - t_1) \tag{8.24}$$

as the probability that there are n individuals at time t_2 given that there were m at time t_1.

The conditional probabilities satisfy the *Chapman-Kolmogorov* equation

$$P_{n/m}(t+h) = \sum_{m'} P_{n/m'}(t) P_{m'/m}(h) \tag{8.25}$$

which is just a statement that at the intermediate time t the system must be in *some* state. The time derivative of the conditional probability is given by

$$\frac{d}{dt} P_{n/m}(t) = \lim_{h\to 0} \frac{1}{h} \left(P_{n/m}(t+h) - P_{n/m}(t)\right) \ . \tag{8.26}$$

We single out the state $m' = m$

$$\frac{d}{dt} P_{n/m}(t) = \lim_{h\to 0} \frac{1}{h} \left(\sum_{m'\neq m} P_{n/m'}(t) P_{m'/m}(h) + (P_{m/m}(h) - 1) P_{n/m}(t) \right) \tag{8.27}$$

and define the transition rate as

$$W(m'|m) = \lim_{h\to 0} \frac{P_{m'/m}(h)}{h} \ . \tag{8.28}$$

The rate at which "something" is happening is

$$\lim_{h\to 0} \frac{1}{h} (1 - P_{m/m}(h)) = \sum_{m'\neq m} W(m'|n) \ . \tag{8.29}$$

Collecting terms we obtain the *Kolmogorov backwards* equation

$$\frac{dP_{n/m}(t)}{dt} = \sum_{m'} W(m'|m)(P_{n/m'}(t) - P_{n/m}(t)) \ . \tag{8.30}$$

For our branching process, the backward equation for 1-initial individual boundary conditions is

$$\frac{dP_{n,1}(t)}{dt} = -(\delta + \beta) P_{n/1}(t) + \beta P_{n/2}(t) + \delta P_{n/0}(t) \ . \tag{8.31}$$

Note that since the 0 individual state is absorbing

$$P_{n/0}(t) = \begin{cases} 1 & n = 0 \\ 0 & n > 0 \end{cases} \ . \tag{8.32}$$

We define the conditional generating functions

$$G_1(z) = \sum_{n=0}^{\infty} P_{n/1}(t) z^n$$

$$G_2(z) = \sum_{n=0}^{\infty} P_{n/2}(t) z^n \tag{8.33}$$

in terms of which the backward equation becomes

$$\frac{\partial G_1(z,t)}{\partial t} = -(\delta+\beta) G_1(z,t) + \beta G_2(z,t) + \delta \ . \tag{8.34}$$

In our model what happens at any branch is independent of what happens at the other branches. Therefore

$$P_{n/2}(t) = \sum_{m=0}^{n} P_{n-m/1}(t) P_{m/1}(t)$$

from which it follows that

$$G_2(z,t) = [G_1(z,t)]^2 \tag{8.35}$$

and we get

$$\frac{\partial G_1(z,t)}{\partial t} = -(\delta+\beta) G_1 + \beta G_1^2(z,t) + \delta,$$

subject to the initial condition

$$G_1(z,0) = z$$

hence

$$G_1(z,t) = \frac{e^{(\delta-\beta)t}(\beta z - \delta) + \delta(1-z)}{e^{(\delta-\beta)t}(\beta z - \delta) + \beta(1-z)} \ . \tag{8.36}$$

We are now in the position to work out properties of the solution. The *survival probability* in a cascade starting with a single individual is

$$P_a(t) = 1 - G_1(0,t) = \frac{\delta - \beta}{\delta e^{(\delta-\beta)t} - \beta} \ . \tag{8.37}$$

For $\beta < \delta$ (death rate higher than the birth rate) the survival probability, as expected, goes to zero in the long time limit. In the opposite limit $\beta > \delta$ we find for the long time survival probability

$$P_\infty = 1 - \frac{\delta}{\beta} = \frac{\Delta}{\beta} \ . \tag{8.38}$$

The critical exponent θ mentioned earlier thus has the value $\theta = 1$. In the critical case $\delta = \beta$ we find for the survival probability

$$P_a(t) = \frac{1}{1 + \beta t} \tag{8.39}$$

so the critical exponent $\gamma = 1$ as well.

8.4 Fokker–Planck Equation

Consider the general one step birth and death process with master equation

$$\frac{\partial P(n,t)}{\partial t} = b(n-1)P(n-1,t) + d(n+1)P(n+1,t)$$
$$-(b(n) + d(n))P(n,t) \tag{8.40}$$

where $b(n)$, $d(n)$ are birth and death rates respectively. It is convenient to introduce the *raising and lowering operators*

$$Eg(n) \equiv g(n+1) = \exp\{\frac{\partial}{\partial n}\}g(n)$$
$$E^{-1}g(n) \equiv g(n-1) = \exp\{-\frac{\partial}{\partial n}\}g(n) \tag{8.41}$$

where $g(n)$ can be an arbitrary function of the number of individuals and where the exponentials of the derivative operator are defined in terms of the Taylor series of the exponential function. In terms of these operators the master equation becomes

$$\frac{\partial P(n,t)}{\partial t} = \left((E^{-1} - 1)b(n) + (E - 1)d(n)\right) P(n,t) \ . \tag{8.42}$$

In Section 8.2 we found for our insect infestation example that, except near the critical point, the fluctuations about the mean field were proportional to the square root of the system size parameter A (which in the particular case represented the population at which the predator consumption would start to saturate). Let us now assume that in the general case we can define a system size parameter which we call Ω. This quantity could be any extensive variable, *e.g.*, the total area or volume or a carrying capacity. We next assume that

$$n = \Omega\phi(t) + \sqrt{\Omega}x \ . \tag{8.43}$$

Here $\phi(t) =$ is a *non-fluctuating* function that describes the average population density and will satisfy an ordinary differential equation. The fluctuations around this average density are described by x which is stochastic continuous variable with probability distribution $\Pi(x,t)$. When n fluctuates by an amount Δn, x will fluctuate by $\Delta x = \Delta n/\sqrt{\Omega}$. We find

$$\Pi(x,t) = \sqrt{\Omega}P(\Omega\phi(t) + \sqrt{\Omega}x, t) \tag{8.44}$$

$$\frac{\partial \Pi}{\partial x} = \Omega\frac{\partial P}{\partial n} \tag{8.45}$$

$$\frac{\partial P}{\partial t} = \frac{1}{\sqrt{\Omega}}\frac{\partial \Pi}{\partial t} - \frac{d\phi}{dt}\frac{\partial \Pi}{\partial x} \tag{8.46}$$

$$E - 1 = \frac{1}{\sqrt{\Omega}}\frac{\partial}{\partial x} + \frac{1}{2\Omega}\frac{\partial^2}{\partial x^2} + \cdots \tag{8.47}$$

$$E^{-1} - 1 = -\frac{1}{\sqrt{\Omega}}\frac{\partial}{\partial x} + \frac{1}{2\Omega}\frac{\partial^2}{\partial x^2} + \cdots \ . \tag{8.48}$$

We next substitute into the master equation and convert it to an equation for Π, sorting terms according to order in Ω. After some algebra we obtain

$$\frac{\partial \Pi(x,t)}{\partial t} = \Omega^{1/2}\frac{\partial \Pi}{\partial x}[\,] + \{\,\}\Pi + O(\Omega^{-1/2}) \tag{8.49}$$

where the operators $[\,]$ and $\{\,\}$ are given by

$$[\,] = \frac{d\phi}{dt} - r(\phi) \tag{8.50}$$

$$\{\,\}\Pi = -\frac{r(\phi)}{d\phi}\frac{\partial}{\partial x}(x\Pi) + D(\phi)\frac{\partial^2 \Pi}{\partial x^2} \tag{8.51}$$

and the *average individual reproduction rate* is

$$r(\phi) = \frac{b(\Omega\phi) - d(\Omega\phi)}{\Omega}$$

while

$$D(\phi) = \frac{b(\Omega\phi) + d(\Omega\phi)}{2\Omega}$$

is the *average indiviual transition rate.* To obtain these equations, we have made the observation that, *e.g.*,

$$b(\Omega\phi + \sqrt{\Omega}x) \approx b(\Omega\phi) + x\sqrt{\Omega}\frac{db(\Omega\phi)}{d\phi}\,.$$

For the expansion in powers of Ω to work we must have $[\,] = 0$ and $\{\,\} = 0$ separately. The first condition yields

$$\frac{d\phi}{dt} = r(\phi) \tag{8.52}$$

which we recognize as the mean field rate equation. The solution to (8.52) depends only on the *net reproduction rate* and not on the transition rate. We next require that terms which are independent of Ω in system size expansion add up to zero. The result is the *Fokker–Planck equation:*

$$\frac{\partial \Pi(x,t)}{\partial t} = -\frac{dr(\phi)}{d\phi}\frac{\partial}{\partial x}(x\Pi) + D(\phi)\frac{\partial^2}{\partial x^2}\Pi \tag{8.53}$$

where we need to substitute $\phi(t)$ from solution to the rate equation (8.52).

We note that we can find the mean and variance of x without solving the Fokker–Planck equation explicitly. We define the intensive variable

$$f(\phi) = -\frac{dr(\phi)}{d\phi} \tag{8.54}$$

and obtain for the time derivative of the mean

$$\frac{d\langle x\rangle}{dt} \equiv \int_{-\infty}^{+\infty} \frac{\partial x\Pi(x,t)}{\partial t} dx = \int_{-\infty}^{\infty} x dx \left(f(\phi)\frac{\partial x\Pi}{\partial x} + D(\phi)\frac{\partial^2 \Pi}{\partial x^2}\right) .$$

Integrating by part assuming that $\Pi \to 0$ for $x \to \pm\infty$

$$\frac{d\langle x\rangle}{dt} = -f(\phi)\langle x\rangle . \tag{8.55}$$

Similarly we find

$$\frac{d\langle x^2\rangle}{dt} = -2f(\phi)\langle x^2\rangle + 2D(\phi) . \tag{8.56}$$

It is of interest to study the fluctuations about a steady state $\phi = \phi_0$, where $f(\phi_0) = f_0 =$ const. and $D(\phi_0) = D_0 =$ const. Assume that initially $x = \langle x\rangle = x_i$. The solution to (8.55) and (8.56) is

$$\langle x\rangle = x_i \exp(-f_0 t)$$

$$\sigma^2 = \langle x^2\rangle - \langle x\rangle^2 = \frac{D_0}{f_0}(1 - \exp(-2f_0 t)) . \tag{8.57}$$

We see that for the steady state to be stable we must have $f_0 > 0$. Recall that steady states are characterized by $r = (b-d)/\Omega = 0$, while we saw in

Section 8.2 that at the critical point $f = -dr/d\phi = 0$ as well. In that case (8.57) diverges. This means that near the critical point the system size expansion in powers of $\sqrt{\Omega}$ breaks down. This is as expected since we saw in Section 8.5 that the critical fluctuations were proportional to $\Omega^{1/4}$. In principle one should then make a system size expansion in powers of $\Omega^{1/4}$.

We can solve the Fokker–Planck equation (8.53) for the steady state in which case it becomes

$$0 = \frac{d}{dx}\left(f_0 x\Pi + D_0\frac{d\Pi}{dx}\right) .$$

If we require that the probability distribution be normalized the solution is

$$\Pi_s(x) = \sqrt{\frac{f_0}{2\pi D_0}}\exp\left(\frac{-f_0 x^2}{2D_0}\right) , \tag{8.58}$$

i.e., the steady state distribution is *Gaussian* with variance D_0/f_0. In the next two sections we will generalize these results first to the case of several variables and after that to continuous processes.

8.5 Fokker–Planck Equation with Several Variables: SIR Model

We wish to illustrate the system size expansion approach when there are several stochastic variables by taking an example from epidemiology, the SIR-model, which is possibly the most basic of epidemiological models. The emphasis in the present book is on systems that are at, or approach, equilibrium. The steady states in the SIR model cannot be considered as equilibrium states, but since this does not affect the method of analysis, we include the model here. Ecological and epidemiological models can be separated into two broad types depending on whether a deterministic or a stochastic approach is employed. In the former the description is in terms of ordinary differential equations and the approach is quite similar to the mean field approach of Chapter 4. The systemsize expansion has the advantage that it gives a handle on both approaches.

In the SIR model the population is divided into three broad classes *susceptibles* S, *infected* I and *removed* (or recovered) R. The total number of individual are then

$$N = S + I + R \tag{8.59}$$

and we take our system size parameter Ω to be the mean value of the total population. In our version of the model we assume that the death rates of susceptibles is γ , while new individuals are introduced at the rate $\Omega\gamma$. A fraction ρ of these are infected, while the remaining fraction $1-\rho$ are susceptible. The susceptibles can become infected by contact with an infected individual. The rate for this is taken to be $\beta SI/\Omega$. The infected are removed from the infected population at a rate λ as either dead or immune. We have made the simplification, which would be serious if one wished to confront actual data, of homogeneous mixing. To some extent this can be corrected for by introducing meta-populations, with homogeneous mixing within each group, and reduced contact between groups. An alternative would be to use a two-dimensional spatial model, but in recent years with increased globalization, such models have become less relevant, and have sometimes been replaced by network models of epidemics and computer viruses (see, *e.g.*, [215, 232, 233]).

As in the previous section we work with raising and lowering operators and use the subscripts S and I to indicate susceptibles or infected, respectively. The model is defined through the master equation

$$\frac{\partial P(S,I,t)}{\partial t} = \Big(\Omega(1-\rho)\gamma(E_S^{-1}-1) + \Omega\rho\gamma(E_I^{-1}-1) + \lambda(E_I-1)I$$

$$+\beta\Omega^{-1}(E_S E_I^{-1}-1)SI + \gamma(E_S-1)S\Big)\, P(S,I,t) \tag{8.60}$$

where $P(S,I,t)$ probability that S susceptibles and I infected are present at time t. As in the previous section, we assume that fluctuations in the populations are proportional to the square root of the system size

$$S = \Omega\phi(t) + \sqrt{\Omega}s; \quad I = \Omega\psi(t) + \sqrt{\Omega}i \tag{8.61}$$

and expect that ϕ and ψ satisfy deterministic rate equations while s and i fluctuate and satisfy a Fokker–Planck equation.

Let $\Pi(s,i,t)$ be the probability distribution for s,i. We have

$$\Pi(s,i,t) = \Omega P(\Omega\phi(t) + \sqrt{\Omega}s, \Omega\psi(t) + \sqrt{\Omega}i, t) \tag{8.62}$$

with the factor Ω arising from normalization condition

$$1 = \sum_{S,I} P(S,I,t) = \int\int ds di \Pi(s,i,t) \ .$$

We have

$$\frac{\partial P}{\partial t} = \frac{1}{\Omega}\frac{\partial \Pi}{\partial t} - \frac{1}{\sqrt{\Omega}}\frac{d\phi}{dt}\frac{\partial \Pi}{\partial s} - \frac{1}{\sqrt{\Omega}}\frac{d\psi}{dt}\frac{\partial \Pi}{\partial i} \ .$$

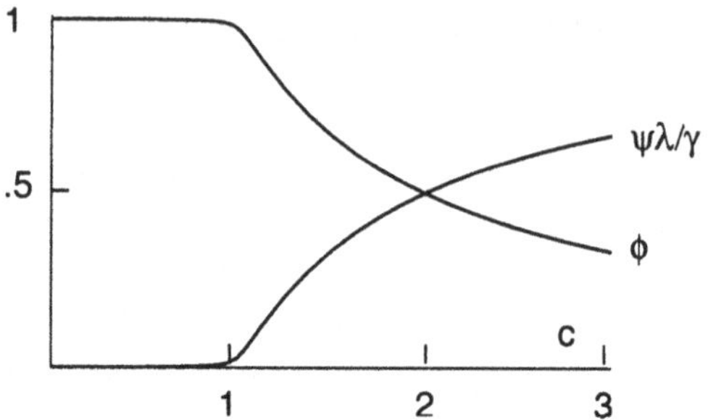

Figure 8.3: Steady state populations of susceptibles and infected as a function of the parameter c, when $\rho << 1$.

Next we expand in powers of $\sqrt{\Omega}$ and in addition to terms of the form (8.47), (8.48) we will need

$$E_S E_I^{-1} - 1 = \frac{1}{\sqrt{\Omega}} \left(\frac{\partial}{\partial s} - \frac{\partial}{\partial i} \right) + \frac{1}{2\Omega} \left(\frac{\partial^2}{\partial i^2} + \frac{\partial^2}{\partial s^2} - 2 \frac{\partial^2}{\partial s \partial i} \right) \quad . \tag{8.63}$$

Substitution and solving for $\frac{\partial \Pi}{\partial t}$ again yields an expression of the form

$$\frac{\partial \Pi}{\partial t} = \sqrt{\Omega}\{\} + [] + O[\frac{1}{\sqrt{\Omega}}] \; .$$

The term $\propto \sqrt{\Omega}$ must vanish:

$$\{\} = \frac{\partial \Pi}{\partial s} \left(\frac{d\phi}{dt} - (1-\rho)\gamma + \beta\psi\phi + \gamma\phi \right) + \frac{\partial \Pi}{\partial i} \left(\frac{d\psi}{dt} - \rho\gamma - \beta\psi\phi + \lambda\psi \right)$$

and we obtain the rate equations

$$\frac{d\phi}{dt} = (1-\rho)\gamma - \beta\psi\phi - \gamma\phi$$

$$\frac{d\psi}{dt} = \rho\gamma + \beta\psi\phi - \lambda\psi \quad . \tag{8.64}$$

From the condition $[] = 0$, we obtain the Fokker–Planck equation

$$\frac{\partial \Pi}{\partial t} = + \frac{\partial}{\partial i}(\lambda i - \beta\psi s - \beta\phi i)\Pi + \frac{\partial}{\partial s}(\gamma s + \beta\psi s + \beta\phi i)\Pi$$

$$+ \frac{1}{2}([1-\rho]\gamma + \beta\psi\phi + \gamma\phi)\frac{\partial^2 \Pi}{\partial s^2}$$

$$+\frac{1}{2}(\rho\gamma + \lambda\psi + \beta\phi\psi)\frac{\partial^2\Pi}{\partial i^2} - \beta\phi\psi\frac{\partial^2\Pi}{\partial s\partial i} \tag{8.65}$$

where the solution to the rate equations has to be substituted for ϕ and ψ. We can obtain the macroscopic steady state by putting the time derivatives of the rate equations to zero:

$$0 = (1-\rho)\gamma - \beta\psi\phi - \gamma\phi$$

$$0 = \rho\gamma + \beta\psi\phi - \lambda\psi \ .$$

The stable solution of these equations is given by

$$\phi_0 = \frac{1}{2c}(1 + c - \sqrt{(1-c)^2 + 4c\rho})$$

$$\psi_0 = \frac{\gamma}{2c\lambda}(c - 1 + \sqrt{(1-c)^2 + 4c\rho}) \tag{8.66}$$

where $c = \beta/\lambda$. If $\rho = 0$ (no immigration of infected individuals) the expressions simplify to

$$\phi_0 = \begin{cases} 1 & \text{for } c < 1 \\ \frac{1}{c} & \text{for } c > 1 \end{cases}$$

$$\psi_0 = \begin{cases} 0 & \text{for } c < 1 \\ \frac{\gamma}{\lambda}(1 - \frac{1}{c}) & \text{for } c > 1 \end{cases} \ .$$

In Figure 8.3 we plot ϕ and $\lambda\psi/\gamma$ for $\rho = .001$, as function of the parameter $c = \beta/\lambda$.

When $c < 1$ and $\rho << 1$ (intermittent regime) only a very small number of individuals succumb to the disease and an epidemic caused by immigration of an infectious individual will die out. The population will remain disease-free until the next infectious individual arrives. For $c > 1$ the disease takes hold and there will be a finite fraction of infectious individuals in the population at all times (endemic regime). For the Ω expansion to be valid we must require $\Omega\psi_0 >> 1$, $\Omega\phi_0 >> 1$. The first condition is the hardest to satisfy.

- For $c > 1$ and $\Omega >> 1$ we may take $\rho = 0$.
- For $c < 1$ and $\rho << 1 - c$ we must have $\gamma\rho\Omega/(\lambda(1-c)) >> 1$.
- For $c = 1$ we must have $\Omega\sqrt{\rho}/\lambda >> 1$.

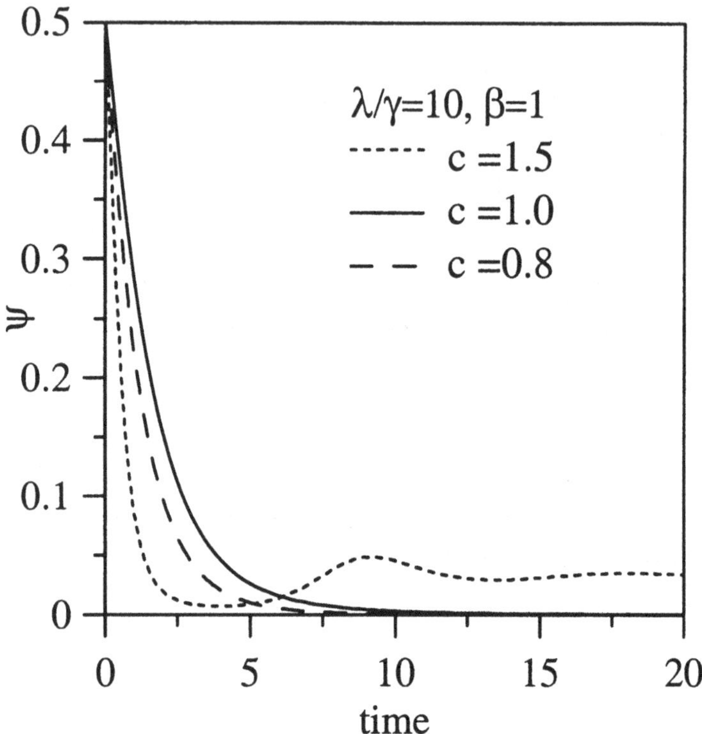

Figure 8.4: Time evolution of approach to the steady state of the infected population for some parameter values.

We interpret ψ, ϕ as an *ensemble average* of many realizations.

We show in figure 8.4 the approach to steady state for some parameter values assuming that initially half of the population infected $\rho << 1$. We note that in the endemic regime the approach to the steady state is sometimes through damped oscillations. We leave it as an exercise to work out the exact conditions for an oscillatory approach to equilibrium. The time dependence of the mean, variance and correlation $\langle si \rangle - \langle s \rangle \langle i \rangle$ for fluctuations can be calculated as in Section 8.4. We leave it as an exercise to work out the time-dependence of these quantities when the endemic steady state is disturbed. The generalization of our approach here to the case of many stochastic variable is, in principle, straightforward, but the algebra tends to be somewhat tedious.

8.6 Jump Moments for Continuous Variables

So far we have only discussed the Fokker–Planck equations for birth and death processes, in which case it was natural to make an expansion in terms of system size. We next turn to continuous processes. Important examples are the motion of particles in a viscous medium, and diffusion problems in general. We will still be dealing with a Fokker–Planck equation of the general form[1]

$$\frac{\partial P(x,t)}{\partial t} = -\frac{\partial}{\partial x}A(x)P(x,t) + \frac{\partial^2}{\partial x^2}D(x)P(x,t) \ . \tag{8.67}$$

In the present context this equation is also called the *Smoluchowski* equation or *second Kolmogorov* equation. The structure of this equation becomes clearer if we rewrite it on the form

$$\frac{\partial P(x,t)}{\partial t} = -\frac{\partial J}{\partial x} \tag{8.68}$$

where

$$J = A(x)P(x,t) - \frac{\partial}{\partial x}D(x)P(x,t) \tag{8.69}$$

is the *probability current*, $A(x)P(x,t)$ is commonly referred to as the *drift* or *transport* term, and $\frac{\partial}{\partial x}D(x)P(x,t)$ is the *diffusion* term. This terminology is in analogy with the customary treatment of macroscopic currents. Suppose $n(x)$ is the concentration of some conserved quantity, *e.g.*, particle number

$$\int n(x,t)dx = \text{const} \ .$$

Then

$$\frac{\partial n(x,t)}{\partial t} = -\nabla\cdot\mathbf{j}$$

where $\mathbf{j}$ is the particle current density. This current density is often obtained from a *constitutive relation*, *e.g.*,

$$\mathbf{j} = \mu\mathbf{f}n - D\nabla n$$

where μ is the *mobility* which for a heavily damped system is defined in terms of the drift velocity (see also Section 12.4.2) $\langle\mathbf{v}\rangle = \mu\ \mathbf{f}$, with $\mathbf{f}$ the *force* (*e.g.*

[1]We will restrict ourselves in the following discussion to a single stochastic variable x. The generalization of (8.67) to higher dimensions is $\partial P(\mathbf{x},t)/\partial t = -\nabla\cdot\mathbf{A}(\mathbf{x},t)P(\mathbf{x},t) + \nabla^2 D(\mathbf{x}P(\mathbf{x},t)$. The generalization of (8.68) is clearly $\partial P(\mathbf{x},t)/\partial t = -\nabla\cdot\mathbf{J}$.

due to gravity or an electric field) and D the *diffusion constant.* The main difference in the Fokker–Planck approach is that we now deal with the probability distribution rather than the particle density.

By definition the Fokker–Planck equation is linear in P. However, the Fokker–Planck equation commonly referred to as linear only if $A(x)$ is linear in x and $D(x) = \text{const}$. Processes governed by linear Fokker–Planck equations

$$\frac{\partial P(x,t)}{\partial t} = -\frac{\partial}{\partial x}[A_0 + A_1 x]P + D\frac{\partial^2 P}{\partial x^2}$$

are called *Ornstein-Uhlenbeck* processes. For such a process to relax towards a stationary steady state we require that $D > 0$. The time dependent Fokker–Planck equation can be solved explicitly for linear processes, while many numerical methods are available for nonlinear processes.

The steady state solution of this equation can be found for arbitrary $A(x)$, $D(x)$, since in the steady state the probability current is zero

$$0 = A(x)P_s(x) - \frac{d}{dx}D(x)P_s(x)$$

with solution

$$P_s(x) = \frac{const}{D(x)} \exp\left[\int^x \frac{A(y)}{D(y)}dy\right] \tag{8.70}$$

where the constant depends on the choice of lower limit in the integral, and is determined by the normalization condition. We next derive the Fokker–Planck equation using Planck's original approach starting from a continuous version of the master equation

$$\frac{\partial P(q,t)}{\partial t} = \int dq'[W(q|q')P(q',t) - W(q'|q)P(q,t)] \ .$$

We rewrite the equation in terms of the "jump" $r = q - q'$ and the jump transition rate $w(q',r) = W(q|q')$

$$\frac{\partial P(q,t)}{\partial t} = \int drw(q-r,r)P(q-r,t) - P(q,t)\int drw(q,-r) \ . \tag{8.71}$$

We expand $w(q-r,r)P(q-r,t)$ with respect to r in the first argument in w. This converts the first integral in (8.71) to the series

$$\int drw(q,r)P(q,t) - \frac{\partial}{\partial q}\int drrw(q,r)P(q,t)$$
$$+\frac{1}{2}\frac{\partial^2}{\partial q^2}\int drr^2 w(q,r)P(q,t) + \cdots \ .$$

We now define the *jump moments*

$$a_n = \int dr r^n w(q,r) \ . \tag{8.72}$$

If we truncate the expansion of (8.71) at 2nd order we recover the Fokker–Planck equation

$$\frac{\partial P(q,t)}{\partial t} = -\frac{\partial}{\partial q}[a_1(q)P(q,t)] + \frac{1}{2}\frac{\partial^2}{\partial q^2}[a_2(q)P(q,t)] \ . \tag{8.73}$$

An added bonus of this derivation is that we have derived the functions $A(q)$, $B(q)$ from the master equation which *defines* the process. We argue in Section 8.8.1 that the constituent relations can be ambiguous for heterogeneous systems. On the other hand, we will find that it is often convenient to obtain the moments from phenomenological considerations.

Why stop at second order? If we include all terms in the expansion we obtain the *Kramers-Moyal expansion*

$$\frac{\partial P(q,t)}{\partial t} = \sum_{n=1}^{\infty} \frac{(-1)^n}{n!}\frac{\partial^n}{\partial q^n}[a_n(q)P(q,t)] \ . \tag{8.74}$$

The usefulness of this expansion is limited by *Pawula's theorem* which states that, if we truncate the expansion at any finite order beyond 2nd, $P(q,t)$ is no longer positive definite, while the expansion to all order is equivalent to the master equation. Another way of looking at this is to say that when the system size expansion works, jump moments of higher than second order are of higher order in $\Omega^{-1/2}$. We refer to the book by van Kampen [309] for more detail on these rather delicate questions.

8.6.1 Brownian motion

We wish to illustrate the above results by considering the motion of a mesoscopic particle in a stationary fluid. A micron-size particle may undergo collisions with the molecules of the fluid at a rate of 10^{12} or more per second, whereas we may be interested in the motion of the mesoscopic particle on the time scale of milliseconds or microseconds. This wide separation of scales allows us to introduce an "infinitesimal" time δt which is long compared to τ_c, the mean time for collisions, but short compared to times t of interest. Let f be an external force acting on the particle and μ the mobility, $\langle v\rangle/f$, where $\langle v\rangle$

is the drift velocity of the particle. During the time δt the particle has moved a distance δx. We find for the jump moments

$$a_1 = \frac{\langle \delta x \rangle}{\delta t} = f\mu \tag{8.75}$$

$$a_2 = \frac{\langle (\delta x)^2 \rangle}{\delta t} \tag{8.76}$$

which we can substitute into the Fokker–Planck equation (8.73). As our first example, consider free Brownian motion with $f = 0$, $a_1 = 0$. We find

$$\frac{\partial P(x,t)}{\partial t} = \frac{\langle (\delta x)^2 \rangle}{2\delta t} \frac{\partial^2 P}{\partial x^2}$$

which is just the diffusion equation. Hence we require

$$D = \frac{\langle (\delta x)^2 \rangle}{2\delta t} \quad . \tag{8.77}$$

This equation does not admit a stationary (steady) solution but we can find the conditional probability

$$P(x,t|0,0) = \frac{\exp\left[-\frac{x^2}{4Dt}\right]}{\sqrt{4\pi D t}} \quad . \tag{8.78}$$

A Gaussian process of this type is called a *Bachelier–Wiener process.* Since we have assumed that $\delta t >> \tau_c$, the process is one in which the particle *moves*, the drift velocity is zero, but the actual velocity is nowhere defined. The theory for such processes was first developed by Bachelier[2] for the fluctuation of market prices, not for Brownian motion.

We now consider the effect of a force $f(x)$, with a Fokker–Planck equation

$$\frac{\partial P(x,t)}{\partial t} = -\mu \frac{\partial}{\partial x} f(x) P + D \frac{\partial^2 P}{\partial x^2} \quad . \tag{8.79}$$

A simple example is the motion of a particle in a gravitational field. In this case, we have $a_1 = -Mg\mu$. For short times the fluctuations in position are typically proportional to the square root of time, while the change in position due to drift will be proportional to time. It is only because the former are random, while the drift is not, that over intermediate times they are of the

[2]Readers interested in the history of science may wish to look up the article *Louis Bachelier* on http://www-groups.dcs.st-and.ac.uk/history/Mathematicians/Bachelier.htm.

same order of magnitude. Thus, the presence of a field is expected to change a_2 only by a negligible amount. The resulting Fokker–Planck equation is

$$\frac{\partial P(x,t)}{\partial t} = Mg\mu\frac{\partial P}{\partial x} + D\frac{\partial^2 P}{\partial x^2} . \tag{8.80}$$

Again, there is no stationary solution on the interval $-\infty < x < +\infty$.

We can, however, solve the time dependent equation by going to a moving frame of reference

$$y = x + Mg\mu t$$

$$P(x,t) = \Pi(x + Mg\mu t, t)$$

and find

$$\frac{\partial P}{\partial t} = \frac{\partial \Pi}{\partial t} + Mg\mu\frac{\partial \Pi}{\partial x} .$$

We are then left with a diffusion equation for Π. This results in the solution

$$P(x,t|0,0) = \frac{\exp - \frac{(x+Mg\mu t)^2}{4Dt}}{\sqrt{4\pi Dt}} . \tag{8.81}$$

We next consider the effect of a *reflecting boundary.* Suppose that the vessel with the diffusing particle has a bottom ($x = 0$) through which it cannot pass. This requires the probability current to be zero for $x = 0$. For the time dependent problem we apply the zero-current condition only at $x = 0$, but in the steady state the probability current is zero everywhere. Let $P_S(x)$ be the steady state probability distribution. The zero-current condition yields the equation

$$J = 0 = -Mg\mu P_s(x) - D\frac{dP_s}{dx}$$

with solution

$$P_s(x) = \text{const.} \exp -[\frac{Mg\mu x}{D}] .$$

At equilibrium we must have

$$P_e(x) = \frac{Mg}{k_B T} \exp -[\frac{Mgx}{k_B T}]$$

from which follows the *Einstein relation*

$$D = \mu k_B T . \tag{8.82}$$

The Einstein relation implies an additional condition, namely that the fluid in which the particle diffuses, is in thermal equilibrium.

8.6.2 Rayleigh and Kramers equations

To this point, we have assumed that the motion is strongly over-damped and taken the "Aristotelian" point of view that it is the velocity, not the acceleration, that is proportional to the force. The macroscopic equation of motion

$$v = \dot{x} = \mu f$$

is not the Newtonian

$$M\frac{dv}{dt} = f - \frac{v}{\mu}.$$

We next wish to consider inertia. The velocity v is now a stochastic variable and we assume that δt is much greater than time between collisions, but that δt is much less than the time for the relaxation of the velocity (to be determined). The macroscopic law is now $M\dot{v} = F - \frac{1}{\mu}v$, where M is the mass of the particle. For a spherical "Stokes" particle, *e.g.*,

$$\mu = \frac{1}{6\pi\eta r}$$

where r is the radius of the particle and η is the viscosity of the fluid through which it is moving. If there is no external force the first jump moment is

$$a_1 = \frac{\delta v}{\delta t} = -\frac{1}{M\mu}v \ . \tag{8.83}$$

We assume that the second jump moment is approximately constant

$$a_2 = \frac{\langle(\delta v)^2\rangle}{\delta t} \tag{8.84}$$

which leads to the Fokker–Planck equation

$$\frac{\partial P(v,t)}{\partial t} = \frac{1}{M\mu}\frac{\partial vP}{\partial v} + \frac{a_2}{2}\frac{\partial^2 P}{\partial v^2}. \tag{8.85}$$

In the stationary state the probability current $J(v) = 0$, and we find

$$\frac{1}{M\mu}vP + \frac{a_2}{2}\frac{\partial P}{\partial v} = 0$$

with solution

$$P_s(v) \propto \exp\left[-\frac{v^2}{a_2 M\mu}\right] \ . \tag{8.86}$$

If the steady state is an equilibrium state, $P_s(v)$ should be the Maxwell-Boltzmann distribution:

$$P_e(v) = \sqrt{\frac{M}{2\pi k_B T}} exp - \left[\frac{Mv^2}{2k_B T}\right] \ .$$

Comparing expressions, we find

$$\frac{a_2}{2} = \frac{k_B T}{M^2 \mu} .$$

Substituting into (8.85), we obtain the *Rayleigh equation*

$$\frac{\partial P(v,t)}{\partial t} = \frac{1}{M\mu}\left[\frac{\partial vP}{\partial v} + \frac{k_B T}{M}\frac{\partial^2 P}{\partial v^2}\right] \ . \tag{8.87}$$

This equation is *linear* according to our terminology and describes an *Ornstein-Uhlenbeck process* with a solution for the conditional probability

$$P(v,t|v_0,0) = \frac{\exp - \left[\frac{M(v-v_0 e^{-t/\tau_v})^2}{2k_B T(1-e^{-t/\tau_v})}\right]}{\sqrt{\frac{2\pi k_B T}{M}(1-e^{-t/\tau_v})}} \tag{8.88}$$

where the initial distribution is given by

$$P(v,0|v_0,0) = \delta(v - v_0)$$

and where $\tau_v = M\mu$ is the *velocity relaxation time.* Using this distribution, we obtain the time dependence of the mean velocity and its mean square fluctuation:

$$\langle v(t)\rangle = v_0 e^{-t/\tau_v};\ \langle v(t)^2\rangle - \langle v\rangle^2 = \frac{k_B T}{M}(1 - e^{-t/\tau_v}) \ . \tag{8.89}$$

We may also calculate the equilibrium *velocity-velocity correlation*

$$\langle v(t)v(0)\rangle = \int v dv \int v_0 dv_0 P(v,t|v_0,0)P_e(v_0)\,, \tag{8.90}$$

and find

$$\langle v(t)v(0)\rangle_e = \frac{k_B T}{M} e^{-t/\tau_v} \ . \tag{8.91}$$

We interpret the above results as implying that Aristotelian mechanics is a good approximation if $f \approx$ const when $\delta t \approx \tau_v$ or

$$\frac{df}{dx} v\tau_v << f \ . \tag{8.92}$$

Substituting $\mu f = v$, $\tau_v = M\mu$ we find that this is equivalent to

$$\frac{df}{dx} << \frac{1}{M\mu^2} \ . \tag{8.93}$$

If this condition is not met we must consider *both* v and x as stochastic variables. The jump moments are now

$$\frac{\langle \delta x \rangle}{\delta t} = v; \quad \frac{\langle \delta v \rangle}{\delta t} = \frac{f}{M} - \frac{v}{\mu}$$

$$\frac{\langle (\delta x)^2 \rangle}{\delta t} = v^2 \delta t \approx 0; \quad \frac{\langle \delta x \delta v \rangle}{\delta t} = v[\frac{\mu f - v}{M\mu}]\delta t \approx 0$$

$$\frac{\langle (\delta v)^2 \rangle}{\delta t} = \frac{k_B T}{M^2 \mu} \tag{8.94}$$

and the Fokker–Planck equation becomes

$$\frac{\partial P(x,v,t)}{\partial t} = -v\frac{\partial P}{\partial x} - \frac{f}{M}\frac{\partial P}{\partial v} + \frac{1}{M\mu}\left[\frac{\partial v P}{\partial v} + \frac{k_T}{M}\frac{\partial^2 P}{\partial v^2}\right] \tag{8.95}$$

and goes under the name of *Kramers's equation.* For a detailed discussion of solutions to (8.95), in a number of different situations, we refer to the book by Risken [256].

8.7 Diffusion, First Passage and Escape

In this section we consider situations in which a particle or a population undergoes diffusive motion towards an absorbing state or boundary. Once the absorbing boundary is reached, the system freezes. In a birth and death problem both the birth $b(n)$ and death rate $d(n)$ are zero on the boundary. If also $d(n), b(n) \to 0$ as $n \to 0$ we say the absorbing boundary is a *natural* boundary, while if

$$\lim_{n=0} d(n), b(n) \neq 0$$

we call the boundary *artificial.* The former case is most straighforward, while for artificial boundaries we have to impose boundary conditions in the continuum limit. Another way of looking at it is that the problem is one of finding the *first passsage time* to the boundary.

8.7.1 Natural boundaries: The Kimura–Weiss model for genetic drift

Kimura and Weiss are among the founders of modern genetics. They argued that to a very large extent genetic evolution is neutral. Many of our genes carry no reproductive advantage or disadvantage, and changes in their occurrence in the population take place through a diffusive process, leading eventually to extinction of some populations.

In the simplest version of the model a gene is expressed as allele a or A. We assume a stable population size in which the individuals randomly pair off. Each pair produces two offspring, and then disappears from the scene. The allowed processes are

$$\begin{aligned}
A + A &\Rightarrow A + A \text{ probability } 1\\
a + a &\Rightarrow a + a \qquad\qquad\quad 1\\
a + A &\Rightarrow a + A \qquad\qquad\quad 1/2\\
&\Rightarrow a + a \qquad\qquad\quad 1/4\\
&\Rightarrow A + A \qquad\qquad\quad 1/4\ .
\end{aligned}$$

Only the last two processes change the composition of the population. We take the rate of the contributing processes to be

$$\beta n(\Omega - n)/\Omega \tag{8.96}$$

where Ω is the size of the population and n number of individuals of type a. The master equation is

$$\frac{\partial P(n,t)}{\partial t} = (E_n + E_n^{-1} - 2)\frac{\beta n(\Omega - n)}{\Omega}P(n,t) \tag{8.97}$$

where E_n and E_n^{-1} are the familiar raising and lowering operators for the variable n. We now go to the continuum limit with macroscopic variables $x = \frac{n}{\Omega}$; $\tau = \frac{t}{\Omega}$. We have

$$\frac{1}{\Omega}\Pi(x,\tau) = P(n,t)\ . \tag{8.98}$$

The mean field rate equation is now trivial. Since $a + A \to 2a$ and $a + A \to 2A$ are equally likely, the concentration will not change in the mean field approximation. If we expand the raising and lowering operators in powers of the system size we get, to lowest order, the Fokker–Planck equation

$$\frac{\partial \Pi(x,\tau)}{\partial \tau} = \beta\frac{\partial^2}{\partial x^2}x(1-x)\Pi(x,\tau)\ . \tag{8.99}$$

We can find solutions by the method of separation of variables

$$\Pi(x,\tau) = \sum_{m=0}^{\infty} \alpha_m e^{-\lambda_m \tau} \psi_m(x) \tag{8.100}$$

and obtain the ordinary differential equation

$$\lambda_m \psi_m(x) = \beta \frac{d^2}{dx^2} x(1-x)\psi_m(x) \ . \tag{8.101}$$

The situation is now quite reminiscent of what happens when one solves the Schrödinger equation for, *e.g.*, the hydrogen atom to get the Legendre and Laguerre equations. In the present case the differential equation is singular at $x = 0$ and $x = 1$. This means that we cannot impose arbitrary boundary conditions. Requiring that the eigenfunctions are well behaved at $x = 0$, $x = 1$ will determine the allowed eigenvalues λ. We let

$$\psi = \sum_{l=0} a_k x^k$$

and obtain the recursion relation

$$a_{k+1} = \left(1 - \frac{\lambda}{\beta(k+2)(k+1)}\right) a_k \ .$$

The radius of convergence of the power series is $x = 1$ and, unless the series is terminated by proper choice of eigenvalues, will diverge as $(1-x)^{-1}$. Therefore, the allowed eigenvalues are

$$\lambda_m = \beta(m+2)(m+1); \quad m = 0, 1, 2 \cdots \tag{8.102}$$

and the eigenfunction $\psi_m(x)$ is a polynomial of degree m.

The lowest eigenvalue is $\lambda_0 = 2\beta$. Asymptotically, the probability that both alleles are found in the population then decays as

$$P_a \propto \exp(-2\beta\tau) = \exp(-\frac{2\beta t}{\Omega}) \ .$$

We note that the time before extinction is proportional to Ω, the population size.

It can be shown that the differential equation for ψ is a special case of the hypergeometric equation. The general solution can then be expressed in terms of Gegenbauer polynomials. For details see Kimura [151].

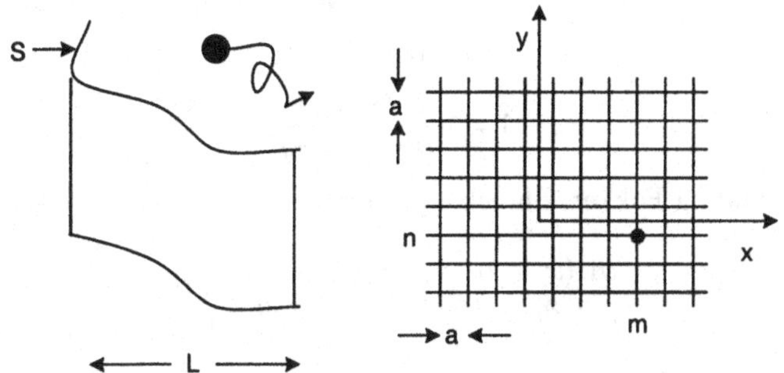

Figure 8.5: Diffusion near a cliff.

8.7.2 Artificial boundaries

When the boundary is artificial, we have to introduce an explicit boundary condition, if we wish to go to the continuum limit. Following van Kampen [309], we do this formally by introducing a *fictitious* state at the boundary S with probability $P = 0$ on S, which we impose as a boundary condition. Instead of arriving at the fictitious state the particle reaching the boundary goes to a *limbo* state denoted by (*). Let $P^*(t)$ be the probability of limbo state. The normalization condition now reads

$$P^*(t) + \sum_{n=0}^{\infty} P(n,t) = 1 \,. \tag{8.103}$$

To see how this works let us consider diffusion near a "cliff". A particle jumps at rate λ a distance a in $\pm x$ and $\pm y$ directions with S an absorbing boundary. The master equation away from the cliff is

$$\frac{\partial P(n,m,t)}{\partial t} = \lambda[E_x + E_x^{-1} + E_y + E_y^{-1} - 4]P(n,m,t) \tag{8.104}$$

where $P(n,m,t)$ denotes the probability that the particle is at $na\hat{\mathbf{x}} + ma\hat{\mathbf{y}}$ at time t (see Figure 8.5).

We introduce the continuous variables $x = na$, $y = ma$ and stipulate that $a \ll L$ where L is a macroscopic length scale. The ratio $\Omega = L/a$ is our dimensionless system size parameter. We now rewrite the master equation in terms of continuous variables:

$$\frac{1}{a^2}\Pi(x,y,t)dxdy = P(n,m,t)$$

$$E_x - 1 = a\frac{\partial}{\partial x} + \frac{a^2}{2}\frac{\partial^2}{\partial x^2}\cdots$$

$$E_y - 1 = a\frac{\partial}{\partial y} + \frac{a^2}{2}\frac{\partial^2}{\partial y^2}\cdots .$$

The corresponding Fokker–Planck equation is the diffusion equation:

$$\frac{\partial \Pi(x,y,t)}{\partial t} = D\left(\frac{\partial^2}{\partial x^2} + \frac{\partial^2}{\partial y^2}\right)\Pi \tag{8.105}$$

where $D = \lambda a^2$. The absorbing boundary condition is $\Pi = 0$ on S. The method of separation of variables then leads to an eigenvalue problem. As an example consider a square with corners at $(0,0),(0,L),(L,L),(L,0)$ and absorbing sides. For this case,

$$\Pi(x,y,t) = \sum_{n,m=1}^{\infty} \alpha_{nm}\psi_{nm}\exp(-\beta_{nm}t)$$

$$\psi_{nm} = \frac{2}{L}\sin\frac{n\pi}{L}\sin\frac{m\pi}{L}$$

$$\beta_{nm} = \frac{D(n^2+m^2)\pi^2}{L^2}$$

$$\alpha_{nm} = \int_0^L dx\int_0^L dy\psi_{nm}(x,y)\Pi(x,y,0) \ .$$

For long times, the smallest eigenvalue $\beta_{11} = 2D\pi^2/L^2$ dominates and the asymptotic formula for survival probability is then

$$\Pi_s(t) \approx \frac{8L\alpha_{11}}{\pi^2}\exp(-\frac{2D\pi^2 t}{L^2}) .$$

8.7.3 First passage time and escape probability

There are a number of important problems in which absorption of a diffusing particle occurs at special sites rather than at a boundary. Examples include the trapping of charge carriers in amorphous photoconductors, chemical reactions that require a diffusing reagent to come in contact with a stationary one, and a predator surprising a prey.[3] A quantity of interest in such situations is *first passage time* distribution which characterizes the lifetime of the diffusing

[3]The book by Redner [253] contains a great deal of information about first passage processes, beyond what we have space for here.

particle. Related to this problem, and treatable by the same methods, is the *escape probability.* In this case, we attempt to calculate the probability that a diffusing particle will ever return to its point of origin.

We consider the case of a particle hopping between neighboring sites on a lattice. For simplicity of notation we take this lattice to be a hypercube of side L in d dimensions. The probability that a particle starting at the origin at time $t = 0$ will reach site $\mathbf{x}$ at time t can be decomposed as

$$P(\mathbf{x}, t|0, 0) = \sum_n \Pi_n(\mathbf{x})\Psi_n(t) \tag{8.106}$$

where $\Pi_n(\mathbf{x})$ is the probability that the particle has gone a distance $\mathbf{x}$ after n steps and $\Psi_n(t)$ is the probability that the particle has taken n steps in time t. With a regular lattice, with constant jump rate λ, the probability distribution $\Psi_n(t)$ for n jumps in time t will, for $t\lambda >> 1$, be Gaussian with mean $\langle n\rangle = \lambda t$ and standard deviation $\sqrt{\lambda t}$. The distributions $P(\mathbf{x}, t|0, 0)$ and $\Pi_{\lambda t}(\mathbf{x})$ are then substantially the same[4], and we will here for convenience focus on the Π_n distribution. We assume that the probability distribution p for each jump is given by

$$\Pi_{n+1}(\mathbf{x}) = \sum_{\mathbf{x}'} p(\mathbf{x}, \mathbf{x}')\Pi_n(\mathbf{x}') \ . \tag{8.107}$$

For a translationally invariant lattice $p(\mathbf{x}, \mathbf{x}') = p(\mathbf{x} - \mathbf{x}')$ and equation (8.107) can be solved using methods of generating functions and Fourier transforms. The generating function is given by

$$G(\mathbf{x}, z) = \sum_{n=0}^{\infty} \Pi_n(\mathbf{x})z^n \ . \tag{8.108}$$

Substituting into (8.107) we obtain

$$G(\mathbf{x}, z) = \delta_{\mathbf{x},\mathbf{0}} + z\sum_{\mathbf{x}'} p(\mathbf{x} - \mathbf{x}')G(\mathbf{x}', z) \ . \tag{8.109}$$

We define the Fourier transforms

$$g(\mathbf{k}, z) = \sum_{\mathbf{x}} \exp(i\mathbf{k}\cdot\mathbf{x})G(\mathbf{x}, z)$$

[4] There are situations (an example is photoconductance in amorphous semiconductors [207]), where occasional long waiting times leads to a qualitative change (subdiffusion). Problems also arise when activity takes place in sudden turbulent bursts (superdiffusion), as happens from time to time in currency and equity markets (see e.g. [190]). For a review of anomalous diffusion see Bouchaud and Georges [50].

$$\lambda(\mathbf{k}) = \sum_{\mathbf{x}} \exp(i\mathbf{k} \cdot \mathbf{x}) p(\mathbf{x}) \tag{8.110}$$

where for periodic boundary conditions

$$\mathbf{k} = \frac{2\pi}{L}(n_1\mathbf{e_1} + n_2\mathbf{e_2} \cdots n_d\mathbf{e_d})$$

where the n_i are integers and $\mathbf{e_i}$ is a unit vector in the i–th direction and $-\frac{L}{a} < n_i \leq \frac{L}{a} - 1$. If the allowed jumps are to nearest neighbors on the lattice,

$$\lambda(\mathbf{k}) = \frac{2}{d} \sum_{i=1}^{d} \cos(\mathbf{k} \cdot a\mathbf{e}_i) \tag{8.111}$$

where the factor of $1/d$ is the probability of hopping to any of the d nearest neighbors. Substituting into (8.109) we obtain

$$g(\mathbf{k}, z) = \frac{1}{1 - z\lambda(\mathbf{k})} \tag{8.112}$$

and performing the inverse Fourier transform we obtain

$$G(\mathbf{x}, z) = \frac{1}{L^d} \sum_{\mathbf{k}} \frac{\exp(-i\mathbf{k} \cdot \mathbf{x})}{1 - z\lambda(\mathbf{k})} \ . \tag{8.113}$$

We are interested in the *first passage time distribution* $\Phi_n(\mathbf{x})$, the probability that the particle reaches $\mathbf{x}$ for the first time on the nth step. We have

$$\Pi_n(\mathbf{x}) = \sum_{n'=1}^{n} \Phi_{n'}(\mathbf{x}) \Pi_{n-n'}(0) \ . \tag{8.114}$$

The generating function for Φ is

$$\Gamma(\mathbf{x}, z) = \sum_{n=1}^{\infty} \Phi_n(\mathbf{x}) z^n \ .$$

Multiplying (8.114) by z^n and summing from 1 to ∞, we obtain

$$\begin{aligned} G(\mathbf{x}, z) &= \delta_{\mathbf{x},0} + \sum_{n=1}^{\infty} \sum_{n'=1}^{n} z^{n'} \Phi_{n'}(\mathbf{x}) z^{n-n'} \Pi_{n-n'}(0) \\ &= \delta_{\mathbf{x},0} + G(0, z)\Gamma(\mathbf{x}, z) \ . \end{aligned} \tag{8.115}$$

We find:

$$\Gamma(0, z) = 1 - \frac{1}{G(0, z)}$$

$$\Gamma(\mathbf{x}, z) = \frac{G(\mathbf{x}, z)}{G(0, z)}; \quad \mathbf{x} \neq 0 \ . \tag{8.116}$$

This completes the formal solution for the first passage problem. We note that we may now obtain $\Phi_n(\mathbf{x})$ for any n from

$$\Phi_n(\mathbf{x}) = \frac{1}{2\pi} \int_0^{2\pi} d\theta e^{-in\theta} \Gamma(\mathbf{x}, e^{i\theta}) \ .$$

We will return to the first passage problem in the special case $d = 1$ at the end of this section. We first discuss the escape probability.

The probability that a particle will return to its starting point ($\mathbf{x} = \mathbf{0}$) *at any time* is

$$\Gamma(\mathbf{0}, 1) = \Phi_1(\mathbf{0}) + \Phi_2(\mathbf{0}) + \Phi_3(\mathbf{0}) + \cdots \ .$$

If in the limit of an infinite lattice, $G(\mathbf{0}, 1) < \infty$, there is a finite *escape probability*, while if $G(\mathbf{0}, 1) = \infty$ the particle will "always" return home, even in the limit of an infinite lattice. In this thermodynamic limit $L \to \infty$, the sum over $\mathbf{k}$ in (8.113) can be replaced by an integral

$$\frac{1}{L^d} \sum_k \Rightarrow \frac{1}{2\pi^d} \int d^d k \ .$$

The integration on $\mathbf{k}$ is over the first Brillouin zone of the hypercubic lattice, *i.e.*, the maximum value of $\mathbf{k}$ will be finite ($\approx 1/a$). The convergence or divergence of the integral therefore depends on the behavior of the integrand for *small k* where

$$1 - \lambda(\mathbf{k}) \propto k^2 a^2 \ .$$

We conclude that, since $d^d k \propto k^{d-1} dk$, the integral diverges for $d \leq 2$. Hence, for $d > 2$ a particle starting at the origin always escapes. For $d \leq 2$ diffusing particles continue to return to their starting points. For notational convenience we have shown this only for a cubic lattice, but the result is quite general. $G(\mathbf{0}, 1)$ has been evaluated analytically for the body centered, face centered and simple cubic lattices. The return probabilities are found to be:

$$0.256318237 \cdots \ (fcc)$$

$$0.282229983 \cdots \ (bcc)$$

$$0.340537330 \cdots \ (sc) \tag{8.117}$$

To find the asymptotic behavior of the first return distribution for $d \leq 2$ we examine how $G(\mathbf{0}, \lambda)$ diverges as $\epsilon = \lambda - 1 \Rightarrow 0$:

$$G(\mathbf{0}, 1 + \epsilon) \propto \epsilon^{-1/2} \ \text{for } d = 1$$

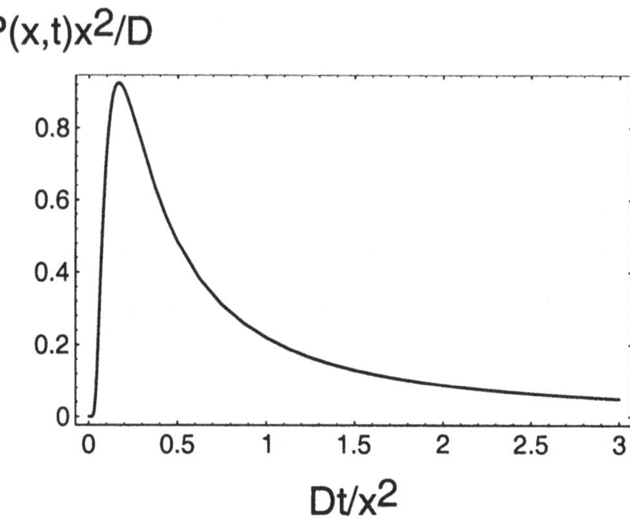

Figure 8.6: First passage time distribution for the one-dimensional Bachelier-Wiener process.

$$G(\mathbf{0}, 1+\epsilon) \propto \ln \frac{1}{\epsilon} \text{ for } d = 2 \, .$$

One can show that this implies that the probability that a walk has not yet returned home after n steps is proportional to $n^{-1/2}$ for $d = 1$ and to $1/\ln n$ for $d = 2$ in the large n limit.

In one dimension we can explicitly calculate the first passage time distribution for $x \neq 0$, in the continuum limit. We take x to be in the range: $x_1 < x < x_2$. A random walk in one dimension from x_1 to x_2 must pass through x at least once. Let $P(x_2 - x_1, t)$ be the probability of motion from x_1 to x_2 in time t. $\Phi(x - x_1, \tau)$ is the probability of arriving *for the first time* at x at time τ. Therefore,

$$P(x_2 - x_1, t) = \int_0^t P(x_2 - x, t - \tau)\Phi(x - x_1, \tau) d\tau \, . \tag{8.118}$$

Let $L(x, u)$ be the Laplace transform of $P(x, t)$ and $\Lambda(x, u)$ be the Laplace transform of $\Phi(x, \tau)$. We find

$$L(x_2 - x_1, u) = L(x_2 - x, u)\Lambda(x - x_1, u) \, .$$

For an unbiased, Gaussian random walk,

$$P(x,t) = \frac{\exp(-\frac{x^2}{4Dt})}{\sqrt{4\pi Dt}}$$

with Laplace transform

$$L(x,u) = \frac{\exp(-\sqrt{x^2u/D})}{\sqrt{4uD}} \ .$$

We find

$$\Lambda(x - x_1, u) = \exp(-\sqrt{(x - x_1)^2 u/D}) \ .$$

Performing an inverse Laplace transform we finally find

$$\Phi(x,t) = \frac{1}{t}\sqrt{\frac{x^2}{4\pi Dt}}\exp[-\frac{x^2}{4Dt}] \ . \tag{8.119}$$

We plot the distribution (8.119) in Figure 8.6. This distribution goes under the name of the *Levy-Smirnov* distribution, and has the curious property that, given x, for any realization the outcome t is finite. The time distribution is normalized, and for most realizations $t \approx x^2/D$. However, both the mean and the variance of $\langle t \rangle$ are infinite.[5] Note that since we have taken x to be continuous we cannot put $x = 0$ in (8.119).

8.7.4 Kramers escape rate

We consider an over-damped particle moving in a free energy potential well with a local minimum and an "abyss", as shown in Figure 8.7. The motion of the particle is due to thermal fluctuations and, as long as $k_BT \ll U(d) - U(b)$, we expect that the particle will remain in the vicinity of the point b for a very long time. Nevertheless, it will eventually escape and our goal in this section is to estimate this escape rate, known as the *Kramers escape rate.*

[5] A similar situation occurs in the Petersburg game first discussed by Daniel Bernoulli in 1730. In a single trial a true coin is tossed until the outcome is "tails". If this occurs after n throws the player receives 2^n ducats. The mean profit is

$$\frac{2}{2} + \cdots \frac{2^n}{2^n} \cdots = \infty \ .$$

In any realization the outcome is finite and the probability distribution for the allowed outcomes can be normalized (see e.g. [227]).

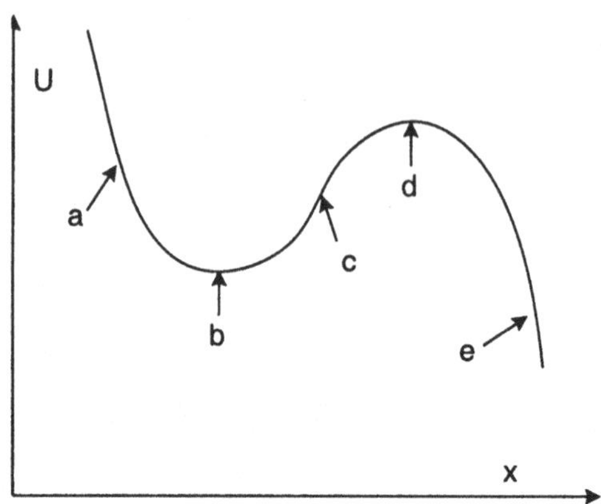

Figure 8.7: Particle moving in a potential well near a metastable state.

We formulate this problem in terms of a Fokker–Planck equation. The particle, at location x is subject to a force $f(x)$

$$f(x) = -\frac{dU(x)}{dx} \tag{8.120}$$

and we assume that the mobility is given by the Einstein equation (8.82). With this assumption, the Fokker–Planck equation is

$$\frac{\partial P(x,t)}{\partial t} = -\frac{D}{k_B T}\frac{\partial}{\partial x} f(x)P(x,t) + D\frac{\partial^2 P(x,t)}{\partial x^2} \ . \tag{8.121}$$

We furthermore assume that if the particle reaches the point e, it is lost. This means that e is an absorbing boundary. The probability current $J(x,t)$ is

$$\begin{aligned} J(x,t) &= \frac{D}{k_B T} f(x)P(x,t) - D\frac{\partial P(x,t)}{\partial x} \\ &= -D\exp(-\frac{U(x)}{k_B T})\frac{\partial}{\partial x}\exp(\frac{U(x)}{k_B T})P(x,t) \ . \end{aligned} \tag{8.122}$$

This current is nonzero because, as mentioned above, the particle must eventually escape. We now make the additional assumption that $J(x,t) = J$ is independent of x. We multiply both sides of (8.122) by $\exp(\frac{U(x)}{k_B T})$ and integrate from b to e with $P(e,t) = 0$ (absorbing boundary). This yields

$$D\exp\left(\frac{U(b)}{k_B T}\right) P(b,t) = J\int_b^e dx \exp\left(\frac{U(x)}{k_B T}\right) \ .$$

To obtain the escape rate r, we divide the probability current J by the probability p that the particle is in the vicinity of b. Near b, we have

$$P(x,t) \approx P(b,t) \exp\left(-\frac{U(x) - U(b)}{k_B T}\right)$$

and, therefore,

$$p = P(b,t) \exp\left(\frac{U(b)}{k_B T}\right) \int_a^c \exp\left(-\frac{U(x)}{k_B T}\right) dx$$

where the limits on the integral are arbitrary, but irrelevant. We obtain for the escape rate

$$r = \frac{J}{p} = \frac{D}{\int_a^c dx \exp\left(\frac{-U(x)}{k_B T}\right) \int_b^e dy \exp\left(\frac{U(y)}{k_B T}\right)} \ . \qquad (8.123)$$

The integrals in the expression above can be evaluated numerically to arbitrary accuracy. In order to get a simple and easy to remember expression we note that the dominant contribution to the first integral in the denominator comes from the region near the minimum. Expanding around $x = b$, we have

$$U(x) \approx U(b) + \frac{1}{2} k_B T (x - b)^2 \alpha \ .$$

The second integral is dominated by contributions from the region near the maximum, where

$$U(y) \approx U(d) - \frac{1}{2} k_B T (y - d)^2 \beta \ .$$

Extending the limits of integration to $(-\infty, \infty)$ and evaluating the Gaussian integrals, we finally obtain Kramers' *escape rate*

$$r = \frac{D\sqrt{\alpha\beta}}{2\pi} \exp(-\frac{\Delta U}{k_B T}) \ . \qquad (8.124)$$

This formula is valid when the escape rate is small compared to the prefactor of the exponential. We refer to the book of Risken [256] for low order correction terms. This formula exhibits the curious feature that the jump rate only depends on the energy gap and the curvature of the potential at the extrema and not on the detailed form of $U(x)$. For this result to hold the diffusion constant D must not depend on x. The case of a spatially varying diffusion constant will be discussed next.

8.8 Transformations of the Fokker–Planck Equation

8.8.1 Heterogeneous diffusion

Consider a Fokker–Planck equation of the form

$$\frac{\partial P(x,t)}{\partial t} = -\frac{\partial}{\partial x}\left[A(x)P(x,t) - \frac{\partial}{\partial x}D(x)P(x,t)\right] . \tag{8.125}$$

We recall from Section 8.6 that this equation is called *linear* if D is constant and if the drift term is of the form $A_0 + A_1 x$. The equation then describes an *Ornstein-Uhlenbeck* process. We have seen that the generalization to a nonlinear drift term is (in principle) straightforward. We will now consider the case where the diffusion term $D(x)$ is not constant, and refer to this situation as *heterogeneous diffusion.*

We first show how to construct a coordinate transformation which makes the diffusion term constant while modifying the drift term.

$$x \Rightarrow y = y(x); \quad y' = \frac{dy}{dx} . \tag{8.126}$$

Under this transformation, we have

$$P(x,t) = \frac{dy}{dx}\Pi(y,t) = y'\Pi(y,t) \tag{8.127}$$

and we also define the functions $a(y) = A(x)$ and $\delta(y) = D(x)$. The Fokker–Planck equation in the new coordinate system becomes

$$\frac{\partial \Pi(y,t)}{\partial t} = -\frac{\partial}{\partial y}\left[a(y)y'\Pi(y,t) - y'\frac{\partial}{\partial y}\delta(y)y'\Pi(y,t)\right] . \tag{8.128}$$

Using

$$y'\frac{dy'}{dy} = \frac{d^2y}{dx^2} = y''$$

and defining

$$y'\frac{d\delta}{dy} = D'$$

we find

$$\frac{\partial \Pi}{\partial t} = -\frac{\partial}{\partial y}[y'(a - D') - \delta y'']\Pi(y,t) - (y')^2\delta\frac{\partial^2 \Pi}{\partial y^2} . \tag{8.129}$$

We have not yet specified the coordinate transformation, but we now choose the transformation to satisfy

$$(y')^2\,\delta(y) = \left(\frac{dy}{dx}\right)^2 D(x) = \Delta = \text{const.} \tag{8.130}$$

with solution

$$y = \sqrt{\Delta}\int^x \frac{dz}{\sqrt{D(z)}} \tag{8.131}$$

where the lower limit of integral can be chosen freely. We finally obtain a Fokker–Planck equation of the form

$$\frac{\partial \Pi(y,t)}{\partial t} = -\frac{\partial f\Pi}{\partial y} + \Delta\frac{\partial^2 \Pi}{\partial y^2} \tag{8.132}$$

where we have a new "effective" force

$$f(y) = y'(a - D') - \delta y'' \ . \tag{8.133}$$

The original problem with heterogeneous diffusion has been transformed into a problem with constant diffusion coefficient, but a new effective drift term. There are physical effects associated with this phenomenon. One example is the *thermoelectric effect* of Section 12.4.3 in which a temperature gradient gives rise to an electromotive force and an electric field can give rise to heat flow. A further example is the *thermomolecular effect* [272] in which a temperature gradient in a rarified gas gives rise to a pressure gradient.

We illustrate this method with the following example which may seem somewhat artificial. However, it has the virtue of simplicity. Suppose

$$A(x) = -\frac{dU(x)}{dx} = -\eta \sin x$$

$$D(x) = \xi(1 + \alpha \sin x)\,, \tag{8.134}$$

so that the Fokker–Planck equation is

$$\frac{\partial P(x,t)}{\partial t} = \eta\frac{\partial}{\partial x}\sin(x)P(x,t) + \xi\frac{\partial^2}{\partial x^2}(1 + \alpha\sin x)P(x,t) \ .$$

When making the coordinate transformation (8.130), we are free to choose the value of the constant Δ. We take $\Delta = \xi$ and find, for $y(x)$

$$y(x) = \int^x \frac{dz}{\sqrt{1 + \alpha \sin z}} \ . \tag{8.135}$$

We also need the following expressions

$$y' = \frac{dy}{dx} = \frac{1}{\sqrt{1+\alpha\sin x}}$$
$$y'' = -\frac{\alpha\cos x}{2(1+\alpha\sin x)^{3/2}}$$
$$D' = 2\xi\alpha\cos x \ . \tag{8.136}$$

The effective force $f(y)$ is given by

$$f(y) = y'(a - D') - \delta y'' \equiv F(x) \tag{8.137}$$
$$= \frac{-\eta\sin x - \frac{\xi\alpha}{2}\cos x}{\sqrt{1+\alpha\sin x}}$$

and the associated potential is

$$U = -\int^{y} dz f(z) = -\int^{x} dx'\, y'(x')F(x')$$
$$= \int^{x} \frac{\eta\sin x' + \frac{\xi\alpha}{2}\cos x'}{1+\alpha\sin x'} \ . \tag{8.138}$$

We now specialize to the case $\alpha << 1$ and, to first order in α obtain

$$U(x) \approx \int dx[\eta\sin x - \alpha(\eta\sin^2 x - \frac{\xi}{2}\cos x)]$$
$$= -\eta\cos x + \frac{\alpha}{2}[\eta(-x + \sin x\cos x) - \xi\sin x] + \text{const.} \tag{8.139}$$

To this order in α, $U(x)$ has local minima at

$$x = n2\pi - \frac{\xi\alpha}{2\eta}; \quad n = 0, \pm 1, \pm 2 \cdots$$

and local maxima at

$$x = (2n+1)\pi - \frac{\xi\alpha}{2\eta}; \quad n = 0, \pm 1, \pm 2 \cdots \ .$$

The difference in potential between successive minima is

$$U[2n\pi - \xi\alpha/2\eta] - U[(2n+2)\pi - \xi\alpha/2\eta] = \pi\alpha\eta$$

so that $U(x)$ has the "washboard" form shown in Figure 8.8. The potential energy barriers to the left and right of a given minimum are

$$\Delta U_L = \eta(2 + \frac{\alpha\pi}{2}); \quad \Delta U_R = \eta(2 - \frac{\alpha\pi}{2}) \ .$$

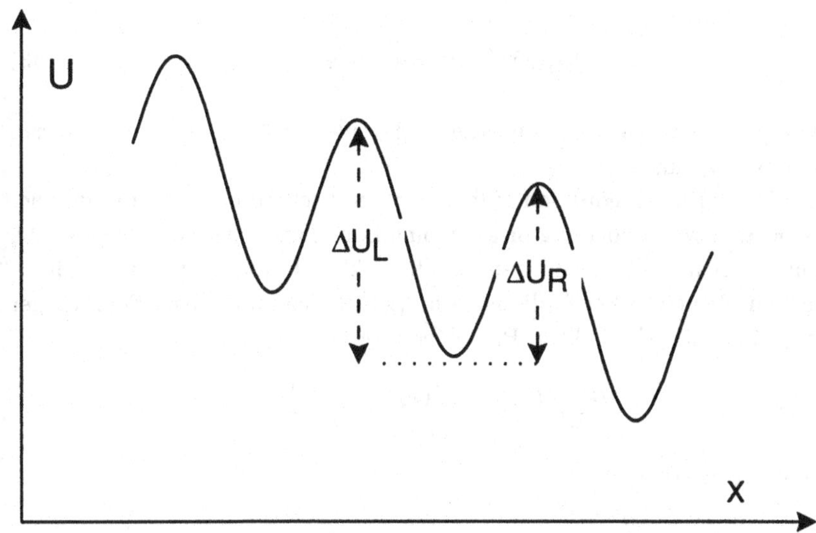

Figure 8.8: The potential $U(x)$ from equation (8.139).

The Kramers escape rates are then

$$r_{L,R} = \eta \exp[-\frac{\eta}{\xi}(2 \pm \frac{\alpha\pi}{2})] \ . \tag{8.140}$$

The net velocity of a particle in the potential is thus

$$\langle v \rangle = 2\pi(r_R - r_L) = 4\pi\eta e^{-2\eta/\xi} \sinh(\frac{\alpha\eta\pi}{2\xi}) \ . \tag{8.141}$$

We conclude that a spatially dependent diffusion constant may introduce an effective force causing particle drift.

8.8.2 Transformation to the Schrödinger equation

A linear differential equation

$$L\phi_\lambda = \rho(x)\lambda\phi_\lambda$$

with eigenvalues λ and eigenfunctions ϕ_λ defined on an interval (a, b) is *self adjoint* if

$$\int_a^b f^*(x)Lg(x)\rho(x)dx = \left(\int_a^b g*(x)Lf(x)dx\right)^* \tag{8.142}$$

for arbitrary functions $f(x)$, $g(x)$. Here L is a linear operator and $\rho(x)$ is a non-negative weighting function. The eigenfunctions are then orthogonal

$$\int_a^b \phi_i^*(x)\phi_j(x)\rho(x)dx = 0 \text{ unless } \lambda_i = \lambda_j \tag{8.143}$$

and the eigenvalues are real. In case of degeneracy, the eigenfunctions can always be orthogonalized.

The Fokker–Planck equation is in general *not* self-adjoint. The eigenfunctions associated with reflecting or absorbing boundaries therefore need not be orthogonal, which can be inconvenient. The Fokker–Planck equation can, however, be transformed to a self-adjoint form, (which looks a lot like a Schrödinger equation). Consider the Fokker–Planck equation

$$\frac{\partial P(x,t)}{\partial t} = -\frac{\partial f(x)P}{\partial x} + \xi\frac{\partial^2 P}{\partial x^2}$$

where f is a conservative force

$$f = -\frac{dU}{dx} \ . \tag{8.144}$$

We wish to solve this equation by the method of separation of variables

$$P(x,t) = \sum_n \Pi_n(x)\exp(-\lambda_n t) \tag{8.145}$$

where the eigenfunctions $\Pi_n(x)$ satisfy

$$-\lambda_n\Pi_n(x) = -\frac{df(x)\Pi_n(x)}{dx} + \xi\frac{d^2\Pi_n(x)}{dx^2} \ . \tag{8.146}$$

We introduce a new function $\Psi(x)$

$$\Pi = \exp\left(-\frac{U(x)}{2\xi}\right)\Psi(x) \ . \tag{8.147}$$

Then Ψ satisfies the "Schrödinger" equation

$$-\xi\frac{d^2\Psi}{dx^2} + \left[\frac{1}{4\xi}\left(\frac{dU}{dx}\right)^2 - \frac{1}{2}\frac{d^2U}{dx^2}\right]\Psi = \lambda\Psi \tag{8.148}$$

with the expression inside the square bracket playing the role of an *effective potential.* This equation is easily seen to be self-adjoint and can be solved in a manner similar to the way one treats the real Schrödinger equation.

8.9 Problems

8.1. *Probability Distribution for the Spruce Budworm Model.*

(a) Verify (8.13).

(b) Evaluate the integral numerically (8.12) for a value of the parameter q and locate the values of u and r for which the maxima of the probability distribution $P_s(u)$ will be the same. Will the free energy (4.51) be the same for these parameter values?

8.2. *Stability and Approach to Equilibrium in the SIR Model.*

(a) Show that in the limit $\rho \to 0$ the steady state solutions (8.60) approach $s = 1,\ i = 0$ for $c < 1$. Show that this solution is stable and that it is approached uniformly in the long time limit.

(b) Find the stable solution for $\rho = 0,\ c > 1$.

(c) Under what conditions will the solution found in (**b**) approach the steady state through damped oscilations? Express your answer in terms of the ratio λ/γ.

d) Discuss the time dependence of the approach to the steady state for $\rho = 0,\ c = 1$.

8.3. *Steady state probability distribution for a birth and death process.* In a certain population the death rate is (in appropriate units) $d(n) = n$, and the birth rate $b(n) = .9n$ and new individuals immigrate into the population at a rate 0.1. Plot the steady state probability distribution for the population size.

8.4 *Brownian particle in oscillator potential.* An overdamped particle with diffusion constant D is moving at temperature T in a harmonic oscillator potential

$$U = \frac{kx^2}{2} \ .$$

Find the probability distribution $P_{x/x'}(t)$.

8.5 *Kramers Equation with a Harmonic Force.* A particle is subject to a harmonic force with force constant k and is in contact with a heat bath at temperature T. The mobility of the particle is μ and it has position x_0, velocity v_0 at time $t = 0$.

(a) Write down Kramers equation for the problem.

(b) Find $\langle v(t)\rangle$ and $\langle x(t)\rangle$.

(c) Find the covariance matrix with components

$$\langle x(t)^2\rangle - \langle x(t)\rangle^2;\ \langle x(t)v(t)\rangle - \langle x(t)\rangle\langle v(t)\rangle;\ \langle v(t)^2\rangle - \langle v(t)\rangle^2 \ .$$

8.6 An overdamped pendulum is subject to a weak constant torque τ and satisfies the Fokker–Planck equation

$$\frac{\partial P(\theta,t)}{\partial t} = \frac{1}{\gamma}\left[\frac{\partial}{\partial\theta}(Mgl\sin\theta - \tau)P + k_BT\frac{\partial^2 P}{\partial\theta^2}\right] \ .$$

Here θ is the angle of the pendulum with respect to the vertical, M is the mass of the pendulum, l its length and γ a suitably chosen damping term. The torque is weak enough that the pendulum will not rotate without thermal noise. In the steady state the probability current will not be zero. Instead the steady state is characterized by periodic boundary conditions $P_s(\theta + 2\pi) = P_s(\theta)$. Estimate the average rate at which the pendulum rotates.

8.7 *The Chemical Reaction* $X + X \to 0$. Consider a chemical reaction in which a certain species X is produced at a steady rate $\beta\Omega$. If two molecules collide they produce a new inert substance which is removed from the system. Let the collision rate be $\alpha n(n-1)/\Omega$, where n is the number of X molecules present. The reactions are irreversible and we write the master equation of the form

$$\frac{\partial P(n,t)}{\partial t} = \Omega\beta(E^{-1} - 1)P(n,t) + \frac{\alpha}{\Omega}(E^2 - 1)n(n-1)P(n,t) \ .$$

Where E is the raising operator.

(a) Carry out a system size expansion of the master equation to derive a Fokker–Planck equation for $\Pi(x,t)$ where

$$n = \phi\Omega + x\sqrt{\Omega} \ .$$

(b) Find the steady concentration ϕ and the variance of x to leading order in Ω.

(c) The problem is actually exactly solvable as shown by Mazo [196]. Show that the results from the Fokker–Planck equation and the exact solution agree to leading order.

8.8 *Kimura Weiss model*

(a) Carry out the necessary transformations to make the Fokker–Planck equation (8.101) linear and self-adjoint.

(b) Modify the Kimura-Weiss master equation (8.97) to allow immigration of individuals with allele A with probability p and a with probability $1-p$. Whenever a new individual is introduced, another, picked randomly from the population, is removed, so that the population stays constant. Find the resulting steady state and the variance of the concentration x.

Chapter 9

Simulations

With the continually increasing speed and power of modern computers, the simulation of large but finite systems has become a more and more important tool in condensed matter physics. Conceptually, simulation techniques are quite simple — one attempts to construct representative states of the finite system and then to extrapolate the relevant thermodynamic quantities to the thermodynamic limit. Our aim in this chapter is only to provide the reader with the basic ideas underlying these methods. Since the object of these techniques is to simulate the largest possible systems, given finite computational resources, very sophisticated programming methods are necessary for state-of-the-art calculations. We will generally not discuss these aspects but rather refer the reader to one of the more specialized texts on the subject.

We begin by discussing the two most commonly used techniques: *molecular dynamics* and the *Monte Carlo* method. In molecular dynamics we consider a system of classical (non-quantum) particles interacting through a set of forces. We numerically integrate the equations of motion, and averages of the appropriate state-variables are then obtained as time averages over the trajectory of the system in phase space. The earliest versions of the molecular dynamics methods were based on the microcanonical ensemble. In Hamiltonian dynamics, with time-independent potentials, the total energy is conserved and the system executes a trajectory on the constant energy surface. The *ergodic assumption* (see Section 2.1) is crucial in order that the results obtained through a time averaging process be equivalent to those that would be obtained from the microcanonical probability density.

Later, molecular dynamics algorithms that simulate the canonical ensemble were developed. These involve dissipation of energy through an explicit friction term (Brownian dynamics) and addition of energy through appropriately chosen random kicks or exchange of energy between the system of interest and a massive test particle (Nosé-Hoover dynamics). The microcanonical methods are preferable if dynamical properties are of interest; for purely static thermodynamic quantities, the canonical methods are advantageous because of better stability of the algorithms.

The Monte Carlo method, in contrast, attempts to directly simulate one of the distributions we have considered, most often the canonical distribution. There is no dynamics; instead the computer generates states of the finite system with a weight proportional to the canonical or grand canonical probability density. In order to lay the ground-work for establishing this point we present our discussion of the Monte Carlo method by a brief outline of the theory of discrete time Markov processes. We then describe some important techniques for extracting thermodynamic information from data obtained by simulations. Finally, we present two examples of how simulations have had an impact in non-traditional areas of research by discussing briefly simulated annealing and a simple model of neural networks.

9.1 Molecular Dynamics

Let us consider a system of classical particles interacting through central two-body forces derivable from a pair potential $u(r_{ij})$ where $r_{ij} = |\mathbf{r}_i - \mathbf{r}_j|$. The force $\mathbf{f}_{ij}(\mathbf{r}_{ij})$ on particle i, due to particle j, is then

$$\mathbf{f}_{ij} = \nabla_j u\left(|\mathbf{r}_i - \mathbf{r}_j|\right) = -\nabla_i u\left(|\mathbf{r}_i - \mathbf{r}_j|\right) .$$

If the mass of each particle is m, the equation of motion for the ith particle is

$$m\frac{d^2\mathbf{r}_i}{dt^2} = \sum_{j\neq i} \mathbf{f}_{ij}(\mathbf{r}_{ij}) . \tag{9.1}$$

We now specialize to the Lennard–Jones or 6–12 potential (5.7) for which the x-component of the force is given by

$$f_{ij,x}(\mathbf{r}_{ij}) = \frac{24\epsilon}{\sigma^2}(x_i - x_j)\left[2\left(\frac{\sigma}{r_{ij}}\right)^{14} - \left(\frac{\sigma}{r_{ij}}\right)^{8}\right] \tag{9.2}$$

with analogous expressions for the other components. For this potential it is instructive to rewrite the equations of motion in dimensionless form. The natural unit of length is the quantity σ and we define $\mathbf{r}_i = \sigma\tilde{\mathbf{r}}_i$ for all i. Substituting into (9.1) and (9.2) we see that the natural unit of time is $t_0 = \sqrt{m\sigma^2/\epsilon}$. Substituting $t = t_0\tau$, we obtain the final form of the equation of motion:

$$\frac{d^2\tilde{r}_{i,\alpha}}{d\tau^2} = 24\sum_j (\tilde{r}_{i,\alpha} - \tilde{r}_{j,\alpha})\left[2\tilde{r}_{ij}^{-14} - \tilde{r}_{ij}^{-8}\right] \tag{9.3}$$

where α labels the components of the position vectors $\tilde{\mathbf{r}}_i$. For a system of Argon atoms, appropriate parameters in the Lennard–Jones potential are $\sigma = 0.34$ nm, $\epsilon/k_B = 120$ K and $m \approx 6.7 \times 10^{-26}$ kg. Thus the basic unit of time is $t_0 \approx 2.15 \times 10^{-12}$ s. Since the equations of motion (9.3) are highly nonlinear it is generally necessary to integrate them numerically using time steps that are very small compared to t_0. It is clear that the iteration of the equations of motion for even one nanosecond is a significant computational task, if the number of particles is large.

Before discussing the integration of (9.3) we note that the time-consuming part of a molecular dynamics calculation is not the integration of the equations but rather the calculation of the force on each of the particles. In the crudest scheme the calculation of the right-hand side of (9.3) requires N^2 steps at each time if the system consists of N particles. Since the Lennard–Jones potential is short-ranged (effectively zero for distances greater than $\approx 2.5\sigma$), a given particle interacts with only a small number of others at any instant. There exist very efficient algorithms (cell methods or neighbor tables) that reduce the calculation of the forces to an operation requiring of the order of N steps. These are described in detail in the book by Allen and Tildesley [11].

We now briefly describe representative computational techniques for both conservative (constant E) and canonical (constant T) molecular dynamics. We then discuss briefly the calculation of thermodynamic or time-dependent quantities.

9.1.1 Conservative molecular dynamics

The task at hand is to integrate approximately the Newtonian equations of motion

$$\frac{dv_{i\alpha}}{dt} = \frac{f_{i\alpha}(t)}{m} = a_{i\alpha}(t) \tag{9.4}$$

$$\frac{dr_{i\alpha}}{dt} = v_{i\alpha}(t) \ . \tag{9.5}$$

This is accomplished through one of a number of finite difference schemes involving discrete steps δt in time. Since we wish to conserve energy, the finite difference scheme must mimic the continuous evolution as closely as possible but, in order to be efficient, must also entail a relatively small number of operations per time step, *i.e.* a compromise is required. A commonly used algorithm is the leapfrog or velocity-Verlet algorithm which we now motivate.

The simplest iteration of equations (9.4) is the following:

$$v_{i\alpha}(t+\delta t) = v_{i\alpha}(t) + \delta t a_{i\alpha}(t) \tag{9.6}$$

$$r_{i\alpha}(t+\delta t) = r_{i\alpha}(t) + \delta t v_{i\alpha}(t) \ . \tag{9.7}$$

This scheme has errors in both the change in position and velocity that are of order $(\delta t)^2$. To see this, consider the velocity equation over the time interval $(t, t+\delta t)$:

$$\begin{aligned} v_{i\alpha}(t+\delta t) &= v_{i\alpha}(t) + \int_t^{t+\delta t} dt' a_{i\alpha}(t') \\ &= v_{i\alpha}(t) + \int_t^{t+\delta t} dt' \left[a_{i\alpha}(t) + (t'-t)a'_{i\alpha}(t) + \frac{1}{2}(t'-t)^2 a''_{i\alpha}(t) \ldots \right] \\ &= v_{i\alpha}(t) + \delta t a_{i\alpha}(t) + \frac{1}{2}(\delta t)^2 a'_{i\alpha}(t) + \ldots \end{aligned} \tag{9.8}$$

where we have simply expanded the acceleration in a Taylor series around point t. A similar calculation for the second equation shows that the errors in the change of position are also of order $(\delta t)^2$.

A very simple, computationally efficient improvement is the following algorithm:

$$v_{i\alpha}(t+\delta t) = v_{i\alpha}(t) + \frac{\delta t}{2}[a_{i\alpha}(t) + a_{i\alpha}(t+\delta t)] \tag{9.9}$$

$$r_{i\alpha}(t+\delta t) = r_{i\alpha}(t) + \delta t v_{i\alpha}(t) + \frac{(\delta t)^2}{2} a_{i\alpha}(t) \tag{9.10}$$

which has errors of at most order $(\delta t)^3$ in both positions and velocities. The first of these equations requires the forces at both time t and $t + \delta t$. The implementation is straightforward. The positions are updated using velocities and forces at time t and the velocities are partially updated using accelerations at time t. The positions at time $t + \delta t$ are then used to calculate the forces at time $t + \delta t$ and the velocity update is completed. As mentioned above, the time consuming step in the process is the calculation of the forces and it is important that, except for one extra calculation of the forces at $t = 0$, each complete update of the positions and velocities requires only one calculation of the forces. This algorithm has proven to be suitable for many applications.

An appropriate choice of δt insures adequate conservation of energy for a large number of time steps.

It is worth pointing out that *any* finite difference scheme for the integration of equations of the type (9.3) is intrinsically unstable. This is due to the fact that trajectories in phase space are extremely sensitive to small changes in initial conditions. Nearby trajectories separate from each other exponentially at long times. Since changing the size of the time step δt is essentially equivalent to changing an initial condition, there is no possibility of integrating the equations with arbitrary accuracy for longer than some finite interval. While the separation of nearby trajectories may appear disturbing at first glance it is not of crucial importance. In a molecular dynamics simulation we wish to sample the constant energy surface of the system. As long as the total energy is conserved, the actual trajectory of the point in phase space is not of great importance as far as expectation values of thermodynamic quantities is concerned. Energy conservation is, however, very important. It turns out that finite difference schemes also fail in this regard after some time. The best that one can hope for is that this time is longer than the relaxation time of any of the quantities of interest. A number of "rules of thumb" that limit the amount of fluctuation that is tolerable in the total energy have been developed. For a discussion of this point and of other finite-difference algorithms (predictor corrector, gear, etc.) the reader is referred to [11].

9.1.2 Brownian dynamics

As mentioned above, there are also molecular dynamics techniques that simulate the canonical distribution. One of these is the Brownian dynamics technique which we first illustrate for a very simple case, a free particle in a viscous medium. The viscous medium both damps the motion of the particle and provides random 'kicks' through collisions. The process is commonly modeled with the following Langevin equation:

$$\frac{d\mathbf{v}(t)}{dt} + \gamma\mathbf{v}(t) = \boldsymbol{\eta}(t) \tag{9.11}$$

where the damping coefficient $\gamma = 1/\tau$ is an inverse relaxation time related to the viscosity of the liquid and $\eta(t)$ is a *Gaussian noise* term with the properties:

$$\langle\eta(t)\rangle = 0 \qquad \langle\eta_\alpha(t)\eta_\beta(t')\rangle = \lambda\delta(t-t')\delta_{\alpha\beta}\,. \tag{9.12}$$

We will relate the *noise strength* λ to the damping coefficient γ below.

A formal solution of equation (9.11) is given by

$$\mathbf{v}(t) = \int_{-\infty}^{t} dt' \boldsymbol{\eta}(t') e^{-\gamma(t-t')} \tag{9.13}$$

with the property, in view of (9.12), $\langle \mathbf{v}(t) \rangle = 0$. However, the velocity-velocity correlation function is nontrivial:

$$\begin{aligned}\langle v_\alpha(t) v_\beta(t') \rangle &= \int_{-\infty}^{t} dt_1 \int_{-\infty}^{t'} dt_2 \langle \eta_\alpha(t_1) \eta_\beta(t_2) \rangle e^{\gamma(t_1+t_2-t-t')} \\ &= \frac{\lambda \delta_{\alpha\beta}}{2\gamma} e^{-\gamma(t-t')} \end{aligned} \tag{9.14}$$

where we have taken $t > t'$. Setting $t' = 0$, we have

$$\langle v_\alpha(t) v_\beta(0) \rangle = \frac{\lambda \delta_{\alpha\beta}}{2\gamma} e^{-\gamma t}$$

and $\langle v_\alpha^2(0) \rangle = \lambda/(2\gamma) = k_B T/m$ where m is the mass of the Brownian particle and where, in the last step, we have invoked equipartition. The noise strength λ is therefore related to the temperature and damping coefficient through $\lambda = 2\gamma k_B T/m$. The velocity autocorrelation function is related to the diffusion constant for the Brownian particle through [125]

$$D = \frac{1}{3} \int_0^\infty dt \langle \mathbf{v(t)} \cdot \mathbf{v(0)} \rangle = \frac{k_B T}{m\gamma} .$$

The relation between D and the velocity autocorrelation function is general. It is an example of a Green-Kubo relation of the type also discussed in Chapter 12. The specific result $D = k_B T/m\gamma$ is, of course, only valid for our simple model.

By using a damping coefficient and a properly chosen noise distribution, we may construct molecular dynamics algorithms that generate canonical distributions in phase space. We begin with the equations of motion for particle i

$$\begin{aligned}\frac{d\mathbf{v}_i}{dt} &= -\gamma \mathbf{v}_i + \frac{\mathbf{f}_i}{m_i} + \boldsymbol{\eta}_i(t) \\ \frac{d\mathbf{r}_i}{dt} &= \mathbf{v}_i \end{aligned} \tag{9.15}$$

where the noise function $\boldsymbol{\eta}_i$ has the property (9.12) and is related to the damping coefficient in the same way as for the single particle. The force $\mathbf{f}_i$ is the total force on particle i due to the other particles, as in conservative molecular dynamics. When these equations are discretized in time, the integration over a finite time interval δt converts both the velocity equation and the coordinate equation into stochastic equations with correlated noise. We refer the reader to the book by Allen and Tildesley [11] for the details.

9.1.3 Data analysis

A typical molecular dynamics calculation consists of the following steps. One starts with a random assignment of positions and velocities (making sure that the center of mass velocity is zero) and iterates the equations for some time to allow the system to equilibrate. This is followed by a 'production run' of a number of time steps to collect data. For an example of a calculation that is carried out as outlined see [313].

The question now arises, how can we obtain thermodynamic or dynamic properties from the molecular dynamics trajectories? Consider as a first example a constant energy simulation. In this case, the temperature is a fluctuating quantity whose mean value can be calculated from a running average of the kinetic energy using equipartition:

$$\frac{1}{2}\langle\sum_i m{v_i}^2\rangle = \frac{3}{2}Nk_B\langle T\rangle$$

or

$$\langle T\rangle = \frac{2\langle\text{kinetic energy}\rangle}{3Nk_B}$$

for a system of N particles in three dimensions. We next turn to the pressure. The quantity

$$\mathcal{V} = \sum_i \mathbf{p}_i \cdot \mathbf{r}_i$$

is called the *virial*. In equilibrium the average of the virial must be independent of time. Therefore

$$\begin{aligned}\left\langle\frac{d\mathcal{V}}{dt}\right\rangle &= 0 = \sum_i\langle\mathbf{p}_i\cdot\dot{\mathbf{r}}_i\rangle + \sum_i\langle\dot{\mathbf{p}}_i\cdot\mathbf{r}_i\rangle \\ &= \sum_i\langle mv_i^2\rangle + \sum_i\langle\mathbf{r}_i\cdot\mathbf{f}_i\rangle\,. \end{aligned} \tag{9.16}$$

The first term in (9.16) is $3Nk_BT$ from the previous calculation. The force $\dot{\mathbf{p}}_i$ on the ith particle can be split into an external force, due to the pressure P

exerted by the walls of the container, and an internal force, due to the other particles. We have for the external force

$$\left\langle \sum_i \mathbf{r}_i \cdot \mathbf{f}_i^{ext} \right\rangle = -P \int \mathbf{r} \cdot \mathbf{dA}$$

where $d\mathbf{A}$ is an area element normal to the wall and directed outward. From Gauss' theorem

$$\left\langle \sum_i \mathbf{r}_i \cdot \mathbf{f}_i^{ext} \right\rangle = -P \int (\nabla \cdot \mathbf{r}) d^3 r = -3PV$$

and from Newton's third law

$$\sum_i \mathbf{r}_i \cdot \mathbf{f}_i^{int} = \sum_{i \neq j} \mathbf{r}_i \cdot \mathbf{f}_{ij} = \frac{1}{2} \sum_{i \neq j} \mathbf{r}_{ij} \cdot \mathbf{f}_{ij}$$

where the last step uses $\mathbf{f}_{ij} = -\mathbf{f}_{ji}$. Collecting terms we get

$$PV = Nk_B T + \frac{1}{6} \left\langle \sum_{i \neq j} \mathbf{r}_{ij} \cdot \mathbf{f}_{ij} \right\rangle . \tag{9.17}$$

Equation (9.17) is called the *virial equation of state*. Since one needs to keep track of the coordinates and forces during the simulation, it is again straightforward to compile a running average of the pressure. Equation (9.17) can be rewritten in terms of the pair distribution $g(\mathbf{r})$ of Section 5.2.1 as

$$PV = Nk_B T \left[1 - \frac{n}{6k_B T} \int d^3 r \, (\mathbf{r} \cdot \nabla u(\mathbf{r})) \, g(\mathbf{r}) \right] . \tag{9.18}$$

Equation (9.18) is known both as the *virial* and as the *pressure* equation of state. Both (9.18) and (9.17) are exact for a system of classical particles with pairwise forces between them. The compressibility obtained by differentiating (9.18) should therefore agree with the result obtained from (5.33). This will, of course, hold when an exact pair distribution function is used. However, it turns out to be difficult to achieve agreement between the two expressions when an approximate form of $g(\mathbf{r})$ is used. Comparison of the two equations thus offers a useful check on the validity of approximate calculations.

It is also possible to obtain other thermodynamic quantities. The expectation value of the potential energy is an obvious byproduct of the calculation of the temperature for the case of conservative dynamics. The specific heat C_V

can be obtained from the fluctuations of the kinetic energy or, equivalently, of the temperature T. The appropriate expression is [169]

$$\frac{\langle T^2\rangle - \langle T\rangle^2}{\langle T\rangle^2} = \frac{2}{3N}\left(1 - \frac{3Nk_B}{2C_V}\right) .$$

We also note that quantities such as the specific heat, that are calculated by means of the fluctuation of another quantity, can generally not be determined to nearly the same accuracy as quantities like the potential energy.

It is also possible to obtain transport coefficients and other dynamical properties from a molecular dynamics simulation. We have already mentioned the self-diffusion coefficient and its relation to the velocity-velocity correlation function. A similar Green-Kubo formula involving the stress-stress correlation function yields the shear viscosity of a fluid or gas from an equilibrium MD simulation.

For reasons of economy one generally wants to run the simulation on systems with at most a few thousand particles, although simulations for 10^6 particles have been carried out. For small systems finite size effects can be important and these are discussed in more detail below. One type of finite size effect that can give rise to inaccuracies is the perturbation due to the surface. This can be reduced by using periodic boundary conditions.

9.2 Monte Carlo Method

In Monte Carlo simulations one does not attempt to simulate the dynamics of the system; instead the idea is to generate states $i, j, \ldots$ by a stochastic process such that the probability $\pi(i)$ of state i is that given by the appropriate distribution (canonical, grand canonical, etc.). In a 'production run' of a simulation $\mathcal{N}$ states are generated and the desired quantity x_i (energy, magnetization, pressure etc.) is calculated for each state. If the probabilities are correct

$$\langle x\rangle = \lim_{\mathcal{N}\to\infty} \frac{1}{\mathcal{N}} \sum_i x_i . \tag{9.19}$$

Most often one simulates the canonical ensemble

$$\pi(i) = \frac{1}{Z} e^{-\beta E(i)}, \quad Z = \sum_i e^{-\beta E(i)} .$$

Two questions then arise:

> How does the computer generate states?
> How can I make sure that the probabilities are right?

9.2.1 Discrete time Markov processes

Let us consider a finite classical system with a finite number M of microstates. For example, for an Ising model consisting of N spins the number $M = 2^N$ and it is easy to calculate the energy E_i of any state i. If we wish to calculate expectation values according to the canonical ensemble we could use a random number generator to assign the values of all the spins, weight the contribution of that microstate by $\exp\{-\beta E\}$ and repeat the process until the relevant expectation values have converged. This would be an extremely inefficient procedure since all states appear with equal probability, including those whose weight is so small that, in effect, they do not contribute to the thermodynamic average. To make a sampling process effective we must focus on those states that make the dominant contribution to the quantities of interest. This can be accomplished by generating a sequence of states according to a *discrete time Markov process*. The difference between what we are doing here and Section 8.1 is that we are looking at transitions taking place on equally spaced time steps. The frequency of occurrence of different microstates in a particular instance of the process will then mirror the probability distribution in the steady state.

Suppose that the system is in a given microstate i. The next state j in the sequence is selected with a transition probability $P_{j\leftarrow i}$ that does not depend on the previous history of the system. Under fairly general conditions such a process, after the passage of a transient, produces states with a *unique* steady-state probability distribution. This steady-state probability $\pi(j)$ is an eigenvector with eigenvalue 1 of the transition matrix:

$$\pi(j) = \sum_i P_{j\leftarrow i}\pi(i) \; . \tag{9.20}$$

The steady state probabilities are unique if the matrix $P_{j\leftarrow i}$ is *regular*, which means that for some integer n all elements of $(P_{j\leftarrow i})^n$ are positive and nonzero. Physically, this restriction implies that it is always possible to go from any of the states to any other state in a finite number of steps. Exceptions are matrices that are block diagonal, *e.g.*

$$\begin{bmatrix} P_{1\leftarrow 1} & P_{1\leftarrow 2} & 0 & 0 \\ P_{2\leftarrow 1} & P_{2\leftarrow 2} & 0 & 0 \\ 0 & 0 & P_{3\leftarrow 3} & P_{3\leftarrow 4} \\ 0 & 0 & P_{4\leftarrow 3} & P_{4\leftarrow 4} \end{bmatrix} .$$

Since there is no way of going from states 1 or 2 to 3 or 4 the stationary probability distribution will depend on whether one started with one of the first two states or one of the last two.

Example of a Markov process

A student is practising 'random walk' in the house shown in Figure 9.1. At regular intervals he changes rooms and uses, with equal probability, any of the doors leaving the room that he currently occupies. Assuming that the student persists for a long time, the task is to find the fraction of the time that he will spend in each room. Suppose we run a simulation in which at some instant the student is in room 2. With probability $\frac{1}{2}$ he will move to room 3 and with probability $\frac{1}{2}$ to room 1. If the student is in room 3 he will move with probability $\frac{1}{3}$ to any of the three other rooms. We can represent the transition probability by a matrix in which the row index represents the final state and the column index the initial state:

$$P_{j\leftarrow i} = \begin{bmatrix} 0 & \frac{1}{2} & \frac{1}{3} & 0 \\ \frac{1}{2} & 0 & \frac{1}{3} & 0 \\ \frac{1}{2} & \frac{1}{2} & 0 & 1 \\ 0 & 0 & \frac{1}{3} & 0 \end{bmatrix} .$$

The steady state probabilities $\pi(i)$ that the student is in a given room can then be determined by solving the eigenvalue problem (9.20) subject to the normalization condition

$$\sum_{i=1}^{4} \pi(i) = 1 .$$

The eigenvector associated with the eigenvalue 1 is $\pi(1) = \pi(2) = \frac{1}{4}, \pi(3) = \frac{3}{8}$, $\pi(4) = \frac{1}{8}$.

9.2.2 Detailed balance and the Metropolis algorithm

We illustrate the Monte Carlo method by describing the simulation of an Ising system at temperature T. Consider a set of sites $\{\alpha\}$ on each of which there is a spin σ_α that can take on the values +1 or −1. A configuration (or microstate) i is specified by the set of values of σ_α for all α. We now wish to determine a transition matrix $P_{i\leftarrow j}$ so that the steady-state distribution is

$$\pi(i) = \exp\{-\beta E(i)\}/Z \tag{9.21}$$

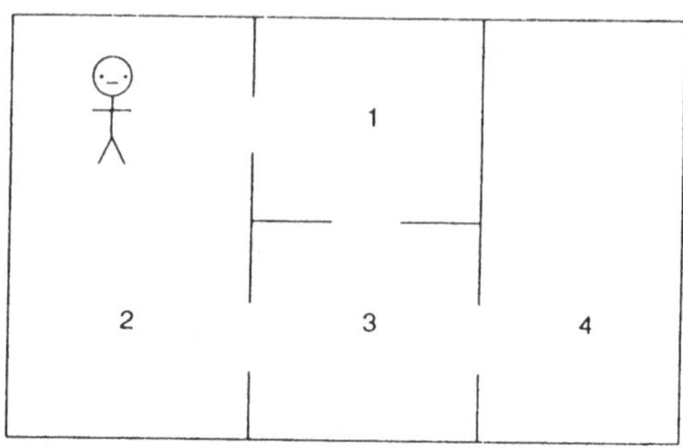

Figure 9.1.

where Z is the partition function. A possible method of generating a sequence of states from an initial state i is to pick a site α randomly and attempt to *flip* (change the sign of) its spin. The resulting state (which may be the same state i if the attempt to flip σ_α fails) we call j. Let $P_{j\leftarrow i}$ be the transition probability $i \to j$. After n steps the transition probability $P_{f\leftarrow i}(n)$ is given by

$$P_{f\leftarrow i}(n) = \sum_{i_1, i_2 \cdots i_{n-1}} P_{f\leftarrow i_{n-1}} P_{i_{n-1}\leftarrow i_{n-2}} \cdots P_{i_1\leftarrow i} \; .$$

As discussed earlier, after many steps the system will approach a limiting distribution

$$\pi(f) = \lim_{n\to\infty} P_{f\leftarrow i}(n)$$

independently of the initial configuration, if the transition matrix is regular. We achieve the desired distribution (9.21) by requiring the probability distribution to be normalized and to satisfy

$$\frac{\pi(m)}{\pi(j)} = \exp[-\beta\{E(m) - E(j)\}]$$

for all pairs of states m, j. We now also require the transition probabilities to be normalized

$$\sum_j P_{j\leftarrow m} = 1 \tag{9.22}$$

and to obey

$$\frac{P_{j\leftarrow m}}{P_{m\leftarrow j}} = \frac{\pi(j)}{\pi(m)} = \exp[-\beta\{E(j) - E(m)\}] \; . \tag{9.23}$$

We find that

$$\pi(m) = \sum_j P_{j\leftarrow m}\, \pi(m) = \sum_j P_{m\leftarrow j}\, \pi(j) \ . \tag{9.24}$$

The first step in (9.24) follows from normalization (9.22) while the second step involves substituting (9.23). From (9.20) we see that (9.24) implies that $\pi(m)$ is a stationary probability distribution of the process. Equation (9.23) is called the *principle of detailed balance* and we see that it is a sufficient condition for arriving at the correct limiting probability distribution, provided that our process for selecting moves does not contain any traps, *i.e.* it should always be possible to get from any given microstate to any other microstate.

The simplest and most frequently used method of achieving detailed balance is the *Metropolis algorithm:*

(i) Pick a site α randomly.
(ii) Compute the energy change $\Delta E = E(f) - E(i)$ that would occur if the spin at site α were flipped.
(iii) If $\Delta E \leq 0$, flip the spin at site α; if $\Delta E > 0$ flip this spin with probability $\exp(-\beta\Delta E)$.
(iv) repeat steps (i) to (iii) until enough data is collected.

An alternative to (iii) that is sometimes used is to flip the spin σ_α with probability $[\exp(\beta\Delta E) + 1]^{-1}$ regardless of the sign of ΔE. It is easy to see that (9.23) is satisfied in both cases for all possible pairs of states. The allowed states therefore will occur with the correct frequency if the simulation is run long enough to reach the steady state.

It is easy (in principle) to generalize this procedure to other systems. For particles in a box, each move could be an attempt to shift a randomly selected particle a random fraction of some maximum distance in an arbitrary direction. According to the principle of detailed balance it should not matter how the maximum distance is selected, although the speed with which the system approaches equilibrium does depend on this choice. If this distance is too long the acceptance rate of the moves is very low, if it is too short the system moves through configuration space very slowly. For non-spherical molecules one must also allow rotation in addition to translations.

One can also generalize to systems at constant pressure rather than volume or constant chemical potential rather than particle-number by modifying the Boltzmann factor to that of the appropriate ensemble and allowing moves that change the overall volume or number of particles. Thus in an NPT ensemble in which the pressure is fixed but the volume is allowed to fluctuate, we simply replace the energy of each microstate by the enthalpy $H(i) = E(i) + PV(i)$

and incorporate attempts to randomly change the volume in our definition of a move.

In the general case a Monte Carlo simulation consists of the following steps:

(1) Choose initial configuration.
(2) Select a move.
(3) Accept or reject the move using a criterion based on detailed balance.
(4) Repeat steps (2) and (3) until enough data is collected.

Typically the data from the early part of a run is discarded since the system will not have had enough time to reach equilibrium.

The Monte Carlo techniques of this section can also be extended to the microcanonical ensemble by the introduction of a 'demon' [65]. Recently some 'cluster methods' have been developed that in some cases significantly shorten the time required to reach equilibrium and the time that it takes for correlations to decay in equilibrium [115]. Cluster methods are particularly attractive for systems with short range interactions near a continuous phase transitions where they can help overcome difficulties associated with "critical slowing down".

At low temperatures one encounters another problem namely extremely high rejection rates, when almost all possible transitons are energetically unfavorable. A better approach may then be to directly simulate the master equation of the dynamics (Section 8.1). To do this we first determine the *rates* r_α at which all possible allowed events take place. Suppose *rand*() is a random number which is uniformly distributed between 0 and 1. A sequence of numbers

$$\tau_\alpha = \frac{-\ln(rand())}{r_\alpha}$$

for each allowed event α, will then generate possible times for each event, that is consistent with the master equation. One then arranges the generated event times in a sequence and executes the event which occurs first. This may change the rates for some events whose positions in the sequence need to be updated, but typically most rates will not be affected. There are data structures such as binary trees and "heaps" which allow efficient management of the sequence of events. The details of how to do this is beyond the present text. Bortz *et al.* [49] developed such a method for Ising spins, while similar approaches to chemical reaction systems have been developed by Gillespie [109][110]. Chemical reactions of interest are commonly driven and out of equilibrium. It is then not possible to appeal to detailed balance and a "real time" simulation may be the only available approach, assuming that the reaction rates are known.

9.2.3 Histogram methods

We continue our discussion of the Monte Carlo method with a few comments on attempts to optimize the method. Suppose that in a production run of a simulation a total number of $\mathcal{N}$ states are generated. Suppose further that we are monitoring a variable x that takes on the value $x(i)$ in microstate i. If the probabilities $\pi(i)$ are correct the mean value of x is

$$\langle x\rangle \simeq \frac{1}{\mathcal{N}}\sum_{i=1}^{\mathcal{N}} x(i) .$$

In the canonical ensemble the frequency with which the different states occur is given by the probability distribution

$$\pi(i) = \frac{1}{Z}\exp[-\beta E(i)] .$$

Usually, the calculated quantity $\langle x\rangle$ is required not just for one value of $\beta = 1/k_BT$, but for a range of temperatures.

The question then arises: Is it necessary to repeat the simulation for each temperature? As we shall show [92], it is in principle possible to use only a single very long run at one temperature. To see this, let $\pi(i)$ be the correct probability for the ith microstate at the actual simulation temperature T and $\pi'(i)$ the probability at another temperature T'. We have

$$\frac{\pi'(i)}{\pi(i)} = \frac{Z}{Z'}\exp[-(\beta'-\beta)E(i)] .$$

If the microstates are generated with the correct frequencies at temperature T

$$\langle x(T)\rangle = \frac{1}{\mathcal{N}}\sum_{i=1}^{\mathcal{N}} x_i . \tag{9.25}$$

At temperature T', the same microstates that appear in (9.25) occur with a weight

$$w_i(T') = \frac{e^{-(\beta'-\beta)E_i}}{\sum_{j=1}^{\mathcal{N}} e^{-(\beta'-\beta)E_j}} \tag{9.26}$$

and

$$\langle x(T')\rangle = \sum_{j=1}^{\mathcal{N}} w_j(T')x_j . \tag{9.27}$$

Equation (9.27) therefore implies that we can calculate the temperature dependence of the desired variable from a single run at a fixed temperature. This 'optimized Monte Carlo method' is not without its pitfalls. It is important that the temperature difference $|T - T'|$ not be too large. If it is, the exponent in $\exp[-(\beta' - \beta)E_i]$ may sometimes be so large that the sample becomes distorted by a too strong dependence on rare events, *i.e.* microstates that are important at temperature T and dominate the expectation value (9.25) may play only a minor role at temperature T'. Another weakness is that the method does not allow easily for numerical error estimates. Generally this method works well for small systems but the allowable temperature range shrinks drastically as the system size is increased.

Histogram methods may also be employed to construct the dependence of the free energy on an order parameter [112] in a spirit somewhat similar to our discussion in Chapter 3. Let h be the field conjugate to the order parameter m and consider the ensemble in which h and T are specified. Let x_m be a microstate for which the order parameter is m, and for simplicity assume discrete variables. We write for the partition function

$$Z = \sum_m \sum_{x_m} e^{-\beta H(x_m)} = \sum_m e^{-\beta G_m} \tag{9.28}$$

where G_m is the "free energy" of the subset of states in which the order parameter takes on the value m. The probability of finding the system in a state with order parameter m is therefore

$$p(m) = \frac{1}{Z} e^{-\beta G(m)} \ . \tag{9.29}$$

We can obtain an approximate probability distribution (9.29) by running a Monte Carlo simulation at a given temperature and keeping track of how frequently the order parameter is in the range m, $m + dm$. In principle we could also extrapolate to different temperatures using (9.26). A plot of $-\beta^{-1} \ln p(m)$ vs. m then gives the dependence of the free energy $G(m)$ on the order parameter and one obtains results near a phase transition similar to Figures 3.8, 3.9. In particular, this method can be used to distinguish between first and second order phase transitions. In the case of a first order transition, $p(m)$ has (at least) two peaks for all temperatures at or below the critical temperature, whereas these peaks merge near T_c in the case of a continuous transition. For a discussion of the finite-size scaling properties of the method see [170].

9.3 Data Analysis

In the previous section we have described the basic Monte Carlo procedure. We now turn to the analysis of the data resulting from such a calculation.[1] There are two important aspects to this. First of all, since the system simulated is of finite size, it is necessary to make an extrapolation of the relevant thermodynamic quantities to the thermodynamic limit. This is quite straightforward if the system is not too close to the critical point of the infinite system. Since the correlation length ξ is finite in such a situation, it may even be possible to simulate systems with linear dimensions greater than ξ. On the other hand, in the more interesting case of a system close to criticality, methods based on finite-size scaling (see Section 6.4) are essential. We will discuss these methods further below. We first discuss the more basic and universal topic of how to extract reliable information from a noisy set of data.

9.3.1 Fluctuations

In a system of finite size there are limitations on the accuracy with which thermodynamic quantities can be calculated because of fluctuations. Consider, for example, the energy of a system simulated by a Monte Carlo method appropriate for the canonical ensemble. Since we are fixing the temperature, the energy fluctuates. An example of this is shown in Figure 9.2 where the energy is plotted as function of 'time' in the steady-state regime of a typical Monte Carlo run. In Chapter 2 we have shown how fluctuations are related to response functions. In the case of the energy the fluctuations are related to the specific heat:

$$\langle E^2\rangle - \langle E\rangle^2 = \langle(\Delta E)^2\rangle = k_B T^2 C_V \ . \tag{9.30}$$

This formula is very useful in computer simulations. First, it allows us to obtain an estimate of the specific heat from an average of $\langle E^2\rangle - \langle E\rangle^2$. Since the energy is an essential component of the Monte Carlo procedure, this quantity is essentially obtained as a byproduct of the simulation. Secondly, it can help to provide a check on whether or not the system is properly equilibrated as we can also obtain C_V by carrying out the calculation at two different temperatures:

$$C_V \simeq \frac{\langle E(T+\delta) - E(T-\delta)\rangle}{2\delta} \ .$$

If the results of these two estimates do not agree, we have an indication that the simulation has not run long enough. Another check on the quality of a

[1] For further reading on the material in this section see [38] and [41].

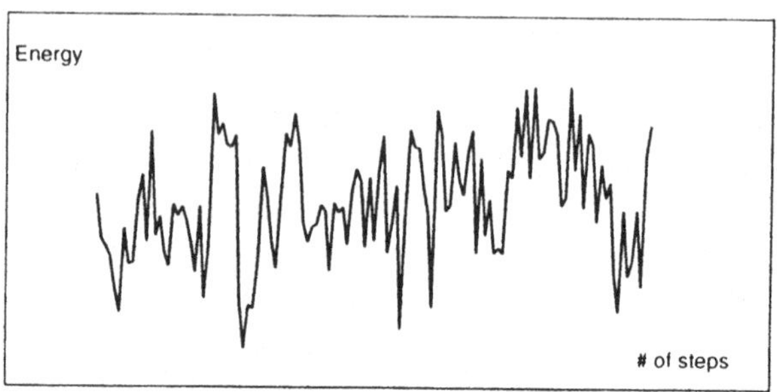

Figure 9.2: Energy fluctuations in a Monte Carlo simulation of the canonical ensemble.

simulation can be obtained by making a histogram of the energy sampled at regular intervals. This should, by the central limit theorem of statistics, be approximately Gaussian. Figure 9.2 illustrates another aspect of computer simulations. The larger energy fluctuations often last many time steps. One does not gain additional information by sampling the energy or other quantities more often than the time scale of the fluctuations, since more frequent measurements are not independent. We will return to this question below.

Clearly, similar considerations apply to any other quantity measured in a Monte Carlo (or molecular dynamics) simulation. For example, we can use the volume fluctuations in the NTP ensemble to measure the isothermal compressibility. Consider the partition function in the NTP ensemble

$$Z_{NPT} = \sum_{\alpha} e^{-\beta H_\alpha}$$

where $H_\alpha = E_\alpha + PV_\alpha$ is the enthalpy of the system in microstate α. Since

$$\langle V \rangle = -k_B T \frac{\partial \ln Z_{NPT}}{\partial P}$$

$$\langle V^2 \rangle = \frac{(k_B T)^2}{Z_{NPT}} \frac{\partial^2 Z_{NPT}}{\partial P^2}$$

we find for the volume fluctuations

$$(\Delta V)^2 = \langle V^2 \rangle - \langle V \rangle^2 = k_B T \langle V \rangle K_T$$

where K_T is the isothermal compressibility (1.50). By monitoring the volume fluctuations in a simulation run at constant NPT we can thus measure K_T. As in the case of the energy, one can obtain a second estimate of K_T by carrying out the simulation at two slightly different pressures. Completely analogous relations obtain for the susceptibility of a magnetic system. Similarly, in an open system at fixed chemical potential μ for which the appropriate weighting function for state α is the grand canonical Hamiltonian $H_\alpha = E_\alpha - \mu N_\alpha$ we have a connection between fluctuations in the particle number and the compressibility (2.65):

$$\langle(\Delta N)^2\rangle = \frac{k_B T K_T \langle N\rangle^2}{V} .$$

9.3.2 Error estimates

Each Monte Carlo step generally only causes a small change in the system. Near phase transitions large systems approach equilibrium very slowly. This phenomenon, which is sometimes called *critical slowing down*, is a major obstacle in simulations of large systems. Suppose $x(s)$ is the value of some observable after s Monte Carlo steps and $x(s+t)$ is the value after t further steps. In a properly functioning simulation the memory of the initial state should decay

$$\lim_{t\to\infty} \langle x(s+t)x(s)\rangle - \langle x\rangle^2 = 0$$

and one can often associate a time scale τ, the *correlation time*, with this decay. If the system is not near a critical point or at coexistence, the decay is approximately exponential

$$\frac{\langle x(s+t)x(s)\rangle - \langle x\rangle^2}{\langle x^2\rangle - \langle x\rangle^2} \simeq e^{-t/\tau} .$$

It is only after approximately τ steps that one has an independent measurement of x. The consequence of this is that one cannot improve the quality of the data by sampling more frequently than every τ time steps. This is useful to know, particularly if there is a significant computational effort involved in the sampling process. The statistical error of each measurement is typically

$$\sigma = \sqrt{\langle x^2\rangle - \langle x\rangle^2} \tag{9.31}$$

giving for the expected random error of a run of N steps

$$\Delta x \simeq \sigma\sqrt{\frac{\tau}{N}} . \tag{9.32}$$

Just as in experimental physics, it is always good practice to make an error estimate of the quantities calculated. One should also make a few independent long runs to check if the spread in the computed values corresponds to what is expected from (9.32). If the spread in the computed value of $\langle x \rangle$ is significantly larger than expected, it is likely that there is more than one relaxation time for the system. Another check that one can make is to calculate the spread in values from several shorter runs and compare with the spread from parts of equal length of a large run. If there is a discrepancy, one must suspect the random number generator. If the spread in computed values from parts of a single long run is less than expected, this may also be serious, since it may indicate that there is a problem with the Markov process (*e.g.* traps).

One aspect of Monte Carlo simulations that we will not discuss in detail is the dependence of results on the quality of the random number generator. Typically in a Monte Carlo simulation one employs a pseudo-random number generator that provides numbers that are *periodic* with a period that depends on the algorithm used and on the initial number in the sequence. Simple random number generators often have a maximum period of $2^{31}-1$—a quantity related to the number of bits in a single-precision word in most computer architectures. With modern computers it is not too difficult to carry out Monte Carlo calculations that require significantly more random numbers than this. For such calculations one must use a random number generator with a longer period, to be safe, a period many times the maximum number of steps carried out. In addition to the problem of the finite period there is the possibility of correlations between different 'random' numbers in the chain. These issues are by no means settled at this point. For a discussion of the problems associated with random number generators, the reader is referred to [249].

9.3.3 Extrapolation to the thermodynamic limit

One of the most important problems of a simulation, be it molecular dynamics or Monte Carlo, is the extrapolation of thermodynamic functions determined for a system of N particles or spins to the thermodynamic limit $N = \infty$. This is particularly difficult near a critical point where, as we have shown in Chapter 6, the singularities associated with the critical behavior of an infinite system are replaced by smooth behavior which sets in as the correlation length ξ becomes equal to the size L of the system. In this subsection we show how some of these difficulties can be managed for a d-dimensional Ising model on a lattice of size L^d and refer the reader to [42] for a more thorough discussion of the general case.

We first note that, since there is no phase transition in any finite system, the order parameter $m_L(T)$ of a system of size L is zero at all temperature

$T \neq 0$. In terms of the order-parameter probability (9.29) this means that $p_L(m)$ is a doubly-peaked function with maxima at $m = \pm m_0(T)$ for $T < T_c$. For a finite system $p_L(m)$ is finite for all values of m in the range $-1 < m < 1$ and this means that in a simulation the system will mostly be found in one of the two ordered states but will cycle between these states with a frequency that becomes smaller as the size of the system is increased. Nevertheless, if the simulation is run long enough, the expectation value of the order parameter is 0. This problem can be overcome by measuring the expectation value of the square of the order parameter:

$$m_{rms}(T) \equiv \left\langle \left(\frac{1}{N} \sum_i \sigma_i \right)^2 \right\rangle^{1/2} . \tag{9.33}$$

This expectation value is dominated by contributions from the two peaks at $\pm m_0$. The disadvantage of using this function instead of $m(T)$ is that it is nonzero at all temperatures $T > T_c$. Therefore, the determination of T_c and the critical exponent β is a nontrivial problem. We note in passing that if the system has an order parameter with a continuous symmetry, *e.g.* the Heisenberg model, the use of a formula such as (9.33) is even more important than for the Ising model.

The behavior of $m_{rms}(T)$ as function of L is easily determined. For $T > T_c$ we have

$$m^2_{rms}(T) = \frac{1}{N^2} \left\langle \sum_{i,j} \sigma_i \sigma_j \right\rangle = \frac{1}{N} k_B T \chi(L,T)$$

where $\chi(L,T)$ is the susceptibility per spin of the finite system. For $L \gg \xi(T)$ we therefore have $m_{rms}(T) \sim L^{-d/2}$ since the L-dependence of $\chi(L,T)$ can be ignored in this regime. Conversely, in the critical region we must use the finite-size scaling form of χ:

$$\chi(L,T) = L^{\gamma/\nu} Q(L/\xi(T)) . \tag{9.34}$$

For temperatures such that $L \ll \xi(T)$ the magnitude of the order parameter therefore varies with system size as $m_{rms}(T) \sim L^{\gamma/2\nu - d/2} \sim L^{1-\eta/2-d/2}$ where we have used the scaling relation $\gamma = \nu(2 - \eta)$. Knowledge of this dependence of thermodynamic functions on the system size is of great help in the determination of T_c and of the critical exponents from a simulation. A similar analysis can be carried out for any other function measured for a range of L and T.

Another method that can be used to determine T_c is based on the study of $p_L(m)$. For $T > T_c$ this probability distribution is a Gaussian with a width given by the susceptibility $\chi(L,T)$:

$$p_L(m) = \sqrt{\frac{1}{2\pi k_B T\chi(L,T)}} \exp\left\{-\frac{m^2}{2k_B T\chi(L,T)}\right\} . \tag{9.35}$$

For $T < T_c$, the single Gaussian is replaced by a doubly-peaked function that becomes a sum of two Gaussians for low temperatures. On the other hand, near the critical point we can use finite-size scaling theory to draw some useful conclusions. A function that has the correct properties is [42]

$$p_L(m) = L^y \tilde{P}(mL^y, L/\xi(T)) . \tag{9.36}$$

The expectation value of the magnitude of the order parameter is then

$$\begin{aligned} \langle|m|\rangle &= L^y \int_{-\infty}^{\infty} dm|m|\tilde{P}(mL^y, L/\xi(T)) \\ &= L^{-y} \int_{-\infty}^{\infty} dz|z|\tilde{P}(z, L/\xi(T)) = L^{-y}\Phi(L/\xi(T)) \end{aligned} \tag{9.37}$$

which is the usual finite-size scaling form for the order parameter and therefore, $y = \beta/\nu$. A particularly powerful technique for determining T_c is based on the fact that at the critical temperature of the infinite system appropriate ratios of moments of this distribution become universal numbers, independent of L. Consider the expectation values $\langle m^2\rangle$ and $\langle m^4\rangle$ in the critical region. Clearly

$$\langle m^2\rangle = L^{-2/y} \int_{-\infty}^{\infty} dz z^2\tilde{P}(z, L/\xi(T)) \tag{9.38}$$

$$\langle m^4\rangle = L^{-4/y} \int_{-\infty}^{\infty} dz z^4\tilde{P}(z, L/\xi(T)) . \tag{9.39}$$

Therefore, the ratio

$$R = \frac{\langle m^4\rangle}{\langle m^2\rangle^2} = \Psi(L/\xi(T)) . \tag{9.40}$$

As $\xi(T) \to \infty$ the right-hand side of this equation becomes independent of L. Noting that for $T \gg T_c$ the Gaussian distribution (9.35) yields $R = 3$ and that, on physical grounds we expect $R \to 1$ as $T \to 0$, we see that the function

$$U_L(T) = 1 - \frac{R}{3} \tag{9.41}$$

plotted as function of T for different L will yield a set of curves that intersect at the critical point T_c of the *infinite* system. For further discussion and elaboration of this method and finite-size analysis of more complicated situations we refer the reader to [42].

9.4 The Hopfield Model of Neural Nets

During the last decade, the study of neural network models has been a highly active area of statistical physics. Although the original motivation for these studies was a desire to understand the functioning of the human brain—the process of learning, the nature of memory, etc. — applications of this type of model are much broader, ranging from *associative* or *content-addressable* memories to autocatalytic systems and classifier systems. There is also an important connection between this field and the theory of spin-glasses. We briefly discuss this subject here as an example of an emergent field in which many of the important results have been attained through simulations.

The human nervous system contains a complex interconnected network of perhaps 10^{11} *neurons*. For our purposes, we model a neuron by a dynamical variable that can assume only two states. The *active* state represents the situation in which a signal is being transmitted from the cell-body of the neuron via the axon to a synaptic junction. Conversely, the *passive* state is the rest state of the neuron. The synaptic junction is the point at which different neurons interact, *i.e.*, transmit signals to each other. In the human cortex a given neuron typically interacts with roughly 10^4 other neurons. Thus in this highly simplistic picture we have a dynamical system consisting of Ising spins — $s_\alpha = 1 \equiv$ active, $s_\alpha = -1 \equiv$ passive — coupled to each other through a set of synapses with a coupling constant $J_{\alpha\gamma}$ that we take to be symmetric: $J_{\alpha\gamma} = J_{\gamma\alpha}$.

We must now define the rules according to which the state of our system can change. Here we consider the Hopfield[2] model [134] which is one of the prototypical neural network models. We let $J_{\alpha\gamma} > 0$ correspond to *stimulation* and $J_{\alpha\gamma} < 0$ to *inhibition* and update the state of the neurons at regular intervals according to the rule

$$s_\alpha(t+1) = sign(\sum_\gamma J_{\alpha\gamma} s_\gamma(t) - h_\alpha) \ . \tag{9.42}$$

We interpret (9.42) as stating that if the number of stimulations of neuron i exceeds the number of inhibitions by a threshold h_i the neuron will be active in the next time interval. Otherwise the neuron will be inactive. If the network is symmetric, we can make contact with physics by describing the situation in

[2]This model was first proposed by McCulloch and Pitts [199] but is generally known as the Hopfield model in the physics community.

terms of an Ising model with the Hamiltonian

$$H = -\sum_{\alpha<\gamma} J_{\alpha\gamma} s_\alpha s_\gamma + \sum_\alpha h_\alpha s_\alpha \,. \tag{9.43}$$

The update rule (9.42) implies that $s_\alpha \to +1$ if this lowers H, otherwise $s_\alpha \to -1$. This rule is thus equivalent to that of a zero temperature Monte Carlo simulation using the Metropolis algorithm. In a neural network one can update the spins either in random order or all spins at the same time. In the latter case the system acts as a "cellular automaton" and the behavior of the network is deterministic.

What does this model have to do with memory? Any string of letters, or pixels in a picture, can be represented by a sequence of bits. Thus the state of an N-neuron network $\{s_1, s_2, \ldots, s_N\}$ corresponds to one of 2^N possible patterns. Before discussing how to store information, or previously learned patterns in a neural network, we note that the dynamics given by (9.42) provides a mechanism for the retrieval of such information. Imagine that the initial state of the network or the *stimulus* is $i = \{s_\alpha\}$. The network then evolves according to (9.42) until (hopefully) some final state f is reached. If the stimulus is sufficiently similar to a previously "learned" pattern we want a properly functioning memory to return that pattern. Clearly, the only way that our network can learn or store information is to adjust the synaptic couplings every time a new stimulus presents itself. The standard method for doing this is the *Hebb rule*[3]. According to this rule a new pattern i is added to the repertoire of the network by modifying the synapse strength so that

$$J_{\alpha\gamma}{}^{new} = \lambda J_{\alpha\gamma}{}^{old} + \epsilon \sigma_\alpha{}^i \sigma_\gamma{}^i \tag{9.44}$$

with $\lambda > 0$ and $\epsilon > 0$. Here $\sigma_\alpha{}^i$ is the αth bit of the stimulus i. We see that if $\{s_\alpha(t)\}$ is 'similar' to the pattern $\sigma_\alpha{}^i$, bits in $\sigma_\alpha{}^i$ which are positive will be stimulated while negative bits will be inhibited. On the other hand, if $\{s_\alpha(t)\}$ is nearly orthogonal to the remaining patterns, in a way that will be defined more precisely below, they will have little effect. Nevertheless if too many patterns are stored there will be interference between the stored responses. This problem can, to some extent, be overcome by letting $\epsilon > \lambda$ so that old memories are deemphasized or "forgotten" as new ones are learned.

[3]Named after the Canadian psychologist Donald O. Hebb 1904-85. For a readable account of his life and work see Milner [205]. This article somewhat conveniently omits reference to one of the darker episodes of Canadian science, namely the heavy involvement of the Psychology Department of McGill University in CIA-sponsored psychological warfare research during the time Hebb was head of the department.

To see that this actually works, we consider a couple of simple situations. Suppose that we initially have $J_{\alpha\gamma} = 0$ for all $\alpha\gamma$ and choose the constant $\epsilon = 1/N$. After a stimulus by a pattern $\{\sigma_i\}$ the synaptic couplings then have the values

$$J_{\alpha\gamma}(1) = \frac{1}{N}\sigma_\alpha\sigma_\gamma \; . \tag{9.45}$$

Now suppose that the network is presented with a stimulus that is similar to the first pattern in the sense that $N - n$ of the bits are the same and $n \ll N$ of the bits are different. For simplicity we take the threshold field $h_\alpha = 0$ for all α. After one step, spin α has the value

$$\begin{aligned} s_\alpha(1) &= sign\left(\frac{1}{N}\sum_\gamma \sigma_\alpha\sigma_\gamma s_\gamma(0)\right) \\ &= sign\left(\sigma_\alpha \frac{N-2n}{N}\right) \; . \end{aligned} \tag{9.46}$$

By assumption, the quantity $N - 2n > 0$ and therefore, after one step the state of the network corresponds to the learned pattern. Since only one pattern has been learned, the only condition for its retrieval is that more than half of the bits of the stimulus must match those of the stored pattern.

This example is easily generalized to the case when M patterns have been memorized. During the learning process we take $\lambda = 1$ and continue to keep $\epsilon = 1/N$. The synaptic couplings at the end of this learning process are then given by

$$J_{\alpha\gamma} = \frac{1}{N}\sum_{j=1}^{M}\sum_{\alpha\gamma} \sigma_\alpha^j \sigma_\gamma^j \tag{9.47}$$

where $\{\sigma_\alpha^j\}$ is the bit pattern of the jth stimulus. Assume now that a stimulus with bit pattern $\{\sigma_\alpha^l\}$, corresponding to the lth memorized pattern acts on the network. The 'field' seen by neuron i at $t = 0$ is

$$\begin{aligned} H_\alpha &= \frac{1}{N}\sum_{j=1}^{M}\sum_\beta \sigma_\alpha^j\sigma_\beta^j\sigma_\beta^l \\ &= \sigma_\alpha^l + \frac{1}{N}\sum_{j\neq l}\sum_\beta \sigma_\alpha^j\sigma_\beta^j\sigma_\beta^l \; . \end{aligned} \tag{9.48}$$

In the second expression, the sum on the right consists of $N(M-1)$ terms that, if the stored patterns are sufficiently different from one another, are equally likely to be +1 and −1. Therefore, we expect that

$$\frac{1}{N}\sum_{j\neq l}\sum_{\beta}\sigma_{\beta}^{l}\sigma_{\alpha}^{j}\sigma_{\beta}^{j}\approx\sqrt{\frac{M-1}{N}}\ .$$

Thus we see that the sign of the field acting on neuron i, in the limit $M/N \to 0$ is that of the bit in stored pattern l for all i and, in one step, we retrieve this pattern. Similarly, one can show that if the initial stimulus is sufficiently close to one of the learned patterns, the correct pattern is likely to be retrieved.

One can quantify the difference between patterns 1 and 2 by defining the 'distance' d between them through

$$d^2=\frac{1}{2N}\sum_{\alpha=1}^{N}\left(s_{\alpha}^{1}-s_{\alpha}^{2}\right)^2=1-\frac{1}{N}\sum_{\alpha=1}^{N}s_{\alpha}^{1}s_{\alpha}^{2}\ .$$

Two similar states (1) and (2) have large 'overlap':

$$\frac{1}{N}\sum_{\alpha}{s_{\alpha}}^{1}{s_{\alpha}}^{2}\simeq 1$$

whereas dissimilar states tend to be nearly orthogonal:

$$\frac{1}{N}\sum_{\alpha}{s_{\alpha}}^{1}{s_{\alpha}}^{2}\simeq 0$$

and 'near opposites' have

$$\frac{1}{N}\sum_{\alpha}{s_{\alpha}}^{1}{s_{\alpha}}^{2}\simeq -1\ .$$

We will not attempt to derive the conditions that lead to a memory that functions reliably. Amit *et al.* [13] analyzed the problem assuming replica symmetry (see Section 13.4.3), that the number of neurons $N\to\infty$ and that the stored patterns are random. Below 'all' and 'always' mean 'almost always' or with probability 1. There may be a few isolated pathological situations. There are four main possibilities if M is the number of stored patterns and $\epsilon=\lambda=1$:

1: $M/N\to 0$ *as* $N\to\infty$. In this case all learned pattern are retrieved correctly if the stimulus is sufficiently close. There will also be spurious solutions; if a stimulus that doesn't correspond to a learned pattern is presented, the network may invent an answer. The system has no way of saying 'I don't know'.[4]

2: *$M/N \to \alpha$ as $M \to \infty$ with $\alpha < 0.05$.* All stored patterns are retrieved in an almost correct form, if the stimulus is close enough, but some bits may be off.

3: *As in (2) but with $0.05 < \alpha < \alpha_c \approx 0.14$.* Most but not all patterns can be retrieved; there will be occasional errors in a few bits of those patterns that can be retrieved.

4 *As in (2) and (3) but with $\alpha > \alpha_c$.* The response of the system is chaotic.

Following the work of [13] a number of authors have attempted to go beyond the replica symmetric approximation, but the conclusions remain substantially the same. For a critical discussion see Stiefvater *et al.* [289].

Here we have only discussed neural networks in their most primitive form. For practical applications one will have to modify the model to address the particular features of the application. There is by now a large literature on the subject, and a number of applications are in use. In practice one often organizes the network in *layers* with an *input* layer at the bottom and an output layer at the top and one or more *hidden* layers in between. The goal is then to produce a desired output for a given input. We may have at our disposal a training set of input patterns for which we know the desired output. One may then employ special "back propagation" algorithms which minimize measures of the expected error. For a recent introduction to neural nets with easy-to-implement examples see [103] or [209]. The book by Hertz *et al.* [129] is closest to the statistical mechanics approach taken here.

One lesson learned from studies of such models is that the Ising model with random coupling J_{ij} between the spins tends to have very many local minima in the energy, a question we will return to in our discussion of spin glasses in Section 13.4. In the case of neural networks one exploits this fact to store information in the system. In the next subsection, we turn to discrete combinatoric problems where the "energy" is a cost function. In this case the presence of many local minima is the aspect of the problem which makes it very hard to solve.

---

[4]For an attempt to overcome this problem by making the network non-symmetric see [230].

9.5 Simulated Quenching and Annealing

We illustrate by an example how some of the statistical mechanics techniques that we have discussed earlier can be applied to an important class of problems involving combinatorial optimization. A sales agent has to visit N cities $c_1, c_2, ..., c_N$. The distance between cities c_i, c_j is $d(i, j)$ and is available in a table. The problem is to find the shortest trip that visits all cities at least once, and returns to the home city. This constitutes the "traveling salesman problem". There are many common problems which involve scheduling that are similar to this problem, *e.g.* school time tables, assignment of duties of drivers in a metropolitan transit system, etc.

The number of possible trips grows with N as $N!$. While it is in principle possible to check all trips, for finite N, and pick the shortest, this is impractical for much more than 10 cities. A number of special purpose, or 'heuristic', techniques have been developed to attack practical problems of this type and in practice some of these methods form part of the solution. Examples of such methods are

1: If it ain't broke don't fix it.

> This method is used to schedule the timetable of classes at many universities. Courses are given 'slots' which don't change from year to year unless a conflict has developed.

2: Steepest descent.

> In the traveling salesman go first to the closest city, then to the closest city not yet visited, etc.

3: Divide and conquer.

> In the case of the traveling salesman one can divide the territory into regions. One finds the shortest trip in each region and then connects these trips.

Here we describe a method for solving combinatorical problems that uses statistical mechanics. In the case of the traveling salesman we describe a possible trip by an array c_i, where c_i is the label of the city visited after the ith leg of the trip. The "cost" of the trip is then

$$L(c) = \sum_{i=1}^{N-1} d(c_i, c_{i+1}) + d(c_N, c_1) \ . \tag{9.49}$$

The next step is to introduce a fictitious inverse 'temperature' β. The average length of the trip for a given β is

$$\langle L \rangle = \frac{\sum\limits_{trips} L(c) e^{-\beta L(c)}}{\sum\limits_{trips} e^{-\beta L(c)}} \; . \tag{9.50}$$

There are more ways of making long trips than short ones. Therefore if the temperature is high, most trips are long, while at low temperature most trips are short. At equilibrium at $\beta = \infty$ only the shortest trip is allowed. The optimum solution to the problem is therefore equivalent to finding the ground state of the system. The difficulty of the problem lies in the fact that there are often very many local minima in the cost function L.

We can use the Monte Carlo method to find nearly optimum solutions. Each possible trip can be represented by a permutation of the numbers $1, 2, ..., N$. For example in a possible 10-city trip

$$5, 2, 8, 10, 9, 1, 3, 7, 4, 6$$

the eighth city is visited after the third leg of the trip. A possible Monte Carlo step is a transposition or the interchange of two cities. For example transposing the third and seventh leg gives

$$5, 2, 3, 10, 9, 1, 8, 7, 4, 6 \; .$$

All possible trips can be reached from any initial configuration by a series of transpositions. Let L_i be the cost function of a trip before a transposition and L_f the cost afterwards. We can generate representative trips using the Metropolis algorithm:

> If $L_f < L_i$ accept the transposition.
> If $L_f > L_i$ accept the transposition with probability $e^{\beta(L_i - L_f)}$.

In the special case $\beta = \infty$ we only accept moves that shorten the trip and the Monte Carlo method in this case is referred to as a zero temperature quench. If there are many local minima in the cost function, a zero temperature quench usually does not give a very good solution to the problem. Somewhat better results can be obtained using the steepest descent method. Since there are only $N(N-1)/2$ possible moves at any one time, one can calculate the change in the cost function for each possible move. In the method of steepest descent one selects the allowed move that lowers the cost function most.

Further improvements can be obtained using *simulated annealing* [154]. The algorithm starts with a configuration corresponding to a high temperature, and the system is cooled *slowly* until one ideally reaches a global optimum as $\beta \to \infty$. In order to reach this result the annealing must be slow enough—typically [108][66] the number of time steps one must use to reach a given value of β grows exponentially with β. There also exists, however, a 'fast simulated annealing' technique [298]. Suppose we have a measure x of distance in state space (*e.g.* the number of spins which are different in an Ising model). Unlike the conventional Metropolis algorithm in which one does a local search for acceptable changes, the attempted configuration is chosen with a Cauchy/Lorentzian distribution

$$G(x) = \frac{T(\beta)}{T(\beta)^2 + x^2} , \tag{9.51}$$

where $T(\beta)$ is a phenomenological parameter. The attempted transposition is accepted or rejected by a detailed balance criterion as before. The fast simulated annealing technique reduces the danger of shallow traps by occasionally allowing longer jumps and it should in principle allow a cooling schedule that is linear in inverse temperature [298].

In practice, the computing times required to aquire thermal equilibrium at a low temperature may be too long for a pure simulated annealing calculation to be practical. A more heuristic annealing/quenching routine may then proceed as follows:

(1). Select an initial allowed configuration (either randomly or using some simple empirical method).

(2). Perform a zero temperature quench (possibly using steepest descent).

(3). Reheat the system to a suitably chosen temperature β^{-1}.

(4). Slowly cool the system finishing with a zero temperature quench.

(5). Repeat steps 3 and 4 while keeping track of the best solution found to date.

Many variations of this procedure are possible, but we leave those to the imagination of the reader. On the whole, the simulated annealing method is a robust method for obtaining near optimum solutions to combinatorical problems. The two main advantages are the generality of the method and the ease by which it can be programmed for the computer. We should emphasize that for a 'classical problem' such as the traveling salesman, which has fascinated researchers for many years, there are special purpose programs that are more efficient for large N than simulated annealing.

9.6 Problems

9.1. *Billiards on a square table.*

Consider a simulation in which a particle collides with the walls of a square box. At each reflection the tangential component of the momentum is conserved, while the normal component changes sign. In a simulation one must specify an initial value of the position x_0, y_0 of a particle, and the x- and y-components of the velocity v_x and v_y (Figure 9.3). In the microcanonical ensemble the speed of a particle is given by the energy of this particle, but all orientations of the velocity and all positions inside the box are equally likely. In the molecular dynamics simulation the magnitude of v_x and v_y will not change, and there exists an infinite number of periodic orbits that do not sample all parts of the box see Figure 9.3(b).[5]

[5]You can obtain a c-program 'ergode.c' from
http://www.physics.ubc.ca/~birger/equilibrium.htm
This program divides a box of unit size into a number of equally sized bins. At equal time intervals, the particle is located within one of the bins. A histogram of the number of visits in each bin is produced at the end of the run, and can be compared with a Gaussian distribution whose variance is the mean number of visits in each bin. A similar histogram can also be produced by a 'random number generator'.

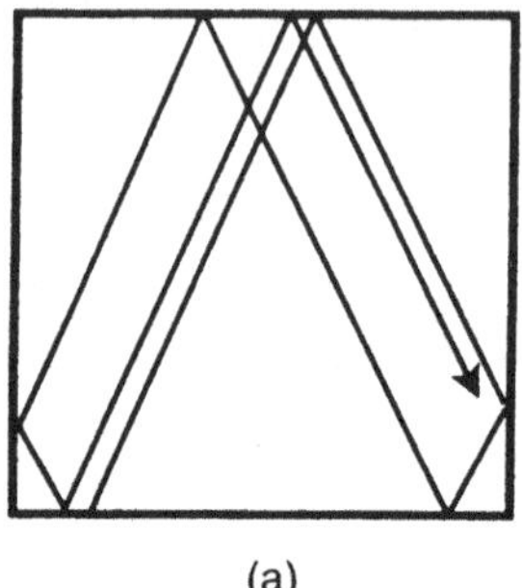

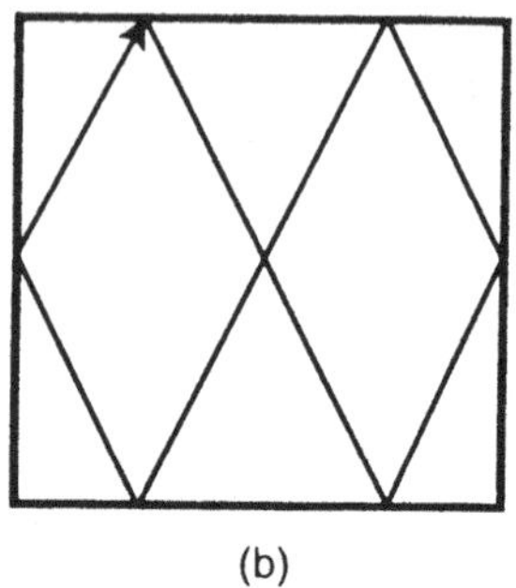

Figure 9.3: Sample trajectories of a particle in a square box. (a) Space-filling trajectory; (b) Periodic orbit.

(a) Verify the existence of periodic orbits for which there are parts of the box that are not visited. What are the characteristics of periodic orbits with short periods?

(b) Show that if the x- or y-components of the velocity is a multiple of the sampling time the sampling distribution will not be uniform.

(c) Select initial conditions so that the orbit, if periodic, has a period which is long compared to the time of a run. How does the distribution of the number of visits in each bin vary with the speed of the particle (given constant sampling frequency)?

9.2. *One-dimensional ideal gas at constant pressure.*

Carry out a simulation of a one-dimensional N-particle system of particles of mass m (Figure 9.4). At one end there is a piston of mass M which is pushed towards the gas with constant force P; the other end is a fixed wall. In collisions with the fixed wall the velocity of the particles changes sign, and the speed remains constant. In collisions with the piston the energy and momentum of particle-piston system is conserved. The instantaneous distance between the piston and the wall is $L(t)$. The gas is 'ideal', *i.e.* the interaction between particles can be neglected.[6]

(a) Check that the enthalpy (PL+kinetic energy) is conserved.

(b) Show that even if the system is started with a non-equilibrium velocity distribution (*e.g.* all the particle moving with constant speed

[6]The source code for this project is available as 'ideal.c' from http://www.physics.ubc.ca/~birger/equilibrium.htm.

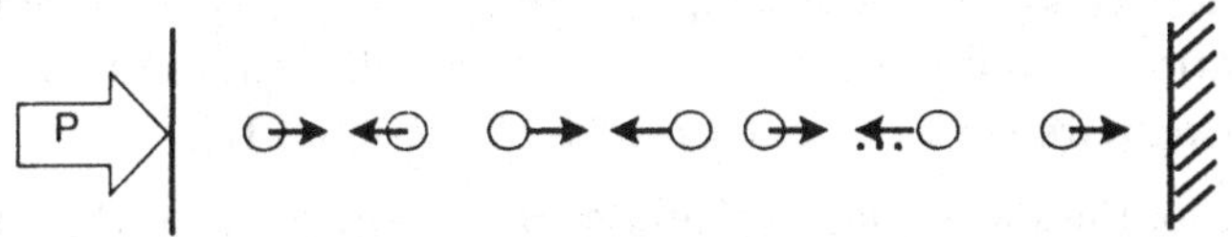

Figure 9.4: One-dimensional ideal gas at fixed pressure.

in the same direction from random positions) the gas, after some time, tends to a Maxwell-Boltzmann velocity distribution

$$P(v) \sim \exp(-\frac{1}{2}\beta m v^2) \tag{9.52}$$

where β^{-1} is the average of $\frac{1}{2}\times$ kinetic energy per particle.

(c) Starting with different initial energies and a fixed P, 'measure' the temperature and average value of L over a production run after allowing the system to thermalize. Show that the system approximately satisfies the ideal gas law.

(d) One may attempt to do a constant $T-P$ simulation by letting the particles that collide with the fixed wall have the velocity distribution (9.52) (such a distribution can be generated using the central limit theorem, by letting a particle undergo a 'random walk' in velocity space). Show that the resulting temperature will be *less* than predicted by (9.52). Why?

9.3. *Another example of a Markov process*

There are initially two white marbles in box A and a black marble in box B. A marble is randomly selected from each box and they are interchanged. What are the steady state probabilities of the two states $\alpha = [\circ\bullet][\circ], \beta = [\circ\circ]\ [\bullet]$?

9.4. *Ising chain*

Consider the one-dimensional Ising chain of N-spins with free ends (3.1).

(a) Carry out a Monte Carlo simulation of the system of spins using the Metropolis algorithm (choose the number of spins according to your computational resources). Compute the average energy and heat capacity at the temperature $T = 2J/k_B$ using both the fluctuation formula (9.30) and a numerical differentiation of the energy. Compare with the exact results of Section 3.1.

(b) Plot the specific heat calculated from a single run at the temperature J/k_B using the fluctuation formula and the histogram method of subsection 9.2.3. Compare with exact results for the specific heat.

9.5. *Traveling salesman by simulated annealing*

Write and test a computer program for the solution of the traveling salesman problem. Distances between a number of cities can be found in most road atlases. If typing in large matrices bores you, generate the coordinates of a number of cities (*e.g.* randomly) and have the computer work out the table. Our experience is that one needs of the order 30 cities before there is any advantage in using simulated annealing over simply performing a large number of zero temperature quenches and selecting the best result.

Chapter 10

Polymers and Membranes

In this chapter we discuss some aspects of the statistical mechanics of linear polymers as well as fluid and tethered membranes. The study of polymers has a long history and continues to be one of great current interest. In our approach to the subject we will focus exclusively on universal properties of these systems at large length scales, ignoring the important microscopic distinctions between different polymers. These microscopic details are of paramount importance to a polymer chemist but play only a minor role in the limit that we will concern ourselves with, namely that of very long chains. Our models will therefore be extremely simple but we will show that even very simple models of linear polymers are capable of producing surprisingly accurate predictions of at least some of the equilibrium properties of these systems.

After a brief introduction we turn in Section 10.1 to a discussion of the simplest model of a macromolecule, namely the Gaussian chain. This will allow us to introduce the concept of *entropic elasticity* which is the counterpart in polymers and rubber (cross-linked polymers) of the Hooke's law of a simple spring. In Section 10.2 we will discuss the effects of self-avoidance on the properties of polymers in good solvents, and in Section 10.3 we will make contact with continuous phase transitions by relating the statistics of a polymer to that on the n-vector model of Section 7.5. Finally, in Section 10.4 we briefly discuss polymers in concentrated solutions. In Section 10.5 we discuss some aspects of the statistical mechanics of membranes. Our simple picture of a membrane will be either a set of particles connected to each other in a two-dimensional network (the *tethered membrane*) or particles free to float on a two-dimensional surface that can stretch, bend and otherwise rearrange itself (the *liquid membrane*). These models are far from the more detailed and sophisticated pictures of membranes that are used by biologists and therefore will allow us only to discuss certain large scale physical properties of such objects. Nevertheless, these models are intrinsically interesting and provide another nice example of the role of dimensionality in statistical physics.

There are a number of excellent monographs on polymers. In our treatment we have leaned heavily on the book by de Gennes [73] and that of Doi and Edwards [76]. An older but still valuable reference is Flory's book [100] and a very thorough modern discussion of the subject is given in Des Cloiseaux and Jannink [75]. An excellent up-to-date reference to polymer physics is Rubinstein and Colby [259]. Widespread interest by the condensed matter physics community in membranes began only in the mid 1980's and we are not aware of any comprehensive treatment of this emerging subject. We will primarily refer to the recent literature on this subject.

10.1 Linear Polymers

Here we will treat polymers as long flexible chains and ignore details on the monomer scale. Consider, for example, polyethylene which is a chain of CH_2 monomers joined at a C–C bond as shown in Figure 10.1(a). The degree of polymerization or the number N of CH_2 units is variable but can be of the order of 10^5 or more. This fact is what makes the use of statistical methods in the description of even a single chain possible, and we shall always assume that we may take the thermodynamic limit $N \to \infty$ for any quantity that we calculate. On a microscopic level, there are several possible configurations of neighboring C–C bonds. The angle θ between nearest neighbor C–C bonds is essentially fixed at $\theta = 68^o$ (see Figure 10.1(b)). However, the bond $C_{n-1} - C_n$ can rotate around the axis defined by the $C_{n-2} - C_{n-1}$ bond. The energy of such a rotation is plotted as function of the azimuthal angle ϕ in Figure 10.1(c). The bond-orientation at the absolute minimum of this energy (at ϕ_0) is referred to as the *trans* configuration; the two orientations at the equivalent higher energy minima at $\phi \approx \phi_0 \pm 2\pi/3$ are called *gauche* configurations. The energy difference between trans and gauche configurations is approximately $\Delta E/k_B \approx 250K$ [100] and it is clear that in equilibrium at room temperature there will be a significant fraction of bonds in the gauche configuration. Of course, the energy $E(\phi)$ depends to some extent on the configuration of other bonds nearby on the chain but this is a correction that does not change the conclusion that at *e.g.* room temperature the configuration of polyethylene will consist of a

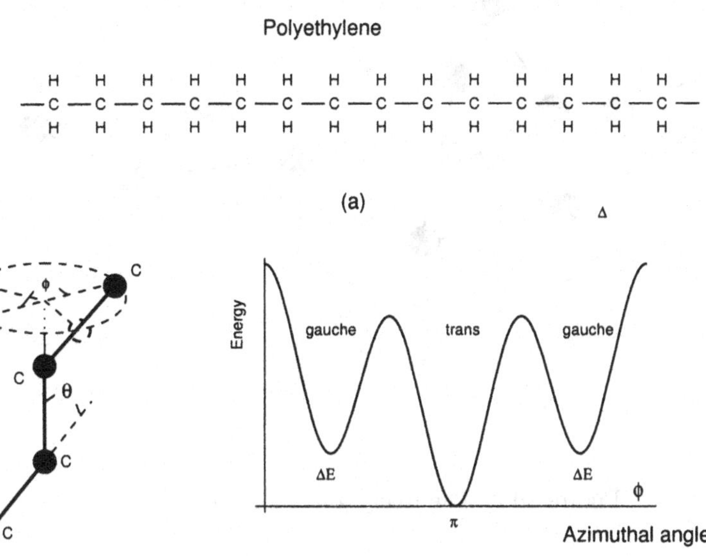

Figure 10.1: (a) Chemical structure of polyethylene. (b) Carbon-carbon bonds on the chain. (c) Energy of a carbon-carbon bond as a function of the azimuthal angle ϕ.

mixture of trans and gauche bonds.

It is clear that there will be a rapid loss of memory as function of distance along the chain and it is this that gives flexible polymers their universal structural properties. There is a length l, sometimes called the *Kuhn length* or *persistence length*, over which the orientations of the bonds become uncorrelated. This length depends on microscopic details such as, in the case of polyethylene, ΔE. As long as l is very much smaller than the total contour length Na, where a nearest neighbor distance, any type of polymer should have physical properties that depend in a universal way on Na/l.

The disorder in the bond orientations makes the configuration of a polymer similar in many respects to a biased random walk and we shall exploit this analogy in the sections that follow. We shall see that the statistical mechanics of flexible polymers, to a first approximation, consists of nothing more than maximizing the entropy of a random walk. However, there is one further important property of flexible chains that makes the polymer problem more

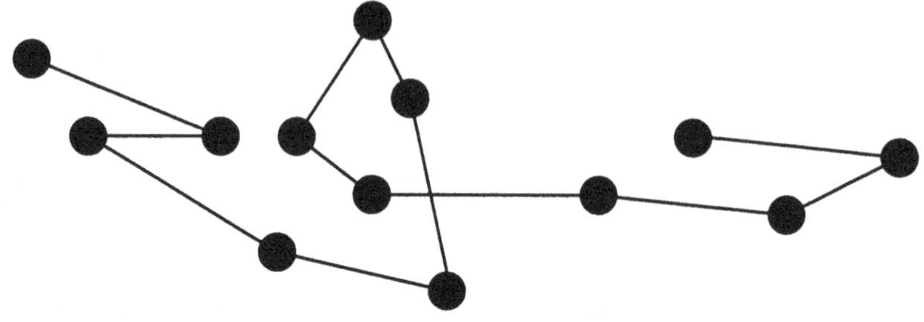

Figure 10.2: The freely jointed chain.

challenging than the random walk problem: the atoms on the chain have hard cores and therefore cannot occupy the same region in space. This causes the chain to expand or swell in a nontrivial way at least for dimensions $d < 4$, which is the *upper critical dimension* for the polymer problem. The appearance of an upper critical dimension is a signal that the polymer problem is in a sense a critical phenomenon problem, and we shall see that this is indeed the case. It should be emphasized that this hard core interaction, although short ranged in d-dimensional space is *long ranged* in terms of distance along the chain, *i.e.* monomers far apart along the chain can interact strongly. It is this feature that makes the polymer problem a highly nontrivial one.

We begin our discussion of the statistical mechanics of polymers by considering, in the next subsection, the simplest model of a polymer, the freely jointed chain.

10.1.1 The freely jointed chain

Consider a system of $N+1$ point-particles, as shown in Figure 10.2, separated by bonds of length a that are free to take on any orientation in the three-dimensional space. The bond-length a is not intended to represent an interatomic distance on *e.g.* the polyethylene chain discussed in the introduction, but rather it is the *persistence length* mentioned above. We denote the location of the particles in space by the set of vectors $\mathbf{R}_0, \mathbf{R}_1, \ldots, \mathbf{R}_N$ with

$\mathbf{R}_i - \mathbf{R}_{i-1} \equiv \mathbf{r}_i$ and $|\mathbf{r}_i| = a$. The assumption that the chain is freely jointed means that

$$\langle \mathbf{r}_i \cdot \mathbf{r}_j \rangle = a^2 \delta_{ij} \ . \tag{10.1}$$

We characterize the configuration of the chain by its end-to-end distance, the square of which is given by

$$\mathcal{S}^2(N) = \left\langle (\mathbf{R}_N - \mathbf{R}_0)^2 \right\rangle \tag{10.2}$$

and by the radius of gyration defined through

$$R_g^2 = \frac{1}{N} \sum_{i=0}^{N} \left\langle (\mathbf{R}_i - \bar{\mathbf{R}})^2 \right\rangle \tag{10.3}$$

where $\bar{\mathbf{R}} = \frac{1}{N+1} \sum_{i=0}^{N} \mathbf{R}_i$ and the angular brackets denote an average over configurations. In the case of this simple model the quantity $\mathcal{S}^2(N)$ is straightforward to evaluate:

$$\begin{aligned}
\mathcal{S}^2(N) &= \left\langle (\mathbf{R}_N - \mathbf{R}_0)^2 \right\rangle \\
&= \left\langle (\mathbf{R}_N - \mathbf{R}_{N-1} + \mathbf{R}_{N-1} - \mathbf{R}_{N-2} + \ldots + \mathbf{R}_1 - \mathbf{R}_0)^2 \right\rangle \\
&= \sum_{i=1}^{N} \langle \mathbf{r}_i^2 \rangle + \sum_{i \neq j} \langle \mathbf{r}_i \cdot \mathbf{r}_j \rangle \\
&= Na^2
\end{aligned} \tag{10.4}$$

where the last result follows from (10.1). Similarly, we may obtain the radius of gyration:

$$\begin{aligned}
R_g^2 &= \frac{1}{N} \sum_{i=0}^{N} \left\langle (\mathbf{R}_i - \bar{\mathbf{R}})^2 \right\rangle = \frac{1}{2N(N+1)} \sum_{i,j=0}^{N} \left\langle (\mathbf{R}_i - \mathbf{R}_j)^2 \right\rangle \\
&= \frac{1}{N(N+1)} \sum_{i>j} \left\langle (\mathbf{r}_i + \mathbf{r}_{i-1} \ldots + \mathbf{r}_{j+1})^2 \right\rangle \ .
\end{aligned} \tag{10.5}$$

If the bonds are freely jointed (10.5) simplifies to

$$R_g^2 = \frac{1}{N(N+1)} \sum_{i>j} (i-j)\, a^2 = \frac{N+2}{6} a^2 \approx \frac{Na^2}{6} \ . \tag{10.6}$$

The fact that the end-to-end distance and the radius of gyration are both proportional to $\sqrt{N}$ will come as no surprise to the reader as this model is nothing more than an N-step random walk with fixed step-length a. More generally, we define an exponent ν that gives the dependence of these characteristic dimensions on the length of a polymer chain:

$$\mathcal{S}(N) \sim R_g(N) \sim N^\nu . \tag{10.7}$$

One of our goals, later in this chapter, will be the calculation of ν for more realistic models.

It is also interesting to obtain the probability distribution for the end-to-end distance for this simple model. The probability distribution for the bonds $\{\mathbf{r}_1, \mathbf{r}_2, \ldots, \mathbf{r}_N\}$ is simply a product of δ functions:

$$P(\mathbf{r}_1, \mathbf{r}_2, \ldots, \mathbf{r}_N) = \prod_{i=1}^{N} \frac{1}{4\pi a^2} \delta\left(|\mathbf{r}_i| - a\right) .$$

Therefore, the probability density $P_N(\mathbf{R})$ that the vector $\mathbf{R}_N - \mathbf{R}_0$ takes on the value $\mathbf{R}$ is given by

$$\begin{aligned} P_N(\mathbf{R}) &= \left(\prod_{i=1}^{N} \frac{1}{4\pi a^2} \int d^3 r_i \delta(|\mathbf{r}_i| - a) \right) \delta(\mathbf{R} - \mathbf{R}_N + \mathbf{R}_0) \\ &= \frac{1}{(4\pi a^2)^N} \left(\prod_{i=1}^{N} \int d^3 r_i \delta(|\mathbf{r}_i| - a) \right) \delta(\mathbf{R} - \mathbf{r}_1 \ldots - \mathbf{r}_N). \end{aligned} \tag{10.8}$$

We evaluate this probability density by first calculating its Fourier transform

$$\begin{aligned} \hat{P}(\mathbf{k}) &= \int d^3 R P_N(\mathbf{R}) e^{-i\mathbf{k}\cdot\mathbf{R}} \\ &= \frac{1}{(4\pi a^2)^N} \prod_{i=1}^{N} \left[\int d^3 r_i e^{-i\mathbf{k}\cdot\mathbf{r}_i} \delta(|\mathbf{r}_i| - a) \right] . \end{aligned} \tag{10.9}$$

The integral over the individual bond lengths is now trivially carried out in spherical coordinates with the result

$$\hat{P}(\mathbf{k}) = \left(\frac{\sin ka}{ka} \right)^N . \tag{10.10}$$

In the thermodynamic limit $N \to \infty$ this function becomes sharply peaked at $k = 0$ and we may approximate it as follows:

$$\ln \hat{P} \approx N \ln\{1 - k^2 a^2 / 6\} \approx -\frac{N k^2 a^2}{6} \tag{10.11}$$

and

$$\begin{aligned} P_N(\mathbf{R}) &= \int \frac{d^3k}{(2\pi)^3} \exp\left[-\frac{Nk^2a^2}{6} + i\mathbf{k}\cdot\mathbf{R}\right] \\ &= \left(\frac{3}{2\pi Na^2}\right)^{3/2} \exp\left[-\frac{3R^2}{2Na^2}\right] \end{aligned} \tag{10.12}$$

which is the characteristic Gaussian distribution for the end-to-end distance of the three-dimensional random walk. Keeping more terms in the expansion (10.11) leads to corrections to (10.12) which are of $\mathcal{O}(1/N)$.

As we have indicated in the introduction, real polymers have a certain amount of freedom in the relative orientation of neighboring bonds but certainly not the complete freedom of the freely jointed chain. We can make our model more realistic by restricting the relative orientation of bonds, *e.g.* by requiring that $\mathbf{r}_i \cdot \mathbf{r}_{i-1} = a^2 \cos\theta$ where θ is fixed. This still allows the bond between particles i and $i-1$ to rotate freely around the axis defined by the bond between $i-2$ and $i-1$ but is clearly more restrictive than the freely jointed model. It is easy to show (Problem 10.1) that the end-to-end distance and radius of gyration still have the characteristic $\sqrt{N}$ dependence on the number of particles but that there is a new "effective bond length" that depends on the angle θ.

10.1.2 The Gaussian chain

We now construct a simple model that is formally equivalent to the freely jointed model in the thermodynamic limit. Assume that the probability distribution $P(\mathbf{r}_i)$ for the vector connecting particles $i-1$ and i is given by

$$P(\mathbf{r}_i) = \left(\frac{3}{2\pi a^2}\right)^{3/2} \exp\left[-\frac{3\mathbf{r}_i^2}{2a^2}\right] \tag{10.13}$$

for all i. Clearly then, the probability distribution for the N vectors $\mathbf{r}_1, \mathbf{r}_2, \ldots, \mathbf{r}_N$ is simply the product of the single bond distributions and the joint probability density for the location of the $N+1$ particles in configuration space is

$$\begin{aligned} P(\mathbf{R}_0, \mathbf{R}_1, \ldots, \mathbf{R}_N) &= \frac{1}{V}\left(\frac{3}{2\pi a^2}\right)^{3N/2} \exp\left[-\frac{3}{2a^2}\sum_{i=1}^{N}\{\mathbf{R}_i - \mathbf{R}_{i-1}\}^2\right] \\ &= \frac{\exp(-\beta H)}{Z} \end{aligned} \tag{10.14}$$

where, in the last step, we have formally written the probability density as a Boltzmann weight and V is the volume. The volume enters into this expression because the center of mass can be anywhere in the container. The "partition function" is the normalization constant $V\left(2\pi a^2/3\right)^{3N/2}$ and the effective Hamiltonian H is

$$H = \frac{3k_BT}{2a^2}\sum_{i=1}^{N}\{\mathbf{R}_i - \mathbf{R}_{i-1}\}^2 \tag{10.15}$$

which is the energy of a set of coupled springs with a "spring constant" linearly proportional to the temperature. This analogy with a set of coupled springs is a useful one as it shows quite transparently that the entropy of the random walk gives rise to an *entropic elasticity.* This entropic elasticity is responsible for many of the striking properties of rubber (an assembly of crosslinked polymers). For example, a rubber band contracts when heated, in contrast to atomic or molecular solids which generically have positive coefficients of expansion.

We now show that the Gaussian model is indeed equivalent to the freely jointed chain, by calculating the end-to-end distance. We must therefore integrate out the positions of particles $1, 2, \ldots, N-1$, keeping the location of particles 0 and N fixed. This can be done in a number of different ways. We note that

$$\begin{aligned}\left(\frac{3}{2\pi a^2}\right)^{3/2}&\int d^3R_1 \exp\left[-\frac{3}{2a^2}\left\{(\mathbf{R}_1-\mathbf{R}_0)^2 + (\mathbf{R}_2-\mathbf{R}_1)^2\right\}\right]\\ &= \left(\frac{1}{2}\right)^{3/2}\exp\left[-\frac{3}{4a^2}\left\{\mathbf{R}_2-\mathbf{R}_0\right\}^2\right]\end{aligned} \tag{10.16}$$

which can be shown by expanding the terms in the exponential and completing the squares in the usual way. Clearly, if we integrate over the position of every second atom beginning with $\mathbf{R}_1$ we find

$$\begin{aligned}&\int d^3R_1\, d^3R_3\cdots d^3R_{N-1}P(\mathbf{R}_0, \mathbf{R}_1, \ldots, \mathbf{R}_N)\\ &= \left(\frac{3}{4\pi a^2}\right)^{3N/4}\exp\left[-\frac{3}{4a^2}\sum_{i=1}^{N/2}\left\{\mathbf{R}_{2i}-\mathbf{R}_{2i-2}\right\}^2\right]\end{aligned} \tag{10.17}$$

i.e. a Gaussian chain of $N/2$ atoms connected by springs with a spring constant reduced by a factor of 2 from that of the N-atom chain. Clearly this process can be continued to completion with the result that

$$P(\mathbf{R}_N, \mathbf{R}_0) = \frac{1}{V}\left(\frac{3}{2N\pi a^2}\right)^{3/2}\exp\left[-\frac{3}{2Na^2}\left\{\mathbf{R}_N-\mathbf{R}_0\right\}^2\right] \tag{10.18}$$

which is equivalent to (10.12). Formula (10.18) is nothing more than the central limit theorem of probability theory. Since $\mathbf{R}_N - \mathbf{R}_0$ is a sum of variables each distributed according to a Gaussian distribution, it must also be governed by a Gaussian distribution.

Before discussing the role of hard core repulsion on the conformations of a polymer we briefly derive a continuum version of the Gaussian model. In (10.15) we may regard $\mathbf{R}_i$ as a variable that depends on a continuous variable i. With $\mathbf{R}_n - \mathbf{R}_{n-1} \to \partial\mathbf{R}(n)/\partial n$ and $\sum_{i=1}^{N} \to \int_0^N dn$ we obtain the equivalent continuum version of (10.15):

$$H = \frac{3k_BT}{2a^2}\int_0^N dn \left(\frac{\partial \mathbf{R}}{\partial n}\right)^2 \tag{10.19}$$

which we shall use in subsequent sections.

10.2 Excluded Volume Effects: Flory Theory

As mentioned in the introduction, one of the important and subtle aspects of polymer statistics is the fact that real chains cannot cross or, equivalently that no two particles can come closer than a minimum hard core distance. The effects of this on polymer conformations are easiest to see in the context of lattice models. One of the most thoroughly studied models of polymers is the self-avoiding random walk on a lattice. We have argued above that on the scale of a persistence length or longer, a polymer can be modeled by a random walk as far as the universal properties are concerned, and walks on a lattice should be in the same universality class as continuum walks. We can easily calculate the first few approximations in a sequence of such approximations by enumerating all self-avoiding walks on our lattice of choice and calculating, for example the quantity $\mathcal{S}^2(N)$ defined in (10.2). The unrestricted random walk, on a square lattice for example, has the property $\mathcal{S}^2(N) \propto N$ which can easily be derived from the multinomial distribution that governs such walks. Conversely, if we enumerate a few short self-avoiding walks we soon see that the mean end-to-end distance is substantially larger (Problem 10.2). Such walks have been enumerated for a number of two- and three-dimensional lattices (see Chapter 4 of [75] for a summary) with the following results:

$$\mathcal{S}^2(N) \sim N^{2\nu(d)} \tag{10.20}$$

where d is the dimensionality of the lattice, $\nu(2) = 0.75$ exactly and $\nu(3) \approx 0.6$. Therefore the swelling of the random walk due to excluded volume is significant, it changes the exponent, not merely the amplitude and has the universal character characteristic of critical behavior — ν depends only on d, not on the type of lattice.

Another way in which self-avoiding walks differ from unrestricted walks is in the total number $\mathcal{N}_N$ of N-step walks. In an unrestricted walk each step can be taken in q different directions, with q the coordination number of the lattice. Clearly, there are $\mathcal{N}_N = q^N$ unrestricted N-step walks. Self-avoiding walks, on the other hand, have the property

$$\mathcal{N}_N \sim \bar{q}^N N^{\gamma-1} \tag{10.21}$$

where $\bar{q} < q$ is an "effective coordination number" that depends on the type of lattice and γ is a second nontrivial *universal* exponent. We will show below that γ is the counterpart of the susceptibility exponent in magnets.

We now describe a simple but very successful approximate theory due to Flory for self-avoiding polymers. We assume that short-range repulsive interactions swell the chain so that the radius of gyration R_g is larger than the Gaussian value R_{g0}. The average concentration of monomers in the region occupied by the chain is therefore proportional to N/R_g^d where d is the dimensionality of the space in which the polymer is free to move. A simple estimate of the interaction energy is then

$$E_{int} = vN\left(\frac{N}{R_g^d}\right) \tag{10.22}$$

where v is a positive number that characterizes the strength of the potential. We can estimate the change in entropy due to the stretching of the polymer by assuming that $H = -TS$ in 10.15 and that the dependence of the entropy on the radius of gyration is the same as for the Gaussian chain

$$\Delta S = \frac{3k_B}{2Na^2}\left(R_{g0}^2 - R_g^2\right) \ . \tag{10.23}$$

Combining (10.22) and (10.23) to form a free energy and minimizing with respect to the parameter R_g we find

$$\frac{\partial A(R_g)}{\partial R_g} = \frac{\partial}{\partial R_g}(E_{int} - T\Delta S) = \frac{\partial}{\partial R_g}\left(v\frac{N^2}{R_g^d} + \frac{3k_BT}{2Na^2}R_g^2\right) = 0 \tag{10.24}$$

or

$$R_g = \left(\frac{vda^2}{3k_BT}\right)^{1/(d+2)} N^{3/(d+2)} = \left(\frac{vda^2}{3k_BT}\right)^{1/(d+2)} N^{\nu_F(d)} \tag{10.25}$$

where the Flory exponent $\nu_F(d) = 3/(d+2)$ has the values $\nu_F(1) = 1$, $\nu_F(2) = 3/4$ and $\nu_F(3) = 3/5$. Of these results, the first two are exact, and the third is an excellent approximation. We also note that $\nu_F(4) = 1/2$ and $\nu_F(d > 4) < 1/2$. The Gaussian chain has $R_g \sim N^{1/2}$ for any dimension d and the result $\nu_F(d > 4) < 1/2$ is an indication that self-avoidance or short-range repulsive interactions are irrelevant for $d > 4$. This is a reflection of the mathematical statement that unrestricted random walks in dimension $d \geq 4$ have zero probability of self-intersection.

The derivation given above of the Flory approximation is very similar to that of the van der Waal's theory of liquids given in Section 3.6. The van der Waal's theory, however, fails in the same way as all other mean field theories in the critical region and does not provide estimates of critical exponents that are even remotely as accurate in the physical dimensions $d = 2, 3$ as the Flory theory does for polymers. The reason for the success of the Flory theory seems to be a spectacular cancellation of errors. This is discussed in considerable detail in Chapter 8 of the book by des Cloizeaux and Jannink [75]. The essence of the matter seems to be that both the estimates (10.22) and (10.23) are incorrect and do not, in fact, give the dominant contribution to the interaction energy or to the change in entropy when the chain swells. These dominant contributions must somehow cancel in (10.24) but this is obviously not something that could be anticipated in advance. Moreover, attempts to systematically improve the Flory theory are therefore likely to fail as spectacularly as the original theory succeeds: the delicate balance needed for this cancellation can easily be destroyed if (10.22) or (10.23) are improved upon. Finally, the Flory method can easily be extended to polymerized membranes (Section 10.4) and one may certainly wonder how reliable it will be in that situation.

We also comment briefly on the parameter v that characterizes the short-range repulsive interaction between different segments on the chain. In our discussion throughout this chapter we have ignored the effect of the solvent—we have essentially treated a chain that is floating freely in space. In reality, the effective interactions between polymer segments are mediated by the solvent and parameters like v will depend on temperature and concentration as well as on the chemistry of the solvent. There is a rough distinction in the polymer literature between *good solvents* and *poor solvents*. Crudely speaking, good solvent conditions correspond to cases in which the interaction between polymer molecules and solvent molecules is more important than the direct interaction between polymer molecules themselves. In other words, we can ignore long range *attractive* interactions between polymer molecules, such as van der Waal's forces. Self-avoidance, on the other hand, cannot be ignored and may be represented by a positive parameter v as we have done above. As the solvent is steadily made poorer the effect of attractive interactions comes into play. Although it is by no means obvious, and will be discussed in more detail below, the effect is to reduce the parameter v to zero at a so-called θ point at which the polymer again becomes Gaussian (although still self-avoiding). In still poorer conditions the chain assumes a collapsed state.

We conclude this section with a brief discussion of a continuum model called the Edwards model that allows more systematic treatments (e.g. perturbation theory, renormalization group) than that given above. To the expression (10.19) for the entropic elasticity we add a simple two-body term:

$$\beta H = K\int_0^N ds\left(\frac{\partial \mathbf{R}(s)}{\partial s}\right)^2 + w\int_0^N ds\int_0^N ds'\delta^d\left(\mathbf{R}(s)-\mathbf{R}(s')\right) \tag{10.26}$$

where s labels the position along the chain and $\mathbf{R}(s)$ is the location of that point in a d-dimensional space. We could add three- and higher-point interactions in the same way through terms of the form

$$\begin{aligned} w_n\int_0^N ds_1\int_0^N ds_2\ldots\int_0^N ds_n\delta^d\left(\mathbf{R}(s_1)-\mathbf{R}(s_2)\right) \\ \times\delta^d\left(\mathbf{R}(s_2)-\mathbf{R}(s_3)\right)\ldots\delta^d\left(\mathbf{R}(s_{n-1})-\mathbf{R}(s_n)\right) \end{aligned} \tag{10.27}$$

but these are all irrelevant for the case of a polymer in a good solvent as we now argue.

We can explore the importance of the various terms in this Hamiltonian by considering a simple scaling argument. We suppose that distances along the chain are rescaled by a factor l and, consistent with the scaling form of the radius of gyration or the end-to-end distance, that vectors $\mathbf{R}(s)$ are rescaled by a factor l^ν where ν is at this point unspecified. The transformations are

$$\begin{aligned} s &= l\bar{s} \\ \mathbf{R}(s) &= l^\nu\mathbf{R}(\bar{s}) \\ ds &= ld\bar{s} \\ \delta^d\left(\mathbf{R}(s)-\mathbf{R}(s')\right) &= l^{-d\nu}\delta^d\left(\mathbf{R}(\bar{s})-\mathbf{R}(\bar{s}')\right) . \end{aligned} \tag{10.28}$$

Substituting into (10.26) we obtain the rescaled dimensionless Hamiltonian

$$\begin{aligned} \beta\tilde{H} &= Kl^{2\nu-1}\int_0^{N/l} d\bar{s}\left(\frac{\partial\mathbf{R}(\bar{s})}{\partial\bar{s}}\right)^2 \\ &\quad + wl^{2-d\nu}\int_0^{N/l} d\bar{s}\int_0^{N/l} d\bar{s}'\delta^d\left(\mathbf{R}(\bar{s})-\mathbf{R}(\bar{s}')\right) . \end{aligned} \tag{10.29}$$

We see that the first term is invariant under rescaling if the exponent $\nu = \frac{1}{2}$ which comes as no great surprise since that term simply represents the Gaussian chain. If we now examine the second term in the vicinity of the Gaussian "fixed point" we see that it grows under rescaling if $d < 4$ and decreases when $d > 4$, *i.e.* w is a *relevant* perturbation when the dimensionality d is less than 4 and irrelevant otherwise. Carrying out the same transformation on the general n-body term (10.27) we find that

$$w_n \rightarrow w_n l^{n-(n-1)d\nu} \tag{10.30}$$

which shows that these terms become relevant in lower dimensions than the two-body term. Of course, in $d < 4$ the Gaussian fixed point is no longer stable and the various n-body terms would have to be re-examined in the vicinity of the nontrivial fixed point.

We also see from (10.29) that straightforward perturbation theory with the Gaussian distribution as starting point will have its difficulties. The appearance of the factor $l^{2-d\nu}$ under rescaling provides a hint that the expansion parameter in a perturbation theory is not w, as might be anticipated, but rather $wN^{2-d\nu} = wN^{1/2}$ in three dimensions. This is indeed the case and makes a perturbation expansion an asymptotic one rather than convergent.

Finally, we will recover the Flory theory from (10.29) in a slightly different way. If we demand that both terms in (10.29) behave in the same manner under rescaling we find

$$l^{2\nu-1} = l^{2-d\nu} \tag{10.31}$$

or

$$\nu = \frac{3}{d+2}$$

which is the Flory result. A little reflection will convince the reader that this approximation is in the same spirit as the free energy minimization used above.

10.3 Polymers and the n-Vector Model

There are a number of different ways in which renormalization group ideas have been applied to the polymer problem and these are described in [75], [102] and [224]. In this section we shall not attempt to carry out a direct renormalization group calculation on the Edwards model (10.26), but rather shall use an argument of de Gennes [72] to make a correspondence between the self-avoiding walk problem and a limiting case of the n-vector model. Once this connection is established we will be able to transcribe the results of Section 7.7 to our situation. We begin with the definition of the classical (continuous spin) n-vector model which has the dimensionless Hamiltonian

$$\mathcal{H} \equiv -\beta H = K \sum_{\langle i,j\rangle,\alpha} S_i^\alpha S_j^\alpha + \sum_{i,\alpha} h_\alpha S_i^\alpha \tag{10.32}$$

where the spins are n-component classical vectors: $\mathbf{S}_i = S_i^1, S_i^2, \ldots, S_i^n$ with fixed length

$$\sum_\alpha (S_i^\alpha)^2 = n\ . \tag{10.33}$$

For this system, the partition function is given by

$$\begin{aligned} Z(K,h) &= \left(\prod_{i,\alpha}\int dS_i^\alpha \delta((\mathbf{S}_i)^2 - n)\right) \exp \mathcal{H}(K,h,\{\mathbf{S}\}) \\ &= n^{N/2}\left(\prod_i \int d\Omega_i\right) \exp \mathcal{H}(K,h,\{\mathbf{S}\}) \end{aligned} \tag{10.34}$$

where, in the second expression, we have replaced the integral over spin components by an integral over the n-dimensional unit sphere of radius $\sqrt{n}$ and where $d\Omega_i$ is the appropriate element of solid angle. Similarly, the thermal average of an arbitrary function of these spin variables $A(\{\mathbf{S}\})$ is

$$\langle A \rangle = \frac{\int \prod_i d\Omega_i A(\{\mathbf{S}\}) \exp \mathcal{H}(K,h,\{\mathbf{S}\})}{\int \prod_i d\Omega_i \exp \mathcal{H}(K,h,\{\mathbf{S}\})}\ . \tag{10.35}$$

In particular, for $\mathbf{h} = 0$ we have

$$\left\langle S_i^\alpha S_i^\beta \right\rangle = \delta_{\alpha\beta} \tag{10.36}$$

which is obvious from symmetry considerations and (10.33). The parameter n has a natural meaning for integer values: $n = 1$ corresponds to the Ising model, $n = 2$ to the XY model etc. However, we will be interested in the value $n = 0$ for which there is an intimate connection to the self-avoiding random walk. In what follows, we will carry out manipulations that are natural for integer n and at appropriate times set n equal to zero. It is important to note that we may insist that (10.36) remain valid at $n = 0$ since, in constructing the average, we have lost the phase-space factor $n^{1/2}$ associated with the magnitude of each of the spins.

We now recall the construction of high-temperature series for magnetic systems developed in Chapter 6. The series for the zero-field susceptibility $\chi_{\alpha\alpha}$ can be written in the form

$$\begin{aligned} k_B T \chi_{\alpha\alpha} &= k_B T \left. \frac{\partial \langle S_i^\alpha \rangle}{\partial h_\alpha} \right|_{\mathbf{h}=0} = \sum_j \langle S_i^\alpha S_j^\alpha \rangle \\ &= 1 + \frac{1}{Z_0} \sum_{j \neq i} \left\langle S_i^\alpha S_j^\alpha \sum_{m=0}^{\infty} \frac{\mathcal{H}^m}{m!} \right\rangle_0 \end{aligned} \tag{10.37}$$

where

$$Z_0 = \left\langle \sum_{m=0}^{\infty} \frac{\mathcal{H}^m}{m!} \right\rangle_0 \tag{10.38}$$

and where the notation $\langle \cdots \rangle_0$ indicates a thermal average taken at infinite temperature. We have indicated, in Section 6.3 how such series can be constructed systematically. In general, the calculation of high-order terms is a technically difficult problem. However, in the limiting case $n = 0$ almost all terms vanish. We demonstrate this using a trick due to de Gennes [73]. Consider the function $f(\mathbf{k})$ defined through

$$f(\mathbf{k}) = \langle \exp\{i\mathbf{k} \cdot \mathbf{S}_i\} \rangle_0 \tag{10.39}$$

where $\mathbf{k}$ is an n-component vector. Clearly, because of the averaging process $f(\mathbf{k}) = f(k)$. Consider

$$\nabla_{\mathbf{k}}^2 f(\mathbf{k}) = \sum_\alpha \frac{\partial^2}{\partial k_\alpha^2} \langle \exp\{i\mathbf{k} \cdot \mathbf{S}_i\} \rangle_0 = -n f(\mathbf{k}) \tag{10.40}$$

where we have used (10.36) in the last step. We now use the fact that $f(\mathbf{k}) = f(k)$ to rewrite the left-hand side:

$$\begin{aligned} \frac{\partial f}{\partial k_\alpha} &= \frac{k_\alpha}{k} \frac{df}{dk} \\ \sum_\alpha \frac{\partial^2 f}{\partial k_\alpha^2} &= \frac{n-1}{k} \frac{df}{dk} + \frac{d^2 f}{dk^2} \, . \end{aligned} \tag{10.41}$$

Combining (10.40) and (10.41), we have

$$\frac{d^2 f}{dk^2} + \frac{n-1}{k} \frac{df}{dk} = -nf \, . \tag{10.42}$$

Finally, taking $n = 0$ in (10.42) we obtain the differential equation

$$\frac{d^2 f}{dk^2} - \frac{1}{k} \frac{df}{dk} = 0 \tag{10.43}$$

which has the general solution $f(k) = a + bk^2$. Expanding (10.39) in powers of $\mathbf{k}$ we also have

$$\begin{aligned} f(\mathbf{k}) &= 1 - \frac{1}{2!}\sum_{\alpha,\beta} k_\alpha k_\beta \langle S_i^\alpha S_i^\beta \rangle + \ldots \\ &= 1 - \frac{1}{2}\sum_\alpha k_\alpha^2 + \ldots = 1 - \frac{1}{2}k^2 + \ldots \end{aligned} \tag{10.44}$$

which holds for arbitrary n. Therefore in the case $n = 0$ we have

$$f(k) = 1 - \frac{1}{2}k^2 \tag{10.45}$$

which has the important consequence that the only nonzero expectation value of products of spin variables is $\langle S_i^\alpha S_i^\alpha \rangle$. For example, the expectation value

$$\left\langle S_i^\alpha S_i^\beta S_i^\gamma S_i^\delta \right\rangle = \frac{\partial^4}{\partial k_\alpha \partial k_\beta \partial k_\gamma \partial k_\delta} f(k)\big|_{k=0} = 0$$

since there are no terms of order k^4 in $f(k)$ when $n = 0$.

Returning now to the susceptibility (10.37) we see that the only terms in the numerator that will survive the averaging process are terms of the form

$$\langle S_i^\alpha S_j^\alpha (S_i^\alpha S_m^\alpha)(S_m^\alpha S_r^\alpha) \cdots (S_t^\alpha S_j^\alpha) \rangle$$

where each site appears exactly two times in the product. Graphically, therefore each term in the expansion corresponds to a self-avoiding walk on the lattice connecting sites i and j since a self-intersection would mean that there would be four spin variables for that site. The factor $1/m!$ in (10.37) is canceled by the number of ways that m distinct nearest-neighbor spin-pairs appear in the expansion of $\mathcal{H}^m$. Furthermore, Z_0, given by (10.38) is, in fact, equal to 1. The only terms that can contribute to the denominator are closed polygons but, in contrast to the numerator in which only the index α appears, the expectation values for spins at the vertices of polygons have to be summed over α. Thus a factor n appears which makes every term except the $m = 0$ term equal to zero. Therefore, we obtain

$$k_B T \chi_{\alpha\alpha} = 1 + \sum_j \sum_N \mathcal{N}_N(i,j) K^N = 1 + \sum_{N=1}^{\infty} \mathcal{N}_N K^N \tag{10.46}$$

where $\mathcal{N}_N(i,j)$ is the number of N-step self-avoiding walks connecting sites i and j and $\mathcal{N}_N$ is the total number of N-step self-avoiding walks.

We are now in a position to make a connection with the polymer problem. Recall that the number of N-step self-avoiding walks on a particular lattice has the asymptotic behavior $\mathcal{N}_N \to \bar{q}^N N^{\gamma-1}$ as N becomes large, where $\bar{q}$ depends on the lattice, and γ is an exponent that depends only on the dimensionality d. From (10.46) we see that the susceptibility will diverge at a critical coupling K_c given by $K_c\bar{q} = 1$ as the susceptibility series converges for any K less than this value. Taking a value of K slightly less than the critical coupling $K = K_c(1-t) \approx K_c e^{-t}$ we may rewrite (10.46) as follows:

$$\begin{aligned} k_B T\chi &= 1 + \sum_{N=1}^{\infty} N^{\gamma-1} e^{-Nt} \approx \int_0^{\infty} dN N^{\gamma-1} e^{-Nt} \\ &= t^{-\gamma}\left[\int_0^{\infty} dx x^{\gamma-1} e^{-x}\right] \end{aligned} \tag{10.47}$$

where, in the last step we have made the substitution $N = t^{-1}x$.

This equation has several important features. First, we see that the exponent γ which characterizes the nontrivial scaling behavior of the number of self-avoiding walks is indeed the susceptibility exponent of the n-vector model in the limit $n \to 0$. Secondly, the relation between the thermodynamic limit $N \to \infty$ in the polymer problem and the approach to a critical point in a magnetic (or other) system is clarified. The point $N^{-1} = 0$ is a critical point for a flexible self-avoiding polymer just as $t = 0$ is a thermodynamic critical point. Finally, we know that for a given t in the magnetic case, there is a characteristic length $\xi(t)$ which diverges according to $\xi(t) \sim |t|^{-\nu}$ as $t \to 0$. Similarly, we have a characteristic length in the polymer problem: For a given N, we have $\mathcal{S}(N) \sim R_g(N) \sim N^{\nu}$. Therefore, it is natural to identify the correlation length in the magnet with the radius of gyration of the polymer and to postulate that the two exponents ν are one and the same.

We may now use our results from Section 7.7 to predict the polymer exponents ν and γ to first order in $\epsilon = 4 - d$. From (7.155) with $n = 0$ we have $\gamma = 1 + \epsilon/8$. Similarly, using (7.155) and (6.98) we find $\nu = 1/2 + \epsilon/16$. Therefore, in three dimensions we have $\nu = 0.5625$ and $\gamma = 1.125$ which are to be compared to the best estimates from series and Monte Carlo methods [259] of $\nu = 0.588 \pm 0.01$ and $\gamma = 1.166 \pm .003$. As is the case for other values of n, the results to order ϵ are not very impressive. However, with the addition of more terms in ϵ the agreement between renormalization group and other methods becomes very good.

10.4 Dense Polymer Solutions

We now discuss the case when interactions between polymers in solution cannot be ignored and construct a mean field theory of this situation known as the Flory-Huggins theory. We consider a lattice-gas model of the solvent and macromolecules and calculate the entropy of mixing and the internal energy. We assume that space is divided into cells of volume a^3 and that each cell contains one particle which may be either a solvent molecule or one of the monomers on a chain. We assume that there are $\tilde{n}$ chains, each containing N monomers, and a total of N_0 cells. The entropy of the system is then entirely due to the configurations of the polymers: once the polymers are distributed among the cells, there is only a single configuration for the indistinguishable solvent particles. To calculate the entropy of the macromolecules, we assume that $j-1$ of these chains have already been placed on the lattice and calculate, in a mean-field way, the number of configurations available to chain j. The first particle in the chain can be placed on any of the available $N_0 - (j-1)N$ sites. The second particle must be placed in one of the q neighboring cells which is empty, on average, with probability $\{N_0 - (j-1)N - 1\}/N_0$. Particle 3 then can occupy one of the $q-1$ neighboring sites of particle 2, again provided that this site is empty. Proceeding in this way until all particles have been distributed we obtain the estimate

$$\begin{aligned} \Omega_j &= [N_0 - (j-1)N]\left[q\frac{N_0-(j-1)N-1}{N_0}\right]\left[(q-1)\frac{N_0-(j-1)N-2}{N_0}\right] \\ &\qquad \ldots\left[(q-1)\frac{N_0-(j-1)N-N+1}{N_0}\right] \\ &= \frac{q(q-1)^{N-2}}{N_0^{N-1}}\frac{\{N_0-(j-1)N\}!}{\{N_0-jN\}!} \end{aligned} \tag{10.48}$$

for the total number of configurations of molecule j. The entropy is then given by

$$\frac{S}{k_B} = \ln\frac{1}{\tilde{n}!}\prod_{j=1}^{\tilde{n}}\Omega_j = \ln\frac{q^{\tilde{n}}(q-1)^{\tilde{n}(N-2)}N_0!}{\tilde{n}!N_0^{\tilde{n}(N-1)}(N_0-\tilde{n}N)!}\,. \tag{10.49}$$

Using Stirling's formula for the factorials and taking $\tilde{n}$, N and N_0 to be large compared to unity, we obtain

$$\frac{S}{N_0 k_B} = -\frac{\Phi}{N}\ln\frac{\Phi}{N} - (1-\Phi)\ln(1-\Phi) + \Phi\left[\ln(q-1) - 1\right] \tag{10.50}$$

where we have defined the quantity $\Phi = \tilde{n}N/N_0$ which is the fraction of cells occupied by chain molecules. We now subtract from this the entropy of a reference state which we take to be $\tilde{n}$ macromolecules confined to $\tilde{n}N$ sites. The calculation proceeds in precisely the same way as in (10.48)–(10.50). The final result for the entropy of mixing (per site) is:

$$\frac{S_{mix}}{N_0 k_B} = \frac{S - S_{ref}}{N_0 k_B} = -\frac{\Phi}{N} \ln \Phi - (1 - \Phi) \ln(1 - \Phi) \, . \tag{10.51}$$

In the same spirit, we now estimate the internal energy of the system. We assume that there are nearest-neighbor interactions between the particles which are characterized by three energies: J_{00} between solvent particles, J_{01} between polymer units and solvent particles and, finally J_{11} between different particles on the chains. The mean field approximation of the internal energy is then

$$\begin{aligned} E &= \sum_{\langle ij \rangle, \alpha\beta} J_{\alpha\beta} \langle n_{i\alpha} \rangle \langle n_{j\beta} \rangle \\ &= \frac{qN_0}{2} J_{00} \left(\frac{N_0 - \tilde{n}N}{N_0} \right)^2 \\ &\quad + N_0(q-2) J_{01} \left(\frac{\tilde{n}N}{N_0} \right) \left(\frac{N_0 - \tilde{n}N}{N_0} \right) + \frac{N_0(q-2)}{2} J_{11} \left(\frac{\tilde{n}N}{N_0} \right)^2 \\ &= N_0 \left[\frac{q}{2} J_{00} (1-\Phi)^2 + (q-2) J_{01} \Phi (1-\Phi) + \frac{q-2}{2} J_{11} \Phi^2 \right] \end{aligned} \tag{10.52}$$

where the factors $q - 2$ in the last two terms come from the fact that a particle on a chain has only $q - 2$ nearest neighbors that are not on the *same* chain. We again subtract from this internal energy the energy of the reference system which is given by

$$E_{ref} = N_0 \left[\frac{q}{2} J_{00} (1 - \Phi) + \frac{q-2}{2} J_{11} \Phi \right] \tag{10.53}$$

to obtain

$$E_{mix} = E - E_{ref} = N_0 \Phi (1 - \Phi) \left[(q-2) J_{01} - \frac{q}{2} J_{00} - \frac{q-2}{2} J_{11} \right] \, . \tag{10.54}$$

We see that the internal energy, in fact, depends only on one parameter χ called the Flory–Huggins parameter:

$$k_B T \chi = (q-2) J_{01} - \frac{q}{2} J_{00} - \frac{q-2}{2} J_{11} \tag{10.55}$$

which characterizes the strength and nature of the interactions. It should be understood that the nearest-neighbor interaction parameters are effective energies which depend in principle on many parameters and, in particular, are expected to be temperature-dependent.

Combining (10.51) and (10.55) we have, for the Helmholtz free energy of the solution

$$\begin{aligned} A &= E_{mix} - TS_{mix} \\ &= N_0 k_B T \chi \Phi (1-\Phi) + N_0 k_B T \left[\frac{\Phi}{N} \ln \Phi + (1-\Phi) \ln (1-\Phi) \right] . \end{aligned} \tag{10.56}$$

It is useful to examine this free energy for small Φ. Expanding the regular terms in powers of Φ we have

$$\frac{\beta A}{N_0} = \frac{\Phi}{N} \ln \Phi + \Phi (\chi - 1) + \frac{\Phi^2}{2} (1 - 2\chi) + \frac{\Phi^3}{6} + \dots . \tag{10.57}$$

In (10.57) the term linear in Φ can be thought of as a chemical potential and the quadratic term represents the effect of monomer-monomer interactions. In the context of mean field theory there will be a qualitative difference between the cases $v = 1 - 2\chi > 0$ and $v < 0$. The case $v > 0$ corresponds to the case of a "good" solvent since the effective interaction between polymer segments is repulsive. Conversely, $v < 0$ corresponds to a poor solvent. The dividing point between these two cases: $v(T = \Theta) = 1 - 2\chi(\Theta) = 0$ is the Flory or Θ temperature. At this point we have cancellation of the two-body interaction.

We now derive an expression for the *osmotic pressure* Π of the solution. The term osmotic pressure refers to the difference in pressure between two compartments separated by a semipermeable membrane. In our case, this membrane will allow the solvent to flow through freely, while restricting the polymer chains to one of the two compartments. In such a situation, the chemical potential of the solvent, which of course depends on the polymer concentration, must be the same in both compartments. In general, the addition of solute lowers the chemical potential of the solvent [318] and results in a flow of solvent into the container with solute until the chemical potentials become equal. This in turn produces the osmotic pressure. In our lattice model we can simply calculate the pressure of our mixture $p = -\partial A_{tot}/\partial V$ where A_{tot} refers to the free energy composed of (10.52) and (10.50) (i.e. without the subtraction of the reference energy and entropy) and subtract from it the corresponding pressure of a lattice filled only with solvent molecules. A little algebra shows that this is equivalent to

$$\Pi = -\left(\frac{\partial A}{\partial V}\right)_{\tilde{n}} = -\frac{1}{a^3}\left(\frac{\partial A}{\partial N_0}\right)_{\tilde{n}} \tag{10.58}$$

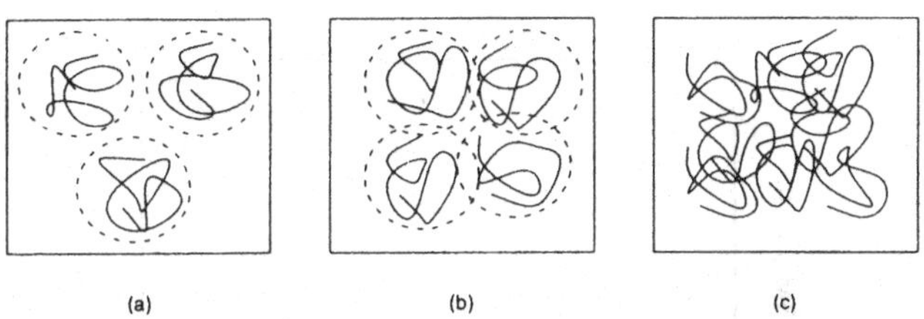

Figure 10.3: Sketch of the three concentration regimes discussed in the text: (a) dilute, (b) semidilute, (c) concentrated. The circles in graphs (a) and (b) represent the characteristic exclusion sphere of radius R_g.

where a^3 is the volume of a cell. This calculation yields

$$\frac{\Pi a^3}{k_B T} = \frac{\Phi}{N} + \frac{\Phi^2}{2}(1 - 2\chi) + \frac{\Phi^3}{3} + \dots \quad (10.59)$$

This equation has several noteworthy features. Firstly, we see that since Φ is a density we have effectively a virial expansion for the osmotic pressure. The leading term $\Phi/N = \tilde{n}/N_0$ is the number of polymers per lattice site reflecting the fact that the chains act as single particles insofar as the pressure is concerned. The second term yields the corrections due to interactions. We see that such corrections become important at concentrations given roughly by $\Phi^2 \sim \Phi/N$ or $\tilde{n}N/N_0 \sim 1/N \sim 10^{-4} - 10^{-6}$ which indicates that the ideal gas law breaks down at extremely low concentrations.

We now discuss these results in light of the scaling picture developed for single chains in the foregoing subsections. As function of concentration, Φ, we can imagine three different regimes. For very small Φ the solution is so dilute that different polymers rarely overlap. In the case of a good solvent, the individual chains will be swollen with a characteristic radius determined to good approximation by the Flory theory. In an intermediate regime, usually called *semidilute* there will be significant overlap of different polymers but not a large amount of *entanglement* which is a term that refers to configurations reminiscent of cooked spaghetti. The third regime is the highly concentrated solution. The situation is sketched in Figure 10.3.

The Flory–Huggins theory described above has deficiencies in the first two concentration ranges. It is clear from the derivation that mean field theory assumes that the density of monomers is uniform in space. In effect, the interaction between particles on the same chain that leads to swelling of an

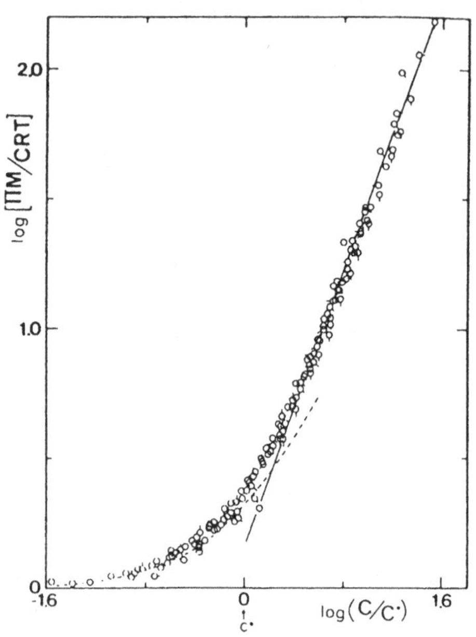

Figure 10.4: Plot of the osmotic pressure Π as function of concentration in the semidilute regime. Here C corresponds to our variable Φ and M corresponds to N. The straight line has slope 1.32, corresponding to $\Pi \sim \Phi^{2.32}$, in good agreement with (10.61). Figure taken from [219].

isolated chain is ignored. We can, at least, use the theory of isolated chains to form an estimate of where the dilute limit of the mean field theory breaks down. We associate a characteristic radius $R_g \sim N^{\nu}$ with each chain. Then chains will begin to have significant overlap with each other when $\tilde{n}R_g^3 \sim N_0$. This signals the end of the ideal gas regime. Using the three-dimensional estimate $R_g \sim N^{3/5}$ we obtain a value $\Phi^* \sim N^{1-3\nu} \sim N^{-4/5}$ for the concentration at which the mixture becomes semi-dilute. In this regime $\Phi^* \ll \Phi \ll 1$ the overlap between different polymers is still not extensive enough that the swelling is destroyed and therefore one might wonder if the second term in the virial form of the osmotic pressure is appropriate. The following scaling argument [73] provides an alternative to the virial expansion and turns out to be correct. We assume

$$\frac{\Pi a^3}{k_B T} = \frac{\Phi}{N} f\left(\frac{\Phi}{\Phi^*}\right) \tag{10.60}$$

where the scaling function $f(x)$ has the following properties. In the dilute limit $x \ll 1$ we must recover the ideal gas limit and therefore $f(x) \to 1$ as $x \to 0$. When $\Phi \gg \Phi^*$ the osmotic pressure should become independent of the degree of polymerization N — it is only the volume fraction Φ that counts, not the length of the individual chains. Since $\Phi^* \sim N^{-4/5}$ we must have

$$f(x) \sim x^{5/4} \ .$$

Inserting this in (10.60) we have the prediction

$$\frac{\Pi a^3}{k_B T} \sim \Phi^{9/4} \tag{10.61}$$

in the semidilute regime. The exponent $\frac{9}{4}$ that appears in (10.61) is not very different from the mean-field prediction of 2. Nevertheless, this difference is measurable. In Figure 10.4 we show measurements of the osmotic pressure of polymer solutions with molecular weight of the constituents varying over more than an order of magnitude. We see that in the regime $\Phi > \Phi^*$ the osmotic pressure scales according to (10.61).

Finally, in the very concentrated regime $\Phi \gg \Phi^*$ we expect that the polymers will become ideal or Gaussian. The reason for this is the "screening" of intrachain repulsion by other polymers that overlap to a significant extent with a given macromolecule. While an exact treatment of this effect is (at this time) not possible, there are approximate calculations [73], [76] that strongly suggest that in dense solutions there is a finite persistence length and that on scales longer than this length chains are Gaussian in the sense of Section 10.1.1.

10.5 Membranes

In this section, we generalize the treatment of dilute polymers in a good solvent to the fluctuation of *solid* or *polymerized* membranes embedded in a higher dimensional space. This is a problem that has received wide attention in the physics community during the last few years, partly because of its relevance to biological materials (e.g. red blood cells) and partly because it is substantially more challenging than the polymer problem. We will also briefly discuss the properties of liquid membranes since these are more common in nature than polymerized membranes.

We begin this section with a discussion of the two-dimensional version of the Gaussian chain, namely the *phantom membrane* for which there is no restriction against self-intersection. In the next subsection we review the extensive numerical studies that have led to the consensus that self-avoiding membranes are *flat* rather than crumpled, at least in three dimensions. In the final subsection, we discuss liquid membranes. We will concentrate on aspects that are likely to be universal — we will not attempt to make our models reasonable from a biological point of view. A discussion of the physical properties of real membranes can be found in [47], see also [175].

10.5.1 Phantom membranes

In Figure 10.5 we show a small section of the simplest model for tethered membranes introduced by Kantor, Kardar and Nelson [148], [149]. Particles occupy the vertices of a triangular network and are joined by flexible strings that have a maximal extension b. In the case of primary interest, the particles have hard cores that prevent two particles from approaching more closely than some minimum distance that we will call σ. In this subsection we will take $\sigma = 0$ so that only the tethering constraint controls the statistical mechanics of the membrane. To begin, we also further approximate the discrete network of tethers by a two-dimensional sheet. In analogy with the Gaussian polymer, we now assume that we may describe the phantom membrane, at large length-scales, by an effective dimensionless Hamiltonian containing only the two-dimensional version of the entropic elasticity:

$$\mathcal{H} = K \int \sum_{\alpha,\beta} d^2x \left(\frac{\partial r_\alpha}{\partial x_\beta}\right)^2 = K \int d^2x \left(\frac{\partial \mathbf{r}}{\partial \mathbf{x}}\right)^2 \tag{10.62}$$

where $\mathbf{r}$ is a vector with components r_α in the d-dimensional space in which the membrane is embedded. In (10.62) the two-dimensional vector $\mathbf{x}$ labels points on the sheet. We may now represent the position $\mathbf{r}(\mathbf{x})$ in terms of a Fourier series:

$$\mathbf{r}(\mathbf{x}) = \frac{1}{L^2} \sum_{\mathbf{k}} \mathbf{A}_{\mathbf{k}} \exp i\mathbf{k} \cdot \mathbf{x} \tag{10.63}$$

where $\mathbf{A}_{\mathbf{k}}$ is a d-dimensional vector, $\mathbf{k}$ is a two-dimensional wave-vector, and L is the linear dimension of the membrane. Substituting and assuming boundary conditions that ensure the orthogonality of the plane waves in (10.63) we have

$$\mathcal{H} = K \sum_{\mathbf{k}} |\mathbf{A}_{\mathbf{k}}|^2 k^2 \ . \tag{10.64}$$

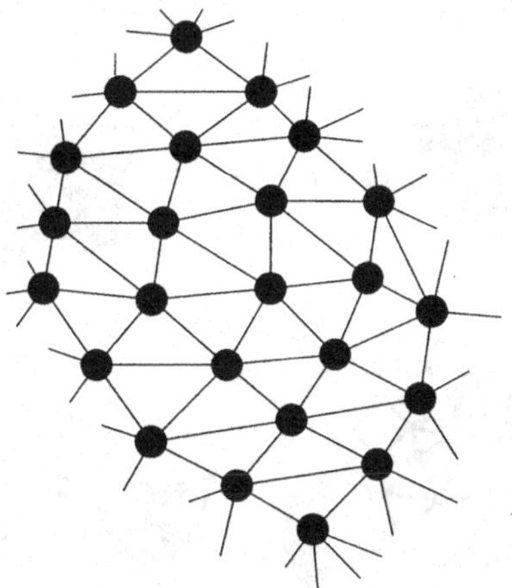

Figure 10.5: Picture of a section of a tethered membrane.

The expectation values of the Fourier components are now simply given by equipartition:

$$\langle A^2_{\mathbf{k},\alpha}\rangle = \frac{1}{2Kk^2} \qquad \alpha = 1, 2, \ldots, d \; . \tag{10.65}$$

We characterize the size of the membrane in the d-dimensional space by its radius of gyration. Choosing our origin so that $\int d^2x\, \mathbf{r}(\mathbf{x}) = 0$ we have

$$\begin{aligned} R_g^2 &= \frac{1}{L^2}\int d^2x\, \langle \mathbf{r}^2(\mathbf{x})\rangle \\ &= \frac{d}{2KL^2}\sum_{\mathbf{k}}\frac{1}{k^2} \sim \frac{d}{4\pi K}\int_{1/L}^{\pi/a}\frac{dk}{k} \end{aligned} \tag{10.66}$$

where, in the last expression, we have introduced the short-distance cutoff a that represents the nearest-neighbor spacing in the discrete case. Clearly, then

$$R_g^2 \sim \frac{d}{4\pi K}\ln\left\{\frac{L}{a}\right\} . \tag{10.67}$$

Since R_g^2 increases more slowly with L for the two-dimensional Gaussian network than for the corresponding polymer ($R_g \sim L^{1/2}$) the former will be much more crumpled. We say that a surface is *crumpled* if the radius of gyration grows slower than linearly with L.

Now imagine that the vertices on the discrete network are occupied by particles of radius σ. The volume required to hold these particles without overlap is of order $L^2\sigma^3$ whereas the volume associated with a typical diameter R_g is of

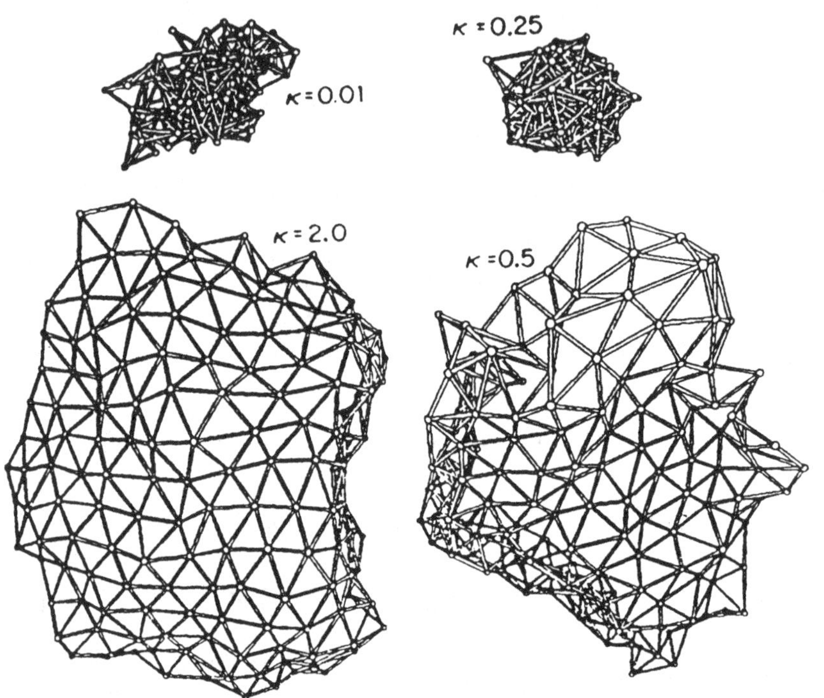

Figure 10.6: Some typical configurations of a phantom membrane with various values of the bending rigidity κ. Taken from [149].

order $(\ln L/\sigma)^{d/2}$ which, in the thermodynamic limit $L \to \infty$ is much less than the total excluded volume in any dimension d. Therefore self-avoidance is always relevant, in contrast to the case of polymers for which the same argument yields the correct result $d_{uc} = 4$. The result (10.67) thus shows that for the phantom membrane the upper critical dimension above which self-avoidance becomes irrelevant is infinity.

In contrast to the case of polymers, the introduction of *bending rigidity* can make a significant difference. In polymers, an energy cost associated with changing the angle between successive links in the chain only renormalizes the persistence length by a finite amount. In phantom membranes, on the other hand, it can change the equilibrium state from crumpled to flat. One can introduce bending rigidity in several ways. For example, one might add to the Hamiltonian a term $-\kappa \sum_{i,j} \hat{\mathbf{n}}_i \cdot \hat{\mathbf{n}}_j$ where $\hat{\mathbf{n}}_i$ is the unit vector normal to triangle i in the network, and j is an adjacent triangle. This energy favors, for positive κ, a phase in which all the normals are parallel, *i.e.* a flat phase. Alternatively, one could imagine a repulsive pair potential between second neighbor particles on the network. This would have the same effect.

When bending rigidity is added to the Hamiltonian, the problem is no longer exactly solvable and approximate or numerical methods must be used. In Figure 10.6 we show the results of Monte Carlo simulations [149] for phantom membranes with bending rigidity for various values of κ. The sample configurations clearly show a change from an open, essentially two-dimensional structure to the very compact spherical structure of the pure phantom network as κ is reduced. Further evidence for a phase transition between these two states is provided by the specific heat which has a clear peak that grows with system size. To date the nature of this transition is not fully understood nor is it known if the analog of this transition exists in the case of self-avoiding membranes.

10.5.2 Self-avoiding membranes

When the excluded volume of the particles on the network of Figure 10.5 is taken into account, the problem becomes much more difficult and exact results are few. One can construct a Flory theory for the crumpled phase, assuming that such a phase exists, but in light of the discussion of Section 10.2 one can certainly question its accuracy. Since the upper critical dimension is infinity, methods such as a standard ϵ-expansion are also not available.

We begin our discussion with the generalized Edwards model:

$$\mathcal{H} = K \int d^D x \left(\frac{\partial \mathbf{r}(\mathbf{x})}{\partial \mathbf{x}}\right)^2 + v \int d^D x \int d^D x' \delta^d \left(\mathbf{r}(\mathbf{x}) - \mathbf{r}(\mathbf{x}')\right) \qquad (10.68)$$

where we have left the *internal* dimension D ($D = 2$ for membranes) unspecified for the time being. In the spirit of Flory theory (Section 10.2) we now assume that if we rescale internal lengths by a factor l, external lengths will be rescaled by a factor l^ν. By making this assumption, we have implicitly assumed that the exponent ν characterizes a crumpled phase rather than a flat phase since a flat phase would have at least two different exponents, corresponding to the directions parallel and perpendicular to the surface. We note that we can bound the exponent ν from the excluded volume constraint: $L^{d\nu} > L^D$ or $\nu > D/d = 2/3$ in the physically relevant case $D = 2$, $d = 3$. A second bound is $\nu < 1$ so that $2/3 < \nu < 1$. The transformation is carried out exactly as in (10.29) and results in the rescaled effective Hamiltonian

$$\begin{aligned}\mathcal{H}(l) &= Kl^{2\nu+D-2}\int d^D\tilde{x}\left(\frac{\partial\tilde{\mathbf{r}}(\tilde{\mathbf{x}})}{\partial\tilde{\mathbf{x}}}\right)^2 \\ &\quad +vl^{2D-d\nu}\int d^D\tilde{x}\int d^D\tilde{x}'\delta^d\left(\tilde{\mathbf{r}}(\tilde{\mathbf{x}})-\tilde{\mathbf{r}}(\tilde{\mathbf{x}}')\right) . \end{aligned} \tag{10.69}$$

Examining (10.69) we find the Gaussian exponent $\nu = (2-D)/2 = 0$ (logarithm) for $D = 2$ if we disregard the second term. Conversely, if we require that the two terms scale with the same exponent we obtain

$$2\nu + D - 2 = 2D - d\nu \qquad \text{or} \qquad \nu = \frac{D+2}{d+2} \tag{10.70}$$

which is the generalization of Flory theory to D-dimensional objects fluctuating in d-dimensional space. Finally, we may obtain one further interesting result. If we examine the scaling behavior of the two-body term at the Gaussian fixed point we see that it varies as

$$vl^{2D-d(2-D)/2}$$

indicating that the two-body term is relevant if $D > 2d/(4+d)$ which defines a *lower critical dimension* $D_{lc}(d)$ for the internal dimension D. This line of lower critical points can be used to construct an ϵ-expansion toward the point of interest in the D, d plane. We shall not discuss this approach but refer the interested reader to [148].

The Flory prediction $\nu = 4/5$ for membranes in three dimensions caused a certain amount of confusion in the early days of this field. Initially, Monte Carlo simulations seemed to be consistent with this prediction [148]. However, Plischke and Boal [244] on the basis of a more detailed analysis of Monte Carlo simulations conjectured that self-avoidance is sufficient to prevent the existence of the crumpled phase and to make the membrane flat but rough. Specifically, they found that if one examines the membrane in the frame of reference defined by the principal axes (in a particular Monte Carlo configuration) one finds an object that is shaped like a pancake. One can find this coordinate system by diagonalizing the inertia tensor, the elements of which are defined to be

$$I_{\alpha\beta} = \frac{1}{N}\sum_i\left[(r_{i,\alpha}-\bar{r}_\alpha)(r_{i,\beta}-\bar{r}_\beta)\right] \tag{10.71}$$

where $\bar{r}_\alpha$ indicates the average of $r_{i,\alpha}$ in a particular configuration. The eigenvalues of $\mathbf{I}$ then provide a measure of how far the object extends in the principal axis directions. The radius of gyration R_g is related to these eigenvalues through

$$R_g^2 = \sum_j \lambda_j \ .$$

Plischke and Boal [244] found that the expectation values of the two largest eigenvalues varied with system-size as $\langle\lambda_1\rangle,\ \langle\lambda_2\rangle \sim L^{2\nu_\parallel}$ where the value of $\nu_\parallel$ found was consistent with $\nu_\parallel = 1$. On the other hand, the smallest eigenvalue was found to scale quite differently with L:

$$\langle\lambda_3\rangle \sim L^{2\zeta}$$

with a roughness exponent $\zeta \approx 0.65$. Thus, while the fluctuations in the transverse directions are large and, indeed, diverge in the thermodynamic limit, the aspect ratio

$$\mathcal{A} = \left\langle \frac{\lambda_3}{\lambda_1} \right\rangle \to 0 \text{ as } L \to \infty \ . \tag{10.72}$$

This constitutes our definition of the flat phase.

The general picture described above was subsequently confirmed by a number of other groups. In particular, Abraham *et al.* [3] carried out extensive molecular dynamics simulations of membranes with hard-core diameters of various sizes. Some of their results are shown in Figure 10.7 in which the radius of gyration is plotted as function of L for various values of σ. The radius of gyration is dominated, for large L, by the largest eigenvalue λ_1 of the inertia tensor and the data, for all $\sigma > 0.4$ are consistent with $\nu_{R_g} = \nu_\parallel = 1.0$. For smaller values of σ there are obvious indications of crossover from phantom behavior at small L to self-avoiding behavior at larger L. Although it is impossible to prove this through simulations, the current consensus is that tethered membranes are flat in $d = 3$ for any non-zero σ. Also shown in Figure 10.8 are some typical snapshots of a membrane with $\sigma = 1$ and 4219 particles. Each configuration is projected onto the planes perpendicular to the eigenvectors of $\mathbf{I}$ for that configuration. Both the extreme anisotropy of the configurations and the roughness characterized by the eigenvalue λ_3 are clearly evident.

It should be mentioned that, while the existence of the flat phase is not under dispute at this time, there is considerable disagreement about the value of the roughness exponent ζ. Computer simulations of tethered membranes are plagued, to a larger extent than is the case for other systems, by long relaxation times and finite-size effects associated with the free-edge boundary conditions most commonly used. Simulations of the basic model described above have yielded values of ζ in the range $0.53 \leq \zeta \leq 0.70$. On the other hand, Lipowsky and Girardet [178] have conjectured that $\zeta = 1/2$ and have simulated large

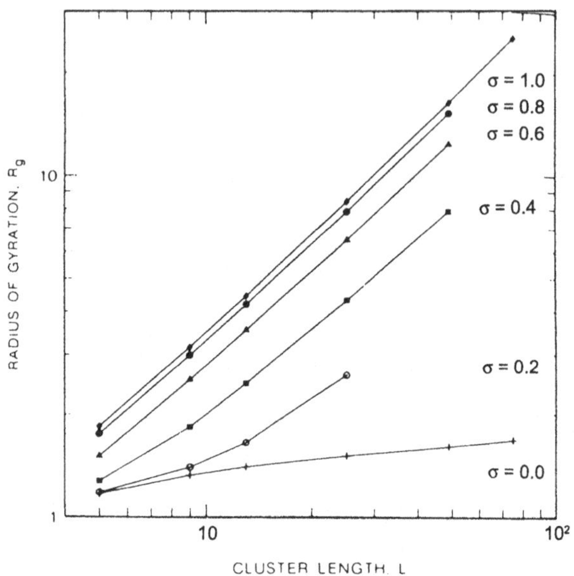

Figure 10.7: Log-log plot of the radius of gyration R_g vs L of self-avoiding tethered membranes for various values of σ. Taken from [3].

infinitely thin elastic sheets, obtaining results consistent with this conjecture. This issue still has to be resolved.

To this point, we have not given a physical argument that explains the stability of the flat phase vis-a-vis the Flory-crumpled phase. The following simple argument due to Abraham and Nelson [4] probably captures the essential physics of the situation. Imagine that the particles on the network interact with a soft repulsive potential rather than an abrupt hard-core repulsion. Consider two elementary triangles that share a side and imagine that a fluctuation that attempts to fold the triangles about their common side occurs (as shown in Figure 10.9). This fluctuation leaves the distance between particles (1) and (2) invariant but brings particles (3) and (4) closer together. In this process, the energy of the system is increased and this increase in energy is effectively the same as if the membrane had a finite bending rigidity. Thus, for any finite repulsion between particles on the network, there will be some effective bending constant κ. It is difficult to produce hard numbers from this type of argument and a much more sophisticated one would be required to show that

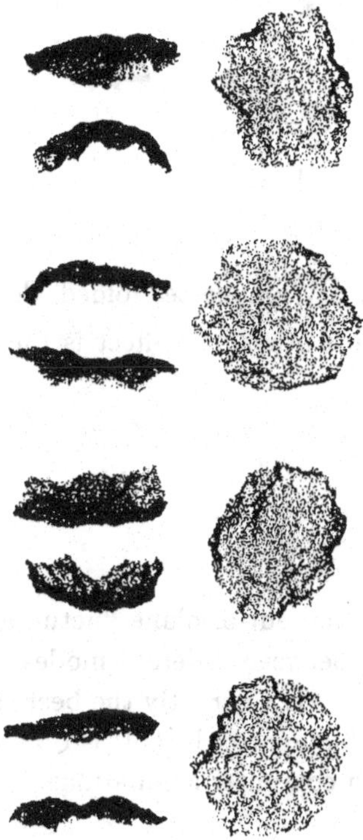

Figure 10.8: Some typical configurations of a self-avoiding tethered membrane consisting of 4219 particles. Taken from [3].

self-avoidance, no matter how small, is sufficient to produce a bending rigidity as large as that required to stabilize the flat phase of phantom membranes.

To conclude this section we briefly discuss some of the analytical work on the tethered membrane problem as well as the experimental situation. Formal theories of the crumpled and flat phases have been based primarily on two different starting points. One can begin with the Edwards model (10.68) and formally construct a perturbation expansion in v away from the line given by the lower critical dimension $D_{lc}(d)$. This approach has already been alluded to above. To date it has not yielded any information about the flat phase. An alternative is to begin with continuum elasticity theory, assuming that there are finite Lamé coefficients as well as a finite bending rigidity [171]. After the phonon degrees of freedom are integrated out one is left with an effective

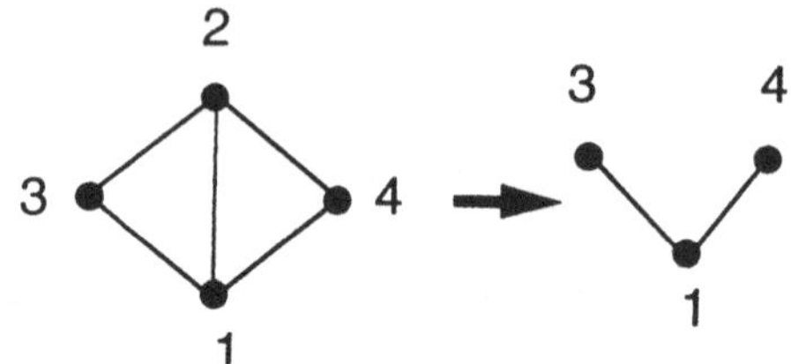

Figure 10.9: Two elementary plaquettes folded along their common line. If particles 3 and 4 repel each other, the effect is the same as if there were an explicit bending rigidity.

free energy functional for the out-of-plane fluctuations. Perturbation theory in the nonlinear coupling between different modes can be recast in the form of an ϵ-expansion in $\epsilon = 4 - D$. Currently the best estimate for the roughness exponent resulting from this approach [171] is $\zeta \approx 0.59$ for $d = 3$ which is in reasonable agreement with computer simulations.

We are aware of two sets of experiments that can be interpreted in terms of the fluctuations of polymerized membranes. Schmidt *et al.* [268] studied the spectrin network of red blood cells by means of light and X-ray scattering. This network, which can be extracted from the cells, consists of long chains anchored to each other in a roughly hexagonal structure. These workers concluded that this network could be characterized as being flat with a roughness exponent $\zeta \approx 0.65$. A quite different system, thin sheets of graphite oxide suspended in solution were studied, also by light scattering, by Wen *et al.* [321]. In contrast to Schmidt *et al.* [268] these workers concluded that their data was consistent with either an isotropically crumpled phase $\nu \approx 0.8$ or a collapsed phase $\nu \approx 2/3$, depending on the pH of the solvent. Assuming that the collapsed phase has indeed been seen, we must conclude that attractive forces between different parts of the graphite oxide sheets play an important role. Attractive forces may be effective in destabilizing the flat phase and causing a transition to either a crumpled or, if strong enough, to a collapsed state. These processes are poorly understood at this time and more experimental as well as theoretical work will be needed to clarify the situation.

10.5.3 Liquid membranes

We conclude this chapter with a very brief discussion of liquid membranes. Physical realizations of liquid membranes that immediately come to mind are microemulsions (surfactant molecules trapped at an interface between oil and water), lipid bilayers in biological membranes and lamellar phases of lyotropic liquid crystals. The important difference between fluid membranes, and the tethered membranes discussed above, is the absence of a shear modulus. The only important physical parameters that govern the conformations of such membranes are the bending energy and self-avoidance. It is easy to show that if there is only bending energy, a fluid membrane will have a finite persistence length. Consider a continuum model, i.e. an infinitely thin sheet, and assume that it is oriented parallel to the $x-y$ plane. If the fluctuations of the sheet are not too large we can specify its height above the $x-y$ plane by a single-valued function $z(x,y)$. The free energy associated with the curvature of the sheet is then given, to lowest order in derivatives of z, by [74]

$$\mathcal{F} = \frac{\kappa}{2}\int d^2r\,\left[\nabla^2 z(\mathbf{r})\right]^2 \tag{10.73}$$

where $\mathbf{r}$ is a vector in the $x-y$ plane and κ the bending rigidity. Writing

$$z(\mathbf{r}) = \frac{1}{L}\sum_{\mathbf{k}} \hat{z}_{\mathbf{k}} e^{i\mathbf{k}\cdot\mathbf{r}}$$

and substituting we find

$$\mathcal{F} = \kappa\sum_{\mathbf{k}} k^4\,|\hat{z}_{\mathbf{k}}|^2\ . \tag{10.74}$$

Therefore, at temperature T, the expectation value of $|z_{\mathbf{k}}|^2$ is given by

$$\langle|\hat{z}_{\mathbf{k}}|^2\rangle = \frac{k_BT}{\kappa k^4}\ . \tag{10.75}$$

The calculation of the mean square width of the interface can now be carried out in strict analogy with the capillary-wave calculation of Section 5.4. Here, we will attempt to calculate the decay of the normal-normal correlation function as function of separation. Defining the position vector of a point on the surface through $\mathbf{R} = (x, y, z(x,y))$ we can find the normal vector $\mathbf{n}(x,y)$ by requiring that $\mathbf{n}\cdot d\mathbf{R} = 0$. This yields in component form

$$\mathbf{n}(\mathbf{r}) = \frac{\{-\partial z/\partial x, -\partial z/\partial y, 1\}}{\sqrt{1+(\nabla z)^2}}\ .$$

We now expand the quantity $\mathbf{n}(\mathbf{r}')\cdot\mathbf{n}(\mathbf{r}'+\mathbf{r})$ in powers of ∇z, keeping only the quadratic terms and obtain

$$\frac{1}{L^2}\int d^2r'\mathbf{n}(\mathbf{r}'+\mathbf{r})\cdot\mathbf{n}(\mathbf{r}')$$
$$\approx 1+\frac{1}{L^2}\int d^2r'\left\{\nabla z(\mathbf{r}'+\mathbf{r})\cdot\nabla z(\mathbf{r}')-\frac{1}{2}[\nabla z(\mathbf{r}+\mathbf{r}')]^2-\frac{1}{2}[\nabla z(\mathbf{r}')]^2\right\} \quad (10.76)$$

Rewriting this in terms of the variables $\hat{z}_{\mathbf{k}}$ and taking the thermal average we find

$$\langle\mathbf{n}(\mathbf{r})\mathbf{n}(0)\rangle \equiv \Gamma(\mathbf{r}) = 1-\frac{k_BT}{\kappa L^2}\sum_{\mathbf{q}}\frac{1-\exp i\mathbf{q}\cdot\mathbf{r}}{q^2}\ . \quad (10.77)$$

Converting the sum over $\mathbf{q}$ to an integral and carrying out the angular integration we have

$$\begin{aligned}\Gamma(\mathbf{r}) &= 1-\frac{k_BT}{2\pi\kappa}\int_{1/L}^{1/a}\frac{dq}{q}[1-J_0(qr)]\\ &= 1-\frac{k_BT}{2\pi\kappa}\int_{r/L}^{r/a}\frac{dy}{y}[1-J_0(y)]\\ &\approx 1-\frac{k_BT}{2\pi\kappa}\ln\left\{\frac{r}{a}\right\}\end{aligned} \quad (10.78)$$

where we have introduced a microscopic cutoff a. The main contribution to the integral in (10.78) for $r/a \gg 1$ comes from values of $y \gg 1$ for which the Bessel function $J_0(y)$ is negligible. This justifies the final step. For values of Γ close to 1 we can write (10.78) in the form

$$\Gamma(\mathbf{r}) \approx \exp\left\{-\ln\left(\frac{r}{a}\right)^{\frac{k_BT}{2\pi\kappa}}\right\} = \left(\frac{a}{r}\right)^{\frac{k_BT}{2\pi\kappa}}$$

which is similar to the power-law decay of correlations seen in the two-dimensional XY model (Section 6.6). Following de Gennes and Taupin [74] we can use this expression to define a persistence length ξ. The correlation function $\Gamma(\mathbf{r}) = \cos\theta(\mathbf{r})$ where $\theta(\mathbf{r})$ is the angle between normals to the surface at two points separated by $\mathbf{r}$. If we take ξ as the distance over which the correlation function (10.78) remains positive, we obtain

$$\xi = a\exp\left\{\frac{2\pi\kappa}{k_BT}\right\} \quad (10.79)$$

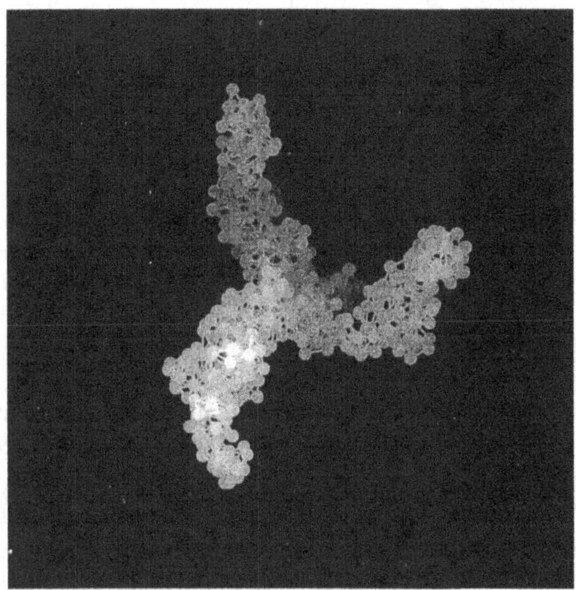

Figure 10.10: Typical configuration of a fluid membrane consisting of 542 particles [46].

i.e., there is a finite persistence length in contrast to the case of tethered self-avoiding membranes which have an infinite persistence length in the flat phase.

In this derivation we have retained only the lowest order terms in the curvature energy. Peliti and Leibler [238] have carried out a renormalization group calculation on a more complete version of (10.73) and found that the nonlinear terms reduce the effective coupling constant κ at long wavelengths and that the persistence length is in fact smaller than the estimate (10.79). This leads to the conjecture that liquid phantom membranes are crumpled at any finite temperature, at least on length-scales large compared to ξ.

This conjecture is supported by computer simulations of tethered membranes in which the particles on the network are allowed to disconnect and reconnect their tethers according to certain constraints. In order to prevent the whole system from evaporating into a three-dimensional gas or condensing into a three-dimensional bulk liquid it is necessary to retain some tethering. Nevertheless, some of the character of a fluid membrane does emerge in these models. In particular, particles diffuse on the network and the persistence length is finite for all κ. The crumpled phase found in these models of fluid

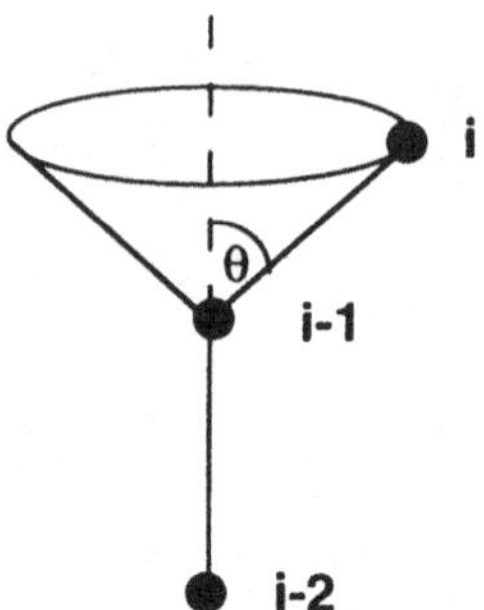

Figure 10.11: The model of problem 10.1.

membranes is, however, in a different universality class than that predicted by Flory theory for tethered membranes. Fluid membranes in three dimensions seem to belong to the universality class of *branched self-avoiding polymers* for which $R_g \sim N^\nu$ with $\nu = 1/2$ (exactly). We will not discuss this work further and simply refer the reader to the original articles [46], [159] for further detail. In Figure 10.10 we show a typical configuration [46] of a finite fluid membrane in which the branched polymer character is clearly visible.

10.6 Problems

10.1. *Effective Bond Length.*

Consider the following model of a polymer. Point particles are located at positions $\mathbf{R}_0, \mathbf{R}_1, \ldots, \mathbf{R}_N$ with $|\mathbf{r}_i| = |\mathbf{R}_i - \mathbf{R}_{i-1}| = a$. The angle θ between successive bonds defined through $\mathbf{r}_i \cdot \mathbf{r}_{i-1} = a^2 \cos\theta$ is fixed but the vector $\mathbf{r}_i$ is otherwise free to rotate around the axis defined by $\mathbf{r}_{i-1}$. The situation is depicted in Figure 10.11.

(a) Write $\mathbf{r}_i = \mathbf{r}_{i-1}\cos\theta + \mathbf{w}_i$ and show that in this model

$$\langle \mathbf{r}_i \cdot \mathbf{r}_{i-n} \rangle = \langle \mathbf{r}_{i-1} \cdot \mathbf{r}_{i-n} \rangle \cos\theta \ .$$

Solve this recursion relation to obtain

$$\langle \mathbf{r}_i \cdot \mathbf{r}_{i-n} \rangle = a^2 \cos^n \theta \ .$$

(b) Show that the mean squared end-to-end distance $S_N^2 = \langle (\sum_{j=1}^N \mathbf{r}_j)^2 \rangle$ is given by

$$S_N^2 = Na^2 \left[\frac{1+\cos\theta}{1-\cos\theta} - \frac{2\cos\theta}{N} \frac{1-(\cos\theta)^N}{(1-\cos\theta)^2} \right]$$

$$\rightarrow \quad N\tilde{a}^2$$

in the thermodynamic limit $N \rightarrow \infty$. The quantity

$$\tilde{a} = a\sqrt{\frac{1+\cos\theta}{1-\cos\theta}}$$

can be thought of as an effective bond length or persistence length. When distances are measured in terms of this length the present model is equivalent to the freely jointed chain discussed in (10.1.1).

10.2. *Normal Modes of the Gaussian Chain.*

Consider the continuum version of the Gaussian chain with temperature-dependent "Hamiltonian" given in (10.19). Since this is in effect a model of a string with no forces exerted on the ends we may use the boundary conditions

$$\left.\frac{\partial \mathbf{R}}{\partial n}\right|_{n=0} = \left.\frac{\partial \mathbf{R}}{\partial n}\right|_{n=N} = 0 \; .$$

(a) Define a set $\mathbf{A}_k$ of normal mode coordinates through the relation

$$\mathbf{R}(n) = \sum_k \mathbf{A}_k \cos kn$$

where the $\mathbf{A}_k$'s are three-dimensional vectors and, consistent with the boundary conditions $k = \frac{\pi}{N}j$, $j = 0, 1, 2, \ldots$. Find the Hamiltonian for the normal modes.

(b) Use equipartition to determine $\langle \mathbf{A}_k^2 \rangle$.

(c) Express the end-to-end distance S_N^2 in terms of these expectation values and show that the result (10.4) is recovered. The following identity may be useful:

$$\sum_{j=0}^{\infty} \left(\frac{1}{2j+1}\right)^2 = \frac{\pi^2}{8} \; .$$

10.3. *Gaussian Chain in an External Potential.*

Suppose that one end of a Gaussian chain with effective Hamiltonian (10.15) is held fixed at $\mathbf{R}_0$ and that a force $\mathbf{F}$ acts at the other end located at $\mathbf{R}_N$. The added potential energy of the chain is given by

$$E_{pot} = -\mathbf{F}\cdot(\mathbf{R}_N - \mathbf{R}_0)$$

(a) Include this term in the effective Hamiltonian and show that:

$$\begin{aligned}\langle \mathbf{r}_i \rangle &= \frac{Fa^2}{3k_BT} \\ \langle \mathbf{r}_i^2 \rangle &= a^2 + \frac{F^2a^4}{9k_B^2T^2} \\ S_N^2 &= \frac{N^2F^2a^4}{9k_B^2T^2} + Na^2 \ .\end{aligned}$$

(b) We may associate a characteristic length $R \sim S_N$ with the end-to-end distance. Express the change in free energy of the chain due to stretching in terms of this length and show that it is of the same form as (10.23), although of opposite sign. Comment on this difference in sign.

10.4. *Self-Avoiding Walks*

(a) Enumerate all open self-avoiding walks with up to 5 steps on the square lattice and calculate the quantity S_N^2 for $N = 1, 2 \ldots, 5$. Note: The safest way to construct the N-step walks for small N is probably to add a single step in all possible ways to each of the $(N-1)$-step walks. The disadvantage of this procedure, on the other hand, is that an error early in the process propagates with exponential consequences.

(b) Estimate the exponent ν, for example by fitting your data to the functional form $S_N^2 = aN^{2\nu}$, or by calculating an effective N-dependent exponent from the formula

$$\nu_{eff}(N) = \frac{\ln\{S_N/S_{N'}\}}{\ln\{N/N'\}}$$

and plotting this as function of N^{-1}.

Chapter 11

Quantum Fluids

In this chapter we begin, in Section 11.1, by discussing in detail one of the most striking consequences of quantum statistics, the condensation of a noninteracting Bose gas. We next turn our attention (Section 11.2) to an interacting Bose system and to the phenomenon of superfluidity. Our treatment of this subject is limited mainly to low-temperature properties and is primarily qualitative. The first part of Section 11.3 is devoted to the Bardeen, Cooper, Schrieffer [25] (BCS) theory of superconductivity, in which fermion pairs undergo a transition that is similar to Bose condensation. We also consider the macroscopic theory of superconductivity due to Ginzburg and Landau. This approach is also applicable to the theory of superfluidity and briefly discussed in that context. We encountered the Landau–Ginzburg formalism previously in Sections 3.10 and 5.4. In the present context we use it to describe some of the important physical properties of superconductors and to indicate why the mean field approach of the BCS theory works so effectively for conventional superconductors. A useful general reference for much of the material of this chapter is Part 2 of the book by Landau and Lifshitz [165]. The recent discovery of high temperature superconductors has rekindled the interest in superconductivity. We will indicate, and give references at appropriate places in the text, when the "unconventional" superconductors differ significantly from those described by BCS theory, but we will not be able to describe the theory of these materials in detail.

11.1 Bose Condensation

Consider a system of noninteracting bosons confined to a cubical box of volume $V = L^3$. We use periodic boundary conditions in the solution of the single-particle Schrödinger equation:

$$\psi(\mathbf{r} + L\hat{a}_i) = \psi(\mathbf{r})$$

where $\hat{a}_i$ is the unit vector in the ith direction.[1] With these boundary conditions the eigenfunctions of the Schrödinger equation are

$$\psi_{\mathbf{k}}(\mathbf{r}) = \frac{1}{\sqrt{V}}\, e^{i\mathbf{k}\cdot\mathbf{r}} \tag{11.1}$$

where $\mathbf{k} = 2\pi(n_1, n_2, n_3)/L$ with $n_i = 0, \pm 1, \pm 2, \ldots$. The single-particle energies are given by

$$\epsilon(\mathbf{k}) = \frac{\hbar^2 k^2}{2m} \tag{11.2}$$

and the logarithm of the grand canonical partition function (2.76) is

$$\ln Z_G = -\sum_{\mathbf{k}} \ln(1 - \exp\{-\beta[\epsilon(\mathbf{k}) - \mu]\}) \tag{11.3}$$

where μ is the chemical potential. The mean occupation number of the state with wave vector $\mathbf{k}$ is

$$\langle n_{\mathbf{k}} \rangle = \frac{1}{e^{\beta[\epsilon(\mathbf{k})-\mu]} - 1}\ . \tag{11.4}$$

For a large system one may attempt to evaluate sums over the closely spaced but discrete values of $\mathbf{k}$ by converting the sum to an integral. Consider first the mean number of particles:

$$\langle N \rangle = \sum_{\mathbf{k}} \frac{1}{e^{\beta[\epsilon(\mathbf{k})-\mu]} - 1} \stackrel{?}{=} \frac{V}{(2\pi)^3} \int \frac{d^3\mathbf{k}}{e^{\beta[\epsilon(\mathbf{k})-\mu]} - 1}\ . \tag{11.5}$$

The integral, in spherical coordinates, is a special case of the Bose–Einstein integral

$$g_\nu(z) = \frac{1}{\Gamma(\nu)} \int_0^\infty dx \frac{x^{\nu-1}}{z^{-1}e^x - 1} = \sum_{j=1}^{\infty} \frac{z^j}{j^\nu} \tag{11.6}$$

[1] A discussion of other boundary conditions may be found in [234] and in the book by Pathria [235].

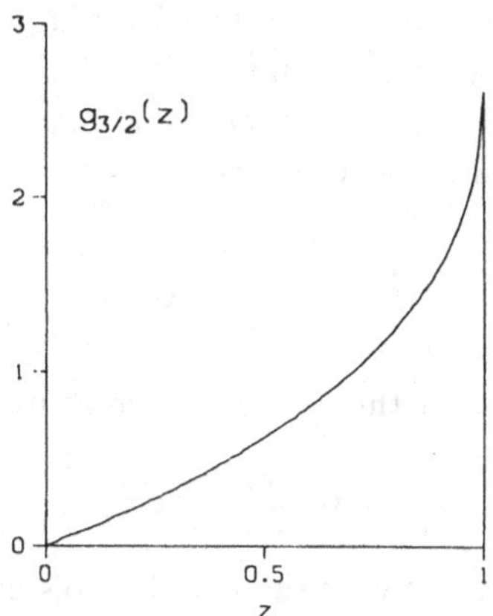

Figure 11.1: The function $g_{3/2}(z)$.

where $\Gamma(\nu)$ is the gamma function. If the replacement of the sum by the integral is legitimate, we obtain, using $\Gamma(\frac{3}{2}) = \frac{1}{2}\pi^{1/2}$,

$$\frac{\langle N \rangle}{V} \stackrel{?}{=} \left(\frac{mk_BT}{2\pi\hbar^2}\right)^{3/2} g_{3/2}(z) \tag{11.7}$$

where $z = e^{\beta\mu}$ is the fugacity. Since occupation numbers cannot be negative or infinite, we must have $\mu < 0$ or $0 \leq z < 1$. The function $g_{3/2}(z)$ is finite in this interval and is depicted in Figure 11.1. The limiting value is $g_{3/2}(1) = \zeta(\frac{3}{2}) = 2.612\ldots$, where $\zeta(x)$ is the Riemann zeta function.

It is clear that the approximation of replacing the sum over $\mathbf{k}$ by an integral is not always valid, since (11.7) yields a solution for z in terms of the density in the allowed region ($z < 1$) only for temperatures greater than

$$k_BT_c = \frac{2\pi\hbar^2}{m}\left[\frac{\langle N \rangle}{\zeta(\frac{3}{2})V}\right]^{2/3} . \tag{11.8}$$

Below this temperature it is impossible to satisfy (11.7) and we use the subscript c to denote that this is the critical temperature of the system. In the replacement of the sum over $\mathbf{k}$ by an integral, we have implicitly assumed that the function $\langle n_{\mathbf{k}} \rangle$ varies smoothly. However, when μ comes close enough to zero [*i.e.*, $\mu \sim \mathcal{O}(1/V)$], the ground state of the system becomes macroscopically occupied [*i.e.*, $\langle n_0 \rangle / V \sim \mathcal{O}(1)$]. Since the spacing between energy levels is $\mathcal{O}(V^{-2/3})$, all the higher-energy states have occupation of at most $\langle n_{\mathbf{k}} \rangle / V \sim \mathcal{O}(V^{-1/3})$. We must therefore single out the ground state in the sum over states in (11.5), while the occupation numbers of all the other states can be summed, in the thermodynamic limit, by replacing the sum by an integral as we did above. For $T > T_c$ there is no macroscopic occupation of the single-particle ground state and we have

$$\frac{\langle N\rangle}{V} = n = \left(\frac{mk_BT}{2\pi\hbar^2}\right)^{3/2} g_{3/2}(z) \qquad T > T_c \tag{11.9}$$

while for $T < T_c$ we consider the gas to be a mixture of two phases,

$$n = n_n(T) + n_0(T) \ . \tag{11.10}$$

The density of particles occupying the $\mathbf{k} \neq 0$ states is

$$n_n(T) = \left(\frac{mk_BT}{2\pi\hbar^2}\right)^{3/2} \zeta\left(\frac{3}{2}\right) \ . \tag{11.11}$$

The density of particles in the ground state adjusts itself to make up for the deficit:

$$n_0(T) = n\left[1 - \left(\frac{T}{T_c}\right)^{3/2}\right] \ . \tag{11.12}$$

The temperature dependence of $n_0(T)$ and of the chemical potential $\mu(T)$ is shown in Figures 11.2 and 11.3. The singular behavior of $n_0(T)$ and $\mu(T)$ at $T = T_c$ is reflected in non-analyticities in all other thermodynamic functions. If we convert (11.3) into an integral, with due attention to the possible macroscopic occupation of the ground state, we find, using (11.6), that

$$\ln Z_G = -\ln(1-z) + \left(\frac{mk_BT}{2\pi\hbar^2}\right)^{3/2} V g_{5/2}(z) \ . \tag{11.13}$$

Since $(1-z) \sim \mathcal{O}(1)$ above T_c and $(1-z) \sim \mathcal{O}(1/V)$ below, we have

$$\frac{1}{V}\ln(1-z) \to 0 \text{ as } V \to \infty \tag{11.14}$$

and find, for the pressure,

$$P = \begin{cases} \left(\dfrac{m}{2\pi\hbar^2}\right)^{3/2} (k_BT)^{5/2} g_{5/2}(z) & T > T_c \\ \left(\dfrac{m}{2\pi\hbar^2}\right)^{3/2} (k_BT)^{5/2} \zeta\left(\dfrac{5}{2}\right) & T \le T_c \end{cases} \tag{11.15}$$

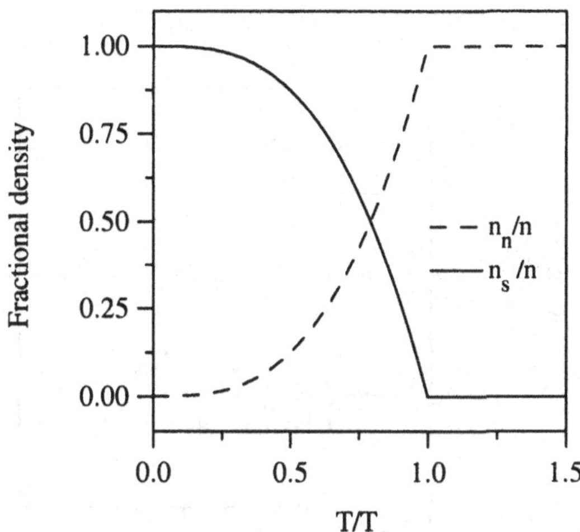

Figure 11.2: Fraction of particles in the $k = 0$ state as function of temperature.

where $\zeta(\frac{5}{2}) = g_{5/2}(1) = 1.341\ldots$. By combining (11.15) with (11.8) and eliminating the temperature, we find that the phase transition line in a Pv diagram with $v = V/\langle N\rangle$ is

$$P_c v_c^{5/3} = \frac{\zeta(\frac{5}{2})}{\left[\zeta(\frac{3}{2})\right]^{5/3}} \frac{2\pi\hbar^2}{m} . \tag{11.16}$$

In Figure 11.4 we plot some of the isotherms and the line of phase transitions in the $P - v$ plane.

The entropy may be obtained from

$$S = \left(\frac{\partial(k_BT\ln Z_G)}{\partial T}\right)_{\mu,v} = \left(\frac{\partial PV}{\partial T}\right)_{\mu,v} . \tag{11.17}$$

Below T_c we have, from (11.15), in the thermodynamic limit,

$$S = \frac{5}{2}k_BV\left(\frac{mk_BT}{2\pi\hbar^2}\right)^{3/2}\zeta\left(\frac{5}{2}\right) = \frac{5}{2}k_Bn_nV\frac{\zeta(\frac{5}{2})}{\zeta(\frac{3}{2})} \tag{11.18}$$

where we have used (11.11) in the last step. This result supports our interpretation, given above, of two coexisting phases below T_c with densities n_n and

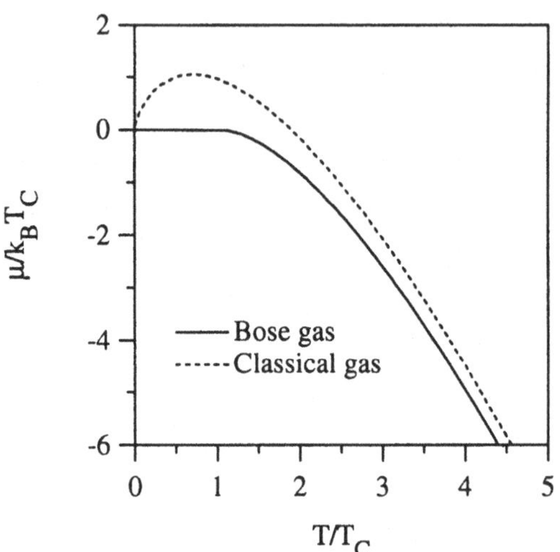

Figure 11.3: Temperature dependence of the chemical potential of an ideal Bose gas. The chemical potential for a classical gas at the same density is shown for comparison.

n_0, respectively. The normal component has an entropy per particle

$$\frac{5}{2}k_B\frac{\zeta(\frac{5}{2})}{\zeta(\frac{3}{2})} \tag{11.19}$$

while the particles condensed into the ground state carry no entropy. The flat parts of the isotherms in Figure 11.4 are thus seen to be coexistence curves. As the normal particles condense there is a latent heat per particle

$$L = T\Delta S = \frac{5}{2}k_B T_c\frac{\zeta(\frac{5}{2})}{\zeta(\frac{3}{2})} \ . \tag{11.20}$$

This equation for the latent heat may also be obtained from the Clausius–Clapeyron equation

$$\frac{dP_c(T)}{dT} = \frac{5}{2}k_B\left(\frac{mk_BT}{2\pi\hbar^2}\right)^{3/2}\zeta\left(\frac{5}{2}\right) = \frac{5}{2}k_B\frac{\zeta(\frac{5}{2})}{\zeta(\frac{3}{2})v_c} = \frac{L}{T\Delta V} \ . \tag{11.21}$$

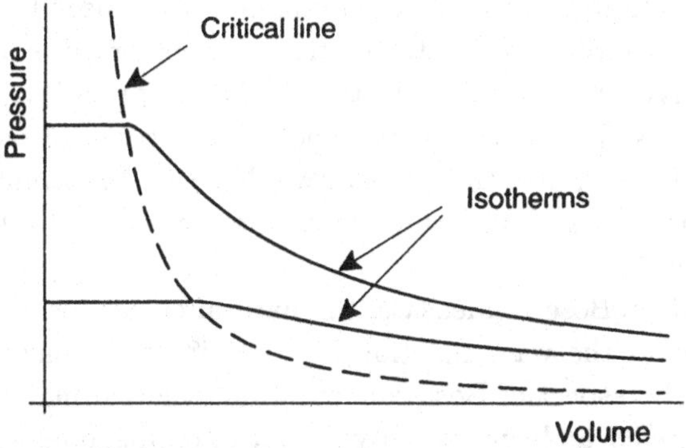

Figure 11.4: PV plane isotherms (solid curves) and critical line (dotted curve).

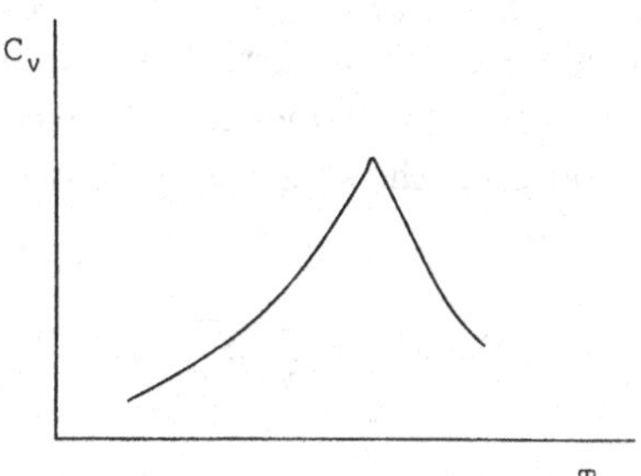

Figure 11.5: Specific heat of the ideal Bose gas.

The interpretation is, once again, in terms of two-fluid coexistence, one part with specific volume v_c, the other with zero specific volume (the condensate).

If, on the other hand, one considers the system at constant volume with a fixed total number of particles, the appropriate specific heat is

$$C_{V,N} = T\left(\frac{\partial S}{\partial T}\right)_{V,N} \quad .$$

We leave it as an exercise (Problem 11.1) for the reader to show that $C_{V,N}$ is continuous at the transition and has a discontinuity in its first derivative (Figure 11.5). For these reasons Bose–Einstein condensation is sometimes called a first-order transition (latent heat!) and sometimes a third-order transition (discontinuous derivative of $C_{V,N}$), depending on the point of view.

We have emphasized the phase-transition aspect of Bose condensation since there are a number of striking similarities with the phenomena of superfluidity and superconductivity. To exploit the analogy, it is useful to comment on the nature of the order parameter of the low-temperature phase. For the properties discussed to this point, the two-fluid description (11.10) is adequate. This description is also used in phenomenological theories of superfluidity and superconductivity, but then there are, as we shall see, important aspects of the low-temperature phases that require description in terms of a two-component order parameter.

In the case of a Bose condensate, the natural choice for a two-component order parameter is the wave function $\psi = |\psi|e^{i\phi}$ of the macroscopically occupied state. The aspect of Bose condensation which formally has an analog in superconductivity and superfluidity is that of off-diagonal long-range order (ODLRO) (see Yang, [332]; Penrose and Onsager, [240]; Penrose, [239]). In the case of the ideal Bose gas, in the momentum space representation, the expectation value $\langle b^\dagger_{\mathbf{k}} b_{\mathbf{k}'}\rangle = \delta_{\mathbf{k},\mathbf{k}'}\langle n_{\mathbf{k}}\rangle$ is given by (11.4). Here $b^\dagger_{\mathbf{k}}, b_{\mathbf{k}}$ are the creation and annihilation operators for a particle in state $\mathbf{k}$ with $\chi^\dagger(\mathbf{r}), \chi(\mathbf{r})$ the corresponding field operators (see the Appendix). The corresponding quantity in the coordinate representation is the single-particle (reduced) density matrix $\rho(\mathbf{r}, \mathbf{r}')$ given by

$$\rho(\mathbf{r}, \mathbf{r}') = \langle\chi^\dagger(\mathbf{r})\chi(\mathbf{r}')\rangle = \frac{1}{V}\sum_{\mathbf{k}}\langle n_{\mathbf{k}} e^{i\mathbf{k}\cdot(\mathbf{r}-\mathbf{r}')}\rangle \ . \tag{11.22}$$

Above T_c all the occupation numbers are of order unity and $\rho(\mathbf{r}-\mathbf{r}') \to 0$ as $|\mathbf{r}-\mathbf{r}'| \to \infty$. Below the transition we have

$$\lim_{|\mathbf{r}-\mathbf{r}'|\to\infty} \rho(\mathbf{r}, \mathbf{r}') = n_0$$

and we may say that the existence of the condensate implies a non-zero limit for the off-diagonal elements of the density matrix, or ODLRO. Similarly, if the condensation were to be into a state with wave vector $\mathbf{k} \neq 0$, we would have

$$\lim_{|\mathbf{r}-\mathbf{r}'|\to\infty} \rho(\mathbf{r}, \mathbf{r}') = \langle n_{\mathbf{k}}\rangle e^{i\mathbf{k}\cdot(\mathbf{r}-\mathbf{r}')} \ .$$

Generalizing to a case in which the condensation is into a state with wave function ψ, we have

$$\rho(\mathbf{r}, \mathbf{r}') \to n_0\psi^*(\mathbf{r})\psi(\mathbf{r}') \ . \tag{11.23}$$

It was conjectured by Yang [332] that the long-range order predicted by (11.22) and (11.23) is retained in the case of an interacting superfluid phase, despite the fact that single-particle momentum states are no longer eigenstates of the Hamiltonian. The concept of off-diagonal long-range order is also seen to fit well with the concepts of the Landau–Ginzburg theory (see (3.10) and later in this chapter).

We now briefly discuss the ideal Bose gas in arbitrary spatial dimension d. We assume that the particles occupy a hypercube of volume L^d ($d = 1, 2, 3, \ldots$) and, as before, impose periodic boundary conditions on the single-particle wave functions. The generalization of equation (11.7) is immediate. Ignoring the possible macroscopic occupation of the ground state, we find using (2.13)

$$\frac{\langle N \rangle}{V} = \left(\frac{mk_BT}{2\pi\hbar^2}\right)^{d/2} g_{d/2}(z) \ . \tag{11.24}$$

Examining the power series (11.6) for $g_\nu(z)$ we see that $g_\nu(z) \to \infty$ as $z \to 1$ for $\nu \le 1$. Thus for $d = 2$ or less, there exists a solution $z < 1$ of equation (11.24) at any finite temperature. The ground state, as a consequence, is macroscopically occupied only at $T = 0$ and there is no finite-temperature Bose condensation for $d \le 2$.

The reason for the divergence of the function $g_{d/2}(z)$ at $z = 1$ is the singularity of the integrand (11.5) near $|\mathbf{k}| = 0$. In spherical coordinates we see that at $z = 1$, in arbitrary dimension d, the contribution to $\langle N \rangle$ of the states near $|\mathbf{k}| = 0$ is

$$\langle N \rangle \sim \frac{V}{(2\pi)^d} \int_{2\pi/L}^{\infty} dk \frac{k^{d-1}}{\left(\frac{\hbar^2\beta}{2m}\right) k^2} \sim \frac{V}{(2\pi)^d} \left(\frac{2mk_BT}{\hbar^2}\right) \int_{2\pi/L}^{\infty} dk\, k^{d-3} \ . \tag{11.25}$$

This integral is finite for $d > 2$ and infinite for $d \le 2$ in the thermodynamic limit. The phase space factor k^{d-1} in low dimensions emphasizes the role of low-lying excitations and destroys the phase transition.

In the case of an interacting Bose gas, the low-lying excitations are, as we shall see, sound-like (*i.e.*, their energy is proportional to k rather than to k^2). The integral corresponding to (11.25) is then convergent, but there is no off-diagonal long-range order in the sense of (11.23) for $d \le 2$. We have encountered this situation previously in connection with our discussion of the Kosterlitz–Thouless theory in Section 6.6.

In recent years, Bose condensation has been observed in dilute gases of trapped alkali atoms cooled to extremely low temperatures. The particles interact very weakly, and this system approximates an ideal Bose gas more closely than liquid ^{4}He discussed in the next Section. This is currently a very active area of research and we will not atempt to provide an overview, and refer the reader instead to a recent text by Pethick and Smithe [241]. One important difference between trapped gases and our discussion of ideal Bose gases is that the traps produce a single-particle potential that is different from the featureless potential $V = 0$ used in our treatment. The effects of the trapping potential is the subject of Problem 11.4.

11.2 Superfluidity

The Bose fluid ^{4}He exhibits a phase transition to a superfluid phase at 2.17 K at atmospheric pressure. Some of the features of this transition are strongly suggestive of Bose condensation, although proper treatment of the ^{4}He liquid requires the inclusion, in the Hamiltonian, of interactions between the particles. In particular, the short-range repulsive interaction turns out to be important. Calculation of the properties of the system, especially near the transition, constitutes a very difficult and, to date, not completely solved problem in statistical physics. Nevertheless, a great deal can be understood about the behavior of ^{4}He at very low temperatures. We discuss, in Section 11.2.1, a number of the most striking features of ^{4}He and attempt to show how they may be understood in terms of the qualitative features of a condensate and the spectrum of excitations in the normal fluid. In Section 11.2.2 we discuss the Bogoliubov theory of the quasiparticle spectrum. We return to this topic from a different point of view in Chapter 12, where we use the formalism of linear response theory to arrive at the same results.

11.2.1 Qualitative features of superfluidity

The feature that has given superfluidity its name is the ability of the liquid to flow through pipes and capillaries without dissipation. We show first that the ideal Bose gas does not have this property. Suppose that our fluid is flowing with velocity $\mathbf{v}$ relative to the laboratory frame of reference, and that an excitation is created in the fluid through its interaction with the walls of the container. In the frame of reference of the fluid the particles are initially at rest. In an ideal Bose gas an excitation from the condensate is single-particle-like, that is, a particle acquires energy $p^2/2m$ through interaction with the "moving" wall. In the laboratory frame of reference the energy of the moving fluid is, in the case of a single excitation,

$$E_L = (N-1)\frac{mv^2}{2} + \frac{(\mathbf{p}+m\mathbf{v})^2}{2m} . \tag{11.26}$$

An excitation with momentum $\mathbf{p}$ (in the moving frame) lowers the total energy of the fluid in the laboratory frame if $E_L < Nmv^2/2$, which occurs if

$$\frac{p^2}{2m} + \mathbf{p} \cdot \mathbf{v} < 0 \,. \tag{11.27}$$

If $\mathbf{p}$ is antiparallel to $\mathbf{v}$ the condition (11.27) becomes simply $p < 2mv$, which can easily be satisfied since the ideal gas allows states of arbitrarily low momentum. The energy of the moving system can thus be dissipated through collisions with the container and the ideal Bose condensate is, therefore, not a superfluid.

The feature that makes ^{4}He a superfluid is the fact that the low-lying excitations of the system are *sound quanta* rather than single-particle excitations. This can be understood intuitively by considering the effect of the strong repulsive interactions between a pair of atoms at short distances. It is obvious that a moving particle will undergo a large number of collisions with other particles and that momentum eigenstates cannot be eigenstates of the Hamiltonian. A cooperative displacement of all the particles, as in a sound wave, is a better candidate for an elementary excitation of the system. The elementary excitation spectrum was postulated by Landau [163], [164] and later justified by Feynman [94]. (Feynman's results were actually obtained much earlier by Bijl [37], whose suggestions unfortunately attracted little attention.) We will not discuss the Bijl-Feynman theory and, at this point, simply note that the excitation spectrum can be obtained experimentally by inelastic neutron scattering (Cowley and Woods [64]). The spectrum is displayed schematically in Figure 11.6.

For low momenta the energy is linear in p, $\epsilon = cp$, with a sound velocity $c = 2.4 \times 10^2$ m/s. There is, moreover, a local minimum in $\epsilon(p)$ at $p/\hbar = 0.19$ nm^{-1} $= p_0$. In the vicinity of the minimum the dispersion relation is given by

$$\epsilon(\mathbf{p}) = \Delta + \frac{(p - p_0)^2}{2m^*} \tag{11.28}$$

with $\Delta/k_B = 8.7$ K and $m^* = 0.16 m_{He}$. Excitations with p close to p_0 are called *rotons* (mainly due to historical accident).

It is now easy to show that a fluid with an excitation spectrum of the general shape shown in Figure 11.6 is a superfluid. Suppose that the fluid is moving through a container and consider, in the rest frame of the fluid, the container to be a classical object with mass M and momentum with an initial value $\mathbf{P}$. The relative velocity of the container with respect to the fluid is thus $\mathbf{v} = \mathbf{P}/M$. It is possible to create excitations of energy $\epsilon(\mathbf{p})$ and momentum

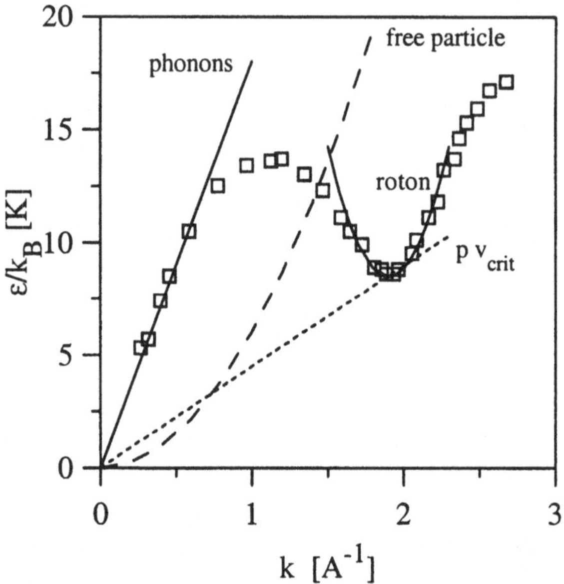

Figure 11.6: Excitation spectrum of ^{4}He below the λ transition.

$\mathbf{p}$ in the fluid if

$$\frac{P^2}{2M} - \frac{(\mathbf{P}-\mathbf{p})^2}{2M} = \epsilon(\mathbf{p}) \ . \tag{11.29}$$

In the limit $M \to \infty$ we see that dissipation of energy is possible only if $v > \epsilon(p)/p$. From Figure 11.6 we see that there is a critical velocity v_{crit} below which it is not possible to dissipate energy through the creation of excitations by contact with the walls of the container through which the fluid is flowing.

The critical velocity for superflow, given by the argument above and the shape of the measured excitation spectrum, is $v_{crit} = 60$ m/s. Experimentally, the critical velocity is much lower than this, and it is strongly dependent on the size and shape of the container. The reason for this is that in most situations turbulent (vortex) excitations occur before the phonon-like excitations considered above.

The preceding argument suggests that further properties of superfluid ^{4}He can be understood by analogy with Bose condensation. To be specific, we will, following Landau, assume that a superfluid can be considered to consist of two coexisting fluids. We write for the density

$$\rho = \rho_s + \rho_n \tag{11.30}$$

where ρ_s is the superfluid mass density and ρ_n is the density of the normal component. We incorporate the idea that there is a macroscopic condensation into a single quantum state by assuming that there exists an order parameter [$\Psi(\mathbf{r}) = (n_0 m)^{1/2}\psi$, where ψ is given by (11.23)] that is proportional to the wave function of this state. We express the order parameter in terms of an amplitude and a phase:

$$\Psi(\mathbf{r}) = a(\mathbf{r})e^{i\gamma(\mathbf{r})} \tag{11.31}$$

and identify the square of the amplitude with the superfluid density:

$$\rho_s = a^2 \ . \tag{11.32}$$

With this interpretation we have for the mass current density of the superflow,

$$\mathbf{j}_s = \frac{1}{2m}\left[\Psi(i\hbar\nabla)\Psi^* - \Psi^*(i\hbar\nabla)\Psi\right] = \hbar a^2 \nabla\gamma(\mathbf{r}) \ . \tag{11.33}$$

The associated velocity field is

$$\mathbf{u}_s = \frac{\hbar}{m}\nabla\gamma(\mathbf{r}) \ . \tag{11.34}$$

A flow pattern in which the velocity field can be expressed as the gradient of a scalar function is referred to as *potential flow*. A characteristic feature of potential flow is that it is irrotational, that is,

$$\nabla\times\mathbf{u}_s = 0 \ . \tag{11.35}$$

One might think that this condition on the flow would prohibit a superfluid from participating in any kind of rotation; for example, if a normal fluid rotates uniformly as a rigid body, we must have

$$\begin{aligned} \mathbf{v} &= \boldsymbol{\omega} \times \mathbf{r} \\ \nabla\times\mathbf{v} &= 2\boldsymbol{\omega} \end{aligned} \tag{11.36}$$

where $\boldsymbol{\omega}$ is the angular velocity. The curl of the velocity field is commonly referred to as the vorticity and in the case of rigid-body rotation is given by the second equation in (11.36). To check the foregoing assumptions, Osborne

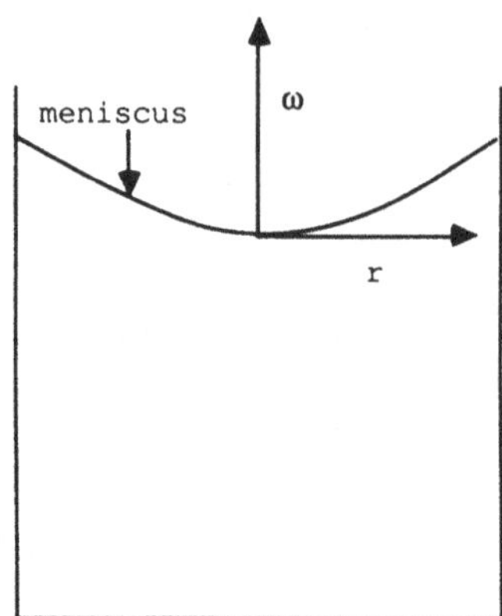

Figure 11.7: Shape of liquid surface in a rotating bucket.

[226] and others conducted a series of "rotating bucket" experiments[2]. In a normal fluid, rotating uniformly, the shape of the upper surface is determined by a balance between the centripetal and gravitational forces (see Figure 11.7), giving

$$z(r) = \frac{\omega^2 r^2}{2g} \ . \tag{11.37}$$

The expectation was that if a bucket of superfluid helium were rotated, only the normal component would participate in the rotation and that the shape of the meniscus would be given by

$$z(r) = \frac{\rho_n \omega^2 r^2}{2\rho g} \ . \tag{11.38}$$

However, the observed shape of the surface conformed with (11.37) rather than (11.38). An observation made originally by Onsager [223] makes it possible to reconcile the outcome of the rotating bucket experiments with our two-fluid picture of superfluidity.

The quantity $\gamma(\mathbf{r})$ differs from an ordinary velocity potential in that it is a phase and is only defined modulo 2π:

$$\gamma(\mathbf{r}) + 2\pi = \gamma(\mathbf{r}) \ .$$

[2]For an early review of the history of the hydrodynamics of rotating superfluid ^{4}He, see Andronikashvili and Mamaladze [19].

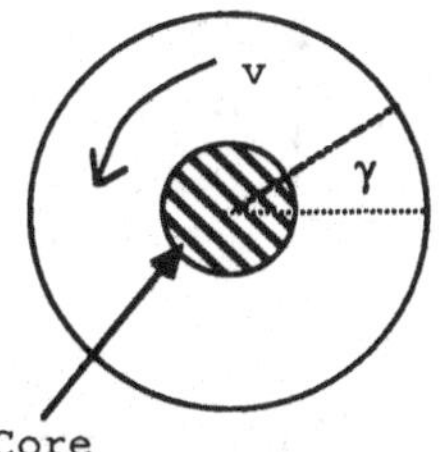

Figure 11.8: Picture of rotating ^{4}He. The superfluid density is zero in the core region.

This does not change the fact that since $\mathbf{u}_s \sim \nabla\gamma$, we must have $\nabla\times\mathbf{u_s} = 0$. It is possible, however, to have a nonzero circulation

$$\oint \mathbf{u}_s \cdot d\mathbf{l} \neq 0$$

without violating $\nabla\times\mathbf{u}_s = 0$ if in the region of phase singularity (γ is undefined at $r = 0$) the superfluid density is zero (see Figure 11.8).

This leads us to consider a new type of excitation in a superfluid called a "vortex". Vortices contain a core of normal fluid surrounded by circulating superfluid. Since the phase of the wave function must have a value (modulo 2π) within the superflow, we must have

$$\oint \nabla\gamma \cdot d\mathbf{l} = 2\pi n \quad n = \text{integer}$$

where the contour in the integration above is any closed path around the core. Equivalently,

$$\oint \mathbf{u}_s \cdot d\mathbf{l} = \frac{nh}{m} \tag{11.39}$$

that is, the circulation around a vortex is *quantized.* It is generally believed that the creation of quantized vortices is the mechanism by which superfluid flow is destroyed when the critical velocity is exceeded. In fact, there is a certain similarity between the breakdown of superfluidity through the formation of vortices and the onset of turbulence from laminar flow in ordinary fluids. It is also worth noting the similarity between the vortices discussed here and in Section 6.6. There is also, in this case, an analogy between (11.39) and

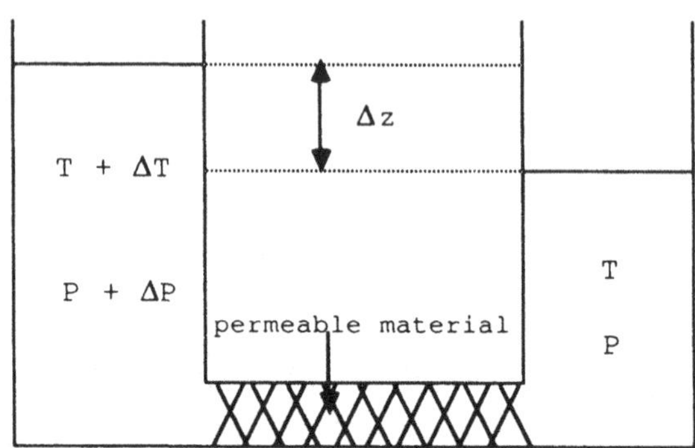

Figure 11.9: Superleak.

Ampère's law in electromagnetic theory which can be used to estimate the physical properties of a simple vortex.

Another important property of superfluidity that can be understood by analogy with the two-fluid model of Bose condensation is that the superfluid component carries no entropy. Combining the Gibbs–Duhem equation

$$E = TS - PV + \mu N \tag{11.40}$$

with the thermodynamic relation

$$dE = TdS - PdV + \mu dN \tag{11.41}$$

we find

$$Nd\mu = -SdT + VdP \ . \tag{11.42}$$

Consider the experimental arrangement of Figure 11.9. Two vessels containing ^{4}He below the λ point are connected by a pipe that is clogged by some permeable obstacle to form a "superleak". The obstacle prevents the flow of normal liquid, but the superfluid component can pass back and forth. We now release a small amount of heat in the container on the left and consider how a new equilibrium state can be established. Since the superfluid can flow freely, the chemical potential must, in the end, be the same on the two sides. On the other hand, since the superfluid component carries no entropy, there is no heat flow and therefore no tendency for the temperature to equilibrate. Similarly, since the pipe is clogged as far as the normal component is concerned, there is nothing to prevent the establishment of a pressure differential. From (11.42) we then obtain

$$\frac{\Delta T}{\Delta P} = \frac{V}{S} \tag{11.43}$$

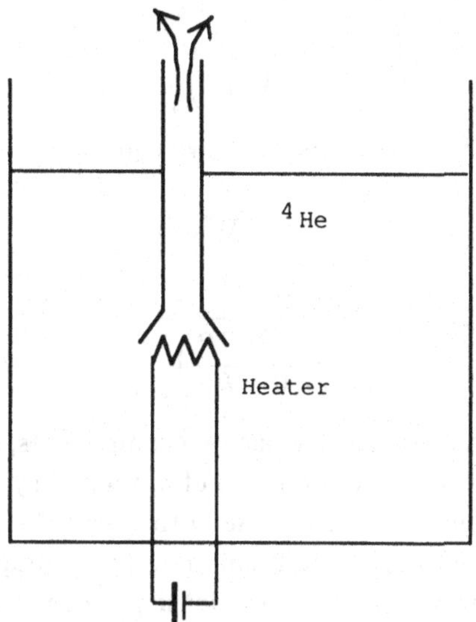

Figure 11.10: Fountain effect.

where ΔT is related to the amount of heat released and to the heat capacity in the normal way. We conclude that in a superfluid a temperature differential is associated with a pressure difference. A dramatic manifestation of this is the fountain effect (Figure 11.10).

We next consider some aspects of the hydrodynamics of a two-component system. Consider first ordinary hydrodynamics. The conditions for mass and momentum balance in an isotropic fluid are

$$\begin{aligned}\frac{\partial \rho}{\partial t} + \nabla\cdot\mathbf{j} = 0 \\ \frac{\partial \mathbf{j}}{\partial t} + \nabla P = 0 \ .\end{aligned} \tag{11.44}$$

In (11.44), ρ is the mass density, $\mathbf{j}$ the mass current, and P the pressure. Taking the time derivative of the first equation, the divergence of the second, and subtracting, we find that

$$\frac{\partial^2 \rho}{\partial t^2} - \nabla^2 P = 0 \ . \tag{11.45}$$

Usually, the pressure variations in a sound wave are sufficiently fast that they can be taken to be adiabatic, so that we can relate pressure and density fluctuations through

$$dP = \left(\frac{\partial P}{\partial \rho}\right)_s d\rho \tag{11.46}$$

which, for small amplitudes, yields the wave equation

$$\frac{\partial^2 \rho}{\partial t^2} - c^2 \nabla^2 \rho = 0 \tag{11.47}$$

with

$$c = \sqrt{\left(\frac{\partial P}{\partial \rho}\right)_s} \; . \tag{11.48}$$

In a superfluid the mass density has two components, ρ_n and ρ_s. It can be shown [165] that there are now two types of motion. In *first sound* the normal and superfluid components are in phase with each other, just as in ordinary sound. In *second sound* the normal and superfluid components beat against each other in such a way that the mass density is constant. Since the normal component carries entropy while the superfluid does not, second sound is an entropy or thermal wave.

Another way of describing second sound is the following. If we think of helium as a degenerate Bose system, the normal component can be identified with collective excitations out of the ground state. These excitations represent first sound. Second sound, on the other hand, can be thought of as a sound wave in the gas of excitations. With this interpretation second sound should be observable in ordinary solids as well as in superfluids. Normally, the damping of sound waves is too large for second sound to be observable, but the phenomenon has indeed been observed in very pure crystals at very low temperatures [201] and in smectic liquid crystals (see de Gennes [71]).

It is remarkable that the foregoing interpretation allowed Landau to make a definite prediction for the velocity of second sound in the low-temperature limit. If $n_{\mathbf{k}}$ is the number of excitations in the mode labeled by $\mathbf{k}$, the energy in this mode is $\hbar\omega n_{\mathbf{k}} = \hbar c_1 k n_{\mathbf{k}}$, where c_1, is the velocity of first sound. The number of longitudinal modes per unit volume is $d^3\mathbf{k}/(2\pi)^3$, giving for the free energy of the gas of excitations,

$$A = \frac{Vk_BT}{2\pi^2 c_1^3} \int_0^\infty d\omega\, \omega^2 \ln\left(1 - \exp\left(-\beta\hbar\omega\right)\right) . \tag{11.49}$$

Comparing $P = -\partial A/\partial V$ and $E = \partial(\beta A)/\partial\beta$ (see Problem 2.8), we find the equation of state

$$PV = \frac{E}{3} \tag{11.50}$$

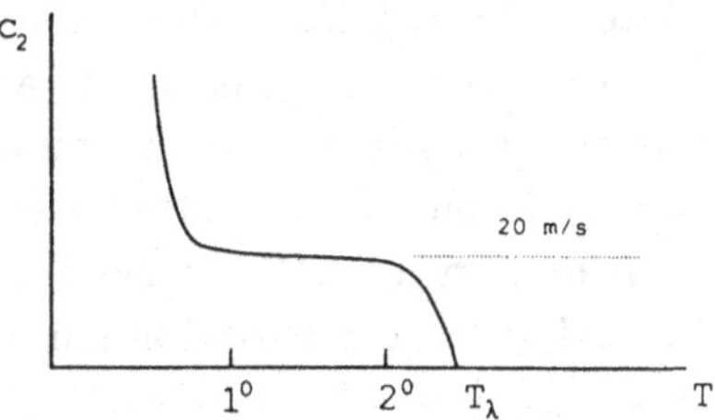

Figure 11.11: Second sound velocity as function of temperature.

where E is the internal energy and P is the pressure. Writing

$$\rho = \frac{E}{c_1^2 V} \tag{11.51}$$

for the density of inertia ("mass density") of the gas, we find from (11.48)

$$c_2 = \frac{c_1}{\sqrt{3}} \; . \tag{11.52}$$

The velocity of second sound found experimentally is sketched in Figure 11.11. The measurements are difficult at very low temperatures, but Landau's prediction about the second sound velocity appears to be essentially correct in the $T \to 0$ limit. On the other hand, over a fairly wide range of temperature, $1\ \mathrm{K} < T < 2\ \mathrm{K}$, the second sound velocity appears to be essentially constant with $c_2 \approx 20$ m/s compared to $c_1 \approx 240$ m/s for first sound.

11.2.2 Bogoliubov theory of the ^{4}He excitation spectrum

We conclude this section with an attempt to make the phonon-like excitation spectrum of superfluid helium in the long-wavelength limit plausible. We do this by developing a theory due to Bogoliubov [48] for a weakly interacting Bose system. For this purpose it is convenient to express the Hamiltonian in the second quantized form (see the Appendix). The grand canonical Hamiltonian in the plane-wave representation is

$$K = H - \mu N = \sum_{\mathbf{k}} \left(\frac{\hbar^2 k^2}{2m} - \mu \right) b_{\mathbf{k}}^\dagger b_{\mathbf{k}} + \frac{1}{2V} \sum_{\mathbf{k},\mathbf{k}',\mathbf{q}} v_{\mathbf{q}} b_{\mathbf{k}+\mathbf{q}}^\dagger b_{\mathbf{k}'-\mathbf{q}}^\dagger b_{\mathbf{k}'} b_{\mathbf{k}} \; . \tag{11.53}$$

We assume that the interaction between the particles, in the long-wavelength limit, may be approximated by a repulsive δ-function potential. In this case the Fourier transform of the potential, $v_{\mathbf{q}}$, is simply a constant, $v_{\mathbf{q}} = u_0$ for all $\mathbf{q}$. This assumption is not realistic in that the interatomic interaction has a hard core, and therefore, no Fourier transform. We are in effect making a "scattering length" approximation [185]. If the system were an ideal gas, the mean occupation number below the condensation temperature would be

$$\langle n_{\mathbf{k}} \rangle = \begin{cases} N_0 \delta_{\mathbf{k},0} & \mathbf{k} = 0 \\ [\exp \beta\epsilon(\mathbf{k}) - 1]^{-1} & \mathbf{k} \neq 0 \end{cases} \tag{11.54}$$

where $N_0 \sim N$ is of the order of Avogadro's number. In this limit the creation and annihilation operators for the lowest single-particle state, $\mathbf{k} = 0$, can be treated approximately as scalars of magnitude $N_0^{1/2}$, that is,

$$\begin{aligned} b_0^\dagger |N_0\rangle &= (N_0 + 1)^{1/2} |N_0 + 1\rangle \approx N_0^{1/2} |N_0\rangle \\ b_0 |N_0\rangle &= N_0^{1/2} |N_0 - 1\rangle \approx N_0^{1/2} |N_0\rangle \ . \end{aligned} \tag{11.55}$$

It is then convenient to identify terms in the Hamiltonian (11.53) in which one or more of the operators act on the $\mathbf{k} = 0$ state. In the two-particle interaction term in (11.53) we find the following combinations of wave vectors that lead to the most important contributions:

$$\begin{array}{lll} \text{(a)} & \mathbf{k} = \mathbf{k}' = \mathbf{q} = 0 & \frac{u_0}{2V} N_0^2 \\ \text{(b)} & \mathbf{k} = \mathbf{k}' = 0, \mathbf{q} \neq 0 & \frac{u_0}{2V} N_0 b_{\mathbf{q}}^\dagger b_{-\mathbf{q}}^\dagger \\ \text{(c)} & \mathbf{q} = -\mathbf{k}, \mathbf{k}' = 0 & \frac{u_0}{2V} N_0 b_{\mathbf{k}}^\dagger b_{\mathbf{k}} \\ \text{(d)} & \mathbf{q} = \mathbf{k}', \mathbf{k} = 0 & \frac{u_0}{2V} N_0 b_{\mathbf{k}'}^\dagger b_{\mathbf{k}'} \\ \text{(e)} & \mathbf{k} = -\mathbf{k}' = -\mathbf{q} & \frac{u_0}{2V} N_0 b_{\mathbf{k}} b_{-\mathbf{k}} \\ \text{(f)} & \mathbf{q} = \mathbf{k} = 0, \mathbf{k}' \neq 0 & \frac{u_0}{2V} N_0 b_{\mathbf{k}'}^\dagger b_{\mathbf{k}'} \\ \text{(g)} & \mathbf{q} = \mathbf{k}' = 0, \mathbf{k} \neq 0 & \frac{u_0}{2V} N_0 b_{\mathbf{k}}^\dagger b_{\mathbf{k}} \ . \end{array} \tag{11.56}$$

All other terms are of lower order in N_0 and will be neglected. The grand canonical Hamiltonian, in this approximation, becomes

$$\begin{aligned} K &= -\mu N_0 + \frac{u_0 N_0^2}{2V} + \sum_{\mathbf{k}}{}' \left[\epsilon(\mathbf{k}) - \mu + \frac{u_0 N_0}{V} \right] b_{\mathbf{k}}^\dagger b_{\mathbf{k}} \\ &\quad + \frac{u_0 N_0}{2V} \sum_{\mathbf{k}}{}' \left(b_{\mathbf{k}}^\dagger b_{-\mathbf{k}}^\dagger + b_{-\mathbf{k}} b_{\mathbf{k}} + b_{\mathbf{k}}^\dagger b_{\mathbf{k}} + b_{-\mathbf{k}}^\dagger b_{-\mathbf{k}} \right) \ . \end{aligned} \tag{11.57}$$

In (11.57) the notation Σ' indicates that the sum excludes terms with wave vector $\mathbf{k} = 0$. The chemical potential, μ must be eliminated through the requirement that the overall density be independent of temperature. This is a nontrivial task and we refer the reader to the book by Fetter and Walecka [93] for a derivation. The result is that for $T \ll T_c$, $\mu = u_0 N_0/V$, which is obtained by minimizing the grand potential with respect to the variational parameter N_0. Assuming this result, we have

$$K = -\frac{u_0 N_0^2}{2V} + \frac{1}{2}\sum_{\mathbf{k}}{}' \left[\epsilon(\mathbf{k})\left(b_{\mathbf{k}}^\dagger b_{\mathbf{k}} + b_{-\mathbf{k}}^\dagger b_{-\mathbf{k}}\right) + \frac{u_0 N_0}{V}\left(b_{\mathbf{k}}^\dagger b_{-\mathbf{k}}^\dagger + b_{-\mathbf{k}} b_{\mathbf{k}} + b_{\mathbf{k}}^\dagger b_{\mathbf{k}} + b_{-\mathbf{k}}^\dagger b_{-\mathbf{k}}\right)\right] . \tag{11.58}$$

This "phonon" Hamiltonian can be diagonalized by means of a canonical transformation known as a *Bogoliubov transformation.* We write

$$\begin{aligned} b_{\mathbf{k}} &= \eta_{\mathbf{k}} \cosh\theta_{\mathbf{k}} - \eta_{-\mathbf{k}}^\dagger \sinh\theta_{\mathbf{k}} \\ b_{-\mathbf{k}} &= \eta_{-\mathbf{k}} \cosh\theta_{\mathbf{k}} - \eta_{\mathbf{k}}^\dagger \sinh\theta_{\mathbf{k}} \end{aligned} \tag{11.59}$$

where $[\eta_{\mathbf{k}}, \eta_{\mathbf{k}}^\dagger] = [\eta_{-\mathbf{k}}, \eta_{-\mathbf{k}}^\dagger] = 1$. This transformation is canonical in that it preserves the commutation relations for any real $\theta_{\mathbf{k}}$:

$$[b_{\mathbf{k}}, b_{\mathbf{k}}^\dagger] = \cosh^2\theta_{\mathbf{k}} - \sinh^2\theta_{\mathbf{k}} = 1 .$$

We determine $\theta_{\mathbf{k}}$ from the condition that the Hamiltonian be diagonal in the quasiparticle operators. Substituting (11.59) into (11.58), we obtain, for the off-diagonal piece

$$K_{o.d.} = -\sum_{\mathbf{k}}{}' \left[\epsilon(\mathbf{k}) + \frac{u_0 N_0}{V}\right] \cosh\theta_{\mathbf{k}} \sinh\theta_{\mathbf{k}} \left(\eta_{\mathbf{k}}^\dagger \eta_{-\mathbf{k}}^\dagger + \eta_{-\mathbf{k}}\eta_{\mathbf{k}}\right) + \frac{u_0 N_0}{V}\sum_{\mathbf{k}}{}' \left(\cosh^2\theta_{\mathbf{k}} + \sinh^2\theta_{\mathbf{k}}\right)\left(\eta_{\mathbf{k}}^\dagger \eta_{-\mathbf{k}}^\dagger + \eta_{-\mathbf{k}}\eta_{\mathbf{k}}\right) . \tag{11.60}$$

We see that $K_{o.d.} = 0$ if

$$\tanh\theta_{\mathbf{k}} = \frac{u_0 N_0/V}{\epsilon(\mathbf{k}) + u_0 N_0/V} \tag{11.61}$$

and note that this equation has a solution for all $\mathbf{k}$ only if $u_0 > 0$. Thus, the repulsive nature of the interparticle potential is crucial for condensation.

Using (11.61), we obtain, for the diagonal part of (11.58),

$$\begin{aligned} K &= -\frac{u_0 N_0^2}{2V} + \sum_{\mathbf{k}}{}' \sqrt{\epsilon^2(\mathbf{k}) + 2\frac{u_0 N_0}{V}\epsilon(\mathbf{k})}\ \eta_{\mathbf{k}}^\dagger \eta_{\mathbf{k}} \\ &\quad + \sum_{\mathbf{k}}{}' \left[\sqrt{\epsilon^2(\mathbf{k}) + 2\frac{u_0 N_0}{V}} - \epsilon(\mathbf{k}) - \frac{u_0 N_0}{V} \right] . \end{aligned} \tag{11.62}$$

Two features of (11.62) are worth noting. We see that, for small $|\mathbf{k}|$, the quasiparticle energy

$$E(\mathbf{k}) = \sqrt{\epsilon^2(\mathbf{k}) + 2\frac{u_0 N_0}{V}\epsilon(\mathbf{k})}$$

is dominated by the second term in the expression under the root and is linear in $|\mathbf{k}|$. We have therefore obtained the phonon dispersion relation for the excitations of a weakly interacting Bose gas. The third term in equation (11.62) is a shift of the ground-state energy that, in fact, diverges. This divergence is due to the contribution from large wave vectors and is an artifact of our approximation $v_{\mathbf{q}} = u_0$ for all $\mathbf{q}$.

The excitation spectrum in this theory, while qualitatively correct at small wave vectors, is not quantitatively accurate. Moreover, it is difficult to improve the theory systematically. For a critique of such "pairing" approximations and a discussion of better theories for superfluid ^{4}He, the reader is referred to Mahan [185].

11.3 Superconductivity

The phenomenon of superconductivity is in some ways similar to superfluidity, and for this reason, we have chosen to discuss it in this chapter. The frictionless flow of superfluid ^{4}H has its analog in the persistent currents in a superconductor, and there are other similarities between the two systems. We begin this section by concentrating on the microscopic theory in order to show how a condensation similar to Bose condensation can occur in a system of interacting fermions. For a more detailed treatment, the reader is encouraged to consult the original papers (Cooper [63], Bardeen *et al.* [25]) or one of the monographs, such as those of De Gennes [72] and Schrieffer [269]. In 1986 Bednorz and Müller [32] reported on a superconductor with a transition temperature substantially higher than any previously reported. The report generated an enormous amount of interest and research activity in the new field of "unconventional superconductors". By now, materials with critical temperatures of well over 100 K have been found. While many aspects of the new materials are understood, no comprehensive theory is currently available. In order to make headway in the large and sometimes confusing literature on unconventional superconductors it is clearly necessary to first understand the theory of "conventional" (BCS) superconductivity. For this reason we will in this section concentrate on the BCS theory.

The crucial feature responsible for conventional superconductivity is an effective attractive interaction between electrons due to "overscreening" by the ions of the metal. The bare Coulomb interaction between electrons is, of course, repulsive. The ions of the system respond to the motion of the electrons and, in certain circumstances, can produce an effective interaction between electrons that is attractive. This discovery is due to Fröhlich [105]. Later, Cooper [63] showed that this attractive interaction could produce a two-electron bound state in the presence of a Fermi sphere of Bloch electrons. These bound pairs have properties that are similar to those of bosons and, at sufficiently low temperatures, condense into the superconducting state. The outline of this section is as follows. We first present the solution of the Cooper problem. In subsection 11.3.2 we solve the BCS ground-state problem and, in Section 11.3.3 discuss the finite-temperature behavior of the system. Finally, in Section 11.3.4, we briefly describe the Landau–Ginzburg theory of superconductivity. For this part of the theory, the books by De Gennes [72] and Part 2 of Landau and Lifshitz [165] are excellent references.

11.3.1 Cooper problem

We consider the problem of two interacting electrons in the presence of a filled Fermi sea. The Hamiltonian of the two-electron system is

$$H = \sum_{\mathbf{k},\sigma} \epsilon(\mathbf{k}) c^{\dagger}_{\mathbf{k},\sigma} c_{\mathbf{k},\sigma} + \frac{1}{2} \sum_{\mathbf{k},\mathbf{k}',\mathbf{q},\sigma,\sigma'} v(\mathbf{q}) c^{\dagger}_{\mathbf{k}+\mathbf{q},\sigma} c^{\dagger}_{\mathbf{k}'-\mathbf{q},\sigma'} c_{\mathbf{k}',\sigma'} c_{\mathbf{k},\sigma} \tag{11.63}$$

where σ, σ' label the spin states of the particles and where the sum over wave vectors is restricted to $|\mathbf{k}| > k_F$, the Fermi wave vector. In a metal, one can show (Fröhlich [105]; see also Chapter 12) that the effective interaction $v(\mathbf{q})$ between electrons can be negative within an energy range $\sim \hbar\omega_D$ near the Fermi energy, where ω_D is the Debye frequency. This "attractive" interaction is due to the overscreening alluded to in the introduction. Instead of using the full potential $v(\mathbf{q})$, we make the simplifying assumption (Cooper [63]) that $v(\mathbf{q}) = -v$ for matrix elements between states with $\epsilon_F \leq \epsilon(k) \leq \epsilon_F + \hbar\omega_D$, and is zero elsewhere. We now attempt to find a two-electron bound state of the Hamiltonian (11.63). We take the zero of energy to be ϵ_F and consider a trial wave function of the form

$$|\psi\rangle = \sum_{\mathbf{k}} \alpha_{\mathbf{k}} c^{\dagger}_{\mathbf{k},\frac{1}{2}} c^{\dagger}_{-\mathbf{k},-\frac{1}{2}} |F\rangle$$

where

$$|F\rangle = \prod_{|\mathbf{k}|<k_F,\sigma} c^{\dagger}_{\mathbf{k},\sigma}|0\rangle$$

is the wave function of the filled Fermi sphere. We demand that $(E-H)|\psi\rangle = 0$ which yields

$$0 = [E - \epsilon(\mathbf{k})]\alpha_{\mathbf{k}} - v\sum_{\mathbf{q}} \alpha_{\mathbf{k}+\mathbf{q}} \,. \tag{11.64}$$

We define the constant Λ to be

$$\Lambda = \sum_{\mathbf{q}} \alpha_{\mathbf{k}+\mathbf{q}} = \int_0^{\hbar\omega_D} d\epsilon \rho(\epsilon)\alpha(\epsilon)$$

and thus obtain the equation

$$\Lambda = v\Lambda \int_0^{\hbar\omega_D} d\epsilon \frac{\rho(\epsilon)}{E - 2\epsilon} \tag{11.65}$$

or

$$1 \approx v\rho(\epsilon_F) \int_0^{\hbar\omega_D} \frac{d\epsilon}{E - 2\epsilon} = v\rho(\epsilon_F) \ln \frac{E - 2\hbar\omega_D}{E} \tag{11.66}$$

where we have assumed that the density of states $\rho(\epsilon)$ does not vary appreciably in the range $(\epsilon_F, \epsilon_F + \hbar\omega_D)$. We may now solve (11.66) for the eigenvalue E to find

$$E = \frac{-2\hbar\omega_D}{\exp\{1/(v\rho(\epsilon_F))\} - 1} \approx -2\hbar\omega_D \exp\left\{\frac{-1}{v\rho(\epsilon_F)}\right\} . \tag{11.67}$$

This equation demonstrates that no matter how weak the effective interaction, v, there always exists a two-particle bound state with energy less than the Fermi energy. This indicates that the Fermi sea is unstable against formation of paired states. We show below that these paired states form a condensate separated from the nearest single-particle states by a gap of width roughly E as given in (11.67). It can also be shown [269] that the paired state with zero center of mass momentum is the lowest-energy paired state. The BCS (Bardeen *et al.*, [25]) theory of superconductivity focuses on such states.

11.3.2 BCS ground state

As we have seen in Section 11.3.1, the normal state of an electron gas is unstable with respect to formation of bound pairs near the Fermi surface. Bardeen *et al.* [25] (BCS) considered an approximate Hamiltonian for the N-electron problem that contains the effective attraction due to overscreening and that contains, besides the Bloch energies, only interactions between paired electrons in singlet states. This "reduced" Hamiltonian is

$$H_{red} = \sum_{\mathbf{k},\sigma} \epsilon(\mathbf{k}) c^{\dagger}_{\mathbf{k},\sigma} c_{\mathbf{k},\sigma} + \frac{1}{V_0} \sum_{\mathbf{k},\mathbf{k}'} V_{\mathbf{k},\mathbf{k}'} b^{\dagger}_{\mathbf{k}} b_{\mathbf{k}'} \tag{11.68}$$

where V_0 is the volume, $c^{\dagger}_{\mathbf{k},\sigma}$,$c_{\mathbf{k},\sigma}$ are creation and annihilation operators for electrons in Bloch states $(\mathbf{k},\sigma)$, and $b^{\dagger}_{\mathbf{k}}$,$b_{\mathbf{k}}$ are creation and annihilation for the pair $(\mathbf{k},+\frac{1}{2};-\mathbf{k},-\frac{1}{2})$ that is, $b^{\dagger}_{\mathbf{k}} = c^{\dagger}_{\mathbf{k},1/2} c^{\dagger}_{-\mathbf{k},-1/2}$ and $b_{\mathbf{k}} = c_{-\mathbf{k},-1/2} c_{\mathbf{k},1/2}$. As in Section 11.3.1 the interaction $V_{\mathbf{k}\mathbf{k}'}$ is assumed to be zero unless both $\epsilon(\mathbf{k})$ and $\epsilon(\mathbf{k}')$ are in a shell of width $\hbar\omega_D$ centered on the Fermi energy. If we restrict our set of basis states to paired states, that is, the single-particle states $(\mathbf{k},+\frac{1}{2})$ and $(-\mathbf{k},-\frac{1}{2})$ are either both occupied or both empty, the Hamiltonian (11.68) can be expressed as

$$H = 2\sum_{\mathbf{k}} \epsilon(\mathbf{k}) b^{\dagger}_{\mathbf{k}} b_{\mathbf{k}} + \frac{1}{V_0} \sum_{\mathbf{k},\mathbf{k}'} V_{\mathbf{k},\mathbf{k}'} b^{\dagger}_{\mathbf{k}} b_{\mathbf{k}'} \ . \tag{11.69}$$

The normal state $|F\rangle$ falls into this category as

$$|F\rangle = \prod_{|\mathbf{k}|<k_F,\sigma} c^{\dagger}_{\mathbf{k},\sigma}|0\rangle = \prod_{|\mathbf{k}|<k_F} b^{\dagger}_{\mathbf{k}}|0\rangle \ .$$

The commutation relations of the pair creation and annihilation operators are

$$[b_{\mathbf{k}}\ , b_{\mathbf{k}'}] = [b^{\dagger}_{\mathbf{k}}\ , b^{\dagger}_{\mathbf{k}'}] = 0$$

$$[b_{\mathbf{k}}\ , b^{\dagger}_{\mathbf{k}'}] = \delta_{\mathbf{k},\mathbf{k}'}\{1 - n_{\mathbf{k},1/2} - n_{-\mathbf{k},-1/2}\} \ . \tag{11.70}$$

If the term $n_{\mathbf{k},1/2}+n_{-\mathbf{k},-1/2}$ in the last commutator were absent, these commutation relations would be those of boson operators. In fact, the commutation relations (11.70) are those of a set of spin-$\frac{1}{2}$ operators with the formal correspondence

$$\begin{aligned} b^{\dagger}_{\mathbf{k}} &= S_{\mathbf{k},x} + iS_{\mathbf{k},y} = S^{+}_{\mathbf{k}} \\ b_{\mathbf{k}} &= S_{\mathbf{k},x} - iS_{\mathbf{k},y} = S^{-}_{\mathbf{k}} \\ 1 - n_{\mathbf{k},1/2} - n_{-\mathbf{k},-1/2} &= 2S_{\mathbf{k},z} \ . \end{aligned} \tag{11.71}$$

This identification of equivalent spin operators holds only for our restricted set of basis states in which the operator $n_{\mathbf{k},1/2}+n_{-\mathbf{k},-1/2}$ has only the eigenvalues 0 and 2. In this spin language the Hamiltonian (11.69) takes the form

$$H = 2\sum_{\mathbf{k}}\epsilon(\mathbf{k})\left(S_{\mathbf{k},z}+\frac{1}{2}\right)+\frac{1}{V_0}\sum_{\mathbf{k},\mathbf{k}'}V_{\mathbf{k},\mathbf{k}'}\left(S_{\mathbf{k},x}S_{\mathbf{k}',x}+S_{\mathbf{k},y}S_{\mathbf{k}',y}\right) \qquad (11.72)$$

which is an XY model with a "Zeeman" energy due to a "magnetic field" in the z direction. The two-dimensional rotational symmetry of the Hamiltonian is obvious in this notation. In magnetic language, if $V_{\mathbf{k},\mathbf{k}'}$ is large enough, the ground state will have some nonzero magnetization $\mathbf{M}=\sum\mathbf{S}_{\mathbf{k}}$, and this vector can be rotated about the z axis without cost of energy. The order parameter is a two-component order parameter and the transition from the normal to the superconducting state will be in the universality class of the XY model. We also see, from this analogy, that conventional long-range order in two-dimensional systems (thin metallic films) will be excluded (see also Section 6.6).

Following BCS we now construct a trial wave function for the ground state and minimize the resulting expression for the energy subject to the constraint that the mean number of electrons is N. The trial wave function

$$|\psi\rangle = \prod_{\mathbf{k}}\frac{1+g_k b_{\mathbf{k}}^{\dagger}}{\sqrt{1+g_k^2}}|0\rangle \qquad (11.73)$$

contains paired states with an indefinite number of particles and is normalized. The function g_k is our variational parameter. The expectation value of the grand canonical Hamiltonian

$$K = H-\mu N$$

in the state (11.73) is

$$K_0 = \langle\psi|K|\psi\rangle = 2\sum_{\mathbf{k}}[\epsilon(\mathbf{k})-\mu]\frac{g_k^2}{1+g_k^2}+\frac{1}{V_0}\sum_{\mathbf{k},\mathbf{k}'}\frac{g_k g_{k'}}{(1+g_k^2)(1+g_{k'}^2)}\ . \qquad (11.74)$$

We note that the normal state corresponds to the value $g_k=\infty$ for $|\,\mathbf{k}\,|<k_F$ and $g_k=0$ for $|\,\mathbf{k}\,|>k_F$. We now define the functions u_k and v_k through

$$\begin{aligned} u_k &= \frac{1}{\sqrt{1+g_k^2}} \\ v_k &= \frac{g_k}{\sqrt{1+g_k^2}} \end{aligned} \qquad (11.75)$$

with $u^2 + v^2 = 1$. In the normal state $u_k = 0$, $v_k = 1$ for $|\mathbf{k}| < k_F$ and $u_k = 1$, $v_k = 0$ for $|\mathbf{k}| > k_F$. The variation of K_0 with respect to u_k and v_k yields

$$\begin{aligned}
\delta K_0 &= 4\sum_{\mathbf{k}}[\epsilon(\mathbf{k}) - \mu]v_k\delta v_k + \frac{2}{V_0}\sum_{\mathbf{k},\mathbf{k}'} V_{\mathbf{k},\mathbf{k}'}(u_{k'}v_{k'})\{u_k\delta v_k + v_k\delta u_k\} \\
&= 4\sum_{\mathbf{k}}[\epsilon(\mathbf{k}) - \mu]v_k\delta v_k + \frac{2}{V_0}\sum_{\mathbf{k},\mathbf{k}'} V_{\mathbf{k},\mathbf{k}'}(u_{k'}v_{k'})\frac{u_k^2 - v_k^2}{u_k}\delta v_k \qquad (11.76)
\end{aligned}$$

where we have used the condition $u_k\delta u_k + v_k\delta v_k = 0$. Requiring the coefficient of δv_k to be zero for all $\mathbf{k}$, we obtain

$$2[\epsilon(\mathbf{k}) - \mu]v_k + \frac{1}{V_0}\sum_{\mathbf{k}'}(u_{k'}v_{k'})\frac{u_k^2 - v_k^2}{u_k} = 0\,. \qquad (11.77)$$

This equation may be simplified by defining new variables. We let

$$u_k v_k = \frac{\Delta_k}{2E_k} \qquad (11.78)$$

where $E_k = \sqrt{\{\epsilon(\mathbf{k}) - \mu\}^2 + \Delta_k^2}$. Using the condition $u_k^2 + v_k^2 = 1$, we may re-express u_k and v_k in terms of the single unknown Δ_k

$$\begin{aligned}
u_k^2 &= \frac{1}{2}\left[1 + \frac{\epsilon(\mathbf{k}) - \mu}{E_k}\right] \\
v_k^2 &= \frac{1}{2}\left[1 - \frac{\epsilon(\mathbf{k}) - \mu}{E_k}\right]\,. \qquad (11.79)
\end{aligned}$$

The choice of sign inside the brackets of (11.79) is determined by the condition $u_k = 0$ for $|\mathbf{k}| < k_F$ in the normal state ($\Delta_k = 0$, $\mu = \epsilon_F$). Substituting into (11.77), we obtain an equation for the zero-temperature "gap", Δ_k

$$\Delta_k = -\frac{1}{V_0}\sum_{\mathbf{k}'} V_{\mathbf{k},\mathbf{k}'}\frac{\Delta_{k'}}{2E_{k'}}\,. \qquad (11.80)$$

At this point we simplify equation (11.80) by taking $V_{\mathbf{k}\mathbf{k}'}$ to be a constant, $-V$, for $|\epsilon(k) - \epsilon_F| < \hbar\omega_D/2$, $|\epsilon(k') - \epsilon_F| < \hbar\omega_D/2$, and $\Delta_k = \Delta$, independent of k in the same region of k-space in which V is nonzero. Converting the sum in (11.80) to an integral over the density of states, $\rho(\epsilon)$, we finally obtain

$$1 = \frac{V}{2}\int_{\epsilon_F - \hbar\omega_D/2}^{\epsilon_F + \hbar\omega_D/2} d\epsilon\rho(\epsilon)\frac{1}{\sqrt{(\epsilon - \mu)^2 + \Delta^2}}\,. \qquad (11.81)$$

It can be shown that the shift in chemical potential is small. Thus, replacing, μ by ϵ_F and assuming that the density of states is essentially constant over the small range $\hbar\omega_D$ near the Fermi surface, we find, from (11.81),

$$1 = V\rho(\epsilon_F)\sinh^{-1}\frac{\hbar\omega_d}{2\Delta} \tag{11.82}$$

or

$$\Delta = \frac{\hbar\omega_D}{2\sinh\left[1/(V\rho(\epsilon_F))\right]} \approx \hbar\omega_D \exp\left\{-\frac{1}{V\rho(\epsilon_F)}\right\} \tag{11.83}$$

in the weak-coupling limit $V\rho(\epsilon_F) < 1$. The reader will note that (11.83) is a singular function of the interaction strength V. Thus any attempt to arrive at the superconducting ground state by perturbative methods would be doomed to failure.

Having obtained an expression for the variational parameter Δ, we must still show that the ground-state energy for nonzero Δ is lower than the normal ($\Delta = 0$) energy. We again take $\mu = \epsilon_F$ and take the zero of energy to be ϵ_F. The difference in energy between the variational ground state and the normal state is given by

$$E_0(\Delta) - E_N = 2\sum_{\mathbf{k}}\left[\epsilon(\mathbf{k}) - \epsilon_F\right]v_k^2 \quad + \quad \frac{1}{V_0}\sum_{\mathbf{k},\mathbf{k}'} V_{\mathbf{k},\mathbf{k}'}(u_k v_k)(u_{k'}v_{k'}) \\ -2\sum_{|\mathbf{k}|<k_F}\left[\epsilon(\mathbf{k}) - \epsilon_F\right] . \tag{11.84}$$

Using the expression (11.79) to substitute for v_k^2, equations (11.78) and (11.80) to eliminate the sum over k' in the second term on the right-hand side, and converting the remaining sums to integrals over the density of states, we find that

$$E_0(\Delta) - E_N = -\frac{V_0}{2}\rho(\epsilon_F)\Delta^2 \tag{11.85}$$

indicating that the variational ground state is of lower energy than the normal state.

In Section 11.3.3 we shall see that the quantities E_k are approximate two-particle eigenvalues of the Hamiltonian. At the Fermi surface they are separated from the nearest excited states (the normal states with no pairing) by a gap in energy of 2Δ. This gap is responsible for the absence of dissipation in current flow.

Finally, we show that the BCS ground state is a state of broken symmetry. To see this we simply calculate the expectation value of the operators $b_{\mathbf{k}}$ and $b_{\mathbf{k}}^\dagger$:

$$\langle\psi_0|b_{\mathbf{k}}|\psi_0\rangle = u_k v_k = \langle\psi_0|b_{\mathbf{k}}^\dagger|\psi_0\rangle . \tag{11.86}$$

In terms of the "pseudo-spin" picture (11.71) this corresponds to the system being "magnetized" in the x direction. The BCS Hamiltonian (11.69) is invariant under the coherent rotation of all the spins, i.e.

$$b_{\mathbf{k}} \rightarrow b_{\mathbf{k}} \exp\{i\phi\}, \quad b^{\dagger}_{\mathbf{k}} \rightarrow b^{\dagger}_{\mathbf{k}} \exp\{-i\phi\} \,.$$

This symmetry is absent in the ground state.

11.3.3 Finite-temperature BCS theory

We found in Section 11.3.2 that the ground state of a superconductor is characterized by broken symmetry and by a nonzero expectation value of the pair creation and annihilation operators $b_{\mathbf{k}}$ and $b^{\dagger}_{\mathbf{k}}$: $\langle b_{\mathbf{k}}\rangle = \langle b^{\dagger}_{\mathbf{k}}\rangle = \Delta_k/(2E_k)$. We now use these results to construct an approximate (but very successful) mean field theory of superconductivity for $T \neq 0$. We confine ourselves to a derivation of the temperature-dependent gap equation. A much more detailed discussion of this aspect of superconductivity can be found in Abrikosov *et al.* [5], Schrieffer [269], or Mahan [185].

We begin, again, with the reduced grand canonical Hamiltonian

$$K = 2\sum_{\mathbf{k}} \epsilon(\mathbf{k}) b^{\dagger}_{\mathbf{k}} b_{\mathbf{k}} + \frac{1}{V_0} V_{\mathbf{k},\mathbf{k}'} b^{\dagger}_{\mathbf{k}} b_{\mathbf{k}'} \tag{11.87}$$

where we have absorbed the chemical potential, μ, into the single-particle energies $\epsilon(\mathbf{k})$. In the spirit of mean field theory (Chapter 3), we now approximate the interaction term in equation (11.87) by the decoupled form

$$\sum_{\mathbf{k},\mathbf{k}'} V_{\mathbf{k},\mathbf{k}'} b^{\dagger}_{\mathbf{k}} b_{\mathbf{k}'} \approx \sum_{\mathbf{k},\mathbf{k}'} V_{\mathbf{k},\mathbf{k}'} \left(b^{\dagger}_{\mathbf{k}} \langle b_{\mathbf{k}'}\rangle + \langle b^{\dagger}_{\mathbf{k}}\rangle b_{\mathbf{k}'} - \langle b^{\dagger}_{\mathbf{k}}\rangle\langle b_{\mathbf{k}'}\rangle \right) \tag{11.88}$$

where the last term in (11.88) has been included to compensate for double counting. The expectation values $\langle b^{\dagger}_{\mathbf{k}}\rangle$, $\langle b_{\mathbf{k}}\rangle$ will be evaluated with respect to the grand canonical probability density and will be temperature dependent. We assume that, as in the ground state, $\langle b^{\dagger}_{\mathbf{k}}\rangle$ is real and hence equal to $\langle b_{\mathbf{k}}\rangle$. This assumption is justified in BCS theory but needs to be modified if the BCS Hamiltonian is replaced by the full electron–phonon interaction. We note, in passing, that the nonzero expectation value of an operator such as $b^{\dagger}_{\mathbf{k}} = c^{\dagger}_{\mathbf{k},1/2} c^{\dagger}_{-\mathbf{k},-1/2}$, that does not conserve the number of particles, is referred to as off-diagonal long-range order (Yang [332]) and is an essential property of superfluids and superconductors (see also Section 11.1 and the Bogoliubov theory of the weakly interacting Bose gas).

In analogy with the ground-state calculations, we define the quantity $\Delta_k(T)$ through

$$\Delta_k(T) = -\frac{1}{V_0}\sum_{\mathbf{k}'} V_{\mathbf{k},\mathbf{k}'}\langle b_{\mathbf{k}'}\rangle \ . \tag{11.89}$$

Using (11.88) and (11.89), we find, for the reduced Hamiltonian,

$$\begin{aligned} K &= \sum_{\mathbf{k}}\Big[\epsilon(\mathbf{k})\left(c^\dagger_{\mathbf{k},1/2}c_{\mathbf{k},1/2} + c^\dagger_{-\mathbf{k},-1/2}c_{-\mathbf{k},-1/2}\right) \\ &\quad -\Delta_k\left(c^\dagger_{\mathbf{k},1/2}c^\dagger_{-\mathbf{k},1/2} + c_{-\mathbf{k},-1/2}c_{-\mathbf{k},-1/2} - \langle b^\dagger_{\mathbf{k}}\rangle\right)\Big] \ . \end{aligned} \tag{11.90}$$

An operator such as (11.90) which is bilinear in fermion operators can always be diagonalized by a Bogoliubov transformation (Section 11.2). Our object is to bring (11.90) into the form

$$K = \sum_{\mathbf{k}} E(\mathbf{k})\xi^\dagger_{\mathbf{k}}\xi_{\mathbf{k}} + \text{constant} \tag{11.91}$$

where the quasiparticle energies $E(\mathbf{k})$ and the new fermion operators are to be determined. Once we have achieved this form, the partition function is simply that of a system of noninteracting fermions. We assume a transformation of the form

$$\begin{aligned} c_{\mathbf{k},1/2} &= \alpha\xi_{\mathbf{k}} + \beta\xi^\dagger_{-\mathbf{k}} \\ c^\dagger_{\mathbf{k},1/2} &= \alpha\xi^\dagger_{\mathbf{k}} + \beta\xi_{-\mathbf{k}} \\ c_{-\mathbf{k},-1/2} &= \gamma\xi_{-\mathbf{k}} + \delta\xi^\dagger_{\mathbf{k}} \\ c^\dagger_{-\mathbf{k},1/2} &= \gamma\xi^\dagger_{-\mathbf{k}} + \delta\xi_{\mathbf{k}} \end{aligned} \tag{11.92}$$

where $[\xi_{\mathbf{k}},\xi^\dagger_{\mathbf{k}'}]_+ = \delta_{\mathbf{k},\mathbf{k}'}$, $[\xi_{\mathbf{k}},\xi_{\mathbf{k}'}]_+ = [\xi^\dagger_{\mathbf{k}},\xi^\dagger_{\mathbf{k}'}]_+ = 0$.

This condition that the c and ξ operators both obey fermion commutation relations, [*i.e.*, that the transformation (11.92) is canonical] produces three equations:

$$\begin{aligned} \alpha^2 + \beta^2 &= 1 \\ \gamma^2 + \delta^2 &= 1 \\ \alpha\delta + \beta\gamma &= 0 \ . \end{aligned} \tag{11.93}$$

The fourth equation for the unknown coefficients $\alpha \ldots \delta$ comes from the requirement that the Hamiltonian should be of the form (11.91). Substituting (11.92) into (11.90), we obtain, for the nondiagonal piece of K,

$$K_{o.d.} = \sum_{\mathbf{k}}[\epsilon(\mathbf{k})(\alpha\beta - \gamma\delta) - \Delta_k(\alpha\gamma + \beta\delta)](\xi^\dagger_{\mathbf{k}}\xi^\dagger_{-\mathbf{k}} + \xi_{-\mathbf{k}}\xi_{\mathbf{k}}) \ . \tag{11.94}$$

If we require this expression to be identically zero, we find, using equation (11.93), that $\gamma = \alpha$, $\delta = -\beta$, and

$$\alpha^2 = \frac{1}{2}\left[1 + \frac{\epsilon(\mathbf{k})}{E(\mathbf{k})}\right]$$
$$\beta^2 = \frac{1}{2}\left[1 - \frac{\epsilon(\mathbf{k})}{E(\mathbf{k})}\right] \tag{11.95}$$

with $E(\mathbf{k}) = [\epsilon(\mathbf{k})^2 + \Delta_{\mathbf{k}}^2]^{1/2}$. The sign convention in (11.95) is such that the quasiparticle energy becomes the usual single-particle energy if $\Delta = 0$ and $k > k_F$. For $k < k_F$ the quasiparticle energy is that of a hole (missing electron). We complete the specification of the coefficients by taking $\alpha,\ \beta > 0$, $\delta < 0$. With these coefficients substituted into (11.90), the remaining diagonal piece of K becomes

$$K = \sum_{\mathbf{k}} \left[E(\mathbf{k})(\xi_{\mathbf{k}}^\dagger \xi_{\mathbf{k}} + \xi_{-\mathbf{k}}^\dagger \xi_{-\mathbf{k}}) + \epsilon(\mathbf{k}) - E(\mathbf{k}) - \Delta_k \langle b_{\mathbf{k}}^\dagger \rangle\right] . \tag{11.96}$$

The lowest-energy state is always the state in which no quasiparticles are present and we see at once that if $\Delta_k = 0$ the ground-state energy of the normal system is obtained.

We are now in a position to calculate Δ_k self-consistently. The expectation value $\langle b_{\mathbf{k}} \rangle$ is given by

$$\langle b_{\mathbf{k}} \rangle = \frac{\mathrm{Tr}\, b_{\mathbf{k}} \exp\{-\beta K\}}{\mathrm{Tr} \exp\{-\beta K\}} = \alpha\delta \frac{\mathrm{Tr}\,(\xi_{\mathbf{k}}^\dagger \xi_{\mathbf{k}} - \xi_{-\mathbf{k}} \xi_{-\mathbf{k}}^\dagger) \exp\{-\beta K\}}{\mathrm{Tr} \exp\{-\beta K\}}$$
$$= \alpha\delta[\langle n_{\mathbf{k}} \rangle + \langle n_{-\mathbf{k}} \rangle - 1] \tag{11.97}$$

where we have used the fact that expectation values of operators such as $\xi_{\mathbf{k}}\xi_{-\mathbf{k}}$ vanish. Since (11.96) is the Hamiltonian of a system of noninteracting "particles," we have

$$\langle n_{\mathbf{k}} \rangle = \frac{1}{\exp\{\beta E(\mathbf{k})\} + 1}$$

and

$$\langle b_{\mathbf{k}} \rangle = \frac{\Delta_{\mathbf{k}}}{2E(\mathbf{k})} \tanh \frac{\beta E(\mathbf{k})}{2} \tag{11.98}$$

and finally,

$$\Delta_{\mathbf{k}}(T) = -\frac{1}{V_0} \sum_{\mathbf{k}'} \frac{V_{\mathbf{k},\mathbf{k}'} \Delta_{\mathbf{k}'}}{2E(\mathbf{k}')} \tanh \frac{\beta E(\mathbf{k}')}{2} . \tag{11.99}$$

If we now assume, as in Section 11.3.2, that $V_{\mathbf{kk}'} = -V$, independent of $\mathbf{k}$ in the shell of width $\hbar\omega_D$ centered on the Fermi energy, then $\Delta_{\mathbf{k}}$ is also independent of $\mathbf{k}$ and given by

$$1 = \frac{V}{2V_0}\sum_{\mathbf{k}} \frac{\tanh\left[\frac{1}{2}\beta\sqrt{\epsilon(\mathbf{k})^2 + \Delta^2}\right]}{\sqrt{\epsilon(\mathbf{k})^2 + \Delta^2}} \tag{11.100}$$

which is the finite-temperature generalization of formula (11.81). From (11.100) we may obtain the critical temperature of the system. As $T \to T_c$, $\Delta \to 0$ and we have

$$1 = \frac{V\rho(\epsilon_F)}{2}\int_{-\hbar\omega_D/2}^{\hbar\omega_D/2} d\epsilon \frac{\tanh\left(\beta_c\epsilon/2\right)}{\epsilon} \ . \tag{11.101}$$

The integral on the right-hand side can be evaluated exactly in the weak-coupling limit [93], with the result

$$k_B T_c = 0.567\hbar\omega_D \exp\left\{-\frac{1}{\rho(\epsilon_F)V}\right\} \ . \tag{11.102}$$

Since $\Delta(T = 0) = \hbar\omega_D \exp\{-1/[\rho(\epsilon_F)V]\}$, we have

$$\frac{\Delta(T = 0)}{k_B T_c} = 1.764 \tag{11.103}$$

which is an important quantitative prediction of the theory. The parameters ω_D,V that appear in the BCS theory are not well known; indeed, the use of the Debye frequency to limit the region in energy in which pairs form is certainly not precise. Nevertheless, experimental data, for a number of weak-coupling superconductors, (mostly elemental), is consistent with (11.103), but significant deviations are sometimes found. These discrepancies are accounted for in the "strong-coupling" theories of superconductivity, that are also based on BCS ideas but incorporate a more realistic form of the electron–phonon interaction and of the Fermi surface. For an introduction to these theories, see Mahan [185].

The theory, as we have presented it above, is quite obviously a form of mean field theory, and it should not be surprising that the order parameter of the system $[\Delta(T)]$ has a typical mean field temperature dependence near the critical point:

$$\frac{\Delta(T)}{\Delta(0)} \approx 1.74\left(1 - \frac{T}{T_c}\right)^{1/2} \ . \tag{11.104}$$

As well, the specific heat has a discontinuity at the critical point.

11.3.4 Landau–Ginzburg theory of superconductivity

The usefulness of the Landau–Ginzburg approach to superconductivity lies in the fact that it provides a direct route to the handling of problems associated with fluctuations and inhomogeneities. The approach also permits a description of the electrodynamics of superconductors and their response to electromagnetic fields. This is, of course, what makes these materials interesting from a practical point of view.

The BCS theory, which we have outlined above, is a mean field theory that exhibits a second-order phase transition. As such, the theory can be described in the language of Chapter 3. In particular, the gap equation (11.100) is analogous to the self-consistent equations encountered in the Weiss molecular field treatment of magnets or in the mean field theory of the Maier–Saupe model. As shown explicitly in subsection 11.3.2, the symmetry of the ordered phase is that of the XY model (i.e., the ordered phase is characterized both by a phase and an amplitude). A description in terms of a one-component order parameter such as the energy gap Δ (or equivalently, by a condensate mass density ρ_s in analogy with the two-fluid model of Bose condensation) therefore cannot account for all of the physics.

The free-energy density must not depend on the phase of the order parameter and we expect that near T_c the free-energy density, $g = G/V_0$, of a bulk superconductor is of the form

$$g = g_n + a(T)|\Psi|^2 + \frac{1}{2}b(T)|\Psi|^4 + \cdots \tag{11.105}$$

where $a(T) = \alpha(T - T_c)$ and where α and $b(T)$ are only weakly temperature dependent. Minimizing this free-energy density, we find that g is a minimum for $T < T_c$ for

$$|\Psi|^2 = \frac{\alpha(T_c - T)}{b} \,. \tag{11.106}$$

The difference in free-energy density between the superconducting and normal states is then

$$g_S - g_N = -\frac{\alpha^2}{2b}(T_c - T)^2 \qquad T < T_c \tag{11.107}$$

and the specific heat has a discontinuity at the transition given by

$$C_S - C_N = \frac{\alpha^2 T_c}{b} \,. \tag{11.108}$$

The normalization of the order parameter is, in principle, arbitrary. In accordance with our intuitive picture of the transition as a condensation of

Cooper pairs, we write, in analogy with (11.31) and (11.32),

$$\Psi = \sqrt{\frac{n_s}{2}} e^{i\gamma} \tag{11.109}$$

where n_s is the superconducting electron density. This density n_s can be determined by subtracting, from the conduction electron density, the inertia of the gas of quasiparticle excitations in much the same way as outlined in Problem 11.3 for a superfluid. The parameters α and b can then be determined by comparing (11.108) and the transition temperature T_c with the predictions of the BCS theory. The results are $\alpha = b\pi^2 T_c/[7\zeta(3)\epsilon_F]$, where ζ is the Riemann zeta function, and $b = \alpha T_c/n$, where $n = k_F^3/(3\pi^2)$ is the conduction electron density.

We are now in a position to discuss inhomogeneities and to derive an expression for the *coherence length.* Recalling our intuitive picture of a condensate of Cooper pairs, each with charge $-2e$ and mass $2m$ we write, in analogy with (11.33), for the superconducting current associated with a spatially varying phase:

$$\mathbf{j}_s = \frac{ie\hbar}{2m}\left(\Psi^*\nabla\Psi - \Psi\nabla\Psi^*\right) . \tag{11.110}$$

We next identify the gradient term in the Landau–Ginzburg free-energy density (3.53) with the kinetic energy density of the moving charges and thus obtain

$$G = G_N + \int d^3r \left(\frac{\hbar^2}{4m}|\nabla\Psi|^2 + a|\Psi|^2 + \frac{1}{2}b|\Psi|^4 + \cdots\right) . \tag{11.111}$$

Remembering that $|\Psi|^2 = \Psi^*\Psi$ and requiring that the functional derivative of G with respect to Ψ^* be zero, we obtain

$$-\frac{\hbar^2}{4m}\nabla^2\Psi + a\Psi + b|\Psi|^2\Psi = 0 . \tag{11.112}$$

We have encountered equations of the form (11.112) earlier in this text, in Section 3.10 and in the discussion of the liquid vapor interface (Section 5.4). We therefore simply note that (11.112) predicts a coherence length (referred to in other contexts as a correlation length)

$$\xi = \frac{\hbar}{\sqrt{8m|a|}} \tag{11.113}$$

for $T < T_c$. This is the natural length scale for spatial variations of the order parameter.

We conclude our discussion of Landau–Ginzburg theory by noting that there is another length scale which is relevant to the properties of superconductors, namely the London penetration depth. Let us consider a superconductor in the presence of an applied magnetic field $\mathbf{B} = \nabla\times\mathbf{A}$. The expression for the current density now becomes

$$\mathbf{j}_s = \frac{ie\hbar}{2m}\left[\Psi^*\left(\nabla + \frac{2ie}{\hbar c}\mathbf{A}\right)\Psi - \Psi\left(\nabla - \frac{2ie}{\hbar c}\mathbf{A}\right)\Psi^*\right] . \tag{11.114}$$

If the length scale over which $\mathbf{A}$ varies is short compared to the coherence length ξ, the magnitude $|\Psi|$ of the order parameter is approximately constant and the expression (11.114) simplifies to

$$\mathbf{j}_s = -\left(\frac{2e^2}{mc}\mathbf{A} + \frac{e\hbar}{m}\nabla\gamma\right)|\Psi|^2 . \tag{11.115}$$

Taking the curl of both sides of (11.115) then leads to the London equation

$$\nabla\times\mathbf{j}_s = -\frac{2e^2}{mc}|\Psi|^2(\nabla\times\mathbf{A}) . \tag{11.116}$$

If we combine (11.116) with Ampère's law,

$$\nabla\times\mathbf{B} = \frac{4\pi}{c}\mathbf{j}_s \tag{11.117}$$

we find

$$\begin{aligned}\nabla^2\mathbf{B} &= \frac{8\pi|\Psi|^2e^2}{mc^2}\mathbf{B} \\ \nabla^2\mathbf{j}_s &= \frac{8\pi|\Psi|^2e^2}{mc^2}\mathbf{j}_s .\end{aligned} \tag{11.118}$$

Equation (11.118) typically has solutions with exponentially decaying currents and fields (Meissner effect). The appropriate length scale is then the London penetration depth, δ, which appears in (11.118):

$$\delta = \sqrt{\frac{mc^2}{8\pi|\Psi|^2e^2}} = \sqrt{\frac{mc^2b}{8\pi e^2\alpha(T_c - T)}} . \tag{11.119}$$

The analysis above assumes that $\xi \gg \delta$. Materials for which this assumption holds are called *type* I *superconductors*. A type I superconductor expels all currents and external fields H except for a thin layer of thickness δ near the surface. The diamagnetic energy cost per unit volume associated with this expulsion is $H^2c^2/8\pi$ and we see immediately from (11.107) that the critical field, above which superconductivity is suppressed, is $H_c = (4\pi\alpha^2/b)^{1/2} (T_c - T)$.

If the coherence length is not large in comparison with the penetration depth, the problem becomes more complicated. Equation (11.112) must then be replaced by

$$-\frac{\hbar^2}{4m}\left(\nabla + \frac{2ie}{\hbar c}\mathbf{A}\right)^2 \Psi + a\Psi + b|\Psi|^2\Psi = 0\,. \tag{11.120}$$

Equation (11.120) together with Ampère's law (11.117) constitute the full Landau–Ginzburg equations. The solution of these equations is beyond the scope of this book and we refer to Landau and Lifshitz [165] for further reading. We only mention that if $\kappa = \delta/\xi > 2^{1/2}$, the Landau–Ginzburg equations can have vortex solutions in analogy with superfluids. A superconductor that allows penetration of magnetic fields by forming vortex lines is called a *type* II *superconductor.*

In Chapter 3 we showed, by means of the Ginzburg criterion, that mean field theories are self-consistent only for spatial dimensionalities $d \geq 4$. Nevertheless, the BCS theory describes real three-dimensional superconductors with very high accuracy, even close to T_c. The reason for this can also be understood in terms of the Ginzburg criterion. By using it to estimate the reduced temperature $(T_c - T)/T_c$ at which the BCS theory should break down in three dimensions, we find (Kadanoff *et al.* [145]) that this will occur at $(T_c - T)/T_c \approx 10^{-14}$, a temperature deviation well beyond current experimental resolution.

We have, in this section, barely scratched the surface of the theory of superconductivity but have demonstrated another phenomenon associated with Bose condensation. Another fermion system in which such condensation occurs is liquid ^{3}He. An introductory discussion of this system can be found in Mahan [185].

11.4 Problems

11.1. *Critical Properties of the Ideal Bose Gas.*

(a) Show that the entropy of an ideal Bose gas above T_c can be written

$$S = \frac{5}{2}aT^{3/2}g_{5/2}(z) - aT^{3/2}g_{3/2}z\ln z = \frac{5}{2}aT^{3/2}g_{5/2}(z) - k_BN\ln z$$

where z is the fugacity and

$$a = V\left(\frac{m}{2\pi\hbar^2}\right)^{3/2} k_B^{5/2} \ .$$

(b) The specific heat $C_{V,N}$ at fixed V and N is

$$C_{V,N} = T\left(\frac{\partial S}{\partial T}\right)_{V,z} + T\left(\frac{\partial S}{\partial z}\right)_{V,T}\left(\frac{\partial z}{\partial T}\right)_{V,N} \ .$$

For $T < T_C$, $(\partial z/\partial T)_{V,N} = 0$ and $C_{V,N} = C_{V,z}$. Show that as $T \to T_c^+$, the second term in the expression for $C_{V,N}$ vanishes and that the specific heat is therefore continuous at the transition.

(c) Show further that $dC_{V,N}/dT$ is discontinuous at T_c.
Hint: It may be useful to derive the limiting form of the function $g_\nu(z)$ for $z \to 1$ and for $\nu \le 1$ as it diverges at $z = 1$ for these values of the index ν. One way of approaching this is to assume that $g_\nu(z) \sim (1-z)^{-\gamma(\nu)}$ and to compare the Taylor expansion of this function with the exact form (11.6) to find the exponent $\gamma(\nu)$ (see also Section 6.2).

(d) Analyze, by the methods of parts (b) and (c), the behavior of the pressure near T_c of an ideal Bose gas at fixed density.

11.2. *Bose Condensation in an Atom Trap.* Consider an atom trap that confines a set of noninteracting bosons to a two-dimensional surface in the trap. The area of this surface is not fixed. Rather, each boson experiences a single-particle potential of the form

$$V(x,y) = \frac{1}{2}m\omega^2(x^2+y^2)\,,$$

i.e., an isotropic two-dimensional harmonic oscillator potential. The energy levels of a particle are therefore

$$\epsilon(j_x, j_y) = \hbar\omega\,(j_x + j_y)\ ,$$

where we have neglected the zero-point energy. If we define $l \equiv j_x + j_y$ then the degeneracy of that level is $g(l) = l + 1 \approx l$, at least for l not too small. For $\hbar\omega/k_BT \ll 1$ the sums over l may be replaced by integrals. With these approximations, show that N identical noninteracting bosons confined in such a trap condense at

$$k_BT_c = \hbar\omega\left[\frac{N}{\zeta(2)}\right]^{\frac{1}{2}} \ .$$

11.3. λ *Point of Helium.*

The theory of superfluidity presented in Section 11.2 was based on an analogy with Bose condensation. This theory does not adequately describe the nature of the transition to the superfluid state. Nevertheless, it is possible to estimate the critical temperature to remarkable accuracy by the following crude argument (Landau and Lifshitz, Part 2 [165]).

Imagine a quasiparticle gas with a Bose distribution function $n(\epsilon(\mathbf{p})-\mathbf{p}\cdot\mathbf{v})$ (i.e., a gas that is moving with velocity $\mathbf{v}$ with respect to the superfluid). The momentum of this gas is, in the low-velocity limit, given by

$$\mathbf{P} = -\sum_{\mathbf{p}} \mathbf{p} n(\epsilon - \mathbf{p}\cdot\mathbf{v}) \approx \sum_{\mathbf{p}} \mathbf{p}(\mathbf{p}\cdot\mathbf{v}) \frac{dn(\epsilon)}{d\epsilon} = \frac{1}{3}\sum_{\mathbf{p}} p^2 \left[-\frac{dn(\epsilon)}{d\epsilon}\right] \mathbf{v} \,.$$

The last step follows if we split $\mathbf{p}$ into components parallel and perpendicular to $\mathbf{v}$ and note that the perpendicular component averages to zero, while the spherical average of $\cos^2\theta = \frac{1}{3}$. The ratio between the momentum and velocity is the inertia or mass of the quasiparticle gas. The inertia of this gas increases with temperature and when the inertia of the gas is equal to the mass of the helium, there is no superfluid component left. This occurs when

$$M_{He} = \frac{1}{3}\sum_{\mathbf{p}} p^2 \left[-\frac{dn(\epsilon)}{d\epsilon}\right] \,.$$

(a) Show that for a phonon gas with velocity of sound c in volume V

$$M_{ph} = \frac{2\pi^2 V T^4 k_B^4}{45\hbar^3 c^5} \,.$$

(b) Near the transition temperature, the thermodynamic properties are dominated not by the phonons but rather by the rotons with energy

$$\epsilon(p) = \Delta + \frac{(p-p_0)^2}{2m^*} \,.$$

Assume that the rotons have a Boltzmann distribution and that m^* is small enough that we may take $p = p_0$ everywhere in thermal averages except in the Boltzmann factor. Show that

$$M_r = \frac{2(m^*)^{1/2} p_0^4 \exp\{-\beta\Delta\}}{3(2\pi)^{3/2}(k_B T)^{1/2}\hbar^3} V \,.$$

(c) Estimate the critical temperature assuming that the density of He is 0.145×10^{-3} kg/m^3, $m^* = 0.16 m_{He}$, $c = 2.4 \times 10^2$ m/s, $p_0/\hbar =$ .19 nm^{-1}, and $\Delta/k_B = 8.7$ K.

(d) Can the argument above be used to calculate the condensation temperature of the ideal Bose gas?

Chapter 12

Linear Response Theory

In this chapter we consider the response of a system to an external perturbation. At this point we are concerned only with perturbations sufficiently weak that a linear approximation of the effect of the perturbation is adequate. In Section 12.1 we define the dynamic structure factor, introduce the concept of generalized susceptibility, and derive an important result, the *fluctuation–dissipation theorem*, which relates equilibrium fluctuations to dissipation in the linear (ohmic) regime. We next show how the thermodynamic properties of a quantum many-body system with particles interacting through two-body potentials can be derived from the response function. Our discussion here is analogous to our treatment of classical liquids in Section 5.2. We illustrate the formalism in Section 12.2 by means of a number of simple examples. We show that a simple mean field theory yields results that are equivalent to those derived in Chapter 11 by means of the Bogoliubov transformation for the weakly interacting Bose gas at low temperatures. We also discuss the dielectric response of an interacting electron gas. We then turn to a discussion of the electron–phonon interaction in metals. The purpose of this discussion is to derive a result originally due to Fröhlich [105] that electrons can interact by exchanging phonons to produce an effective interaction that can be attractive in certain circumstances. This result was used in Chapter 11 in our treatment of the BCS theory of superconductivity.

In Section 12.3 we return to the development of the formalism by considering *steady-state* situations with a current flowing in response to a field. We derive the Kubo relations between conductivities and the appropriate equilibrium current-current correlation functions. On the basis of an assumption of microscopic reversibility, we next demonstrate the Onsager reciprocity theorem. Finally, in Section 12.4, we briefly discuss the Boltzmann equation approach to transport.

12.1 Exact Results

In this section we develop the formalism of linear response theory quite generally and exhibit a number of interesting and useful properties. Within the framework of the linear approximation, the results of this section are exact.

12.1.1 Generalized susceptibility and the structure factor

We consider a system of particles subject to an external perturbation that may be time dependent. The Hamiltonian is

$$H_{tot} = H + H_{ext} = H_0 + H_1 + H_{ext} \tag{12.1}$$

where H_0 contains the kinetic energy and the single-particle potential and H_1 the interparticle potential energy. H_0 will usually represent an ideal Bose or Fermi gas. The pair interaction H_1 can be treated only approximately, and we will generally assume that a theory of the mean-field type will be adequate. We assume that the external perturbation couples linearly to the density:

$$H_{ext} = \int d^3x \; n(\mathbf{x})\phi_{ext}(\mathbf{x}, t) \tag{12.2}$$

where ϕ_{ext} is a scalar function of position and time.

The system responds to the perturbation (12.2) through an induced change in the particle density

$$\langle \delta n(\mathbf{x}, t)\rangle = \langle n(\mathbf{x}, t)\rangle_{H_{tot}} - \langle n(\mathbf{x}, t)\rangle_H \; . \tag{12.3}$$

The expectation value $\langle n(\mathbf{x}, t)\rangle_{H_{tot}}$ in (12.3) is the density at $\mathbf{x}$, at time t, when the system has evolved from an equilibrium state of H (at $t = -\infty$) with the external perturbation switched on for all finite times. This expectation value thus involves a trace over an equilibrium density matrix (at $t = -\infty$) and a modification of the states of the system due to H_{ext}. We assume that H_{ext} is small enough that we can treat it in first-order perturbation theory. This assumption is frequently harmless, for example, if H_{ext} represents an infinitesimal "test field" introduced formally to probe the equilibrium dynamic or static response of the system.

We suppose that we have determined the eigenstates of H in the Heisenberg picture and denote them by $|\psi_H\rangle$. With the perturbation switched on at $t = -\infty$, the time-dependent Schrödinger equation for an arbitrary state $|\psi\rangle$ is

$$i\hbar\frac{\partial}{\partial t}|\psi\rangle = H_{tot}|\psi(t)\rangle \equiv H_{tot}\exp\left\{-\frac{i}{\hbar}Ht\right\}|\psi_H(t)\rangle \,. \tag{12.4}$$

The time dependence in $|\psi_H(t)\rangle$ is due to the external perturbation, and the state $|\psi_H(t)\rangle$ therefore obeys the differential equation

$$i\hbar\frac{\partial}{\partial t}|\psi_H(t)\rangle = H_{ext}|\psi_H(t)\rangle \tag{12.5}$$

where

$$\begin{aligned} H_{ext}(t) &= e^{iHt/\hbar}H_{ext}e^{-iHt/\hbar} \\ &= e^{iKt/\hbar}H_{ext}e^{-iKt/\hbar} \end{aligned} \tag{12.6}$$

and K is the grand canonical Hamiltonian

$$K = H - \mu N \,. \tag{12.7}$$

The last equality in (12.6) follows if the external perturbation conserves particle number (*i.e.*, N commutes with H_{ext}). Solving (12.5) to lowest order in H_{ext} we find

$$|\psi_H(t)\rangle = |\psi_H\rangle - \frac{i}{\hbar}\int_{-\infty}^{t} dt' H_{ext}(t')|\psi_H\rangle \,. \tag{12.8}$$

Thus

$$\begin{aligned} &\langle\psi(t)|n(\mathbf{x})|\psi(t)\rangle \\ &= \langle\psi_H|e^{iHt/\hbar}n(\mathbf{x})e^{-iHt/\hbar}|\psi_H\rangle - \frac{i}{\hbar}\int_{-\infty}^{t} dt'\langle\psi_H| \\ &\quad\times\left(e^{iHt/\hbar}n(\mathbf{x})e^{-iHt/\hbar}H_{ext}(t') - H_{ext}(t')e^{iHt/\hbar}n(\mathbf{x})e^{-iHt/\hbar}|\psi_H\rangle\right) \\ &= \langle\psi_H|n(\mathbf{x},t)|\psi_H\rangle \\ &\quad - \frac{i}{\hbar}\int_{-\infty}^{t}\int d^3x'\phi_{ext}(\mathbf{x}',t')\langle\psi_H|[n(\mathbf{x},t),n(\mathbf{x}',t')]|\psi_H\rangle \end{aligned} \tag{12.9}$$

where we use the notation that operators such as $n(\mathbf{x},t)$ with an explicit time dependence are Heisenberg operators:

$$n(\mathbf{x},t) = e^{iKt/\hbar}n(\mathbf{x})e^{-iKt/\hbar}$$

while $n(\mathbf{x})$ is in the Schrödinger picture and $[A, B] = AB - BA$.

As mentioned above, the system was in equilibrium at $t = -\infty$. Hence we average equation (12.9) over the grand canonical probability density to obtain finally,

$$\langle \delta n(\mathbf{x}, t)\rangle = \frac{i}{\hbar}\int_{-\infty}^{t} dt' \int d^3x' \phi_{ext}(\mathbf{x}', t')\langle[n(\mathbf{x}', t'), n(\mathbf{x}, t)]\rangle_H \ . \tag{12.10}$$

It is convenient to express the response in terms of the retarded density-density correlation function, D^R, defined through

$$D^R(\mathbf{x}, t; \mathbf{x}', t') \equiv -\frac{i}{\hbar}\theta(t - t')\langle[n(\mathbf{x}, t), n(\mathbf{x}', t')]\rangle_H \tag{12.11}$$

where $\theta(t)$ is the Heaviside step function. This correlation function (also called a propagator or Green's function) is independent of the perturbing potential and thus can be used to describe the response of the system to any external perturbation that couples linearly to the density. Substituting in (12.10), we obtain

$$\langle \delta n(\mathbf{x}, t)\rangle = \int d^3x' \int_{-\infty}^{\infty} dt' D^R(\mathbf{x}, t; \mathbf{x}', t')\phi_{ext}(\mathbf{x}', t') \ . \tag{12.12}$$

If the unperturbed system is homogeneous in space and time, $D^R(\mathbf{x}, t; \mathbf{x}', t')$ can only be a function of $\mathbf{x} - \mathbf{x}'$ and $t - t'$ and the integral in (12.12) is simply a four-dimensional convolution. Thus if we define Fourier transforms according to

$$\langle \delta n(\mathbf{x}, t)\rangle = \frac{1}{2\pi V}\sum_{\mathbf{q}}\int_{-\infty}^{\infty} d\omega \, \delta\rho(\mathbf{q}, \omega)e^{i\mathbf{q}\cdot\mathbf{x}}e^{-i\omega t} \tag{12.13}$$

$$D^R(\mathbf{x}, t; \mathbf{x}', t') = \frac{1}{2\pi V}\sum_{\mathbf{q}}\int_{-\infty}^{\infty} d\omega \, \chi^R(\mathbf{q}, \omega)e^{i\mathbf{q}\cdot(\mathbf{x}-\mathbf{x}')}e^{-i\omega(t-t')} \tag{12.14}$$

and

$$\phi_{ext}(\mathbf{x}', t') = \frac{1}{2\pi V}\sum_{\mathbf{q}}\int_{-\infty}^{\infty} d\omega \, \phi_{ext}(\mathbf{q}, \omega)e^{i\mathbf{q}\cdot\mathbf{x}'}e^{-i\omega t'} \tag{12.15}$$

we have

$$\delta\rho(\mathbf{q}, \omega) = \chi^R(\mathbf{q}, \omega)\phi_{ext}(\mathbf{q}, \omega) \ . \tag{12.16}$$

The function χ^R is called the generalized susceptibility.

Before discussing the susceptibility and its relation to the dynamic structure factor, we pause to express the Fourier transformed quantities in second quantized form. The density operator $n(\mathbf{x})$ may be written in the form

$$n(\mathbf{x}) = \psi^\dagger(\mathbf{x})\psi(\mathbf{x}) = \frac{1}{V}\sum_{\mathbf{k},\mathbf{q},\sigma} c^\dagger_{\mathbf{k}-\mathbf{q},\sigma} c_{\mathbf{k},\sigma} e^{i\mathbf{q}\cdot\mathbf{x}} = \frac{1}{V}\sum_{\mathbf{q}} \rho(\mathbf{q})e^{i\mathbf{q}\cdot\mathbf{x}} \tag{12.17}$$

where

$$c^\dagger_{\mathbf{k},\sigma}, c_{\mathbf{k},\sigma} = \begin{cases} a^\dagger_{\mathbf{k},\sigma}, a_{\mathbf{k},\sigma} & \text{fermions} \\ b^\dagger_{\mathbf{k},\sigma}, b_{\mathbf{k},\sigma} & \text{bosons} \end{cases}$$

are the creation and annihilation operators introduced in the Appendix. In the case of spinless bosons, the sum over σ in (12.17) should, of course, be omitted. The time-dependent density operator can then be written in the form

$$\begin{aligned} n(\mathbf{x},t) &= \frac{1}{V}\sum_{\mathbf{q}} e^{iKt/\hbar}\rho(\mathbf{q})e^{-iKt/\hbar}e^{i\mathbf{q}\cdot\mathbf{x}} \\ &= \frac{1}{V}\sum_{\mathbf{q}} \rho(\mathbf{q},t)e^{i\mathbf{q}\cdot\mathbf{x}} \\ &= \frac{1}{2\pi V}\sum_{\mathbf{q}} \int_{-\infty}^{\infty} d\omega \rho(\mathbf{q},\omega)e^{i\mathbf{q}\cdot\mathbf{x}}e^{-i\omega t} \end{aligned} \tag{12.18}$$

and

$$\rho(\mathbf{q},\omega) = \int_{-\infty}^{\infty} dt\; \rho(\mathbf{q},t)e^{i\omega t} = \int d^3x \int_{-\infty}^{\infty} dt\; n(\mathbf{x},t)e^{-i\mathbf{q}\cdot\mathbf{x}}e^{i\omega t}\ . \tag{12.19}$$

The expression (12.16) for the frequency-dependent response is consistent with the conventional definition of the magnetic susceptibility as the ratio of the magnetization to the magnetic field, H (rather than the magnetic induction B). The magnetic field is usually interpreted as being due to an external source. On the other hand, the conventional definition of the electric susceptibility is the ratio of the polarization to the electric field E, not to the source field D.

We now define the *dynamic structure factor*, $S(\mathbf{q},\omega)$:

$$S(\mathbf{q},\omega) = \int_{-\infty}^{\infty} dt e^{i\omega t}\langle\rho(\mathbf{q},t)\rho(-\mathbf{q},0)\rangle\ . \tag{12.20}$$

The generalized susceptibility $\chi^R(\mathbf{q},\omega)$ (12.14) can be expressed in terms of this function as we now demonstrate. We first note that the expectation value

$$\langle n(\mathbf{x},t)n(\mathbf{x}',t')\rangle = \frac{1}{V^2}\sum_{\mathbf{k}_1,\mathbf{k}_2}\langle\rho(\mathbf{k}_1,t)\rho(\mathbf{k}_2,t')\rangle e^{i\mathbf{k}_1\cdot\mathbf{x}}e^{i\mathbf{k}_2\cdot\mathbf{x}'}$$

is a function only of $|\mathbf{x}-\mathbf{x}'|$. Letting $\mathbf{x}=\mathbf{x}'+\mathbf{y}$ and integrating over $\mathbf{x}'$, we obtain

$$\begin{aligned}\frac{1}{V}\int d^3x'\langle n(\mathbf{x}'+\mathbf{y},t)n(\mathbf{x}',t')\rangle &= \frac{1}{V^2}\sum_{\mathbf{k}}\langle\rho(\mathbf{k},t)\rho(-\mathbf{k},t')\rangle e^{i\mathbf{k}\cdot\mathbf{y}}\\ &= \frac{1}{V^2}\sum_{\mathbf{k}}\langle\rho(\mathbf{k},t-t')\rho(-\mathbf{k},0)\rangle e^{i\mathbf{k}\cdot\mathbf{y}}\ .\end{aligned} \tag{12.21}$$

Hence

$$\begin{aligned}\chi^R(\mathbf{q},\omega) &= -\frac{i}{\hbar V^2}\sum_{\mathbf{k}}\int d^3y\int_{-\infty}^{\infty}d\tau\ e^{i(\mathbf{k}-\mathbf{q})\cdot\mathbf{y}+i\omega\tau}\theta(\tau)\langle\rho(\mathbf{k},\tau)\rho(-\mathbf{k},0)\\ &\qquad -\rho(-\mathbf{k},0)\rho(\mathbf{k},\tau)\rangle\\ &= -\frac{i}{\hbar V}\int_{-\infty}^{\infty}d\tau e^{i\omega\tau}\theta(\tau)\langle\rho(\mathbf{q},\tau)\rho(-\mathbf{q},0)\\ &\qquad -\rho(-\mathbf{q},0)\rho(\mathbf{q},\tau)\rangle\ .\end{aligned} \tag{12.22}$$

Using the integral representation for the Heaviside function,

$$\theta(\tau)=\frac{1}{2\pi i}\int_{-\infty}^{\infty}d\omega'\frac{e^{i\omega'\tau}}{\omega'-i\eta} \tag{12.23}$$

where η is a positive infinitesimal, we find

$$\begin{aligned}\chi^R(\mathbf{q},\omega) &= \frac{-1}{2\pi\hbar V}\int_{-\infty}^{\infty}d\tau\int_{-\infty}^{\infty}d\omega' e^{i(\omega'+\omega)\tau}\\ &\quad\times\frac{\langle\rho(\mathbf{q},\tau)\rho(-\mathbf{q},0)-\rho(-\mathbf{q},0)\rho(\mathbf{q},\tau)\rangle}{\omega'-i\eta}\ .\end{aligned} \tag{12.24}$$

Substituting $\omega'=-\omega+\omega''$ and noting, from the definition of the expectation values, that for any pair of operators A, B,

$$\langle A(0)B(t)\rangle=\langle A(-t)B(0)\rangle$$

and, moreover, that

$$\langle\rho(-\mathbf{k},\tau)\rho(\mathbf{k},0)\rangle=\langle\rho(\mathbf{k},\tau)\rho(-\mathbf{k},0)\rangle$$

we have

$$\chi^R(\mathbf{q},\omega)=\frac{1}{\hbar V}\int_{-\infty}^{\infty}\frac{d\omega'}{2\pi}\frac{S(\mathbf{q},\omega')-S(\mathbf{q},-\omega')}{\omega-\omega'+i\eta}\ . \tag{12.25}$$

We now examine the dynamic structure factor in more detail. Let $|n\rangle$ be a complete set of eigenstates of the grand Hamiltonian K:

$$K|n\rangle = K_n|n\rangle \ . \tag{12.26}$$

Using these basis states we find, from the definition (12.20),

$$\begin{aligned}
S(\mathbf{q},\omega) &= \int_{-\infty}^{\infty} dt\, e^{i\omega t} \sum_n \langle n|e^{-\beta K} e^{iKt/\hbar} \rho(\mathbf{q},0) e^{-iKt/\hbar} \rho(-\mathbf{q},0)|n\rangle / Z_G \\
&= \sum_{n,m} \frac{e^{-\beta K_n}}{Z_G} \int_{-\infty}^{\infty} dt e^{i\{\omega+(K_n-K_m)/\hbar\}t} \langle n|\rho(\mathbf{q},0)|m\rangle\langle m|\rho(-\mathbf{q},0)|n\rangle \\
&= 2\pi\hbar \sum_{n,m} \frac{e^{-\beta K_n}}{Z_G} |\langle n|\rho(\mathbf{q},0)|m\rangle|^2 \delta(\hbar\omega + K_n - K_m) \ .
\end{aligned} \tag{12.27}$$

From this equation we see that $S(\mathbf{q},\omega)$ is real and non-negative:

$$S(\mathbf{q},\omega) \geq 0 \tag{12.28}$$

and that

$$S(-\mathbf{q},\omega) = S(\mathbf{q},\omega) = e^{\beta\hbar\omega} S(\mathbf{q},-\omega) \ . \tag{12.29}$$

Equation (12.29) is an expression of the principle of *detailed balance.* We interpret (12.29) in the following way. The quantum-mechanical transition rate between two states is independent of the direction of the transition; there is no distinction between emission and absorption. The dynamic structure factor, on the other hand, is a thermal average of transition rates, as can be seen from (12.27), and hence is the quantum transition rate weighted by the average occupation number of the initial state. This accounts for (12.29).

The function $S(\mathbf{q},\omega)$ represents the frequency spectrum of density fluctuations. On the other hand, the imaginary part of the response function χ^R has a physical interpretation in terms of energy dissipation. Combining (12.25) and (12.29) and using

$$\frac{1}{\omega + i\eta} = P\left(\frac{1}{\omega}\right) - i\pi\delta(\omega) \tag{12.30}$$

where P indicates principal value, we obtain the *fluctuation–dissipation* theorem:

$$\left(1 - e^{-\beta\hbar\omega}\right) S(\mathbf{q},\omega) = -2\hbar V \,\mathrm{Im}\, \chi^R(\mathbf{q},\omega) \ . \tag{12.31}$$

Before proceeding to some examples, we generalize the formalism above by considering the effect of an external field that couples to an arbitrary dynamical variable A rather than specifically to the density. Thus

$$H_{ext} = \phi_{ext} A \ . \tag{12.32}$$

The response of another observable B to this perturbation can be derived, in strict analogy with (12.10), to be

$$\langle \delta B(t) \rangle = \frac{i}{\hbar} \int_{-\infty}^{t} dt' \langle [A(t'), B(t)] \rangle_H \phi_{ext}(t') \tag{12.33}$$

and the appropriate Green's function, or propagator, is given by

$$D^R_{BA}(t,t') = -\frac{i}{\hbar}\theta(t-t')\langle [B(t), A(t')] \rangle_H \tag{12.34}$$

which is the generalization of (12.11). Proceeding as before, we define

$$\begin{aligned} \langle \delta B(t) \rangle &= \int_{-\infty}^{\infty} \frac{d\omega}{2\pi} \langle \delta B(\omega) \rangle e^{-i\omega t} \\ D^R_{BA}(t,t') &= \int_{-\infty}^{\infty} \frac{d\omega}{2\pi} \chi^R_{BA}(\omega) e^{-i\omega(t-t')} \\ \phi_{ext}(t') &= \int_{-\infty}^{\infty} \frac{d\omega}{2\pi} \phi_{ext}(\omega) e^{-i\omega t'} \end{aligned} \tag{12.35}$$

and obtain

$$\langle \delta B(\omega) \rangle = \chi^R_{BA}(\omega) \phi_{ext}(\omega) \ . \tag{12.36}$$

We may, as above, express the response function in terms of an equilibrium correlation function. We let

$$\begin{aligned} S_{BA}(\omega) &= \int_{-\infty}^{\infty} dt \, \langle B(t) A(0) \rangle_H e^{i\omega t} \\ S_{AB}(\omega) &= \int_{-\infty}^{\infty} dt \, \langle A(t) B(0) \rangle_H e^{i\omega t} \end{aligned} \tag{12.37}$$

and after some manipulations similar to (12.20)–(12.25) obtain the result

$$\chi^R_{BA}(\omega) = \frac{1}{2\pi\hbar} \int_{-\infty}^{\infty} d\omega' \frac{S_{BA}(\omega') - S_{AB}(-\omega')}{\omega - \omega' + i\eta} \ . \tag{12.38}$$

Using arguments similar to those leading to (12.29), we may easily relate $S_{AB}(-\omega)$ to $S_{BA}(\omega)$. The result after some algebra is

$$S_{AB}(-\omega) = e^{-\beta\hbar\omega} S_{BA}(\omega) \ . \tag{12.39}$$

To obtain the fluctuation dissipation theorem (12.31), we need to restrict the formalism somewhat. For the case $B = A^\dagger$ we see that S_{AB} is real and non-negative and, using (12.30), we find that

$$(1 - e^{-\beta\hbar\omega})S_{BA}(\omega) = -2\hbar \,\mathrm{Im}\,\chi^R_{BA}(\omega) \;. \tag{12.40}$$

We shall be using this more general formalism in Section 12.2.3 where we derive the magnon spectrum of the Heisenberg ferromagnet.

12.1.2 Thermodynamic properties

In Section 5.2 we expressed the equation of state, the internal energy, and the compressibility of a system of classical particles, interacting through two-body forces, in terms of the pair correlation function, or equivalently, in terms of the static structure factor. We now wish to derive similar relations for a quantum many-body system. For simplicity we limit ourselves to an isotropic system and assume that the Hamiltonian can be written in the form

$$H = H_0 + H_1 = \sum_{i=1}^{N} \frac{p_i^2}{2m} + \sum_{i<j} v(\mathbf{r}_i - \mathbf{r}_j) \;. \tag{12.41}$$

The equipartition theorem allowed us, in Section 5.2, to consider the two terms $\langle H_0\rangle$ and $\langle H_1\rangle$ separately. For a quantum system this is no longer possible and we must take a slightly more roundabout route. Let us first consider the ground-state energy and rewrite the Hamiltonian (12.41) in the form

$$H_\lambda = H_0 + \lambda H_1 \;. \tag{12.42}$$

For the physical system $\lambda = 1$, but we can imagine intermediate values. In the latter case we assume that the ground state $|\lambda\rangle$ has energy E_λ and is normalized so that $\langle\lambda|\lambda\rangle = 1$. Thus

$$E_\lambda = \langle\lambda|H_\lambda|\lambda\rangle \;.$$

Differentiation with respect to λ yields

$$\frac{\partial E_\lambda}{\partial\lambda} = \left(\frac{\partial}{\partial\lambda}\langle\lambda|\right) H_\lambda|\lambda\rangle + \langle\lambda|\frac{\partial H_\lambda}{\partial\lambda}|\lambda\rangle + \langle\lambda|H_\lambda\frac{\partial}{\partial\lambda}|\lambda\rangle \;.$$

The first and third terms combine to give

$$E_\lambda \frac{\partial}{\partial\lambda}\langle\lambda|\lambda\rangle = 0$$

since the state $|\lambda\rangle$ is normalized. We therefore have

$$\frac{\partial E_\lambda}{\partial \lambda} = \langle H_1 \rangle_\lambda$$

where the expectation value of the two-body potential is evaluated for the ground state at coupling strength λ. Integrating, we finally obtain

$$E = E_0 + \int_0^1 d\lambda \langle H_1 \rangle_\lambda \ . \tag{12.43}$$

This argument can easily be generalized to finite temperatures [see the discussion of (3.23)–(3.27)]. We have, for example, for the Helmholtz free energy,

$$A(N,V,T) = A_0(N,V,T) + \int_0^1 d\lambda \langle H_1 \rangle_{\lambda,c} \tag{12.44}$$

where the subscripts indicate that the expectation value is to be evaluated in the canonical ensemble at temperature T, volume V, for an N-particle system. In (12.44), $A_0(N,V,T)$ is the corresponding free energy of the noninteracting system. The interaction energy is related to the static (or geometric) structure factor of Section 5.2 through

$$\langle H_1 \rangle_\lambda = \frac{N}{2V} \sum_{\mathbf{q}\neq 0} v_{-\mathbf{q}} [S_\lambda(\mathbf{q}) - 1] + \frac{N^2}{2V} v_0 \tag{12.45}$$

where $v_{\mathbf{q}}$ is the Fourier transform of the pair potential. The geometric structure factor is, in turn, related to the dynamic structure factor through

$$S(\mathbf{q}) = \frac{1}{N} \int_{-\infty}^{\infty} \frac{d\omega}{2\pi} S(\mathbf{q},\omega) \ . \tag{12.46}$$

It is now a straightforward matter to express the ground-state energy, or the free energy, in terms of the structure factor, and from the free energy one can obtain all other thermodynamic properties. A fairly serious complication is the fact that one needs to know the structure factor at all intermediate coupling strengths $\lambda < 1$. We will return to this point in the mean field approximation of the next section and discuss a case for which the integration over λ is quite straightforward.

12.1.3 Sum rules and inequalities

We next derive some exact relationships that must be satisfied by the response functions. Some of these "sum rules" are useful in checking the validity of approximations, while others offer valuable insights into basic principles.

Let us first consider the consequences of *causality*. Since the response of a stable system cannot precede the disturbance, we required (12.11) that $D^R(\mathbf{x}, t; \mathbf{x}', t') = 0$ for $t < t'$. The theory of analytic functions of a complex variable tells us that $\chi^R(\mathbf{q}, \omega)$ then is an analytic function of the complex variable ω in the upper halfplane and, in particular, can have no poles there. Poles on the real axis, on the other hand, would correspond to nondissipative resonances and can be ruled out if we require that a finite source field give rise to a finite response. These analytic properties, together with the assumption that the response function falls off sufficiently rapidly as $\omega \to \infty$, are sufficient to show that χ must satisfy Kramers–Kronig relations. We simply carry out the contour integral

$$0 = \frac{1}{2\pi i} \oint_C dz \frac{\chi^R(\mathbf{q}, \omega)}{z - \omega}$$

where the contour C is shown in Figure 12.1. Taking the radius R of the large semicircle to infinity, and that of the small one, ρ, to zero we obtain the Kramers–Kronig relations:

$$\begin{aligned} \mathrm{Re}\, \chi^R(\mathbf{q}, \omega) &= P \int_{-\infty}^{\infty} \frac{d\omega'}{\pi} \frac{\mathrm{Im}\, \chi^R(\mathbf{q}, \omega')}{\omega - \omega'} \\ \mathrm{Im}\, \chi^R(\mathbf{q}, \omega) &= -P \int_{-\infty}^{\infty} \frac{d\omega'}{\pi} \frac{\mathrm{Re}\, \chi^R(\mathbf{q}, \omega')}{\omega - \omega'} \end{aligned} \tag{12.47}$$

where P indicates the principal value of the integrals.

Another important relationship, the f-sum rule, holds for systems in which the interparticle potential is independent of velocity. The density operator (12.17) and its Fourier transform commute with H_1 in (12.41) but not with

$$H_0 = \sum_{\mathbf{k},\sigma} \frac{\hbar^2 k^2}{2m} c^\dagger_{\mathbf{k},\sigma} c_{\mathbf{k},\sigma} \,. \tag{12.48}$$

It is easy to verify that

$$\langle [[H, \rho(\mathbf{q})], \rho(\mathbf{q})] \rangle = \langle [[H_0, \rho(\mathbf{q})], \rho(\mathbf{q})] \rangle = \frac{\hbar^2 q^2}{m} \langle N \rangle \,.$$

Moreover, we also have

$$\langle [[H, \rho(\mathbf{q})], \rho(\mathbf{q})] \rangle = \sum_{n,m} |\langle n|\rho(\mathbf{q})|m\rangle|^2 (E_n - E_m)(e^{-\beta E_n} - e^{-\beta E_m}) \,.$$

Using (12.27), (12.31), and (12.46), we find, after some algebra,

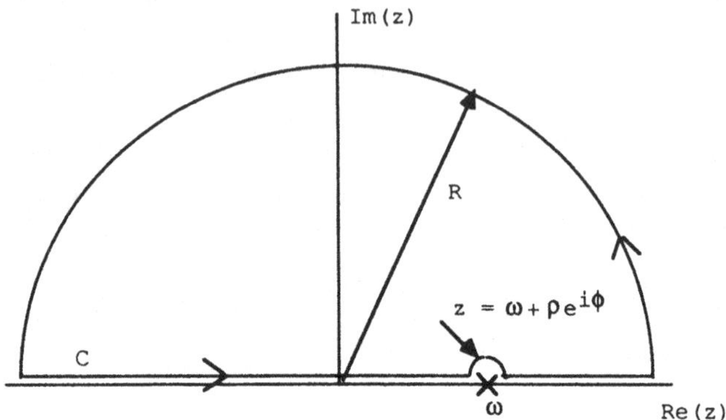

Figure 12.1: The contour used to derive the Kramers-Kronig relations.

$$\int_{-\infty}^{\infty} \frac{d\omega}{\pi} \, \omega \, \mathrm{Im} \, \chi^R(\mathbf{q}, \omega) = -\frac{q^2 \langle N \rangle}{mV} \tag{12.49}$$

which is the f-sum rule.

Another self-consistency condition can be derived from the relationship between the pair distribution function and the response function. Since the pair distribution function has the interpretation of a probability, it must be non-negative. This imposes restrictions on the response function that can be difficult to satisfy in approximate theories. We refer the reader to the texts by Mahan [185] and by Pines and Nozières [243] and references therein for discussion of this question and for further sum rules that can be derived.

12.2 Mean Field Response

To this point our treatment is exact within the linear response approximation. The dynamic structure factor and the response functions can usually not be evaluated exactly since the statistical treatment of a system of interacting particles is in general an unsolved problem. However, the approach of Section 12.1 forms a useful starting point for some of the most successful approximation schemes for many-particle systems. The most common approach is that of mean field theory or generalizations thereof. In its most straightforward and simple form, mean field theory is based on the assumption that a system responds as a system of free particles to an effective potential that is determined self-consistently in terms of the externally applied potential. This method has been applied in many different situations and, consequently, has many different names, such as the random phase approximation, time-dependent Hartree approximation, and Lindhard and self-consistent field approximation. The Vlasov equation of plasma physics is another example of an approximation made in the same spirit. We illustrate this approach through a few simple examples.

12.2.1 Dielectric function of the electron gas

Assume an externally controlled electrostatic potential energy $\phi_{ext}(\mathbf{q},\omega)$ (12.15). The response of the system is formally given by

$$\langle\delta\rho(\mathbf{q},\omega)\rangle = \chi^R(\mathbf{q},\omega)\phi_{ext}(\mathbf{q},\omega) \tag{12.50}$$

where χ^R, in (12.50), is the exact response function. Since a density fluctuation $\langle\delta\rho(\mathbf{q},\omega)\rangle$ produces a Coulomb potential given by

$$\begin{aligned}\phi_{ind}(\mathbf{q},\omega) &= \int d^3r\int d^3r'\frac{e^2}{4\pi\epsilon_0|\mathbf{r}-\mathbf{r}'|}\langle\delta n(\mathbf{r}',\omega)\rangle e^{i\mathbf{q}\cdot\mathbf{r}}\\ &= \frac{e^2}{\epsilon_0 q^2}\langle\delta\rho(\mathbf{q},\omega)\rangle\end{aligned} \tag{12.51}$$

we have an effective potential

$$\phi_{eff}(\mathbf{q},\omega) = \phi_{ext}(\mathbf{q},\omega) + \phi_{ind}(\mathbf{q},\omega)\ . \tag{12.52}$$

In (12.51) we have assumed that each particle has charge e. We define the relative dielectric function $\epsilon(\mathbf{q},\omega)$ through

$$\phi_{eff}(\mathbf{q},\omega) = \frac{\phi_{ext}(\mathbf{q},\omega)}{\epsilon(\mathbf{q},\omega)} \tag{12.53}$$

and obtain

$$\frac{1}{\epsilon(\mathbf{q},\omega)} = 1 + \frac{e^2}{\epsilon_0 q^2}\chi^R(\mathbf{q},\omega) \tag{12.54}$$

which relates the dielectric constant to the exact response function. We note, in passing, that in our formulation the response function is $\epsilon^{-1}(\mathbf{q},\omega)-1$, not $\epsilon(\mathbf{q},\omega)$. For this reason $\epsilon^{-1}(\mathbf{q},\omega)-1$ must satisfy the Kramers–Kronig dispersion relations (12.47). We can imagine exposing our polarizable material to a fixed potential rather than to an external test charge (Figure 12.2). In this case the polarization response is proportional to $\epsilon(\mathbf{q},\omega)-1$ and, as a consequence, the dielectric function itself must satisfy Kramers–Kronig relations.

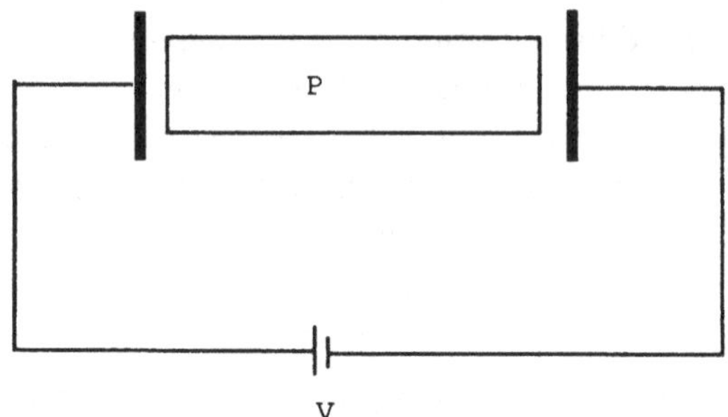

Figure 12.2: Polarization of a sample by a fixed external potential.

However, we can only attach the capacitor plates externally with a macroscopic separation. Causality therefore imposes the Kramers–Kronig relations for the polarization only in the long-wavelength limit. For a more detailed discussion of this point, see Kirzhnits [155] and Dolgov *et al.* [77].

We can easily construct a mean field approximation for $\chi^R(\mathbf{q},\omega)$ by writing

$$\langle\delta\rho(\mathbf{q},\omega)\rangle = \chi_0^R(\mathbf{q},\omega)\phi_{eff}(\mathbf{q},\omega) \tag{12.55}$$

where χ_0 is the response function of a noninteracting system. Using (12.52) and (12.53), we obtain

$$\langle\delta\rho(\mathbf{q},\omega)\rangle = \chi_0^R(\mathbf{q},\omega)\phi_{ext}(\mathbf{q},\omega) + \chi_0^R(\mathbf{q},\omega)\frac{e^2}{\epsilon_0 q^2}\langle\delta\rho(\mathbf{q},\omega)\rangle \tag{12.56}$$

or

$$\langle\delta\rho(\mathbf{q},\omega)\rangle = \frac{\chi_0^R(\mathbf{q},\omega)}{1-e^2\chi_0^R(\mathbf{q},\omega)/\epsilon_0 q^2}\phi_{ext}(\mathbf{q},\omega) \tag{12.57}$$

and the mean field theory response function is given by

$$\chi_{MF}(\mathbf{q},\omega) = \frac{\chi_0^R(\mathbf{q},\omega)}{1-e^2\chi_0^R(\mathbf{q},\omega)/\epsilon_0 q^2} \ . \tag{12.58}$$

Similarly, the mean field version of the dielectric constant is given by

$$\epsilon_{MF}(\mathbf{q},\omega) = 1 - \frac{e^2}{\epsilon_0 q^2}\chi_0^R(\mathbf{q},\omega) \ . \tag{12.59}$$

The mean field approach is clearly not exact. The local potential seen by a particle is not equal to the average potential $\phi_{eff}(\mathbf{q},\omega)$ but rather is modified by exchange and correlation effects. It is customary to write

$$\langle\delta\rho(\mathbf{q},\omega)\rangle = \chi_0^R(\mathbf{q},\omega)\psi_{eff}(\mathbf{q},\omega) \tag{12.60}$$

where the effective potential, as seen by one of the particles, is given by

$$\psi_{eff}(\mathbf{q},\omega) = \phi_{ext}(\mathbf{q},\omega) + \frac{e^2}{\epsilon_0 q^2} f(\mathbf{q},\omega)\langle\delta\rho(\mathbf{q},\omega)\rangle \ . \tag{12.61}$$

The *local field* correction [1] $f(\mathbf{q},\omega)$ (often taken to be independent of frequency) has been estimated in many different ways, the most popular being that of Singwi and co-workers (Vashista and Singwi [311] and references therein; see also Mahan [185]). In this approximation the particles respond to a weak perturbation in the same way as independent particles would respond to an effective potential

$$\psi_{eff}(\mathbf{q},\omega) = \frac{\phi_{ext}(\mathbf{q},\omega)}{1 - e^2/\epsilon_0 q^2 f(\mathbf{q},\omega)\chi_0^R(\mathbf{q},\omega)} \ . \tag{12.62}$$

It is worth noting that the ratio between ϕ_{ext} and ψ_{eff} is not the dielectric function. This function is still given by (12.53). After some straightforward algebra we obtain

$$\epsilon(\mathbf{q},\omega) = 1 - \frac{e^2}{\epsilon_0 q^2}\frac{\chi_0^R(\mathbf{q},\omega)}{1 - (e^2/\epsilon_0 q^2)(1 - f(\mathbf{q},\omega))\chi_0^R(\mathbf{q},\omega)} \ . \tag{12.63}$$

In the absence of local field corrections $f(\mathbf{q},\omega) = 1$ and (12.63) reduces to the simple mean field result (12.59).

12.2.2 Weakly interacting Bose gas

Our discussion, here, of the weakly interacting Bose gas is complementary to the treatment of Section 11.2. The approximations made are in the same spirit, but the formalism is that of linear response theory. We begin by calculating the dynamic structure factor $S_0(\mathbf{q},\omega)$ of the noninteracting (ideal) Bose gas.

At $T = 0$, the only state that contributes to the ensemble average (12.20) is the ground state $|N\rangle$ which has all the particles condensed into the zero-momentum state. In the occupation number representation we have

$$\rho(\mathbf{q}) = \sum_{\mathbf{k}} b^\dagger_{\mathbf{k}-\mathbf{q}} b_{\mathbf{k}} \tag{12.64}$$

[1] The term "local field correction" is also used to describe effects due to the lack of complete translation symmetry of a real metal. Equation (12.60) is then no longer correct and one must include "umklapp" terms, where the wave vector $\mathbf{q}$ on the left-hand side differs from that on the right-hand side of the equation by a reciprocal lattice vector. For a discussion, see Adler [6] and Wiser [328].

where $b^\dagger_{\mathbf{k}}$, $b_{\mathbf{k}}$ are the creation and annihilation operators for particles with momentum $\hbar\mathbf{k}$. Since $b_{\mathbf{k}}|N\rangle = 0$ unless $\mathbf{k} = 0$ we see that the state $\rho(\mathbf{q})|N\rangle$ is one in which a single particle has been excited into a state of energy

$$\epsilon_0(\mathbf{q}) = \frac{\hbar^2 q^2}{2m} \; . \tag{12.65}$$

Using

$$b^\dagger_{-\mathbf{q}} b_0 |N\rangle = \sqrt{N} b^\dagger_{-\mathbf{q}} |N-1\rangle$$

and (12.20), we find for $\mathbf{q} \neq 0$

$$S_0(\mathbf{q},\omega) = 2\pi\hbar N \delta(\hbar\omega - \epsilon_0(\mathbf{q})) \; . \tag{12.66}$$

Similarly,

$$S_0(\mathbf{q},-\omega) = 2\pi\hbar N \delta(\hbar\omega + \epsilon_0(\mathbf{q})) \; . \tag{12.67}$$

Thus, using (12.25), we obtain

$$\begin{aligned} \chi_0^R(\mathbf{q},\omega) &= \frac{N}{V}\int_{-\infty}^{\infty} d\omega' \, \frac{\delta(\hbar\omega' - \epsilon_0(\mathbf{q})) - \delta(\hbar\omega' + \epsilon_0(\mathbf{q}))}{\omega - \omega' + i\eta} \\ &= \frac{N}{V}\frac{q^2/m}{(\omega + i\eta)^2 - q^4/4m^2} \end{aligned} \tag{12.68}$$

or

$$\chi_0^R(\mathbf{q},\omega - i\eta) = \frac{N}{V}\frac{2\epsilon_0(\mathbf{q})}{(\hbar\omega)^2 - \epsilon_0^2(\mathbf{q})} \; . \tag{12.69}$$

Assume next that the particles interact through a pair potential that has a spatial Fourier transform $v_{\mathbf{q}}$. As in Section 12.2.1, we obtain, with $e^2/\epsilon_0 q^2$ replaced by $v_{\mathbf{q}}$,

$$\begin{aligned} \chi_{MF}^R(\mathbf{q},\omega - i\eta) &= \frac{\chi_0^R(\mathbf{q},\omega - i\eta)}{1 - v_{\mathbf{q}}\chi_0^R(\mathbf{q},\omega - i\eta)} \\ &= \frac{N}{V}\frac{\epsilon_0(\mathbf{q})}{\epsilon(\mathbf{q})}\left[\frac{1}{\hbar\omega - \epsilon(\mathbf{q})} - \frac{1}{\hbar\omega + \epsilon(\mathbf{q})}\right] \end{aligned} \tag{12.70}$$

where

$$\epsilon(\mathbf{q}) = \sqrt{\epsilon_0^2(\mathbf{q}) + \frac{2N}{V} v_{\mathbf{q}} \epsilon_0(\mathbf{q})}$$

and, by comparison with (12.66),

$$S(\mathbf{q},\omega) = 2\pi N\hbar \frac{\epsilon_0(\mathbf{q})}{\epsilon(\mathbf{q})} \delta(\hbar\omega - \epsilon(\mathbf{q})) \tag{12.71}$$

is the dynamic structure factor for $\mathbf{q} \neq 0$, in mean field theory, of weakly interacting bosons at $T = 0$. Thus the response of the system to an external perturbation that couples to the density is to create sound waves with energy

$$\epsilon(\mathbf{q}) \sim \sqrt{\frac{2N}{V} v_0 \frac{\hbar^2}{m}} q$$

for small wave vectors, where we have assumed (see also Section 11.2) that the limit as $\mathbf{q} \to 0$ of $v_{\mathbf{q}}$ exists and is finite. The energy spectrum is identical to that found in Section 11.2 by means of the Bogoliubov transformation.

12.2.3 Excitations of the Heisenberg ferromagnet

We consider the anisotropic Heisenberg model at low temperature. The Hamiltonian is

$$H = -\sum_{\langle ij \rangle} [J_z S_{iz} S_{jz} + J_{xy}(S_{ix} S_{jx} + S_{iy} S_{jy})] - mB \sum_i S_{iz} \qquad (12.72)$$

where the sum in the first term is over nearest-neighbor pairs on a lattice with coordination number ν and where $mS\hbar$ is the magnetic moment per atom. The isotropic Heisenberg ferromagnet corresponds to $J_z = J_{xy} > 0$. Here we simply assume that $J_z \geq J_{xy} > 0$ and we have taken the magnetic field to be in the z direction, which, with our choice of coupling constants, is the direction of the ground-state magnetization. The spin operators obey the usual angular momentum commutation relations:

$$\begin{aligned} [S_i^+, S_{jz}] &= -\hbar S_i^+ \delta_{ij} \\ [S_i^-, S_{jz}] &= \hbar S_i^- \delta_{ij} \\ [S_i^+, S_j^-] &= 2\hbar S_{iz} \delta_{ij} \end{aligned} \qquad (12.73)$$

where $S^+ = S_x + iS_y$, $S^- = S_x - iS_y$ are the raising and lowering operators.

We note that the operator $M_z = \sum_i S_{iz}$ commutes with H. The eigenstates of H can therefore be partially indexed by the z component of the magnetization M_z. In particular, the ground state $|0\rangle$ is the state with all spins fully aligned in the positive z direction:

$$S_{iz}|0\rangle = S\hbar|0\rangle \qquad (12.74)$$

for all i.

We now suppose an external perturbation of the form

$$H_{ext}(t) = -\sum_j h_j(t)(S_j^+ + S_j^-) \tag{12.75}$$

and calculate the response of the system. A physical realization of such a perturbation could be a beam of neutrons polarized in the x direction. We apply the formalism (12.32) - (12.39) to this situation. In particular, we shall calculate the correlation functions

$$\begin{aligned} S_{+-}(\mathbf{q},\omega) &= \int_{-\infty}^{\infty} \langle S^+(\mathbf{q},t)S^-(\mathbf{q},0)\rangle e^{i\omega t}dt \\ S_{-+}(\mathbf{q},\omega) &= \int_{-\infty}^{\infty} \langle S^-(\mathbf{q},t)S^+(\mathbf{q},0)\rangle e^{i\omega t}dt \end{aligned} \tag{12.76}$$

where

$$S^+(\mathbf{q},t) = \frac{1}{\sqrt{N}}\sum_j S_j^+(t)e^{i\mathbf{q}\cdot\mathbf{r}_j} \tag{12.77}$$

$$S^-(\mathbf{q},t) = \frac{1}{\sqrt{N}}\sum_j S_j^-(t)e^{-i\mathbf{q}\cdot\mathbf{r}_j} = [S^+(\mathbf{q},t)]^\dagger\ . \tag{12.78}$$

We begin by considering the zero-temperature case and calculate, for $B = 0$,

$$\begin{aligned} HS^-(\mathbf{q},0)|0\rangle &= -\frac{1}{\sqrt{N}}\sum_m e^{-i\mathbf{q}\cdot\mathbf{r}_m} \\ &\times \sum_{\langle ij\rangle}\left[J_z S_{iz}S_{jz} + \frac{J_{xy}}{2}(S_i^+S_j^- + S_j^+S_i^-)\right]S_m^-|0\rangle\ . \end{aligned} \tag{12.79}$$

We examine (12.79) term by term:

$$J_z S_{iz}S_{jz}S_m^- = \begin{cases} J_z S^2\hbar^2 S_m^-|0\rangle & \text{if } i \neq m \text{ and } j \neq m \\ J_z S(S-1)\hbar^2 S_m^-|0\rangle & \text{if } i = m \text{ or } j = m\ . \end{cases}$$

Therefore,

$$-\sum_{\langle ij\rangle} J_z S_{iz}S_{jz}S_m^-|0\rangle = E_0 S_m^-|0\rangle + \nu J_z\hbar^2 S S_m^-|0\rangle \tag{12.80}$$

where $E_0 = -\nu N J_z\hbar^2 S^2/2$ is the zero-field ground-state energy. Also, since i, j are nearest neighbors,

$$J_{xy}S_i^+S_j^-S_m^-|0\rangle = \begin{cases} 0 & i \neq m \\ 2J_{xy}\hbar^2 S S_j^-|0\rangle & i = m\ . \end{cases}$$

Therefore,

$$\sum_{\langle ij \rangle} \frac{J_{xy}}{2}(S_i^+ S_j^- + S_j^+ S_i^-)S_m^-|0\rangle = J_{xy}\hbar^2 {\sum_j}' S_j^- |0\rangle \tag{12.81}$$

where $\sum'$ indicates that the summation extends only over nearest neighbors of site m. Substituting (12.80) and (12.81) into (12.79), we obtain

$$HS^-(\mathbf{q},0)|0\rangle = E_0 S^-(\mathbf{q},0)|0\rangle + \frac{\hbar^2 S}{\sqrt{N}} \sum_{m,\boldsymbol{\delta}} \left(J_z - J_{xy}e^{i\mathbf{q}\cdot\boldsymbol{\delta}}\right) S_m^- e^{-i\mathbf{q}\cdot\mathbf{r}_m}|0\rangle \tag{12.82}$$

where $\boldsymbol{\delta}$ is a nearest-neighbor vector. Therefore,

$$HS^-(\mathbf{q},0)|0\rangle = [E_0 + \epsilon_0(\mathbf{q})]S^-(\mathbf{q},0)|0\rangle \tag{12.83}$$

with

$$\epsilon_0(\mathbf{q}) = \hbar^2 S \sum_{\boldsymbol{\delta}} \left(J_z - J_{xy}e^{i\mathbf{q}\cdot\boldsymbol{\delta}}\right) \tag{12.84}$$

which is the energy of the spin wave or magnon with wave vector $\mathbf{q}$. For the simple cubic lattice, with nearest-neighbor spacing a, the dispersion relation (12.84) takes the form

$$\epsilon_0(\mathbf{q}) = 6\hbar^2 S(J_z - J_{xy}) + 6\hbar^2 S J_{xy}\left[1 - \frac{1}{3}(\cos q_x a + \cos q_y a + \cos q_z a)\right] . \tag{12.85}$$

The anisotropy in the coupling constants produces a gap $\Delta = 6\hbar^2 S(J_z - J_{xy})$ in the spin wave spectrum. In the case of a finite magnetic field in the z direction this gap would be increased by $mB\hbar$, as is easily shown. The gap leads to an exponential dependence of the order parameter on T at low temperatures:

$$\Delta M_z(T) \equiv M_z(0) - M_z(T) \sim e^{-\Delta/k_B T} . \tag{12.86}$$

In the case of the isotropic Heisenberg model, $J = J_z = J_{xy}$, the spectrum is free-particle like for small wave vectors $\mathbf{q}$:

$$\epsilon_0(\mathbf{q}) \approx J\hbar^2 S q^2 a^2 . \tag{12.87}$$

If we assume that the spin waves or magnons are noninteracting bosons, we can then show that at low temperatures and zero applied magnetic field, the temperature dependence of the magnetization is given by (see Problem 12.5)

$$\Delta M_z(T) \sim T^{3/2} . \tag{12.88}$$

The assumption that magnons are noninteracting bosons is only approximately valid but becomes more and more exact as the temperature is lowered, and equation (12.88) can be shown to hold in this limit (see Mattis [194] or the original paper by Dyson [82] for a complete discussion).

We note, in passing, that in contrast to the case of the Ising model, where the statistical mechanics of the antiferromagnet (at least on lattices such as the simple cubic which consist of interpenetrating sublattices, see section 4.1) is identical to that of the ferromagnet, the Heisenberg antiferromagnet poses a much more difficult problem than the ferromagnet. Indeed, not even the ground state is known in three or two dimensions. Mattis [194] provides a thorough discussion of this problem. We mention only that antiparallel ordering on two sublattices produces a phonon-like (linear in q) excitation spectrum rather than the free-particle spectrum (12.87).

Returning to the correlation functions (12.76), we immediately find

$$S_{+-}(\mathbf{q},\omega) = 4\pi\hbar^2 S\,\delta\left(\omega - \frac{\epsilon_0(\mathbf{q})}{\hbar}\right) \tag{12.89}$$

and $S_{-+}(\mathbf{q},\omega) = 0$ for the ground state aligned in the $+z$ direction. The response function or transverse susceptibility is therefore given by (12.38)

$$\chi^R_{+-}(\mathbf{q},\omega) = 2\hbar S\int_{-\infty}^{\infty} d\omega'\,\frac{\delta(\omega' - \epsilon_0\mathbf{q}/\hbar)}{\omega - \omega' + i\eta} = \frac{2\hbar S}{\omega - \epsilon_0(\mathbf{q})/\hbar + i\eta}\,. \tag{12.90}$$

Specializing to a static field, independent of position [*i.e.*, $h_j(t) = h$ in (12.75)], we find, for the static transverse susceptibility,

$$\chi^S_{+-} = -\lim_{\omega,\mathbf{q}\to 0}\frac{2\hbar S}{\omega - \epsilon_0(\mathbf{q})/\hbar + i\eta} = \frac{2\hbar S}{6\hbar S(J_z - J_{xy}) + mB} \tag{12.91}$$

which diverges in the isotropic case as the field in the z direction approaches zero. This divergence is due to the infinite degeneracy of the ground state — there is no energy cost associated with the coherent rotation of all N spins.

12.2.4 Screening and plasmons

As in the case of the weakly interacting Bose gas [Section 12.2.2], we must first calculate the dynamic structure factor of the noninteracting system. Again, at $T = 0$, the only state that contributes to the ensemble average (12.20) is the ground state, which has plane wave states occupied up to a spherical Fermi surface of radius k_F. Consider the states $|n\rangle$ that can be populated through

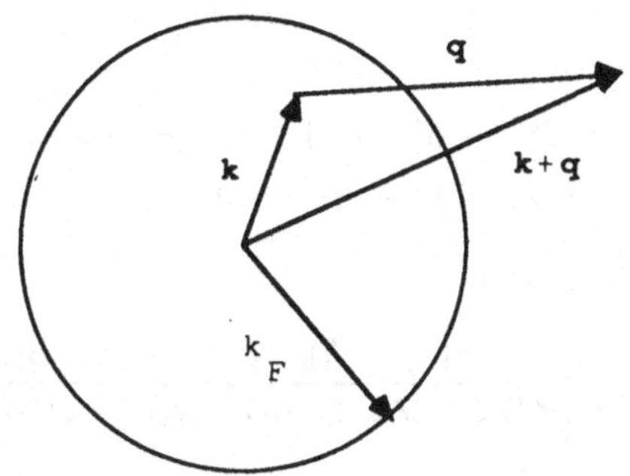

Figure 12.3: Electron-hole excitation from a filled Fermi sphere.

the action of the density operator

$$\rho_{\mathbf{q}} = \sum_{\mathbf{k},\sigma} a^{\dagger}_{\mathbf{k+q},\sigma} a_{\mathbf{k},\sigma} \tag{12.92}$$

on the ground state $|0\rangle$. The state $|n\rangle = a^{\dagger}_{\mathbf{k+q},\sigma} a_{\mathbf{k},\sigma}|0\rangle$ is one in which a particle with wave vector $\mathbf{k}$ inside the Fermi surface is excited to a state of wave vector $\mathbf{k}+\mathbf{q}$ outside the Fermi surface as illustrated in Figure 12.3. The energy of this state is

$$E_n = E_0 + \frac{\hbar^2(\mathbf{k}+\mathbf{q})^2}{2m} - \frac{\hbar^2 k^2}{2m} \tag{12.93}$$

where E_0 is the energy of the filled Fermi sea. Using (12.93) we find, for the dynamic structure factor of the noninteracting system (note a factor of 2 for spin),

$$S_0(\mathbf{q},\omega) = 4\pi\hbar \sum_{\mathbf{k},\mathbf{q}} \theta(k_F - k)\theta(|\mathbf{k}+\mathbf{q}| - k_F)\, \delta\left\{\hbar\omega - \frac{\hbar^2}{2m}[(\mathbf{k}+\mathbf{q})^2 - k^2]\right\} \tag{12.94}$$

where $\theta(x) = 1$ for $x > 0$, 0 for $x < 0$.

The free-particle response function is given by (12.25), and with (12.94), this becomes

$$\begin{aligned} \chi_0^R(\mathbf{q},\omega) &= \frac{2}{V} \sum_{k<k_F,|\mathbf{k}+\mathbf{q}|>k_F} \left\{ \frac{1}{\hbar\omega - (\hbar^2/2m)[(\mathbf{k}+\mathbf{q})^2 - k^2] + i\eta} \right. \\ & \left. \quad - \frac{1}{\hbar\omega + (\hbar^2/2m)[(\mathbf{k}+\mathbf{q})^2 - k^2] + i\eta} \right\} . \end{aligned} \tag{12.95}$$

If we make the substitution $\mathbf{k} = -(\mathbf{p}+\mathbf{q})$ in the second term and then relabel $\mathbf{p} \to \mathbf{k}$ the denominators in (12.95) become identical. Moreover, we use the identity

$$\begin{aligned}\theta(k_F - k)\theta(|\mathbf{k}+\mathbf{q}| - k_F) - \theta(k - k_F)\theta(k_F - |\mathbf{k}+\mathbf{q}|) \\ = \theta(k_F - k) - \theta(k_F - |\mathbf{k}+\mathbf{q}|)\end{aligned} \tag{12.96}$$

to simplify (12.95) and obtain

$$\chi_0^R(\mathbf{q},\omega) = \frac{2}{V}\sum_{\mathbf{k}} \frac{\theta(k_F - k) - \theta(k_F - |\mathbf{k}+\mathbf{q}|)}{\hbar\omega - (\hbar^2/2m)[(\mathbf{k}+\mathbf{q})^2 - k^2] + i\eta} . \tag{12.97}$$

The function $\chi_0^R(\mathbf{q},\omega)$ is commonly referred to as the Lindhard function. If the sum over $\mathbf{k}$ is transformed into an integral, analytical expressions can be found for both the real and imaginary parts. The resulting expressions are rather complicated and we will here exhibit only two special cases. Explicit formulas for the general case can be found in Mahan [185], Fetter and Walecka [93], or Pines and Noziéres [243].

(i) $\omega \approx 0, q \neq 0$. This limit is important for the theory of static screening. It is also the limit in which one usually studies the electron phonon interaction, since lattice vibrational energies are generally small compared with typical electronic excitation energies. Substituting

$$\frac{1}{\hbar\omega + E + i\eta} = P\frac{1}{\hbar\omega + E} - i\pi\delta(\hbar\omega + E) \tag{12.98}$$

where P indicates principal value, we find for $\omega = 0$,

$$\begin{aligned}\mathrm{Im}\,\chi_0^R(\mathbf{q},0) &= -\frac{2\pi}{V}\sum_{\mathbf{k}}(\theta(k_F - k) - \theta(k_F - |\mathbf{k}+\mathbf{q}|)) \\ &\quad \times\delta\left\{\hbar\omega - \frac{\hbar^2}{2m}[(\mathbf{k}+\mathbf{q})^2 - k^2]\right\} = 0\end{aligned} \tag{12.99}$$

since the two step functions cancel when $|\mathbf{k}| = |\mathbf{k}+\mathbf{q}|$. Thus the static susceptibility is real for all $\mathbf{q}$ and

$$\chi_0^R(\mathbf{q},0) = \frac{m}{2\pi^3\hbar^2}\left[P\int_{k<k_F}\frac{d^3k}{k^2 - (\mathbf{k}+\mathbf{q})^2} - P\int_{|\mathbf{k}+\mathbf{q}|<k_F}\frac{d^3k}{k^2 - (\mathbf{k}+\mathbf{q})^2}\right] . \tag{12.100}$$

If we make the substitution $\mathbf{k}+\mathbf{q} = -\mathbf{p}$ in one of the integrals, we see that the two terms give identical contributions and have

$$\chi_0^R(\mathbf{q},0) = -\frac{m}{\pi^3\hbar^2}P\int_{k<k_F}\frac{d^3k}{q^2 + 2\mathbf{k}\cdot\mathbf{q}} . \tag{12.101}$$

In spherical coordinates, with $\mu = \cos\theta$, (12.101) becomes

$$\begin{aligned} \chi_0^R(\mathbf{q},0) &= -\frac{2m}{\pi^2\hbar^2}\int_0^{k_F} dk\, k^2 P\int_{-1}^{1} d\mu \frac{1}{q^2+2kq\mu} \\ &= -\frac{m}{q\pi^2\hbar^2} P\int_0^{k_F} dk\, k \ln\frac{q+2k}{q-2k} \\ &= -\frac{m}{q\pi^2\hbar^2} P\int_{-k_F}^{k_F} dk\, k \ln|q+2k| \;. \end{aligned} \tag{12.102}$$

The last integral can easily be evaluated through integration by parts with the final result:

$$\chi_0^R(\mathbf{q},0) = -\frac{mk_F}{\pi^2\hbar^2}\left(\frac{1}{2} + \frac{4k_F^2-q^2}{8qk_F}\ln\left|\frac{q+2k_F}{q-2k_F}\right|\right) \;. \tag{12.103}$$

In the mean field approximation the static dielectric function is therefore given by (12.59)

$$\epsilon(\mathbf{q},0) = 1 - \frac{e^2}{\epsilon_0 q^2}\chi_0^R(\mathbf{q},0) \;. \tag{12.104}$$

Defining the function

$$u_{\mathbf{q}} = \left(\frac{1}{2} + \frac{4k_F^2-q^2}{8qk_F}\ln\left|\frac{q+2k_F}{q-2k_F}\right|\right) \tag{12.105}$$

we thus have

$$\epsilon(\mathbf{q},0) = 1 + \frac{k_{TF}^2}{q^2}u_{\mathbf{q}} \tag{12.106}$$

where k_{TF} is the *Thomas–Fermi wave vector*

$$k_{TF}^2 = \frac{me^2}{\pi^2\epsilon_0\hbar^2}k_F \;. \tag{12.107}$$

The function $u_{\mathbf{q}}$ is sketched in Figure 12.4. We note that for small q, $u_{\mathbf{q}} \approx 1$ while $u_{\mathbf{q}} \to 0$ as $q \to \infty$. In the vicinity of $2k_F$, $u_{\mathbf{q}}$ varies rapidly and there is a logarithmic singularity in the derivative with respect to q at $2k_F$.

Consider next the effective potential due to a point charge e in the mean field linear response approximation:

$$\phi_{eff}(\mathbf{q}) = \frac{e^2/\epsilon_0}{q^2 + k_{TF}^2 u_{\mathbf{q}}} \;. \tag{12.108}$$

Normally, the spatial dependence of a function at large r will be determined primarily by the behavior of the Fourier transformed function at small q. Since

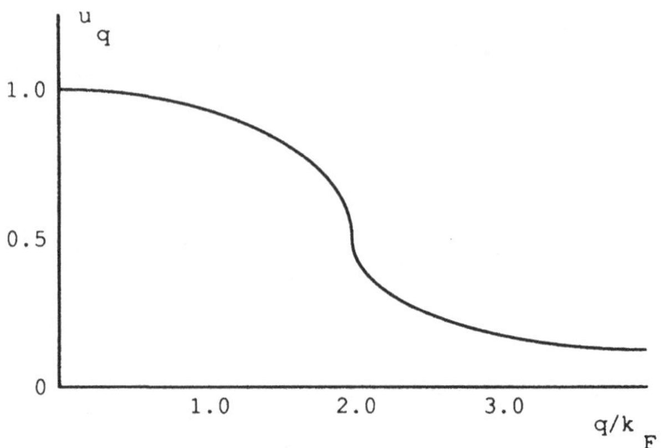

Figure 12.4: The function $u_{\mathbf{q}}$ of equation (12.105).

$u_{\mathbf{q}} \to 1$ as $q \to 0$, one might expect that for large r the effective potential will behave like the Fourier transform of the $q = 0$ limit of (12.108), which is (see Problem 12.1)

$$\phi_{TF}(\mathbf{r}) = \frac{e^2}{4\pi\epsilon_0 r} e^{-k_{TF} r} \ . \tag{12.109}$$

This argument is, however, not correct because of the singularity in $u_{\mathbf{q}}$ at $2k_F$. It can be shown (see, e.g., Fetter and Walecka [93]) that for large r

$$\phi_{eff}(\mathbf{r}) \sim \frac{1}{r^3} \cos 2k_F r \ . \tag{12.110}$$

The oscillations in $\phi_{eff}(\mathbf{r})$ are commonly referred to as *Friedel oscillations.* The result (12.110) is gratifying from the point of view of understanding the cohesion of metals. If the effective interaction of ions in an electron gas had been purely repulsive, as in the Thomas–Fermi approximation (12.109), it would be difficult to see what keeps a metal together. The potential (12.110), however, leads to an ion–ion interaction which is attractive at certain distances, with an energy minimum at the equilibrium ionic positions of the resulting crystal. Unfortunately, this model is too crude to be useful quantitatively for the description of real metals. We next consider another interesting limiting case.

(ii) $|q|$ ***small,*** $\omega \gg \hbar q^2/2m$. In this limit the denominator in (12.97) cannot vanish and we rewrite this equation in the form

$$\chi_0^R(\mathbf{q},\omega) \quad = \quad \frac{2}{V}\sum_{\mathbf{k}} \theta(k_F - k) \left\{ \frac{1}{\hbar\omega - (\hbar^2/2m)[(\mathbf{k}+\mathbf{q})^2 - k^2]} \right.$$

$$- \frac{1}{\hbar\omega + (\hbar^2/2m)[(\mathbf{k}-\mathbf{q})^2 - k^2]} \bigg\} . \tag{12.111}$$

Using $(\mathbf{k} \pm \mathbf{q})^2 - k^2 = q^2 \pm 2\mathbf{k}\cdot\mathbf{q}$ and combining the two terms, we obtain

$$\begin{aligned} \chi_0^R(\mathbf{q},\omega) &= \frac{2}{V}\sum_{k<k_F} \frac{q^2/m}{[\omega - (\hbar/2m)(q^2 + 2\mathbf{k}\cdot\mathbf{q})][\omega + (\hbar/2m)(q^2 - 2\mathbf{k}\cdot\mathbf{q})]} \\ &\approx \frac{q^2 N}{mV}\frac{1}{\omega^2} . \end{aligned} \tag{12.112}$$

Using this result, we find the dielectric function in the mean field approximation

$$\epsilon(\mathbf{q},\omega) = 1 - \frac{e^2}{\epsilon_0 q^2}\chi_0(\mathbf{q},\omega) \approx 1 - \frac{e^2 N}{\epsilon_0 m V} = 1 - \frac{\Omega_p^2}{\omega^2} \tag{12.113}$$

where $\Omega_p = (e^2N/\epsilon_0 mV)^{1/2}$ is the *plasma frequency.* We note that Planck's constant does not appear in the plasma frequency. Indeed, the plasma frequency obtained in this approximation is identical to that of a classical system of charged particles in the Drude model (see e.g. Ashcroft and Mermin [21]). This is easily seen from the following elementary argument. Suppose that a classical free electron system is subject to a time-dependent electric field of the form

$$\mathbf{E}(t) = \mathbf{E}_0 e^{-i\omega t} . \tag{12.114}$$

The resulting force on a particle is $\mathbf{f} = -e\mathbf{E}(t)$ and the mean displacement is $\langle \mathbf{x}(t)\rangle = e\mathbf{E}/\omega^2 m$. Therefore, the induced polarization is given by

$$\mathbf{P} = -\frac{Ne}{V}\langle \mathbf{x}\rangle = -\frac{ne^2}{m\omega^2}\mathbf{E} . \tag{12.115}$$

The dielectric function, in turn, is given by

$$\mathbf{D} = \epsilon\epsilon_0\mathbf{E} = \epsilon_0\mathbf{E} + \mathbf{P} . \tag{12.116}$$

Combining (12.115) and (12.116), we recover the previous result (12.113).

From $\phi_{eff}(\mathbf{q},\omega) = \phi_{ext}(\mathbf{q},\omega)/\epsilon(\mathbf{q},\omega)$, we see that a zero in the dielectric function corresponds to a resonant response. The resulting excitation is a longitudinal charge density wave that for nonzero $\mathbf{q}$ propagates with a frequency $\Omega_{pl}(\mathbf{q})$ where $\Omega_{pl}(\mathbf{q})$ reduces to the plasma frequency Ω_p (12.113) in the long-wavelength limit. It can be seen, by expanding (12.112) to higher order in q, that for small q, $\Omega_{pl}(\mathbf{q})$ differs from Ω_p by an amount proportional to q^2. The corresponding quantum of excitation or quasiparticle is called a *plasmon* and has energy $\hbar\Omega_{pl}(\mathbf{q})$.

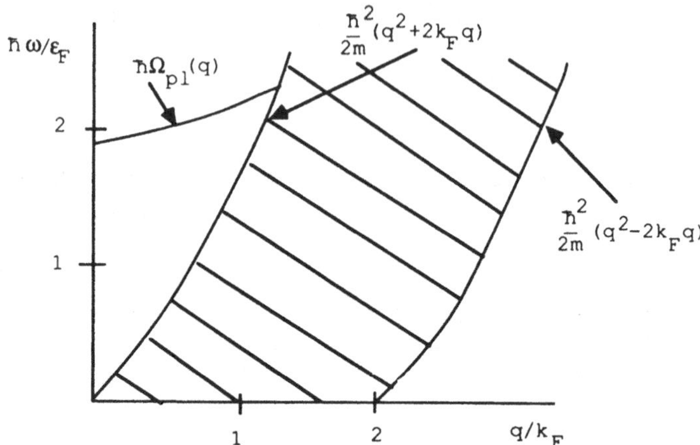

Figure 12.5: Excitation spectrum of an interacting electron gas.

12.2.5 Exchange and correlation energy

We now return to the case of arbitrary values of $\mathbf{q}$ and ω. From the expression (12.94) for the dynamic structure factor $S_0(\mathbf{q},\omega)$ and the inequalities, for $k < k_F$,

$$q^2 + 2k_F q \geq q^2 + 2kq \geq (\mathbf{k}+\mathbf{q})^2 - k^2 \geq q^2 - 2kq \geq q^2 - 2k_F q$$

we see that $S_0(\mathbf{q},\omega)$ is nonzero in the shaded region of Figure 12.5.

If $\chi_0^R(\mathbf{q},\omega)$ is continued analytically to complex values of ω,

$$\chi_0^R(\mathbf{q},\omega) = 2\int \frac{d^3k}{(2\pi)^3} \frac{\theta(k_F - k) - \theta(k_F - |\mathbf{k}+\mathbf{q}|)}{\hbar\omega + \epsilon_{\mathbf{k}} - \epsilon_{\mathbf{k}+\mathbf{q}}} \tag{12.117}$$

and we see that for ω on the real axis,

$$\operatorname{Im}\chi_0^R(\mathbf{q},\omega) = \frac{1}{2i}\left[\chi_0^R(\mathbf{q},\omega+i\eta) - \chi_0^R(\mathbf{q},\omega-i\eta)\right] \tag{12.118}$$

that is, $\operatorname{Im}\chi_0^R(\mathbf{q},\omega)$ is $\frac{1}{2}$ times the discontinuity across a branch cut with extension

$$-\frac{\hbar^2}{2m}(q^2+2k_F q) < \hbar\omega < \frac{\hbar^2}{2m}(q^2+2k_F q) \qquad q < 2k_F$$

$$\frac{\hbar^2}{2m}(q^2-2k_F q) < |\hbar\omega| < \frac{\hbar^2}{2m}(q^2+2k_F q) \qquad q > 2k_F\ .$$

The expression

$$\frac{\chi_0^R(\mathbf{q},\omega)}{1 - v_\mathbf{q}\chi_0^R(\mathbf{q},\omega)} = \frac{1}{\epsilon(\mathbf{q},\omega)}$$

has a branch cut for the same range of ω as well as a pole at the plasmon frequency. Here $v_\mathbf{q} = e^2/\epsilon_0 q^2$. The residue at this pole is largest for $q = 0$ and gradually goes to zero as the pole merges with the particle–hole continuum (shaded region in Figure 12.5).

We next turn to the problem of evaluating the ground-state energy of the electron gas. By combining (12.45), the $T = 0$ limit of (12.40) and (12.104), we find

$$E_{int}(\lambda) = -\sum_\mathbf{q}\left\{\int_0^\infty \frac{\hbar d\omega}{2\pi}\mathrm{Im}\frac{1}{\epsilon_\lambda(\mathbf{q},\omega)} + \frac{Nv_\mathbf{q}}{2V}\right\} . \tag{12.119}$$

From (12.118) we see that the integration over frequency in (12.119) is equivalent to following the contour C in Figure 12.6. We can add a semicircle to this contour since the contribution to the integral vanishes in the limit of large radius. The analytical continuation (12.117) does not have any singularities except on the real axis. We can therefore deform the contour to lie along the imaginary axis (C') and obtain, with $\omega = iu$, in the mean field approximation:

$$E = \frac{3}{5}\frac{\hbar k_F^2}{2m}N - \sum_\mathbf{q}\left\{\frac{Nv_\mathbf{q}}{2V} + \int_{-\infty}^{\infty}\frac{\hbar du}{4\pi}\int_0^1 d\lambda\frac{v_\mathbf{q}\chi_0^R(\mathbf{q},iu)}{1-\lambda v_\mathbf{q}\chi_0^R(\mathbf{q},iu)}\right\} .$$

$\chi_0^R(\mathbf{q},iu)$ turns out to be real and, after performing the integration over λ, we finally obtain

$$E = \frac{3}{5}\frac{\hbar k_F^2}{2m}N - \sum_\mathbf{q}\left\{\frac{Nv_\mathbf{q}}{2V} - \int_{-\infty}^{\infty}\frac{\hbar du}{4\pi}\ln\left[1 - v_\mathbf{q}\chi_0^R(\mathbf{q},iu)\right]\right\} . \tag{12.120}$$

This approximate expression for the ground-state energy was first obtained by Gell–Mann and Brueckner [107]. It can easily be evaluated numerically and turns out to give rise to a significant contribution to the cohesive energy of metals. From the expression for the energy, other quantities, such as the compressibility, can be obtained using thermodynamic identities. We refer the interested reader to the texts by Mahan [185] and Pines and Noziéres [243] for further discussion of the electron liquid.

12.2.6 Phonons in metals

We now consider an idealized model of a metal in which we treat the system as a two-component plasma of N_i ions and $N_e = ZN_i$ electrons. We will calculate

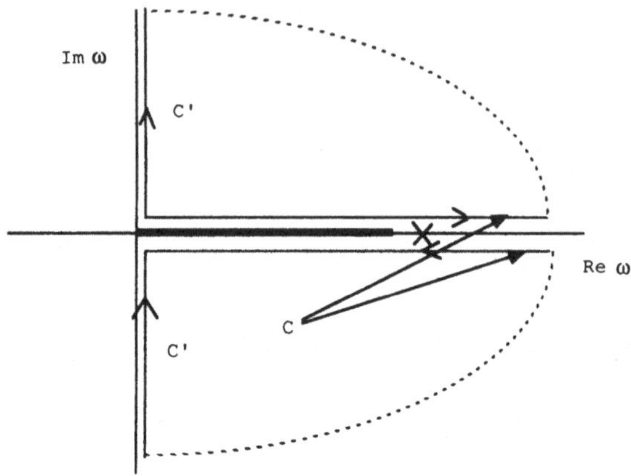

Figure 12.6: Contour used in the integration (12.119). The branch cut is shown as a bold line.

the *charge* response

$$\langle -e\delta\rho^e(\mathbf{q},\omega) + Ze\delta\rho^i(\mathbf{q},\omega)\rangle$$

to an external potential (voltage) $v_{ext}(\mathbf{q},\omega)$ and let $\chi_0^i(\mathbf{q},\omega)$ and $\chi_0^e(\mathbf{q},\omega)$ be, respectively, the free ion and free electron response functions as defined earlier. We find, for the mean field response, suppressing explicit reference to the dependence on $\mathbf{q}$ and ω

$$\langle\delta\rho^i\rangle = \chi_0^i(Zev_{ext} + \frac{Ze}{\epsilon_0 q^2}\langle Ze\delta\rho^i - e\delta\rho^e\rangle) \tag{12.121}$$

with a similar formula for the response of the electrons. After some straightforward algebra, we obtain

$$\begin{aligned}
\langle -\delta\rho^e + Z\delta\rho^i\rangle &= \frac{-e\chi_0^e + Ze\chi_0^i}{1 - e^2/\epsilon_0 q^2(Z^2\chi_0^i + \chi_0^e)} v_{ext} &(12.122)\\
\langle\delta\rho^i\rangle &= \frac{Ze\chi_0^i}{1 - e^2/\epsilon_0 q^2(Z^2\chi_0^i + \chi_0^e)} v_{ext} &(12.123)\\
\langle\delta\rho^e\rangle &= \frac{-e\chi_0^e}{1 - e^2/\epsilon_0 q^2(Z^2\chi_0^i + \chi_0^e)} v_{ext}\,. &(12.124)
\end{aligned}$$

Because of their large mass the ions respond much more slowly than do the electrons. We therefore assume that we are dealing with frequencies at which the high-frequency approximation (12.112) is valid for the ion response function:

$$\chi_0^i(\mathbf{q},\omega) \approx \frac{q^2 N_i}{MV}\frac{1}{\omega^2} \tag{12.125}$$

where M is the mass of an ion. We next assume that the electronic response is sufficiently fast that the static approximation (12.103), (12.105):

$$\chi_0^e(\mathbf{q},0) = -\frac{mk_F}{\pi^2\hbar^2}u_\mathbf{q} \tag{12.126}$$

is valid. With these approximations, the ionic response becomes

$$\langle\delta\rho^i\rangle = \frac{1}{\omega^2-\omega^2(\mathbf{q})}\frac{ZeN_i}{MV}\frac{q^2 v_{ext}(\mathbf{q},\omega)}{q^2+k_{TF}^2 u_\mathbf{q}} \ . \tag{12.127}$$

If it were not for the screening of the ions by the electrons, the lattice vibrational frequency would be the ion plasma frequency $\Omega_i = \sqrt{N_i Z^2 e^2/\epsilon_0 M}$. With screening, we find instead an acoustic mode, the quantum of which is the *phonon*, with frequency given by

$$\omega^2(\mathbf{q}) = \frac{e^2 Z^2 N_i}{(q^2+k_{TF}^2 u_\mathbf{q})\epsilon_0 V}q^2 \ . \tag{12.128}$$

In the long-wavelength limit, $u_\mathbf{q} \to 1$ and (12.128) reduces to $\omega = cq$ where

$$c \approx \left(\frac{Z^2 e^2 N_i}{\epsilon_0 k_{TF}^2 V}\right) \tag{12.129}$$

is the Bohm–Staver sound velocity.

The model that we have used to derive this result is extremely crude. Nevertheless, the resulting velocity of sound is of the right order of magnitude for most metals. We also note that for any nonzero frequency $\chi_0^e(\mathbf{q},\omega)$ will have a small but nonzero imaginary part. The theory thus predicts attenuation of acoustic waves.

From (12.122) we find for the dielectric function in the mean field approximation:

$$\epsilon_{MF}(\mathbf{q},\omega) = 1 - \frac{e^2}{\epsilon_0 q^2}\left[Z^2\chi_0^i(\mathbf{q},\omega) + \chi_0^e(\mathbf{q},\omega)\right] \ . \tag{12.130}$$

Once again, using the approximations (12.125) and (12.126), we obtain

$$\epsilon_{MF}(\mathbf{q},\omega) \approx \left(1+\frac{k_{TF}^2 u_\mathbf{q}}{q^2}\right)\left[1+\frac{\omega^2(\mathbf{q})}{\omega^2-\omega^2(\mathbf{q})}\right] \tag{12.131}$$

and we see that for frequencies less than $\omega(\mathbf{q})$ the dielectric function is negative. This has the consequence that the quasielastic interaction between electrons in states near the Fermi surface is attractive. As discussed in Section 11.3, such an attractive interaction in the BCS theory produces a Fermi surface instability. This is the mechanism generally believed to be responsible for the phenomenon of superconductivity in the traditional low temperature superconductors.

12.3 Entropy Production, the Kubo Formula, and the Onsager Relations for Transport Coefficients

In this book we are concerned primarily with equilibrium properties of matter. For systems sufficiently close to equilibrium that linear response theory is adequate, it is possible to relate *transport properties* to equilibrium correlation functions. We begin the discussion of transport by deriving the Kubo formula for the special case of the electrical conductivity. In Section 12.3.2 we generalize the concepts of currents and fields and also introduce the notions of microscopic reversibility and local equilibrium. In Section 12.3.3 we use these ideas to derive the Onsager relations between transport coefficients.

12.3.1 Kubo formula

Consider a system of particles, each with charge e and with positions specified by $\mathbf{r}_m$. The electrical current density at point $\mathbf{x}$ is given by

$$\mathbf{j}(\mathbf{x}) = \sum_m e\mathbf{v}_m \delta(\mathbf{x} - \mathbf{r}_m) \tag{12.132}$$

where $\mathbf{v}_m$ is the velocity of particle m. We assume an external electric field, $\mathbf{E}(t)$, that does not vary with position. In this case the current density will also be time but not position dependent and the perturbing Hamiltonian is

$$H_{ext}(t) = -e\sum_m \mathbf{r}_m \cdot \mathbf{E}(t) \ . \tag{12.133}$$

Using (12.33) with $B = j_\alpha$, the αth component of the current density, we have

$$j_\alpha(t) = -\frac{i}{\hbar V}\int_{-\infty}^{t} dt' \sum_{m,n,\gamma} \langle [r_{m\gamma}(t'), \dot{r}_{n\alpha}(t)] \rangle_H e^2 E_\gamma(t') \tag{12.134}$$

where V is the volume of the system. Substituting

$$\langle\phi\rangle = \mathrm{Tr}\,\rho\phi$$

where ρ is the density matrix of the unperturbed system, and using the fact that the trace of a product of operators is invariant under cyclic permutation of the operators, we may write (12.134) in the form

$$j_\alpha(t) = \frac{i}{\hbar}\int_{-\infty}^{t} dt'\mathrm{Tr}\, j_\alpha(t-t')\sum_{m\gamma}[r_{m\gamma},\rho]eE_\gamma(t') \,. \tag{12.135}$$

We simplify this expression by rewriting the commutator. For an arbitrary operator A we have

$$Z_G[A,\rho] = [A, e^{-\beta K}] \equiv e^{-\beta K}\phi(\beta) \tag{12.136}$$

where

$$\phi(\lambda) = e^{\lambda K}Ae^{-\lambda K} - A \,. \tag{12.137}$$

Differentiating (12.137) with respect to λ and integrating from 0 to β, we obtain

$$\phi(\beta) = \int_0^\beta d\lambda\, e^{\lambda K}[K,A]e^{-\lambda K}$$

or, in (12.135),

$$[r_{m\gamma},\rho] = \rho\int_0^\beta d\lambda\, e^{\lambda K}[K,r_{m\gamma}]e^{-\lambda K} = \rho\int_0^\beta d\lambda \frac{\hbar}{i}\dot{r}_{m\gamma}(-i\hbar\lambda) \tag{12.138}$$

where we have used the Heisenberg equation of motion

$$[K,A] = \frac{\hbar}{i}e^{-iKt/\hbar}\frac{dA}{dt}e^{iKt/\hbar} \,.$$

Finally,

$$\begin{aligned} j_\alpha(t) &= -V\sum_\gamma\int_{-\infty}^{t} dt'\,\mathrm{Tr}\, j_\alpha(t-t')\rho\int_0^\beta d\lambda\, j_\gamma(-i\hbar\lambda)E_\gamma(t') \\ &= V\sum_\gamma\int_{-\infty}^{t} dt'\int_0^\beta d\lambda\langle j_\gamma(-i\hbar\lambda)j_\alpha(t-t')\rangle E_\gamma(t') \,. \end{aligned} \tag{12.139}$$

If we specialize to a time-independent field $\mathbf{E}(t') = \mathbf{E}$ and let $t' = t-\tau$, we have

$$j_\alpha(t) = j_\alpha = -V\sum_\gamma\int_0^\infty d\tau\int_0^\beta d\lambda\langle j_\gamma(-i\hbar\lambda)j_\alpha(\tau)\rangle = \sum_\gamma \sigma_{\alpha\gamma}E_\gamma \tag{12.140}$$

or

$$\sigma_{\alpha\gamma} = -v \int_0^\infty d\tau \int_0^\beta d\lambda \langle j_\gamma(-i\hbar\lambda) j_\alpha(\tau)\rangle \ . \tag{12.141}$$

Thus we have expressed the conductivity tensor, σ in terms of the *equilibrium* current-current correlation function. In the classical limit $\hbar \to 0$, which is equivalent to the high-temperature limit $\beta \to 0$, we may approximate (12.141) by the formula

$$\sigma_{\alpha\gamma} = -\frac{v}{k_B T} \int_0^\infty d\tau \langle j_\gamma(0) j_\alpha(\tau)\rangle \ . \tag{12.142}$$

Equations (12.141) and (12.142) are known as Kubo formulas and it is clear that similar expressions can be derived for other transport coefficients. The Kubo formulation of transport theory is complementary to the Boltzmann equation approach. We refer the interested reader to the original work of Kubo [160] and the book of Mahan [185].

12.3.2 Entropy production and generalized currents and forces

We now wish to extend our formalism to a wider class of transport phenomena. As before, we limit ourselves to nonequilibrium phenomena that are sufficiently close to equilibrium that the macroscopic intensive variables such as pressure, temperature, chemical potential, and electrostatic or vector potential are well defined locally. We further assume that these variables are related to the densities associated with the conjugate extensive variables through the equilibrium equations of state. In equilibrium the temperature, pressure, and chemical potential must be constant throughout the system. Here we consider instead the situation where a gradient in these variables is maintained, resulting in currents of the corresponding density variables. To develop a systematic approach, we must first select appropriate variables. In equilibrium we have, from (1.25), the differential for the entropy density

$$ds = \frac{1}{T} de - \frac{1}{T} \sum_i X_i d\xi_i - \frac{1}{T} \sum_j \mu_j dn_j = \sum_k \phi_k d\rho_k \ . \tag{12.143}$$

In (12.143), e is the energy density, X a generalized force, ξ_i the density that couples to this force, and n_j the particle density of species j. For notational convenience we have combined these terms into a set of generalized potential variables ϕ_k and density variables ρ_k.

We now consider the effect of spatially varying potentials ϕ_k. These in turn generate current densities $\mathbf{j}_k$ and we assume that there is a conservation law associated with each density ρ_k so that

$$\frac{\partial \rho_k}{\partial t} + \nabla \cdot \mathbf{j}_k = 0 \, . \tag{12.144}$$

We may also associate an entropy current density $\mathbf{j}_s$ with the currents $\mathbf{j}_k$, that is,

$$\mathbf{j}_s = \sum_k \phi_k \mathbf{j}_k \, . \tag{12.145}$$

The rate of change in the entropy density is then given by

$$\frac{ds}{dt} = \frac{\partial s}{\partial t} + \nabla \cdot \mathbf{j}_s \tag{12.146}$$

where

$$\frac{\partial s}{\partial t} = \sum_k \phi_k \frac{\partial \rho_k}{\partial t} \, . \tag{12.147}$$

It is a straightforward consequence of (12.144)–(12.147) that

$$\frac{ds}{dt} = \sum_k \nabla \phi_k \cdot \mathbf{j}_k \, . \tag{12.148}$$

Equation (12.148) specifies the appropriate conjugate generalized *force fields* $\nabla \phi_k$ and *current densities* $\mathbf{j}_k$.

We now specialize to systems in which there is a steady-state current in response to a static field. The reader will be familiar with a number of phenomenological laws that are used to describe this type of situation. We have already come across Ohm's law for the electrical conductivity. Other examples are Fick's law for diffusion and the Seebeck and Peltier relations for the thermoelectric effects discussed in the next section. In general, in linear response theory, we assume relationships of the form

$$\mathbf{j}_i = \sum_k L_{ik} \nabla \phi_k + O(\nabla \phi_k)^2 \tag{12.149}$$

where the coefficients $L_{ik}(\rho_k)$ are called *kinetic coefficients* and where we ignore the second- and higher-order terms. In the next section we shall discuss Onsager's analysis of these coefficients which led him to the conclusion, that under the very reasonable assumption of *microscopic reversibility*, these coefficients must satisfy the Onsager reciprocity relations:[2]

$$L_{ik} = L_{ki} \ . \tag{12.150}$$

12.3.3 Microscopic reversibility: Onsager relations

In the following discussion we give a derivation very similar to that in Onsager's [220] original articles. Consider first the specific situation of heat flow in an insulator. We denote the heat currents in the three Cartesian directions by j_1, j_2, and j_3 and write

$$j_i = \sum_j L_{ij} \left(\nabla T^{-1}\right)_j \tag{12.151}$$

where L_{ij} are transport coefficients (the thermal conductivity tensor) which are independent of the temperature gradients in the linear approximation. Note that (12.143) implies that it is T^{-1}, not T, that is the field conjugate to the energy density. The coefficients L_{ij} are not all independent. If, for example, the crystal structure has a 3-fold symmetry axis with $i = 3$ denoting the axis, the most general form of equation (12.151) is

$$\begin{aligned} j_1 &= L_{11}\left(\frac{\partial T^{-1}}{\partial x}\right) + L_{12}\left(\frac{\partial T^{-1}}{\partial y}\right) \\ j_2 &= -L_{12}\left(\frac{\partial T^{-1}}{\partial x}\right) + L_{11}\left(\frac{\partial T^{-1}}{\partial y}\right) \\ j_3 &= L_{33}\left(\frac{\partial T^{-1}}{\partial z}\right) \end{aligned} \tag{12.152}$$

where the relations $L_{21} = -L_{12}$ and $L_{22} = L_{11}$ are a consequence of the symmetry (Problem 12.7) of the lattice. The Onsager relations that we shall derive for this particular example are $L_{ij} = L_{ji}$, which have the further consequence that $L_{12} = 0$.

Following Onsager [220], we consider a fluctuation in the energy density. This fluctuation, in general, results in nonzero values for the components of

[2] We have assumed that the generalized force field is the gradient of a scalar potential (*i.e.*, a polar vector). The magnetic field $\mathbf{B} = \nabla\times\mathbf{A}$, on the other hand, is an axial vector and odd under time reversal. For this reason, the Onsager relations in the presence of a magnetic field are $L_{ij}(\mathbf{B}) = L_{ji}(-\mathbf{B})$.

the first moments

$$\alpha_i = \int d^3r\, r_i \epsilon(\mathbf{r}) \tag{12.153}$$

where $\epsilon(\mathbf{r})$ is the energy density and where all coordinates are measured from the center of the crystal. One expects the thermal averages of these moments to have the properties

$$\langle \alpha_i \rangle = 0 \qquad \langle \alpha_i \alpha_j \rangle = \delta_{ij} \langle \alpha^2 \rangle \tag{12.154}$$

for a spherical crystal. The second of these equations is important for what follows and basically contains the statement that the equilibrium fluctuations are spatially isotropic, whereas the thermal conductivity, which depends on the connectivity of the lattice, may not be isotropic. Later in this section we formulate (12.154) more generally.

Suppose next that at a given time t variable α_1 takes on a specific value $\alpha_1(t)$. The correlation functions that characterize the decay of this fluctuation are the expectation values of the functions

$$\alpha_1(t+\Delta t)\alpha_1(t) \qquad \alpha_2(t+\Delta t)\alpha_1(t) \qquad \alpha_3(t+\Delta t)\alpha_1(t)\ .$$

Consider, for example, the correlation function

$$\begin{aligned} \langle \alpha_2(t+\Delta t)\alpha_1(t)\rangle &= \langle \alpha_2(t)\alpha_1(t)\rangle + \Delta t \langle \dot{\alpha}_2(t)\alpha_1(t)\rangle \\ &= \Delta t \langle \dot{\alpha}_2(t)\alpha_1(t)\rangle \end{aligned} \tag{12.155}$$

where we have used (12.154) to eliminate the first term. The time derivative $d\alpha_2/dt$ is proportional to the heat current j_2 and we may now use (12.151) to obtain

$$\langle \alpha_2(t+\Delta t)\alpha_1(t)\rangle = \Delta t L_{21} \left\langle \left(\frac{\partial T^{-1}}{\partial x}\right) \alpha_1(t) \right\rangle\ . \tag{12.156}$$

Moreover, $\partial T^{-1}/\partial x = -C\alpha_1/T^2$ and we have

$$\langle \alpha_2(t+\Delta t)\alpha_1(t)\rangle = -\frac{\Delta t L_{21} C\langle \alpha^2\rangle}{T^2}\ . \tag{12.157}$$

Similarly,

$$\langle \alpha_1(t+\Delta t)\alpha_2(t)\rangle = \frac{\Delta t L_{21} C\langle \alpha^2\rangle}{T^2} \tag{12.158}$$

where we have used the 3-fold rotation symmetry (12.152).

We now define a joint probability $P(\alpha_2', t+\Delta t | \alpha_1', t)$ which is the probability that variable α_2 takes on the specific value α_2' at time $t+\Delta t$ and that α_1 has value α_1' at time t. In terms of this function, the expectation value (12.157) is given by

$$\langle \alpha_2(t+\Delta t)\alpha_1(t)\rangle = \int d\alpha_1 \int d\alpha_2 \alpha_1\alpha_2 P(\alpha_2, t+\Delta t|\alpha_1, t) \ . \tag{12.159}$$

The principle of microscopic reversibility states that

$$P(\alpha_1, t+\Delta t|\alpha_2, t) = P(\alpha_2, t+\Delta t|\alpha_1, t) \ . \tag{12.160}$$

This equation is very plausible. If the velocities of all particles in a given configuration with a specific value of α_2, resulting from a configuration with value α_1 at time $t-\Delta t$ are reversed, then at time $t+\Delta t$ we will once again have the configuration of time $t-\Delta t$. Equation (12.160) states that these velocity-reversed configurations are equally probable. Clearly, then,

$$\langle \alpha_1(t+\Delta t)\alpha_2(t)\rangle = \langle \alpha_2(t+\Delta t)\alpha_1(t)\rangle \tag{12.161}$$

or

$$L_{12} = 0 \ . \tag{12.162}$$

We generalize the foregoing discussion by considering an arbitrary fluctuation in an isolated system. From the discussion of the preceding section we see that an appropriate choice of these fluctuating variables are the densities $\rho_k(\mathbf{r})$ and the entropy is then a functional of these variables. To avoid the complications of functional differentiation we assume that the variables $\rho_k(\mathbf{r})$ have been expressed in terms of a set of discrete variables $\alpha_i, i = 1, 2, \ldots$ that are zero in the equilibrium state of maximum entropy. The number of such variables is, in principle, equal to the total number of degrees of freedom of the system less the number of extensive bulk thermodynamic variables. The variables α_i could, for example, be coefficients of a Fourier expansion or of an expansion in a set of orthogonal polynomials. The probability that the system is in a state with specific values of these variables is

$$P(\alpha_1, \alpha_2, \ldots \alpha_n) = \frac{\exp\{S(\alpha_1, \alpha_2, \ldots \alpha_n)/k_B\}}{\int d\alpha_1, d\alpha_2 \ldots d\alpha_n \exp\{S(\alpha_1, \alpha_2, \ldots \alpha_n)/k_B\}} \ . \tag{12.163}$$

Since $(\partial S/\partial \alpha_i)|_{\{\alpha\}=0} = 0$, we have, quite generally,

$$P(\alpha_1, \alpha_2, \ldots \alpha_n) = C \exp\left\{-\frac{1}{2}\sum_{j,m} \frac{g_{jm}\alpha_j\alpha_m}{k_B}\right\} \tag{12.164}$$

where C is a normalizing constant and where the coefficients g_{jm} are symmetric: $g_{jm} = g_{mj}$. This expression immediately implies that $\langle \alpha_j \rangle = 0$. Thermodynamic stability also implies that the determinant of the matrix $\mathbf{g}$ is greater than zero and that all its eigenvalues are positive.

We now obtain a simple expression for the correlation functions. Noting that

$$\frac{\partial P(\{\alpha\})}{\partial \alpha_i} = \frac{1}{k_B}\frac{\partial S}{\partial \alpha_i}P$$

we find

$$\begin{aligned}\left\langle \alpha_i \frac{\partial S}{\partial \alpha_i}\right\rangle &= \int d\alpha_1 d\alpha_2 \ldots d\alpha_n \alpha_i \frac{\partial S}{\partial \alpha_i} P(\{\alpha\}) \\ &= k_B \int d\alpha_1 d\alpha_2 \ldots d\alpha_n \alpha_i \frac{\partial P}{\partial \alpha_i}\end{aligned} \tag{12.165}$$

and integrating by parts,

$$\left\langle \alpha_i \frac{\partial S}{\partial \alpha_i}\right\rangle = -k_B \tag{12.166}$$

and for $i \neq m$,

$$\left\langle \alpha_i \frac{\partial S}{\partial \alpha_m}\right\rangle = 0 \;. \tag{12.167}$$

Thus

$$\sum_m g_{im}\langle \alpha_m \alpha_j \rangle = k_B \delta_{ij} \;. \tag{12.168}$$

We now return to the transport equations. The time derivatives of the variables α_i are proportional to observable currents such as the heat currents considered in the first part of this subsection. We assume again that a linearized equation

$$j_i = \frac{d\alpha_i}{dt} = \sum_j L_{ij} X_j \tag{12.169}$$

describes the response of the system to generalized forces X_j. These generalized forces can be expressed in terms of the entropy through $X_j = \partial S/\partial \alpha_j$. Thus

$$\frac{d\alpha_i}{dt} = \sum_j L_{ij}\left(\frac{\partial S}{\partial \alpha_j}\right) . \tag{12.170}$$

Consider, once again, the expectation value

$$\begin{aligned}\langle \alpha_i(t+\Delta t)\alpha_j(t)\rangle &= \langle \alpha_i(t)\alpha_j(t)\rangle + \Delta t \langle \dot{\alpha}_i(t)\alpha_j(t)\rangle \\ &= \langle \alpha_i(t)\alpha_j(t)\rangle + \Delta t \sum_m L_{im}\left\langle \frac{\partial S}{\partial \alpha_m}\alpha_j \right\rangle \\ &= \langle \alpha_i(t)\alpha_j(t)\rangle - \Delta t L_{ij} k_B \;.\end{aligned} \tag{12.171}$$

Similarly,

$$\langle\alpha_j(t+\Delta t)\alpha_i(t)\rangle = \langle\alpha_j(t)\alpha_i(t)\rangle - \Delta t L_{ji} k_B \tag{12.172}$$

which implies, by the principle of microscopic reversibility, that

$$L_{ij} = L_{ji} \ . \tag{12.173}$$

Thus, quite generally, we have arrived at the symmetry relations of the linear transport coefficients.

12.4 The Boltzmann Equation

In this section we discuss transport theory from the point of view of the Boltzmann equation. This approach lacks the general validity of the Kubo formalism. On the other hand, the different terms in the Boltzmann equation have a straightforward physical interpretation and the approach leads to explicit results. Our discussion will be rather brief and we will limit ourselves to situations in which linear response theory is applicable. We refer the reader to Ziman [334] and Callaway [54] for a more extensive discussion. We develop the formalism in Section 12.4.1, discuss DC conductivity in Section 12.4.2, and thermoelectric effects in Section 12.4.3.

12.4.1 Fields, drift and collisions

We assume, in the spirit of mean field theory, that the system of interest can be adequately described in terms of the single-particle distribution, $f_{\mathbf{p}}(\mathbf{r})$, which is the density of particles with momentum $\mathbf{p}$ at position $\mathbf{r}$. The distribution is normalized (in accordance with our phase space definitions of Section 2), so that

$$\frac{1}{h^3}\int d^3p \int d^3r f_{\mathbf{p}}(\mathbf{r}) = N \tag{12.174}$$

where N is the number of particles.

In the case of a quantum system, such as electrons in a crystalline solid, a particle with momentum $\mathbf{p}$ is represented as a Bloch function with wave vector $\mathbf{k} = \mathbf{p}/\hbar$, energy $\epsilon_{\mathbf{k}}$, and velocity $\mathbf{v_p} = \partial\epsilon_{\mathbf{k}}/\partial\hbar\mathbf{k}$. Because of the uncertainty principle, we must assume that $f_{\mathbf{p}}(\mathbf{r})$ is coarse grained over a sufficiently large volume that $\mathbf{r}$ can be considered to be a *macroscopic* variable. This is consistent with the fundamental assumption, made elsewhere in this chapter, that densities and fields are sufficiently slowly varying that they are well defined locally. We will take a *semiclassical* approach to quantum systems, that is, we assume that in the presence of macroscopic electric and magnetic fields the equation of motion for particles of charge e is

$$\dot{\mathbf{p}} = \hbar\dot{\mathbf{k}} = e\left(\mathbf{E} + \mathbf{v_k} \times \mathbf{B}\right) \ . \tag{12.175}$$

If the acceleration of the particles due to the fields were the only effect causing changes in the distribution function, we would have

$$f_{\mathbf{p}}(t + \delta t) = f_{\mathbf{p}-\delta\mathbf{p}}(t)$$

or

$$\left.\frac{\partial f_{\mathbf{p}}(\mathbf{r})}{\partial t}\right|_{field} = -e\left(\mathbf{E} + \mathbf{v_k} \times \mathbf{B}\right) \cdot \frac{\partial f_{\mathbf{p}}}{\partial \mathbf{p}} \ . \tag{12.176}$$

If we take the fields to be slowly varying, we can visualize the states as wave packets that are accelerated by the fields according to the classical equations of motion. This assumption is difficult to justify rigorously and we will not attempt to do so. In some cases, such as in inhomogeneous semiconductors, near surfaces, or in insulators subjected to intense fields, the electric fields are strong enough to cause *tunneling*. The semiclassical approach is then not appropriate.

Particles from time to time undergo *collisions* with obstacles in their path. Electrons in solids are scattered by impurities, vacancies, dislocations, and phonons. Because of screening, the interactions responsible for scattering are generally short-ranged and this implies that collisions should be treated quantum mechanically. Let $W(\mathbf{k},\mathbf{k}')$ be the transition rate from state $\mathbf{k}$ to state $\mathbf{k}'$. This transition rate is typically calculated approximately by means of the golden rule (Problem 12.8). The distribution function then changes in time due to transitions *into* and *out* of state $\mathbf{k}$ and we may write

$$\left(\frac{\partial f_{\mathbf{k}}}{\partial t}\right)_{coll} = \sum_{\mathbf{k}'}[f_{\mathbf{k}'}(1 - f_{\mathbf{k}})W(\mathbf{k}',\mathbf{k}) - f_{\mathbf{k}}(1 - f_{\mathbf{k}'})W(\mathbf{k},\mathbf{k}')] \ .$$

From the principle of detailed balance, we have, for elastic processes, $W(\mathbf{k},\mathbf{k}') = W(\mathbf{k}',\mathbf{k})$ and

$$\left.\frac{\partial f_{\mathbf{p}}(\mathbf{r})}{\partial t}\right|_{coll} = \sum_{\mathbf{k}'}\{f_{\mathbf{k}'} - f_{\mathbf{k}}\}\, W(\mathbf{k}',\mathbf{k}) \ . \tag{12.177}$$

In the rest of this section we consider only a system of fermions and label states by wave vectors $\mathbf{k}$ rather than momenta $\mathbf{p}$, as we have done in (12.177). In what follows we also limit ourselves to the simplest treatment of collisions and work within the relaxation time approximation.

We assume that the external fields produce only a small change in the distribution function and write

$$f_{\mathbf{k}} = f^0_{\mathbf{k}} + g_{\mathbf{k}} \tag{12.178}$$

where $f^0_{\mathbf{k}}$ is the equilibrium (zero-field) distribution given, as appropriate by the Boltzmann, Bose–Einstein, or, in our case, the Fermi–Dirac distributions. In the relaxation-time approximation one assumes that if the external fields were switched off, the nonequilibrium part of the distribution function would decay exponentially with time:

$$g_{\mathbf{k}}(t) = g_{\mathbf{k}}(0)e^{-t/\tau}$$

where τ is the relaxation time. We thus obtain

$$\left.\frac{\partial f^0_{\mathbf{k}}}{\partial t}\right|_{coll} = 0 \qquad \frac{\partial g_{\mathbf{k}}}{\partial t} = -\frac{g_{\mathbf{k}}}{\tau}\;. \tag{12.179}$$

If the distribution is inhomogeneous, it will change in time due to *drift*. If the particles are not subject to any forces,

$$f_{\mathbf{k}}(\mathbf{r}, t+\delta t) = f_{\mathbf{k}}(\mathbf{r} - \mathbf{v}_{\mathbf{k}}\delta t, t) \tag{12.180}$$

and hence

$$\left.\frac{\partial f_{\mathbf{k}}}{\partial t}\right|_{drift} = -\mathbf{v}_{\mathbf{k}}\cdot\frac{\partial f_{\mathbf{k}}}{\partial \mathbf{r}}\;. \tag{12.181}$$

Combining the various terms, we obtain the *Boltzmann equation* for the distribution function:

$$\frac{df_{\mathbf{k}}}{dt} = \left.\frac{\partial f_{\mathbf{k}}}{\partial t}\right|_{field} + \left.\frac{\partial f_{\mathbf{k}}}{\partial t}\right|_{coll} + \left.\frac{\partial f_{\mathbf{k}}}{\partial t}\right|_{drift}\;. \tag{12.182}$$

In a steady-state situation we require $df_{\mathbf{k}}/dt = 0$.

12.4.2 DC conductivity of a metal

We next assume that for a weak electric field the nonequilibrium part, $g_{\mathbf{k}}$, is proportional to the field and linearize the Boltzmann equation. This yields

$$-\frac{e\mathbf{E}}{\hbar}\cdot\frac{\partial f^0_{\mathbf{k}}}{\partial \mathbf{k}} - \frac{g_{\mathbf{k}}}{\tau} = 0$$

or

$$g_{\mathbf{k}} = -\frac{e\mathbf{E}\tau}{\hbar}\frac{\partial f^0_{\mathbf{k}}}{\partial \epsilon_{\mathbf{k}}}\cdot\frac{\partial \epsilon_{\mathbf{k}}}{\partial \mathbf{k}} = -\tau e\mathbf{E}\cdot\mathbf{v}_{\mathbf{k}}\frac{\partial f^0_{\mathbf{k}}}{\partial \epsilon_{\mathbf{k}}}\;. \tag{12.183}$$

Note that we have taken the charge of an electron to be e. For a metal at not too high a temperature,

$$\frac{\partial f^0}{\partial \epsilon} \sim -\delta(\epsilon - \epsilon_F) \ . \tag{12.184}$$

The electrical current density is given by

$$\mathbf{j} = \frac{2}{V}\sum_{\mathbf{k}} e\mathbf{v}_{\mathbf{k}} g_{\mathbf{k}} \ . \tag{12.185}$$

Let S_ϵ be a constant energy surface, that is, the surface in k-space for which $\epsilon_{\mathbf{k}} = \epsilon$ and let $\hat{n}$ be the unit vector normal to S_ϵ. We then have

$$\sum_{\mathbf{k}} = \frac{V}{(2\pi)^3}\int d^3k = \frac{V}{(2\pi)^3}\int d\epsilon \int dS_\epsilon \left(\hat{n}\cdot\frac{\partial \mathbf{k}}{\partial \epsilon_{\mathbf{k}}}\right) \tag{12.186}$$

and obtain, for the current density,

$$\mathbf{j} = \frac{e^2\tau}{4\pi^3\hbar}\int \frac{dS_F}{|v_{\mathbf{k}}|}\mathbf{v}_{\mathbf{k}}(\mathbf{v}_{\mathbf{k}}\cdot\mathbf{E}) \tag{12.187}$$

where S_F is the Fermi surface. This in turn yields the conductivity (in dyadic form):

$$\sigma = \frac{e^2\tau}{4\pi^3\hbar}\int dS_F \frac{\mathbf{v}_{\mathbf{k}} : \mathbf{v}_{\mathbf{k}}}{|\mathbf{v}_{\mathbf{k}}|} \ . \tag{12.188}$$

This formula illustrates that the conductivity is, in general, a tensor. In component form

$$j_i = \sum_m \sigma_{im} E_m \tag{12.189}$$

and we have, for example,

$$\sigma_{xx} = \frac{e^2\tau}{4\pi^3\hbar}\int dS_F \frac{(\mathbf{v}_{\mathbf{k}})_x^2}{|\mathbf{v}_{\mathbf{k}}|} \ . \tag{12.190}$$

In discussions of systems that lack spherical symmetry, one finds that some authors use anisotropic relaxation times $\tau(\mathbf{k})$. The relaxation time must then be kept inside the integral. In the references listed at the beginning of this section it is pointed out that there are serious difficulties with such an approach.

In an isotropic system the conductivity tensor is diagonal and

$$\sigma = \frac{e^2\tau}{12\pi^3\hbar}\int dS_F v_F \ . \tag{12.191}$$

We write $v_F = hk_F/m^*$ for the Fermi velocity, where m^* is the effective mass and, using $n = N/V = k_F/3\pi^2$, obtain

$$\sigma = \frac{ne^2\tau}{m^*} \tag{12.192}$$

which is of the same form as the conductivity of the simple Drude model (see, *e.g.*, Ashcroft and Mermin, [21]).

It is useful to interpret the result

$$f_{\mathbf{k}} = f^0_{\mathbf{k}} + g_{\mathbf{k}} = f^0_{\mathbf{k}} - \tau e(\mathbf{E}\cdot\mathbf{v}_{\mathbf{k}})\frac{\partial f^0_{\mathbf{k}}}{\partial \epsilon_{\mathbf{k}}} \tag{12.193}$$

in a different way. To first order in E, we can rewrite (12.193) in the form

$$f_{\mathbf{k}} = f^0_{\mathbf{k}}\{\epsilon_{\mathbf{k}} - \tau e(\mathbf{E}\cdot\mathbf{v}_{\mathbf{k}})\}\ . \tag{12.194}$$

The right-hand side of this equation is simply the equilibrium distribution of the system with all energies shifted by an amount

$$\delta\epsilon_{\mathbf{k}} = \tau e(\mathbf{E}\cdot\mathbf{v}_{\mathbf{k}}) \tag{12.195}$$

i.e., by precisely the amount expected classically for particles moving with constant velocity $\mathbf{v}$ for a time τ in a force field $e\mathbf{E}$. The extra energy gained in this way can be interpreted in terms of a drift velocity $\delta\mathbf{v}_{\mathbf{k}}$ in the direction of the field so that

$$\delta\mathbf{v}_{\mathbf{k}}\cdot\frac{\partial \epsilon_{\mathbf{k}}}{\partial \mathbf{v}_{\mathbf{k}}} = e\tau(\mathbf{v}_{\mathbf{k}}\cdot\mathbf{E})\ . \tag{12.196}$$

If

$$\frac{\partial \epsilon_{\mathbf{k}}}{\partial \mathbf{v}_{\mathbf{k}}} = \mathbf{p}_{\mathbf{k}} = m^*\mathbf{v}_{\mathbf{k}} \tag{12.197}$$

we obtain

$$\delta\mathbf{v}_{\mathbf{k}} = \frac{e\tau}{m^*}\mathbf{E}\ . \tag{12.198}$$

For n particles per unit volume, we have for the current,

$$\mathbf{j} = ne\delta\mathbf{v} \tag{12.199}$$

and we recover (12.192) for the conductivity.

In the case of a metal the drift velocities are typically very small compared to the Fermi velocity v_F, mainly because the electric fields inside a metal tend to be small. In a semiconductor one sometimes deals with fields which are large enough that nonohmic effects are important. It is then not adequate to linearize the Boltzmann equation and one must consider the nonlinear problem. In such situations collisions often occur so frequently that one cannot describe them as independent events and the whole Boltzmann approach becomes suspect.

12.4.3 Thermal conductivity and thermoelectric effects

To this point we have only considered a situation in which the distribution function was spatially uniform. To give an example of the use of the Boltzmann equation when the drift term comes into play, we discuss the case of a time-independent temperature gradient maintained across a metallic sample. From Fourier's law we expect that, in analogy with Ohm's law, there will be a heat current whenever there is a temperature gradient:

$$\mathbf{j}_Q = L_{QQ}\nabla\left(\frac{1}{T}\right) = -\kappa\nabla T \tag{12.200}$$

where κ is the thermal conductivity and L_{QQ} the kinetic coefficient defined in Section 12.3. We also allow for an electric field $\mathbf{E}$, with corresponding scalar potential $\phi(\mathbf{r})$. From thermodynamics we have

$$Tds = du - \mu' dn \tag{12.201}$$

where $\mu' = \mu + e\phi(\mathbf{r})$ is the electrochemical potential at point $\mathbf{r}$ and where s, u, n are the entropy, energy, and particle densities. Hence

$$\mathbf{j}_Q = \mathbf{j}_\epsilon - \mu'\mathbf{j}_N \ . \tag{12.202}$$

We assume that the heat current is due entirely to the motion of electrons and neglect the lattice thermal conductivity. In the presence of the electrostatic potential $\phi(\mathbf{r})$ the electronic energies will be "locally" shifted by an amount $e\phi(\mathbf{r})$ so that the *energy current* density is then given by

$$\mathbf{j}_\epsilon(\mathbf{r}) = \frac{2}{(2\pi)^3}\int d^3k\,[\epsilon_\mathbf{k} + e\phi(\mathbf{r})]\,\mathbf{v}_\mathbf{k} f_\mathbf{k}(\mathbf{r}) \tag{12.203}$$

and the particle current density is

$$\mathbf{j}_N(\mathbf{r}) = \frac{2}{(2\pi)^3}\int d^3k\,\mathbf{v}_\mathbf{k} f_\mathbf{k}(\mathbf{r}) \tag{12.204}$$

which, of course, also implies an electrical current density $\mathbf{j}_c = e\mathbf{j}_N$. Thus the heat current (12.202) is given by

$$\mathbf{j}_Q(\mathbf{r}) = \frac{2}{(2\pi)^3}\int d^3k[\epsilon_\mathbf{k} - \mu]\mathbf{v}_\mathbf{k} f_\mathbf{k}(\mathbf{r}) \ . \tag{12.205}$$

As before we write

$$g_\mathbf{k}(\mathbf{r}) = f_\mathbf{k}(\mathbf{r}) - f^0_\mathbf{k}$$

and, in addition, assume that the thermal gradient is small enough that it is meaningful to talk about a local temperature and a local chemical potential. With these assumptions, the Boltzmann equation becomes

$$-\mathbf{v}_{\mathbf{k}}\cdot\frac{\partial f_{\mathbf{k}}}{\partial \mathbf{r}} - \frac{e}{\hbar}\mathbf{E}\cdot\frac{\partial f_{\mathbf{k}}}{\partial \mathbf{k}} + \left.\frac{\partial f_{\mathbf{k}}}{\partial t}\right|_{coll} = 0 . \tag{12.206}$$

We take $f^0_{\mathbf{k}}(\mathbf{r})$ to be the equilibrium distribution with the local temperature $T(\mathbf{r})$ and the local chemical potential $\mu'(\mathbf{r})$ controlling the density at point $\mathbf{r}$. Noting that $\epsilon_{\mathbf{k}} + e\phi(\mathbf{r}) - \mu'(\mathbf{r}) = \epsilon_{\mathbf{k}} - \mu(\mathbf{r})$, we have

$$f^0_{\mathbf{k}} = f^0\{\epsilon_{\mathbf{k}}, \mu(\mathbf{r}), T(\mathbf{r})\} = \left[\exp\left\{\frac{\epsilon_{\mathbf{k}} - \mu(\mathbf{r})}{k_B T(\mathbf{r})}\right\} + 1\right]^{-1} \tag{12.207}$$

and hence

$$\frac{\partial f^0_{\mathbf{k}}}{\partial \mathbf{r}} = \frac{\partial f^0_{\mathbf{k}}}{\partial T}\frac{\partial T}{\partial \mathbf{r}} + \frac{\partial f^0_{\mathbf{k}}}{\partial \mu}\frac{\partial \mu}{\partial \mathbf{r}} . \tag{12.208}$$

We next make the relaxation-time approximation (12.180) and in the spirit of the linearized Boltzmann equation, neglect terms such as

$$\frac{\partial g}{\partial \mathbf{r}} \quad \text{and} \quad \frac{e}{\hbar}\mathbf{E}\cdot\frac{\partial g}{\partial \mathbf{k}} .$$

Using

$$\frac{\partial f^0_{\mathbf{k}}}{\partial T} = -\frac{\epsilon_{\mathbf{k}} - \mu}{T}\frac{\partial f^0_{\mathbf{k}}}{\partial \epsilon_{\mathbf{k}}} \qquad \frac{\partial f^0_{\mathbf{k}}}{\partial \mu} = -\frac{\partial f^0_{\mathbf{k}}}{\partial \epsilon_{\mathbf{k}}}$$

and collecting terms, we obtain

$$\frac{1}{\tau}g_{\mathbf{k}} = -\frac{\partial f^0_{\mathbf{k}}}{\partial \epsilon_{\mathbf{k}}}\mathbf{v}_{\mathbf{k}}\cdot\left[-\frac{\epsilon_{\mathbf{k}} - \mu}{T}\frac{\partial T}{\partial \mathbf{r}} + \left(e\mathbf{E} - \frac{\partial \mu}{\partial \mathbf{r}}\right)\right] . \tag{12.209}$$

The potential difference measured by, say, a voltmeter is not

$$\int \mathbf{E}\cdot d\mathbf{s}$$

but rather the quantity

$$\Psi = \int \left(\mathbf{E} - \frac{1}{e}\nabla\mu\right)\cdot d\mathbf{s} .$$

We therefore introduce the "electromotive field" or "observed" field

$$\mathcal{E} = \mathbf{E} - \frac{1}{e}\nabla\mu = -\frac{1}{e}\nabla\mu' . \tag{12.210}$$

Clearly, $\mathcal{E}$ is of more interest than the electric field $\mathbf{E}$ itself. We now define the kinetic coefficients L_{CC}, L_{CQ}, L_{QC}, and L_{QQ} through

$$\begin{aligned} \mathbf{j}_C &= L_{CC}\mathcal{E} + L_{CQ}\nabla\left(\frac{1}{T}\right) \\ \mathbf{j}_Q &= L_{QC}\mathcal{E} + L_{QQ}\nabla\left(\frac{1}{T}\right) . \end{aligned} \tag{12.211}$$

Using (12.204) and (12.205), we see that the kinetic coefficients can all be expressed in terms of the integral

$$I_\alpha = \int d\epsilon \left(-\frac{\partial f^0}{\partial \epsilon}\right)(\epsilon - \mu)^\alpha \sigma(\epsilon) \tag{12.212}$$

where

$$\sigma(\epsilon) = e^2\tau \int \frac{d^3k}{4\pi^3}\delta(\epsilon - \epsilon_{\mathbf{k}})\mathbf{v}_{\mathbf{k}} : \mathbf{v}_{\mathbf{k}} \tag{12.213}$$

is the generalized energy-dependent form of the conductivity tensor (12.188). We shall evaluate I_α for conditions appropriate to a metal and in this case

$$-\frac{\partial f^0}{\partial \epsilon} = \frac{\beta \exp\{\beta(\epsilon - \mu)\}}{[\exp\{\beta(\epsilon - \mu)\} + 1]^2} \tag{12.214}$$

can be taken to be nonzero only in a narrow energy range of order k_BT around ϵ_F. We introduce the new variable $z = \beta(\epsilon - \mu)$ and expand

$$\sigma(k_BTz + \mu) = \sigma(\mu) + k_BTz\frac{\partial \sigma}{\partial \mu} + \dots .$$

Substituting in (12.213), we then have the transport coefficients expressed in terms of

$$I_\alpha \approx (k_BT)^\alpha \int_{-\infty}^{\infty} dz \frac{z^\alpha e^z}{(1+e^z)^2}\left[\sigma(\mu) + k_BT\frac{\partial \sigma}{\partial \mu}\right] . \tag{12.215}$$

Defining

$$Q_j = \int_{-\infty}^{\infty} dz \frac{z^j}{(e^z+1)(e^{-z}+1)}$$

we have $Q_0 = 1$, $Q_1 = 0$, $Q_2 = \pi^2/3$, and $Q_3 = 0$. Taking $\mu \approx \epsilon_F$, we thus obtain

$$L_{CC} = \sigma(\epsilon_F) = \sigma \tag{12.216}$$

$$L_{CQ} = TL_{QC} = \frac{\pi^2}{3e^2}k_B^2T^3\left.\frac{\partial \sigma(\epsilon)}{\partial \epsilon}\right|_{\epsilon=\epsilon_F} \tag{12.217}$$

$$L_{QQ} = \frac{\pi^2}{3e^2} k_B^2 T^3 \sigma \ . \tag{12.218}$$

We see that in our approximate treatment we obtain $L_{CQ} = TL_{QC}$, which is an Onsager relation, as can easily be shown (Problem 12.9). We note also that for electrons ($e = -|e|$), L_{QC} and L_{CQ} are negative. If these coefficients are found experimentally to be positive, it is an indication that the charge carriers are holes.

To obtain the thermal conductivity we require that there be no electric current, or from (12.211),

$$\mathcal{E} = -L_{CC}^{-1} L_{CQ} \nabla \left(\frac{1}{T}\right) \ . \tag{12.219}$$

Substituting into the second equation of (12.211), we have for the thermal conductivity

$$\kappa = \frac{L_{QQ} - L_{QC} L_{CC}^{-1} L_{CQ}}{T^2} \ . \tag{12.220}$$

We now argue that in a metal the second term in (12.220) is small compared to the first. We first note that the second law of thermodynamics implies that κ is positive or $L_{CC}L_{QQ} > L_{CQ}L_{QC}$. To obtain an order-of-magnitude estimate, we make the approximation (on dimensional grounds) $\partial\sigma/\partial\epsilon \sim \sigma/\epsilon$ (for free electrons $\partial\sigma/\partial\epsilon = 3\sigma/2\epsilon$) and thus have

$$\frac{L_{CQ}L_{QC}}{L_{CC}L_{QQ}} \approx \frac{\pi^2}{3}\left(\frac{k_B T}{\epsilon_F}\right)^2 \sim 10^{-4}$$

for a typical metal at room temperature. Neglecting the second term in (12.220), we therefore find

$$\kappa = \frac{L_{QQ}}{T^2} = \frac{\pi^2}{3e^2} k_B^2 T \sigma \ . \tag{12.221}$$

This result is known as the Wiedemann–Franz law and is often derived from more elementary considerations (see, e.g., Ashcroft and Mermin, [21]). However, the present derivation suggests that this law should hold under quite general circumstances. The main assumption made above was the use of the relaxation-time approximation. This relies, in turn, on the assumption of elastic scattering. The dominant inelastic processes are emission and absorption of phonons and it is interesting that the most significant deviations from the Wiedemann–Franz law occur near the Debye temperature. At lower temperatures most electron–phonon processes are frozen out, while at room temperature or higher, phonon energies are small compared to k_BT, and the scattering processes can be taken to be "quasi elastic."

The coupled transport equations (12.211) suggest other thermoelectric effects such as the Seebeck and Peltier effects. We refer to the literature for a discussion of these.

12.5 Problems

12.1. *Thomas-Fermi and Debye Screening.*

If an external potential energy $\phi(\mathbf{r})$ varies sufficiently slowly the chemical potential must satisfy

$$\mu = \mu_0(n(\mathbf{r})) + \phi(\mathbf{r}) = \text{constant}$$

where μ_0 is the chemical potential of a system of particles of density $n(\mathbf{r})$ in the absence of the external field.

(a) Consider an electron gas at $T = 0$ with electron states filled up to a Fermi level $\epsilon_F = \hbar^2 k_F^2/2m$, where k_F is the Fermi wave vector. Show that in these circumstances the free particle static susceptibility is given by

$$\chi_0(\mathbf{q}) = -\frac{\partial n_0}{\partial \epsilon_F} = -\frac{1}{2\pi^2}\left(\frac{2m}{\hbar^2}\right)^{3/2} \epsilon_F^{1/2} .$$

(b) Show that the mean field dielectric function can be expressed in the form

$$\epsilon(\mathbf{q}) = 1 + \frac{k_{TF}^2}{q^2}$$

where $k_{TF} = [me^2 k_F/(\pi^2 \epsilon_0 \hbar^2)]^{1/2}$ is the Thomas-Fermi wave vector.

(c) Show that if $\phi_{ext} = -e^2/4\pi\epsilon_0 r$ is the potential energy due to an external point charge, the mean field effective potential becomes

$$\phi_{ind}(\mathbf{r}) = -\frac{e^2}{4\pi\epsilon_0 r} e^{-k_{TF} r} .$$

(d) Consider a classical gas of charged particles at temperature T. Show that with the same approximations

$$\phi_{tot}(\mathbf{r}) = -\frac{e^2}{4\pi\epsilon_0 r} e^{-r/\lambda_D}$$

where λ_D is the Debye screening length given by $\lambda_D = (n_0 e^2/\epsilon_0 k_B T)^{1/}$

12.2. *Pair Distribution Function of an Ideal Bose Gas.*

(a) Compute the dynamic structure factor $S(\mathbf{q}, \omega)$ for an ideal Bose gas at $T \neq 0$.

(b) Find the expression for the geometric structure factor (i) by integrating the expression for $S(\mathbf{q}, \omega)$ over frequency and (ii) by evaluating the expression for the geometric structure factor directly.

(c) Evaluate numerically the pair distribution function $g(r)$ for an ideal Bose gas for a temperature (i) above and (ii) below the Bose condensation temperature (see Section 11.1). Show that in the latter case $g(r)$ does not approach unity as $r \to \infty$.

12.3. *Mean Field Approximation for the Bose Gas.*

(a) Show that the mean field response function (12.70) satisfies the f-sum rule.

(b) The mean field theory for the weakly interacting Bose gas may give rise to a pair distribution function that is quite unphysical at short distances. Show that (12.71) yields a form for $g(r)$ which diverges as $r \to 0$ unless the pair potential $v_{\mathbf{q}}$ falls off faster than $1/q$ for large q.

12.4. *Dispersion of Plasma Oscillations.*

Extend the calculation of the mean field response of an electron gas at $T = 0$ for $\omega \gg \hbar^2 q^2/2m$ to next order in q and show that

$$\Omega_{pl}(q) = \Omega_{pl}(0) \left[1 + \frac{3q^2 v_F^2}{10\Omega_{pl}^2(0)} + \cdots\right]$$

where $v_F = \hbar k_F/m$ is the Fermi velocity and $\Omega_{pl}^2(0) = e^2 n/\epsilon_0 m$.

12.5. *Screening in Two Dimensions.*

Consider a two-dimensional system of electrons distributed on a surface of area $\mathcal{A}$ at $T = 0$ and with a compensating uniform positive background ensuring overall neutrality.

(a) Show that the two-dimensional Fourier transform of the Coulomb potential is $v_{\mathbf{q}} = e^2/(2q\epsilon_0)$.

(b) Show that the static free-particle susceptibility is

$$\chi_0(\mathbf{q},0)\begin{cases} = -\frac{m}{\hbar^2\pi} & q < 2k_F \\ = -\frac{m}{\hbar^2\pi}\left[1-\sqrt{1-\frac{4k_F^2}{q^2}}\right] & q > 2k_F\,. \end{cases}$$

(c) Consider a point charge e a distance z above the layer of electrons. Show that the two-dimensional Fourier transform of the bare Coulomb potential from this charge in the electron gas is $-e^2\exp\{-qz\}/(2q\epsilon_0)$.

(d) Find expressions for the mean field screened potential and for the screening charge distribution in the electron layer.

12.6. *Low-Temperature Properties of the Heisenberg Model.*

Consider a spin system described by the isotropic Heisenberg model with spin wave excitation spectrum $\epsilon(\mathbf{q}) = J\hbar^2 Sa^2q^2$. Assume that the mean number of spin waves with wave vector $\mathbf{q}$ obeys the Bose-Einstein distribution function. The total magnetization is reduced from its saturation value $N\hbar S$ by $\hbar$ for each spin wave that is excited.

(a) Show that with these assumptions the low-temperature spontaneous magnetization is of the form

$$M(T) = M(0)(1 - \text{const.}T^{3/2})$$

in three dimensions. Find the value of the multiplicative constant.

(b) Show that the integral which gives the deviation of the spontaneous magnetization from the saturation value diverges in one and two dimensions. This result is an indication of the absence of long-range order in the Heisenberg model in one and two dimensions. For a proof not based on spin wave theory, see Mermin and Wagner [203].

12.7. *Symmetry of Transport Coefficients.*

Consider the thermal conductivity tensor L defined in (12.151).

(a) Show that in the case of a cubic crystal the symmetry of the lattice requires that this tensor be diagonal with $L_{11} = L_{22} = L_{33}$.

(b) Show that for a crystal with a 3-fold axis of rotation $L_{12} = -L_{21}$, $L_{11} = L_{22}$, and $L_{i3} = L_{3i} = 0$ for $i = 1, 2$, where 3 represents the direction of the symmetry axis.

12.8. *Estimate of Relaxation Time for Impurity Scattering.*

Consider a situation in which the transition rate for impurity scattering can be described by plane-wave matrix elements of a spherically symmetric impurity potential $u(\mathbf{r})$:

$$W(\mathbf{k},\mathbf{k}') = \frac{2\pi}{\hbar}|\langle\mathbf{k}'|u(\mathbf{r})|\mathbf{k}\rangle|^2\delta(\epsilon_{\mathbf{k}'} - \epsilon_{\mathbf{k}})$$

and let

$$U_{\mathbf{q}} = \int d^3re^{-i\mathbf{q}\cdot\mathbf{r}}u(\mathbf{r}) \ .$$

(a) Consider a free electron gas at $T = 0$ with n_i impurities per unit volume. Show that substitution of

$$g_{\mathbf{k}} = -\tau e\mathbf{E}\cdot\mathbf{v}_{\mathbf{k}}\frac{\partial f^0_{\mathbf{k}}}{\partial\epsilon_{\mathbf{k}}}$$

into the linearized Boltzmann equation yields

$$\frac{1}{\tau} = \frac{n_i}{4\pi^2 v_F\hbar^2}\int d^2S_F U^2_{\mathbf{k}-\mathbf{k}'}\left(1 - \frac{\mathbf{k}\cdot\mathbf{k}'}{k_F^2}\right)$$

where $\mathbf{k}$ lies on the Fermi surface, the integration over $\mathbf{k}'$ extends over the Fermi surface, and $v_F = \hbar k_F/m$ is the Fermi velocity.

(b) Estimate the low-temperature resistivity in Ωm of a sample of 0.1% Mg in Na. Since Mg has a valence $Z = 2$ (one more than Na), the impurity potential can be taken to be that of a Thomas-Fermi screened point charge e (use Lindhard screening if you have easy access to a computer). The Fermi wave vector of Na is $9.2nm^{-1}$.

12.9. *Onsager Relation for the Thermoelectric Effect.*

To obtain the Onsager relation (12.217) in the situation in which there is a thermal gradient as well as an electric field, one writes, as in (12.201),

$$Tds = du - \mu' dn$$

where $\mu' = \mu + e\phi(\mathbf{r})$ and where μ is the chemical potential in the absence of an electric field. The energy and particle currents, defined as in Section 12.3.2, then obey the phenomenological equations

$$\begin{aligned}\mathbf{j}_E &= L_{EE}\nabla\left(\frac{1}{T}\right) + L_{EN}\nabla\left(-\frac{\mu'}{T}\right)\\ \mathbf{j}_N &= L_{NE}\nabla\left(\frac{1}{T}\right) + L_{NN}\nabla\left(-\frac{\mu'}{T}\right)\end{aligned}$$

with an Onsager relation $L_{NE} = L_{EN}$. Show that the relation (12.217) $L_{CQ} = TL_{QC}$ follows.

Chapter 13

Disordered Systems

Real materials are seldom, if ever, the idealized pure systems that we have discussed to this point. Magnetic crystals invariably contain defects and non-magnetic impurities. Liquids, which we have generally taken to be composed of a single component, invariably have impurities dissolved in them. Even liquid helium will have a certain amount of isotopic disorder. Thus it is important to understand the effect of disorder on the properties of materials and to what extent the theoretical framework that we have constructed remains valid when we attempt to describe real materials.

The physics of disordered systems is a vast subject with an extensive literature. In one chapter we will hardly be able to give a comprehensive treatment of the material, and the reader may wish to consult, for example, the book by Ziman [335] or one of the review articles that we shall mention when we discuss specific topics.

The effects of disorder begin at the microscopic level. The energy levels and wave functions of a particle in a disordered medium can be substantially different from those in a pure material. In our discussion of electrons in metals we have, to this point, assumed translational invariance. This assumption is, for the statistical mechanics of a pure metal, quite adequate. The periodic potential due to the ions changes the shape of the Fermi surface and is responsible for the detailed properties of individual materials. The incorporation of such effects into our statistical formalism presents us only with minor technical problems; the basic physical picture is the same as that of the idealized system. Similarly, the use of realistic force constants, rather than springs between nearest neighbors, in the vibrational energy of a crystal is a conceptually trivial change. The translational symmetry of the Hamiltonian makes both of these problems in principle very simple, although technical difficulties may arise in actual computations.

The situation is dramatically different in a disordered material. In a one-dimensional disordered material the electronic wave functions are *localized* while they are *extended* in a periodic potential. In two and three dimensions we believe that some, if not all, of the states of a disordered system are localized. This has important implications for transport coefficients. Intuitively, one expects that if the Fermi level falls into a range of localized states, the conductivity will decrease as the temperature is lowered—in contrast to the behavior of a pure material in which the resistivity decreases due to the freezing out of phonon scattering. We discuss the properties of single-particle states in Section 13.1.

In our discussion of phase transitions we have noted that a true phase transition can occur only in the thermodynamic limit. Any finite system has a partition function that is a smooth function of its variables. Consider now a crystal that has a certain concentration of magnetic atoms interacting with each other through short-range exchange interactions. Clearly, if the concentration of magnetic atoms is too small, they will be for the most part isolated, and even in the thermodynamic limit, no infinite cluster of interacting magnetic atoms will exist. Thus one of the important aspects of disordered systems is the geometry or connectivity of random clusters. The question as to when a system of randomly occupied lattice sites forms an infinite connected cluster and what the dimensionality and other properties of this cluster are, is the subject of percolation theory, which is discussed in Section 13.2.

Another issue that one can consider is the effect of randomness on the critical behavior of a system. Are the critical exponents the same as in a pure material? Does the phase transition remain well defined (*i.e.*, occur at a unique critical temperature T_c), or is it smeared out over some temperature interval? We will briefly discuss these questions in Section 13.3.

To this point we have mentioned effects of disorder that are partially understood and for which a certain number of analytical results exist, at least for simplified models. A subject that is less well understood is the statistical mechanics of glassy or amorphous materials. A related area that has seen much activity in recent years is the physics of spin glasses—a set of materials that display singularities in some thermodynamic properties at well-defined temperatures but without simply-ordered low-temperature phases. We discuss some aspects of such systems in Section 13.4.

Before beginning our treatment of these topics we note that, from a statistical point of view, disorder is often classified as *annealed* or *quenched.* By quenched disorder we mean that randomly distributed impurities or defects do not rearrange themselves. An example of this is a solid solution of magnetic and nonmagnetic atoms at low temperature. The diffusion of the two atomic species is then a very slow process, sometimes involving time constants comparable to the age of the universe, and usually slower than other processes, such as the demagnetization of the sample due, for example, to excitation of spin waves as the temperature is increased. Annealed disorder refers to the opposite situation where the distribution that describes the disorder is also changing in the temperature or time interval of interest. An example of this might be the case of β-brass (Section 4.1) near the order–disorder transition. If one is interested in the electronic properties of CuZn near 740K, one must take the variation of the atomic distribution into account, and in a fundamental approach to this system, one would of course attempt to derive the effective interactions that drive the order–disorder transition from the band structure of the system. In the following sections we take disorder to be quenched unless we state explicitly that it is annealed.

13.1 Single-Particle States in Disordered Systems

To have a concrete model to work with, we consider the following simple tight-binding Hamiltonian (a description of the tight-binding method can be found in most solid-state physics texts, see, *e.g.*, [21]).

$$H = \sum_{\langle ij \rangle} t_{ij} \left(c_i^\dagger c_j + c_j^\dagger c_i \right) + \sum_j \epsilon_j c_j^\dagger c_j \tag{13.1}$$

where the c's are the usual fermion creation and annihilation operators (see the Appendix) and where we have ignored the electron spin. The sum in (13.1) is over the sites of a perfect lattice, and disorder is introduced by taking either, or both, of the coefficients ϵ_j or t_{ij} to be random variables subject to some distribution. The second term in (13.1) represents a set of "atomic" levels with energies ϵ_j and corresponding Wannier functions centered on sites j. The first term represents a covalent lowering of the energy due to overlap of atomic orbitals on neighboring sites. In what follows we take the disorder to be "site diagonal" and take $t_{ij} = t$ to be a constant for i, j nearest neighbors and zero otherwise. The on-site energies ϵ_j will be ϵ_A with probability p_A

and ϵ_B with probability p_B independent of the site j. In this case (13.1) is a crude model for a substitutionally disordered binary alloy. In situations in which the Wannier representation is appropriate, the independent electron approximation is generally unrealistic. In particular, electron correlations are essential for an understanding of the origin of magnetic effects. However, we wish to concentrate on the effects of disorder and ignore these complications. Also, in a more realistic model the hopping matrix element t_{ij} would depend on the nature of the atoms occupying sites i and j and we would also expect to see a certain amount of clustering of atoms in solid solutions (see Section 4.1). Hence at least the probabilities p_{AA}, p_{AB}, and p_{BB} for the occurrence of AA, AB, and BB pairs should be specified in order to make the model reasonably realistic. However, even with our simplified version we will be able to demonstrate some of the effects of disorder.

13.1.1 Electron states in one dimension

Consider the two limiting cases $p_A = 1$ and $p_B = 1$. In this case the Hamiltonian (13.1) can be written in the form

$$H = -t\sum_j \left(c_j^\dagger c_{j+1} + c_{j+1}^\dagger c_j\right) + \epsilon_{A,B}\sum_j c_j^\dagger c_j \ . \tag{13.2}$$

We easily obtain a diagonal form by making the canonical transformation

$$\begin{aligned} c_j^\dagger &= \frac{1}{\sqrt{N}}\sum_k c_k^\dagger e^{ikja} \\ c_j &= \frac{1}{\sqrt{N}}\sum_k c_k e^{-ikja} \end{aligned} \tag{13.3}$$

with $k = 2\pi m/(Na)$, $m = 0, \pm1, \pm2, \ldots \pm (N/2-1), N/2$ for a chain of length Na with periodic boundary conditions. Substituting, we find

$$H = \sum_k \epsilon(k) c_k^\dagger c_k = \sum_k \epsilon(k) n_k \tag{13.4}$$

where n_k is the occupation number (0 or 1) of state k with energy given by

$$\epsilon(k) = \begin{cases} \epsilon_A - 2t\cos ka & \text{for } p_A = 1 \\ \epsilon_B - 2t\cos ka & \text{for } p_B = 1 \end{cases} \ . \tag{13.5}$$

The eigenstate corresponding to the eigenvalue $\epsilon(k)$ is simply

$$|\psi(k)\rangle = c_k^\dagger\,|0\rangle = \frac{1}{\sqrt{N}}\sum_j e^{-ikja} c_j^\dagger\,|0\rangle \ . \tag{13.6}$$

Defining $|j\rangle = c_j^\dagger|0\rangle$, we see that the probability amplitude for the electron to be on site j is

$$\langle j|\psi(k)\rangle = \frac{1}{\sqrt{N}}e^{-ijka} \tag{13.7}$$

and hence the probability of finding the electron on site j is simply $1/N$ independent of j and k. The eigenstates are therefore *extended.* This property is independent of dimensionality, as is the diagonalization procedure. In three dimensions, for the simple cubic lattice, one simply has

$$\epsilon_{A,B}(\mathbf{k}) = \epsilon_{A,B} - 2t(\cos k_x a + \cos k_y a + \cos k_z a) \ . \tag{13.8}$$

In one dimension it is also possible to find an analytic expression for the density of states per site. The result for spinless fermions is

$$n_{A,B}(E) = \frac{1}{\pi}\frac{1}{\sqrt{4t^2 - (E - \epsilon_{A,B})^2}} \tag{13.9}$$

for $\epsilon_{A,B} - 2t \leq E \leq \epsilon_{A,B} + 2t$ and $n_{A,B}(E) = 0$ outside this interval. The proof of this result is left as an exercise.

We note that in the pure system the Hamiltonian has the property $H(x + ja) = H(x)$ where j is any integer. This immediately implies Bloch's theorem and the classification of eigenstates in terms of wave vectors k [21].

13.1.2 Transfer matrix

We now consider the more general case of a one-dimensional disordered chain and assume that $\epsilon_m = \epsilon_A$ with probability p_A and $\epsilon_m = \epsilon_B$ with probability p_B for all m. Since the system is not periodic we no longer have Bloch's theorem to help us classify the eigenstates. Intuitively, we expect that the energy levels, for any configuration, will be confined to the range $\epsilon_A - 2t \leq E \leq \epsilon_B + 2t$ for $\epsilon_A < \epsilon_B$, and this is indeed the case (see [335] for further discussion on this point). We can formulate the one-electron problem in terms of a transfer matrix in a manner similar to our treatment of the one-dimensional Ising model (Section 3.6). We assume that

$$|\psi\rangle = \sum_{j=1}^{N} A_j\,|j\rangle = \sum_{j=1}^{N} A_j c_j^\dagger\,|0\rangle \tag{13.10}$$

is an eigenstate of (13.1) with eigenvalue E. Then

$$\langle j|\,H\,|\psi\rangle = EA_j = \epsilon_j A_j - t(A_{j+1} + A_{j-1}) \tag{13.11}$$

for $j = 1, 2, \ldots, N$ and where $A_{N+i} = A_i$, in the case of periodic boundary conditions. Defining the vector

$$\phi_j = \begin{bmatrix} A_{j+1} \\ A_j \end{bmatrix}$$

we immediately have the recursion relation

$$\phi_j = \mathbf{T}_j \phi_{j-1} \tag{13.12}$$

where

$$\mathbf{T}_j(\epsilon_j, E) = \begin{bmatrix} \frac{\epsilon_j - E}{t} & -1 \\ 1 & 0 \end{bmatrix} \tag{13.13}$$

is the transfer matrix. Thus the solution of the Schrödinger equation is reduced to finding E such that

$$\phi_N = \mathbf{T}_N(\epsilon_N, E)\mathbf{T}_{N-1}(\epsilon_{N-1}, E) \cdots \mathbf{T}_1(\epsilon_1, E)\phi_N \tag{13.14}$$

or

$$\prod_{j=1}^{N} \mathbf{T}_j(\epsilon_j, E) = \mathbf{1} \ . \tag{13.15}$$

We first show that this equation reproduces the eigenstates of the pure chain. If $\epsilon_j = \epsilon_A$ for all j, we can easily show (Problem 13.2) that (13.15) reduces to the requirement that the two eigenvalues of $\mathbf{T}$, λ_+ and λ_-, be complex conjugate of each other ($\lambda_- = \lambda_+^*$), and that $|\lambda_+| = |\lambda_-| = 1$. We have

$$\begin{aligned} \lambda_+ &= \frac{\epsilon_A - E}{2t} + i\sqrt{1 - \left(\frac{\epsilon_A - E}{2t}\right)^2} = e^{i\theta} \\ \lambda_- &= \frac{\epsilon_A - E}{2t} - i\sqrt{1 - \left(\frac{\epsilon_A - E}{2t}\right)^2} = e^{-i\theta} \end{aligned} \tag{13.16}$$

which yields

$$E = \epsilon_A - 2t\cos\theta \ .$$

The periodic boundary condition ($\lambda^N = 1$) produces $\theta = 2\pi j/N$, $j = 1, \cdots, N$. In the case of real eigenvalues ($\lambda \neq \pm 1$) (*i.e.*, $|\epsilon_A - E| > 2t$) it is impossible to satisfy the periodic boundary conditions, and this corresponds to an energy cutoff in the spectrum of the translationally invariant chain.

In the disordered situation we have, in (13.15), a product of transfer matrices which are either of the form $\mathbf{T}_j = \mathbf{T}_A$ or $\mathbf{T}_j = \mathbf{T}_B$ depending on whether site j is occupied by an A or B atom. One would expect that if E is a forbidden energy of *both* the pure A and pure B system, it will be impossible to satisfy (13.15). Similarly, if E lies in the forbidden region of one of the materials, say A, then each time the transfer matrix $\mathbf{T}_A$ appears in the product (13.15), the vector ϕ will in general increase in modulus since the larger of the two eigenvalues of $\mathbf{T}_A$ is greater than 1. Thus it seems unlikely that (13.15) could be satisfied for any E that lies outside the region of allowed energies in any of the two pure materials. However, extensive numerical work (see Ishii [139] for a review) has shown that there are, in fact, eigenstates of H in this region but that these wave functions are *localized.* We discuss localization further below.

The determination of energy levels in the disordered case requires the solution of the eigenvalue problem (13.15) for a particular realization of a chain of length N. If we are interested only in a rough determination of the location of the eigenvalues, or the average density of states, we can use the following simple method, which is applicable only in one dimension. Quite generally, the eigenfunction corresponding to the mth eigenvalue has $m-1$ nodes (this is not strictly true for chains with periodic boundary conditions). A node between sites j and $j+1$ in our case corresponds to the ratio A_{j+1}/A_j being negative and this ratio can easily be expressed in terms of the matrix elements of $\mathbf{T}_j$ and the ratio A_j/A_{j-1}. Thus if we fix E and the initial ratio A_2/A_1 and simply count the number of negative ratios of successive coefficients A_j for a particular configuration of the N potentials, we have, equivalently, a count of the number of energy levels below E. Averaging over many realizations of the random chain, a process easily carried out on a computer, we find the average integrated density of states:

$$\mathcal{N}(E) \equiv \int_{-\infty}^{E} dE'\, n(E') = S_N(E)$$

where $S_N(E)$ is the total number of nodes in the wave function at energy E. In Figures 13.1 and 13.2 we display the density of states of a 500-atom chain obtained by numerical differentiation of $\mathcal{N}(E)$ for $\epsilon_A = 2$, $\epsilon_B = 3$, $t = 1$ and two different concentrations of the constituents. The density of states is plotted as a histogram instead of as a smooth curve since we have no information as to how $n(E)$ varies between the points at which the nodes were counted.

A number of features of Figures 13.1 and 13.2 are of interest. In the strongly disordered alloy (Figure 13.1, $p_A = p_B = \frac{1}{2}$) no trace of the square-root sin-

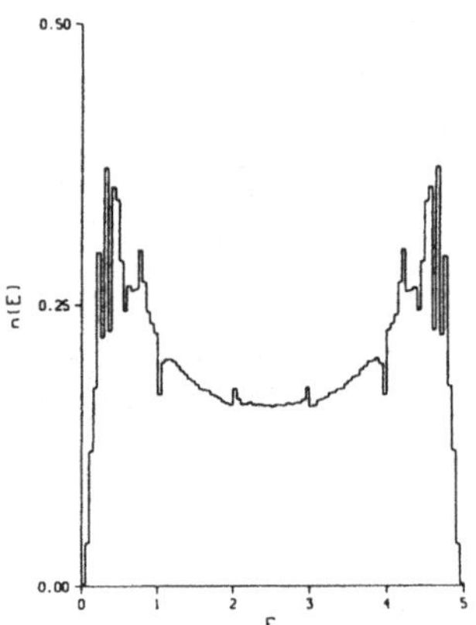

Figure 13.1: Density of states $n(E)$ for the disordered one-dimensional chain with $\epsilon_A = 2$, $\epsilon_B = 3$, and $p_A = p_B = 0.5$.

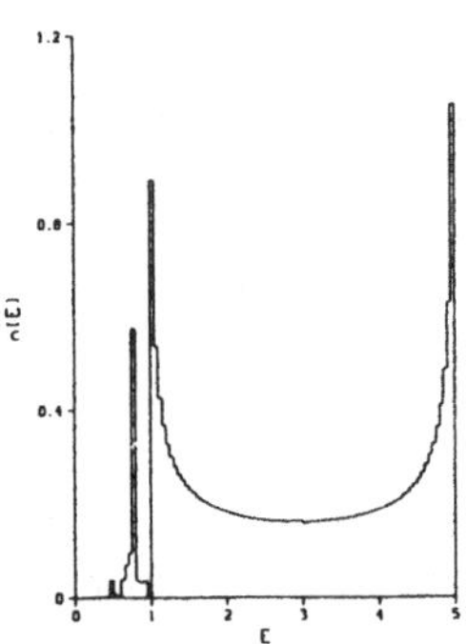

Figure 13.2: Density of states $n(E)$ for the disordered one-dimensional chain with $p_A = 0.05$, $p_B = 0.95$ and the same energy parameters as in Figure 13.1.

gularities (13.9) near the upper and lower band edges remains. On the other hand, $n(E)$ is quite noisy in these regions. This is not a statistical artifact. A measure of statistical error is the degree of asymmetry of the density of states around the average energy $E = 2.5$ and we see that this is negligible. The noisy structure near the band edges is due to the existence of gaps in the spectrum at arbitrarily fine energy resolution (see Gubernatis and Taylor [122] for a graphic demonstration of this fine structure). We also note that the density of states seems to extend essentially to the pure system band edges $\epsilon_A - 2$ and $\epsilon_B + 2$, although it drops rather precipitously near these points. These low density regions are known as Lifshitz tails. One can understand that states with energy very close to the lower limit $\epsilon_A - 2$ will exist. In a random system there is a finite probability that a very long island, say of length L, of pure A material will occur. At least one eigenstate of the complete system should therefore be very close to the lowest eigenstate of a pure A system of length L. A similar argument applies at the upper band edge.

Figure 13.2 shows the density of states for a weakly disordered chain ($p_A = 0.05$). In this case we see almost the pure B density of states with a noisy impurity band essentially separated from the main part of the spectrum at lower energies.

We now return to the question of localization of the eigenstates. It has been rigorously proven that all eigenstates (except possibly a set of measure zero) of a one-dimensional disordered system with site-diagonal disorder are localized. The proof requires theorems on the properties of random matrices that are beyond the scope of this book and the reader is referred to Matsuda and Ishii [192] for the rigorous mathematical arguments. We present here a weaker demonstration of localization that should at least make the result plausible.

We consider the quantity

$$\phi_{N+1}^\dagger \phi_{N+1} = A_{N+1}^2 + A_N^2 = \phi_0^\dagger \left(\mathbf{T}_1^\dagger \mathbf{T}_2^\dagger \cdots \mathbf{T}_N^\dagger \mathbf{T}_N \mathbf{T}_{N-1} \cdots \mathbf{T}_1 \right) \phi_0 \,. \quad (13.17)$$

The Hermitian matrix $\mathbf{M}_{N-i} = \mathbf{T}_i^\dagger \mathbf{T}_{i+1}^\dagger \cdots \mathbf{T}_N^\dagger \mathbf{T}_N \mathbf{T}_{N-1} \cdots \mathbf{T}_i$, will be of the form

$$\mathbf{M}_{N-i} = \begin{bmatrix} a_{N-i} & b_{N-i} \\ b_{N-i} & c_{N-i} \end{bmatrix} \,. \quad (13.18)$$

We can easily derive a set of recursion relations for the matrix elements. We have

$$\mathbf{M}_N = \mathbf{T}_1^\dagger \mathbf{M}_{N-1} \mathbf{T}_1 = \begin{bmatrix} \frac{\epsilon_1 - E}{t} & 1 \\ -1 & 0 \end{bmatrix} \begin{bmatrix} a_{N-1} & b_{N-1} \\ b_{N-1} & c_{N-1} \end{bmatrix} \begin{bmatrix} \frac{\epsilon_1 - E}{t} & -1 \\ 1 & 0 \end{bmatrix} \tag{13.19}$$

and find

$$\begin{aligned} a_N &= \left(\frac{\epsilon_1 - E}{t}\right)^2 a_{N-1} + 2\left(\frac{\epsilon_1 - E}{t}\right) b_{N-1} + c_{N-1} \\ b_N &= -\left(\frac{\epsilon_1 - E}{t}\right) a_{N-1} - b_{N-1} \\ c_N &= a_{N-1} \,. \end{aligned} \tag{13.20}$$

We now average these recursion relations over the probability distribution of the atomic potentials. This yields the expectation value of the quantity $\phi_{N+1}^\dagger \phi_{N+1}$, rather than the probability distribution of this quantity which is needed for a proper proof of localization. Once the average of (13.20) is obtained, the difference equations can be solved by the ansatz

$$\begin{bmatrix} a_N \\ b_N \\ c_N \end{bmatrix} = \begin{bmatrix} x_1 \\ x_2 \\ x_3 \end{bmatrix} \lambda^N \tag{13.21}$$

which then yields a 3×3 eigenvalue problem. An extended eigenstate of the system corresponds to an eigenvalue λ of magnitude 1 and it is easy to show that in the pure case ($p_A = 1$ or $p_B = 1$) we recover the usual dispersion relations for the energy levels.

Since the secular equation for this eigenvalue problem is cubic and rather unenlightening, we simply plot, in Figure 13.3, the quantity

$$L^{-1} = \lim_{N\to\infty} \frac{1}{N} \ln \frac{\phi_{N+1}^\dagger \phi_{N+1}}{\phi_0^\dagger \phi_0} \tag{13.22}$$

as a function of E for the two sets of parameters for which we have plotted the average density of states in Figures 13.1 and 13.2, namely $\epsilon_A = 2$, $\epsilon_B = 3$ and $p_A = 0.5$ and 0.05. We also take as boundary condition $A_0 = 0$. The quantity $L(E)$ clearly has a physical interpretation as a mean localization length. Referring to Figure 13.3, we see that in both cases the localization length is a smooth function of the energy E and that it decreases sharply near the band

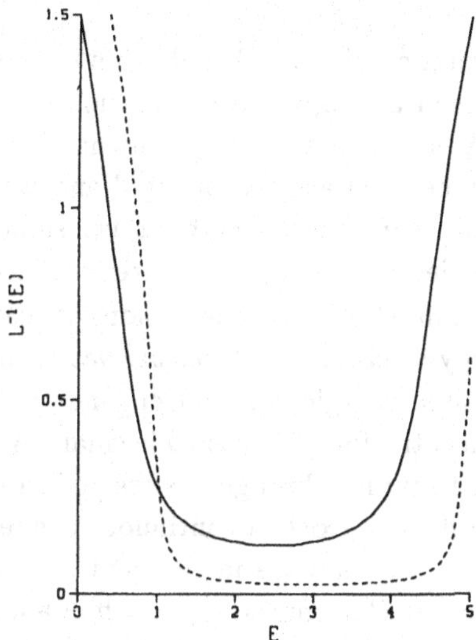

Figure 13.3: Plot of the inverse localization length (13.22) as a function of E for the disordered one-dimensional alloys with the densities of state of Figures 13.1 and 13.2. Solid curve, $p_A = 0.5$; dashed curve, $p_A = 0.05$.

edges but remains finite throughout the entire energy range. Similar results hold in two dimensions except that the states may be "weakly localized", *i.e.*, have a power law rather than an exponential decay. For further details on this point we refer to the review of Lee and Ramakrishnan [174]. The feature of strong localization near band edges is thought to hold in three dimensions as well as in one dimension. The three-dimensional case is distinguished from that of the one-dimensional by the existence of a *mobility edge* which separates a region of extended states from localized states.

The foregoing discussion for the case of a disordered tight binding model holds as well for the vibrational properties of an isotopically disordered chain (random masses, fixed spring constants), for a disordered Kronig–Penney model (Problem 13.2) and, in general, for any disordered one-dimensional system. An extensive discussion of both numerical and analytic results can be found in the review by Ishii [139].

13.1.3 Localization in three dimensions

The notion that wave functions in a disordered three-dimensional system might be localized is due to Anderson, who, in a classic paper, [15] considered the

Hamiltonian (13.1) with a continuous distribution of energies ϵ_j characterized by a width W. Anderson attempted a calculation of the probability $a_j(\infty)$ for an electron to be at site j at $t = \infty$, given that it is there at $t = 0$. The mathematical methods he used are complicated and we shall not attempt to repeat his arguments. His conclusion was that given sufficiently strong disorder (W large enough), the electronic eigenstates are localized.

One can partially understand this conclusion, at least in the pathological limit in which the energy levels of the atoms are vastly different. Suppose that we have only two possible energies ϵ_A and ϵ_B and that $|\epsilon_A - \epsilon_B| \gg 1$ and, moreover, that the concentration of B atoms is small. An electron, initially on a B site, will be unable to tunnel through A sites and can escape the vicinity of its initial location only if there exists a continuous connected path of B atoms extending to infinity. As we shall see in the next section, such a path exists only if the concentration of B atoms is greater than a critical value called the percolation concentration. Thus we have a situation in which at least some of the electronic states are localized. Once we accept this possibility, it is not hard to see that there might be a transition, as a function of electron energy, from localized to extended wave functions.

The picture that we now believe to be correct is shown schematically in Figure 13.4 and is supported by extensive numerical work. For a given band the localized states extend from the band edges to mobility edges at energies E_1 and E_2. The localized states have wave functions with an exponential envelope. There is some evidence [166], [176] that there may be a region in which the "localized" states of a three-dimensional system have a power law envelope rather than an exponential envelope as well. The notion of sharp mobility edges separating localized and extended states seems to be due to Mott [208]. Much of the numerical work on this topic is discussed by Thouless [302].

It is quite clear that the effect of localized states on the conductivity will be quite interesting and theoretically complicated. At zero temperature we expect that the conductivity will be zero if the Fermi level lies in the region of localized states. Wave functions of localized states will in general not overlap with other localized states of the same energy and hence there can be no tunneling between different states without thermal activation (hopping). At small but finite temperature this argument no longer holds, but there are now two length scales in the problem—the extent of the localized wave function L and the inelastic mean free path L_i. If $L_i \ll L$ the electron is scattered many times while traversing the distance L, and hence the information as to which localized state it started in is lost. Localization is irrelevant in this case

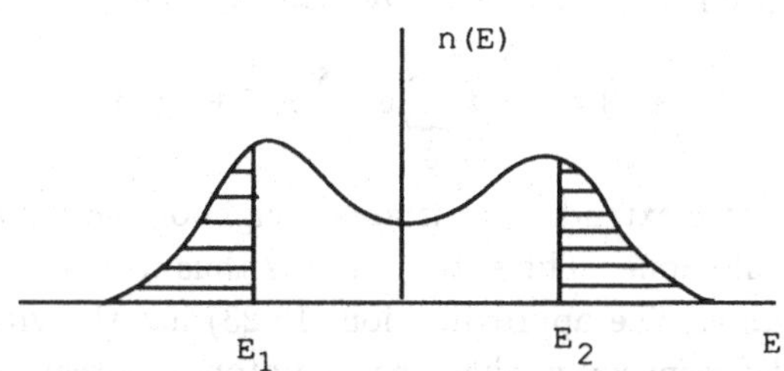

Figure 13.4: Density of states of a disordered alloy with localized states in the shaded regions ($E < E_1$, $E > E_2$), extended states for $E_1 < E < E_2$.

and resistivity is dominated by the usual scattering processes. Conversely, if $L \ll L_i$, the effect of localization becomes important and this manifests itself in added dependence of the resistivity of thin wires and films on the physical dimensions of the sample. These ideas have been used by Thouless and co-workers and Wegner to develop scaling theories of localization which are reviewed, together with the experimental situation, by Lee and Ramakrishnan [174].

13.1.4 Density of states

The equilibrium thermodynamics of a disordered system can be expressed entirely as a functional of the density of states or spectral density (see also Section 5.5). The response to external perturbations (conductivity, thermal conductivity, etc.) depends in greater detail on the nature of the wave functions and presents more difficult problems. The density of states is understood in much more detail and reliable approximations have been developed which allow us to determine this function in many cases. The simplest method for calculating the spectral density is the *rigid band* or *virtual crystal* approximation; the most successful is the *coherent potential approximation.* We briefly discuss them below.

(i) Virtual crystal approximation. Consider, again, the Hamiltonian (13.1) with $\epsilon_j = \epsilon_A$ with probability $p_j = p_A$ and ϵ_B with probability $p_j = p_B$. If we average the Hamiltonian over the impurity distribution function, before attempting to diagonalize it, we obtain the translationally invariant effective Hamiltonian

$$\bar{H} = -t\sum_{\langle ij\rangle}\left(c_i^\dagger c_j + c_j^\dagger c_i\right) + \bar{\epsilon}\sum_i c_i^\dagger c_i \tag{13.23}$$

with $\bar{\epsilon} = p_A\epsilon_A + p_B\epsilon_B$. Using Bloch's theorem, we easily find the energy levels to be

$$\epsilon(\mathbf{k}) = \bar{\epsilon} - t\sum_{\boldsymbol{\delta}} e^{i\mathbf{k}\cdot\boldsymbol{\delta}} \equiv \bar{\epsilon} + w(\mathbf{k}) \tag{13.24}$$

where the sum over $\boldsymbol{\delta}$ extends over nearest-neighbor vectors and where $\mathbf{k}$ lies in the first Brillouin zone. The entire band is thus displaced uniformly without change in shape. The approximation (13.23) has the virtue of simplicity but is not realistic, especially when the two atomic potentials ϵ_A and ϵ_B are substantially different.

(ii) Coherent potential approximation. A more sophisticated and far more accurate approximation is the coherent potential approximation [280]. The essential idea is to replace each atom by an "effective" atom so that on the average no scattering takes place on each site. To be more precise, consider a single impurity of type B at site i in an otherwise type A crystal. The Hamiltonian then is

$$H = H_A + (\epsilon_A - \epsilon_B)c_i^\dagger c_i = H_A + U \ . \tag{13.25}$$

We now define the resolvent operator (or Green's function) $G(z)$

$$\begin{aligned} G(z) &= (z-H)^{-1} = (z - H_A - U)^{-1} \\ &= (z-H_A)^{-1} + (z-H_A)^{-1}U(z-H)^{-1} \ . \end{aligned} \tag{13.26}$$

The last equation can easily be shown to be correct by premultiplying by $(z - H_A)$ and postmultiplying by $(z - H)$. Using Wannier basis functions and defining

$$G_{mj}(z) = \langle m|(z-H)^{-1}|j\rangle \qquad G^0_{mj}(z) = \langle m|(z-H_A)^{-1}|j\rangle$$

we obtain, on iterating (13.26),

$$\begin{aligned} G_{mj}(z) &= G^0_{mj}(z) + G^0_{mi}(z)U_{ii}G_{ij}(z) \\ &= G^0_{mj}(z) + G^0_{mi}(z)U_{ii}G^0_{ij}(z) + G^0_{mi}(z)U_{ii}G^0_{ii}(z)U_{ii}G^0_{ij}(z) + \cdots \\ &= G^0_{mj}(z) + G^0_{mi}(z)U_{ii}\left(1 - G^0_{ii}(z)U_{ii}\right)^{-1} G^0_{ij}(z) \ . \end{aligned} \tag{13.27}$$

The operator $T = U(1 - G^0U)^{-1}$ is known as the T-matrix of the potential U and has, in the particular case of a single impurity (13.25), only diagonal matrix elements. The generalization of (13.26) and (13.27) for an arbitrary perturbation U is, in operator form,

$$G(z) = G^0(z) + G^0(z)UG(z) = G^0(z) + G^0(z)T(z)G^0(z) \ . \tag{13.28}$$

The Green's function $G(z)$ yields the density of states, as we now show. Suppose that the eigenstates of H are $|\phi_m\rangle$ with energies E_m and consider

$$\begin{aligned}\operatorname{Tr} G(E+i\eta) &= \sum_m \langle\phi_m|(E-H+i\eta)^{-1}|\phi_m\rangle \\ &= \sum_m (E-E_m+i\eta)^{-1} . \end{aligned} \tag{13.29}$$

Using

$$\lim_{\eta\to 0} \frac{1}{E-E_m+i\eta} = P\frac{1}{E-E_m} - i\pi\delta(E-E_m) \tag{13.30}$$

we see that

$$n(E) = \sum_m \delta(E-E_m) = -\frac{1}{\pi}\operatorname{Im}\operatorname{Tr} G(E+i0^+) . \tag{13.31}$$

Since the trace in (13.31) can be evaluated in any basis, we are free to use our Wannier states $|m\rangle$ or the Bloch states $|\mathbf{k}\rangle$ to calculate the density of states. It is only necessary to find the diagonal matrix elements of the operator $G(z)$.

In the coherent potential approximation one determines the Green's function G by assuming that there exists an effective complex translationally invariant potential such that the averaged T-matrix vanishes for each site. Thus one writes the Hamiltonian in the form

$$\begin{aligned} H &= \sum_{\mathbf{k}} [w(\mathbf{k})+\Sigma(E)] c_{\mathbf{k}}^\dagger c_{\mathbf{k}} + \sum_j [\epsilon_j - \Sigma(E)] c_j^\dagger c_j \\ &= H_0 + \sum_j U_j(E) \end{aligned} \tag{13.32}$$

where we have used a mixed $\mathbf{k}$-space and real space representation and where $w(\mathbf{k})$ is defined in (13.24). Using $\langle\mathbf{k}|j\rangle = N^{-1/2}\exp\{-i\mathbf{k}\cdot\mathbf{r}_j\}$, we have, for the diagonal matrix elements of the Green's function G^0,

$$G_{ii}^0(z) = \langle i|(z-H_0)^{-1}|i\rangle = \frac{1}{N}\sum_{\mathbf{k}} \frac{1}{z-\Sigma(z)-w(\mathbf{k})} . \tag{13.33}$$

The T-matrix of a single "impurity" potential at site i is then diagonal and is given by

$$T_i(z) = [\epsilon_i - \Sigma(z)]\left\{1-[\epsilon_i-\Sigma(z)]\,G_{ii}^0(z)\right\}^{-1} . \tag{13.34}$$

Averaging over the atomic distribution, we have

$$\begin{aligned} \langle T_i(z)\rangle &= p_A\,[\epsilon_A-\Sigma(z)]\left\{1-[\epsilon_A-\Sigma(z)]\,G_{ii}^0(z)\right\}^{-1} \\ &+ p_B\,[\epsilon_B-\Sigma(z)]\left\{1-[\epsilon_B-\Sigma(z)]\,G_{ii}^0(z)\right\}^{-1} = 0 \end{aligned} \tag{13.35}$$

which determines the unknown function $\Sigma(z)$. Rearranging, we find the simpler form

$$\Sigma(z) = \bar{\epsilon} - [\Sigma(z) - \epsilon_A]\,[\Sigma(z) - \epsilon_B]\, G^0_{ii}(z) \tag{13.36}$$

which can be solved numerically for the complex function $\Sigma(z)$. The CPA approximation for the average density of states per site is then

$$\begin{aligned} n(E) &= -\frac{1}{\pi N}\mathrm{Im}\ \mathrm{Tr}\, G^0(E+i0) \\ &= -\frac{1}{\pi N}\mathrm{Im} \sum_{\mathbf{k}} \frac{1}{E + i0 - w(\mathbf{k}) - \Sigma(E+i0)}\ . \end{aligned} \tag{13.37}$$

In Figures 13.5 and 13.6 we plot this function for the one-dimensional disordered alloy for the same parameters as used to obtain the average density of states by the exact node counting method (Figures 13.1 and 13.2). As we see, the gross features of the density of states are well reproduced by the coherent potential approximation, and indeed, near the center of the band, the two functions are essentially identical. What is missing in the CPA density of states is the fine structure near the band edges and the low-density tails, which in the exact calculation, extend to the edges of the pure system bands. This is not surprising since the CPA deals with the impurity distribution in a strictly local manner. The structure in the Lifshitz tails is due to islands of one species and the probability of such islands never enters into the CPA formalism.

One can understand the success of the coherent potential approximation in the weak scattering limit, which we characterize by $|G^0_{jj}(z)T_j(z)| \ll 1$. Using (13.28) and expanding the full T matrix, we have

$$\begin{aligned} G_{mj} &= G^0_{mj} + \sum_i G^0_{mi}T_iG^0_{ij} + \sum_{k\neq i} G^0_{mi}T_iG^0_{ik}T_kG^0_{kj} \\ &+ \sum_{k\neq i,\, n\neq k} G^0_{mi}T_iG^0_{ik}T_kG^0_{kn}T_nG^0_{nj} \\ &+ \sum_{k\neq i,\, n\neq k,\, s\neq n} G^0_{mi}T_iG^0_{ik}T_kG^0_{kn}T_nG^0_{ns}T_sG_{sj} + \cdots . \end{aligned} \tag{13.38}$$

Taking G^0 to be the CPA Green's function and averaging over the impurity distribution, we see that because of the restrictions on successive indices in the summations in (13.38), the first few terms vanish and

$$\langle G_{mj}\rangle = G^0_{mj} + \sum_{k\neq i} G^0_{mi}\langle T_iG^0_{ik}T_kG^0_{ki}T_iG^0_{ik}T_k\rangle G^0_{kj} + \mathcal{O}(T^6)\ . \tag{13.39}$$

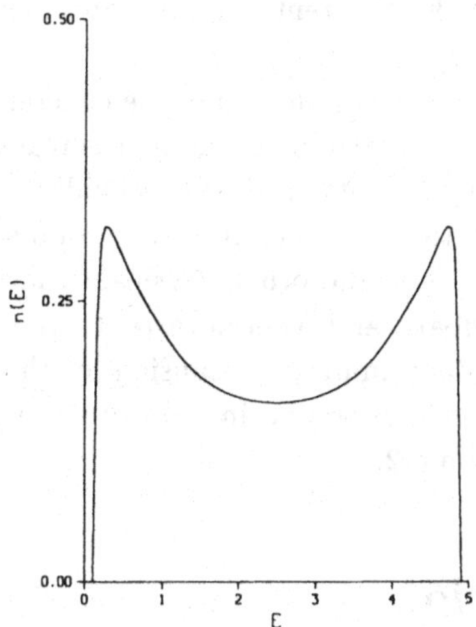

Figure 13.5: Density of states from the coherent potential approximation for the one-dimensional disordered binary alloy. Parameters are the same as in Figure 13.1.

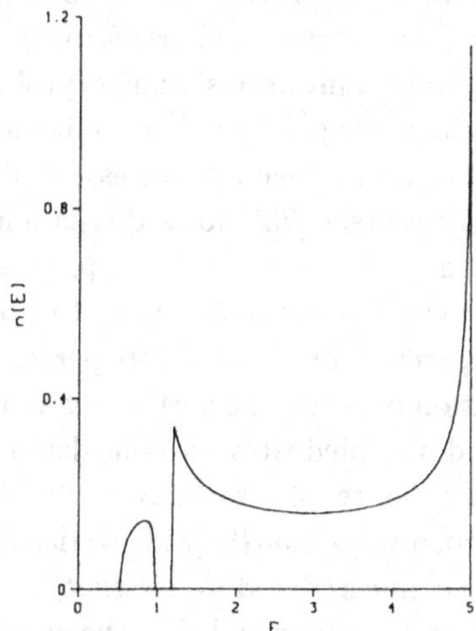

Figure 13.6: Density of states from the coherent potential approximation for the one-dimensional disordered binary alloy. Parameters as in Figure 13.2.

Thus the corrections to the coherent potential approximation are of order T^4 and higher.

For a much more extensive discussion of the coherent potential approximation for the case of the disordered binary alloy, the reader is referred to Velicky *et al.* [312]. The review of Elliott *et al.* [84] provides a thorough discussion of applications of the CPA to vibrational properties of disordered materials, spin waves in disordered magnets, and other physical situations.

As shown by Matsubara and Yonezawa [191], the expansion (13.27) can be used to construct a diagrammatic expansion for the density of states. The resulting moment expansion is similar in form to the high-temperature expansions discussed in Section 6.2.

13.2 Percolation

In our brief discussion of localization in three dimensions, we have noted that the nature of connected clusters of equivalent atoms in a disordered binary alloy may be important in determining the nature of electronic states. In dilute magnetic alloys where the fraction of magnetic atoms is p, the existence of a finite-temperature phase transition depends on whether or not there exists an infinite connected cluster (of suitably high dimensionality) of interacting magnetic atoms. A random resistor network formed of elements with finite resistance (probability p) and infinite resistance (probability $1-p$) conducts only if the network of conducting elements is continuous across the sample. Many other physical situations depend in an essential way on the geometric properties of random clusters (see [287] for a discussion of forest fires) and, in particular, on the existence of a connected cluster that spans the system in question. The study of such clusters is the subject of percolation theory.

There are two basic percolation models: *site* percolation and *bond* percolation. In the site percolation problem the vertices of a lattice are occupied with a given probability p, and occupied sites are considered to be connected if they are nearest neighbors. In Figure 13.7 we show a 20×20 segment of a square lattice occupied with probability 0.5575 (223 particles). The particles in the largest cluster have been connected and we see that it spans the lattice in both the horizontal and vertical directions and that the connections are tenuous. In fact, the existence of a spanning cluster at this probability of occupation is a finite-size effect. The site percolation probability (*i.e.*, the probability at which the infinite cluster forms in the thermodynamic limit) on the square lattice is known to be $p_c = 0.5927\ldots$.

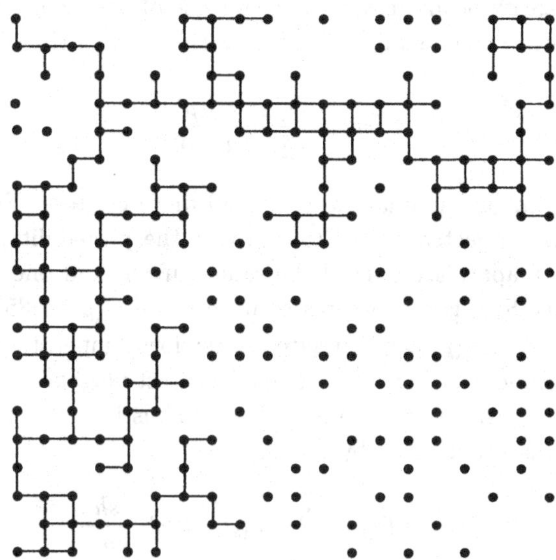

Figure 13.7: Example of site percolation cluster on 20×20 square lattice at concentration 0.5575. The largest cluster has been connected.

Bond percolation, on the other hand, is defined in terms of the probability of occupation of nearest-neighbor bonds on a lattice, and this model is clearly relevant to conduction in inhomogeneous media. The site and bond problems are closely related and one can show, for example, that on any lattice the critical probability for bond percolation $p_c^B \leq p_c^S$, the site percolation probability (see, *e.g.* [85]).

We now define the quantities of interest in a percolation problem. Let $n_s(p)$ be the number of clusters per lattice site of size s. The probability that a given site is occupied and part of a cluster of size s is simply $sn_s(p)$. Now let $P(p)$ be the fraction of occupied sites that belong to the spanning cluster. Clearly, $P(p) = 1$ for $p = 1$ and $P(p) = 0$ for $p < p_c$ (in the thermodynamic limit). In this sense $P(p)$ is similar to the order parameter of a system that undergoes an ordinary thermal phase transition. Indeed, percolation is often referred to as a geometric phase transition. Clearly, we have the relation

$$\sum_s sn_s(p) + pP(p) = p \tag{13.40}$$

where the summation extends over all finite clusters.

Another quantity of interest is the mean size of finite clusters, denoted by $S(p)$. This quantity can also be related to $n_s(p)$ through the relation

$$S(p) = \frac{\sum_s s^2 n_s(p)}{\sum_s s n_s(p)} \tag{13.41}$$

where, once again, the summation is over all finite clusters. Finally, one can define the "pair connectedness" $C(p,r)$ to be the probability that occupied sites a distance r apart are part of the same cluster, and one can then show that the quantity $S(p)$ can be expressed in terms of $C(p,r)$ [85], [180].

We also note that $S(p)$ can be related to the low-temperature susceptibility of a dilute Ising model. For $k_BT \ll J$ all nearest-neighbor magnetic atoms will be in the same state, either up or down. Thus in a magnetic field h the magnetization per site is given by

$$\langle \sigma \rangle = pP(p) + \sum_s s n_s(p) \tanh \frac{sh}{k_BT} \tag{13.42}$$

where the first term on the right is the contribution from the infinite cluster. Differentiating, we find for the zero-field susceptibility:

$$\chi(0,T) = \frac{1}{k_BT} \sum_s s^2 n_s(p) = \frac{p\,[1-P(p)]}{k_BT} S(p) . \tag{13.43}$$

The functions $P(p)$, $S(p)$, and $C(p,r)$ can be calculated by most of the methods that we have developed for other statistical problems, such as series expansions, Monte Carlo simulations, and renormalization group methods.

Also of interest and important for a number of physical processes is the geometry of the percolating cluster close to p_c. If the finite clusters are removed from the system, we are left with a tenuous network which, at p_c, does not fill space *i.e.*, $N_\infty(L,p_c)/L^d \to 0$ as $L \to \infty$. Here N_∞ is the number of particles on the percolating cluster on a lattice of linear dimension L. In fact, the spanning cluster is a fractal with fractal dimension D. This is illustrated in Figure 13.8 where the largest cluster on a 100×100 lattice with site occupation probability $p = 0.6$ is displayed. There are holes of many different sizes as is typical of a fractal. In two dimensions, $D = 43/24$ exactly; in three dimensions, $D \approx 2.5$.

Referring back to Figure 13.7 or to Figure 13.8, we see that the percolating cluster contains a large number of dangling bonds, *i.e.*, bonds that can be

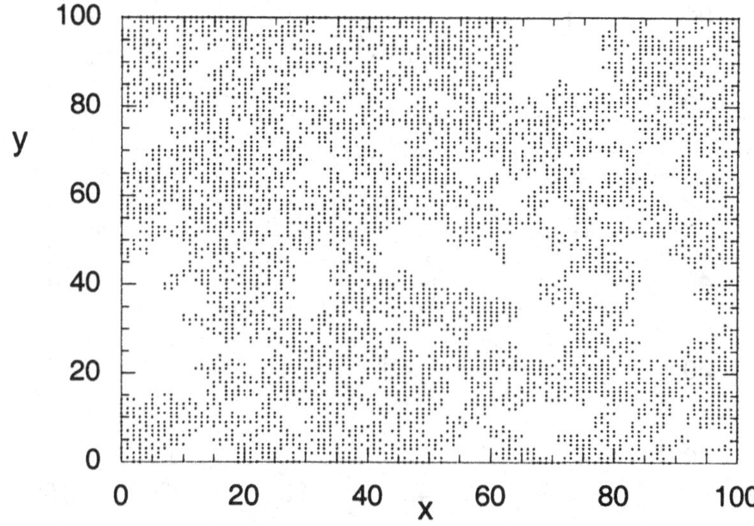

Figure 13.8: Percolating cluster on a square lattice with sites occupied with probability $p = 0.6$. The sites not connected to this cluster have been removed.

removed without destroying percolation. If all of these bonds are removed, we are left with what is called the *backbone* (see e.g. Figure 13.9). It is also a fractal with its own fractal dimension $D_B \approx 1.64$ in two dimensions and $D_B \approx 1.8$ in three dimensions. The backbone of the percolating cluster of Figure 13.8 is shown in Figure 13.9. It is clear that most of the mass has disappeared with the dangling ends. The backbone is clearly the only relevant feature for the conductivity of a percolating network of conductors — no current flows in the dangling ends. There are a number of other geometric features that are of interest. We refer the reader to [8] for further discussion. To date the percolation problem in two and three dimensions remains unsolved, although in two dimensions the percolation probabilities for several lattices as well as many of the critical exponents are known exactly. The one-dimensional percolation problem is, of course, trivial and both the bond and site percolation probability can be calculated exactly for a Bethe lattice (see, e.g. [302]). We will first exploit the analogy between percolation and phase transitions to postulate a simple scaling theory.

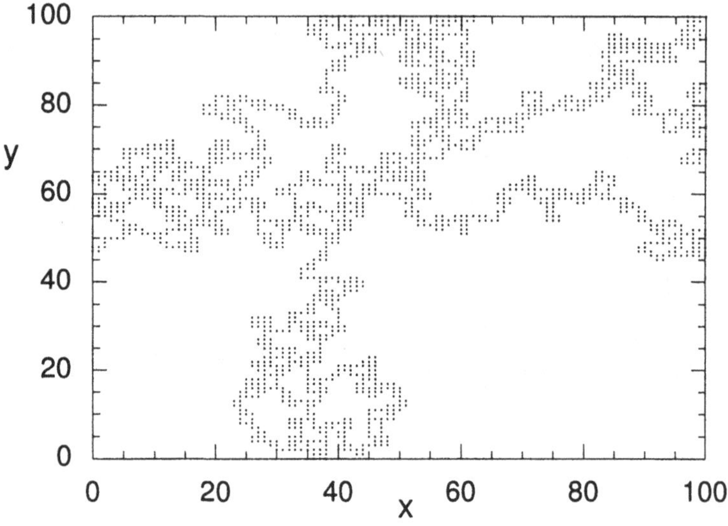

Figure 13.9: Backbone of the percolating cluster shown in Figure 13.8.

13.2.1 Scaling theory of percolation

The analogy between a thermal phase transition and percolation alluded to above can be put on a rigorous basis by means of an argument due to Kasteleyn and Fortuin [150], who showed that the bond percolation problem is isomorphic to the q-state Potts model in the limit $q \to 1$ (see also [180] or [58] for a discussion of this point). The probability of occupation of a bond is related to a Boltzmann weight in the Potts model and is therefore like the temperature variable in the scaling theory of Chapter 6 (the analog of the magnetic field can be introduced as well but does not have as direct an interpretation in the percolation problem). The correspondence between thermodynamic functions and the geometric quantities defined above is as follows. The analog of the free energy per site is the total number of clusters per site, that is,

$$G(p) = \sum_s n_s(p) . \tag{13.44}$$

The role of an order parameter is played by the probability $P(p)$, as is intuitively obvious. Similarly, the mean size of finite clusters $S(p)$ is equivalent to the susceptibility and the pair connectedness $C(p, r)$ is equivalent to the thermal pair correlation function.

We therefore expect these quantities to have power law singularities at the percolation probability and can define the usual set of exponents :

$$G(p) \sim |p - p_c|^{2-\alpha} \tag{13.45}$$

$$P(p) \sim (p - p_c)^{\beta} \tag{13.46}$$

$$S(p) \sim |p - p_c|^{-\gamma} \tag{13.47}$$

$$C(p, r) \sim \frac{\exp\{-r/\xi(p)\}}{r^{d-2+\eta}} \tag{13.48}$$

where, moreover, the correlation length ξ is expected to diverge as

$$\xi(p) \sim |p - p_c|^{-\nu} \ . \tag{13.49}$$

If universality (Section 6.5) holds, the percolation exponents α, β, γ, ... will depend on dimensionality but not on details such as the type of lattice or whether it is bond or site percolation that is being considered.

The basic assumption of the scaling theory [286] is that for p near p_c, or for a given ξ in (13.48), there is a typical cluster size s_ξ which produces the dominant contribution to the functions (13.45)–(13.47). This cluster size must diverge as $p \to p_c$ and we assume that the divergence is of the power law form:

$$s_\xi \sim |p - p_c|^{-1/\sigma} \tag{13.50}$$

which defines the exponent σ. We further assume that the number of clusters of size s at probability p can be related to s/s_ξ and to $n_s(p_c)$ through

$$n_s(p) = n_s(p_c) f\left(\frac{s}{s_\xi}\right) \tag{13.51}$$

where $f(x) \to 0$ as $x \to \infty$ and $f(x) \to 1$ as $x \to 0$ but is otherwise unspecified. It is known from Monte Carlo simulations [286] that for large s, $n_s(p_c) \sim s^{-\tau}$, where τ is an exponent that depends on the dimensionality d. Using this form and (13.51), we have

$$n_s(p) = s^{-\tau} \phi\left(s|p - p_c|^{1/\sigma}\right) \tag{13.52}$$

which, as in the case of thermodynamic scaling (Section 6.3), allows us to express the percolation exponents in terms of the two independent exponents σ and τ. For example,

$$\begin{aligned} G(p) &= \sum_s n_s(p) = \sum_s s^{-\tau} \phi\left(s|p - p_c|^{1/\sigma}\right) \\ &\approx |p - p_c|^{\frac{\tau-1}{\sigma}} \int dx x^{-\tau} \phi(x) \end{aligned} \tag{13.53}$$

where $x = s|p - p_c|^{1/\sigma}$. Assuming that the integral over x converges, we have $\alpha = 2 - (\tau - 1)/\sigma$. Similarly, we obtain $\gamma = (3 - \tau)/\sigma$ and $\beta = (\tau - 2)/\sigma$ and we find the familiar scaling relation,

$$\alpha + 2\beta + \gamma = 2 \ .$$

The pair connectedness can also be included in the scaling formalism. If we integrate $C(p, r)$ over the volume of the sample, we should obtain the mean cluster size $S(p)$, that is,

$$S(p) \sim |p - p_c|^{-\gamma} \sim \int dr r^{d-1} \frac{e^{-r/\xi(p)}}{r^{d-2+\eta}} \sim |p - p_c|^{-\nu(2-\eta)} \tag{13.54}$$

and hence $\gamma = \nu(2 - \eta)$. The hyperscaling equation $d\nu = 2 - \alpha$ can also be obtained with one further assumption. The dominant cluster size at a given p, from (13.50), obeys the relation $s \sim |p - p_c|^{-1/\sigma}$ and the concentration of these clusters (13.52) is $n_s \sim |p - p_c|^{\tau/\sigma}$. Assuming that this concentration is inversely proportional to the volume $\xi^d(p)$ occupied by these clusters, we find that

$$|p - p_c|^{\tau/\sigma} = |p - p_c|^{d\nu}$$

or $d\nu = 2 - \alpha$.

We also note that the fractal dimension D of the percolating cluster can be expressed in terms of these exponents. By definition, the number of particles within a distance R of a given particle scales as $N(R) \sim R^D$ and

$$N(R, p) \propto \int_0^R dr r^{d-1} C(p, r) \sim R^{2-\eta}$$

at p_c. Therefore, $D = 2 - \eta = \gamma/\nu$.

In the review of Stauffer [286], the reader will find a critical discussion of numerical tests of the scaling theory. Although the exponents obtained from Monte Carlo simulations and series expansions are not extremely accurate, they are consistent with the scaling theory. There is also no evidence for violation of the universality hypothesis—percolation exponents seem to depend only on the dimensionality. In two dimensions it is known (see [39]) that the critical exponents are given by the rational numbers $\alpha = -\frac{2}{3}$, $\beta = \frac{5}{36}$, $\gamma = \frac{43}{18}$, $\nu = \frac{4}{3}$, and $\eta = \frac{5}{24}$. In three dimensions $\beta \approx 0.41$, $\gamma \approx 1.82$, and $\nu \approx 0.88$ [263].

13.2.2 Series expansions and renormalization group

The application of Monte Carlo methods to the percolation problem is straightforward and we shall not discuss it here. Series expansions can be constructed

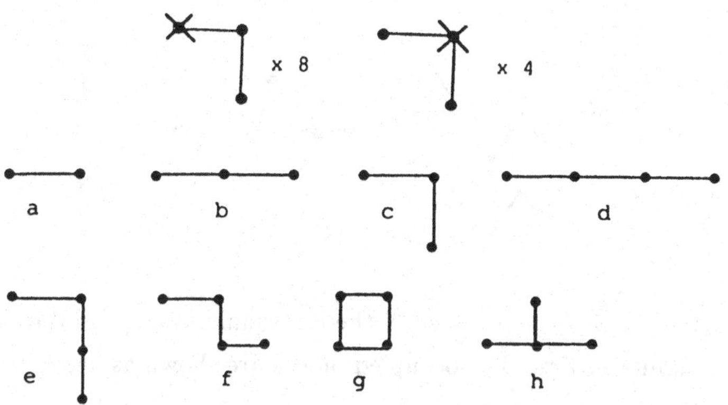

Figure 13.10: Graphs for percolation series.

for all the relevant functions discussed in Section 13.2.1 and analyzed by the standard techniques (Section 6.2) used for thermodynamic functions. Consider, for example, the quantity $S(p)$ (13.41), the mean size of finite clusters. We wish to obtain a power series in p for this function, and since we are expanding around $p = 0$, the denominator in (13.41) is simply p from (13.40). We write

$$S(p) = \frac{1}{p}\sum_s s^2 n_s(p) = \sum_{j=0}^{\infty} a_j p^j \ . \tag{13.55}$$

One can systematically calculate the coefficients a_j by enumerating clusters of increasing size and calculating the probability of occurrence. To be specific, consider the square lattice. The probability that a given site is occupied and isolated is simply $p(1-p)^4 = n_1(p)$. Consider now the next few clusters shown above. Cluster (a) will contribute terms of order p^2 and higher, (b) and (c) terms of order p^3, and the remaining five graphs terms of order p^4 and higher.

It is easiest to calculate the quantity $sn_s(j,p)$, where j refers to the label of the graph. Taking a site, say 0, in the lattice and associating it with each of the inequivalent vertices of the graph j, we count the number of configurations in which site 0 is part of cluster j. For graph (c) in the table below, we may place one of the two equivalent points at the end of the arms on site 0. There are then 8 different configurations for this case. If the central point is placed on site 0, we obtain another 4 configurations. Thus

$$3n_3(c,p) = 12p^3(1-p)^7 \ .$$

Carrying out this calculation for the remaining graphs and expanding the factor

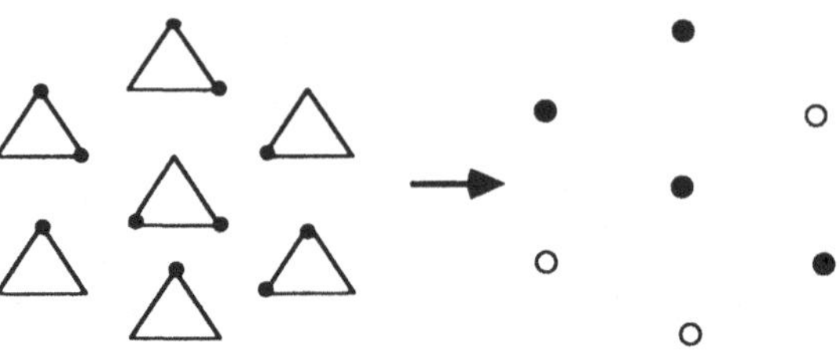

Figure 13.11: Example of one step in the renormalization procedure described for the triangular lattice. The occupied blocks are shown as heavy dots on the right.

$(1-p)^t$, where t is the number of perimeter sites, we find, through order p^3,

$$S(p) = 1 + 4p + 12p^2 + 24p^3 + \cdots \ . \tag{13.56}$$

It is clear that as in the case of high-temperature series for the Ising model (Section 6.2), the calculation of high-order terms becomes a formidable problem of graph enumeration. In the review of Essam [85] the coefficients a_j, in (13.55), are tabulated for a number of two- and three-dimensional lattices (to $j = 11$ for the square lattice). The reader may wish to derive one or two more terms to add to (13.56) or to analyze the longer series by applying the ratio methods of Section 6.2. As is the case with high-temperature expansions, one can obtain well-converged estimates of p_c and to a lesser extent of the critical exponents by the series expansion method [80].

We now briefly discuss a renormalization group approach to percolation. The renormalization group has been used primarily in two different ways to attack the percolation problem. The first method makes use of the formal equivalence between percolation and the $q \to 1$ limit of the q-state Potts model, which we have already mentioned. One can carry out an ϵ-expansion (around $d = 6$, the upper critical dimension for percolation) and attempt to evaluate the critical exponents at $\epsilon = 3$ [127], [250].

A technically simpler approach is to apply real space renormalization group techniques directly. The essential idea is that at the percolation probability, the system should be invariant under rescaling by an arbitrary length. Thus, in a simple case (Figure 13.11), we divide the triangular lattice into blocks of three sites and attempt to calculate the block occupation probability in terms of the site occupation probability. We assume that the block is occupied if the

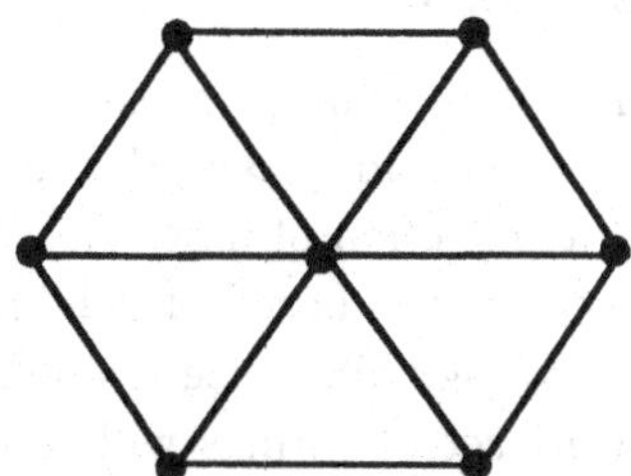

Figure 13.12: Seven site cluster on triangular lattice.

majority of its sites are occupied. Thus the probability for occupation of the blocks is given by

$$p' = R_{\sqrt{3}}(p) = p^3 + 3p^2(1-p) \ . \tag{13.57}$$

This recursion relation has trivial fixed points at $p = 0$ and $p = 1$, and a nontrivial fixed point at $p^* = 0.5$, which we identify with the percolation concentration and which is, fortuitously, exact. The correlation length exponent ν can be determined immediately from the recursion relation. Recall that $\xi(p) \sim |p - p_c|^{-\nu}$ and that $\xi'(p') = \xi(p)/L \sim |p' - p_c|^{-\nu}$, where $L = \sqrt{3}$ is the rescaling length. Thus

$$p' - p_c = L^{1/\nu}(p - p_c)$$

near the fixed point. From (13.57) we therefore obtain $\nu = 1.35$, which is in excellent agreement with the exact result $\nu = \frac{4}{3}$.

One can easily generalize these results to larger clusters. For the seven site block in Figure 13.12, the recursion relation is

$$p' = R_{\sqrt{7}}(p) = 35p^4 - 84p^5 + 70p^6 - 20p^7 \tag{13.58}$$

which, once again, has a nontrivial fixed point at $p^* = p_c = 0.5$. In this case the correlation length exponent is given by

$$\nu = \frac{\ln 7}{2 \ln \frac{7}{3}} = 1.243$$

in respectable agreement with $\nu = \frac{4}{3}$.

It is clear how to generalize this procedure to arbitrary lattices and that the Monte Carlo renormalization group methods described in Section 7.5 can also be easily adapted to the percolation problem. We refer to the reviews of Binder and Stauffer [39], Stanley *et al.* [285], and Kirkpatrick [153] for further discussion.

13.2.3 Rigidity percolation

We have seen that the percolation point p_c separates two distinct geometrical phases, a connected infinite cluster for $p > p_c$ that percolates from one side of a system to the other and a collection of finite clusters for $p < p_c$. It is clear that if the percolating cluster is composed of conductors then a current will flow across the sample if a voltage difference is applied. It is much less clear that close to p_c a percolating set of springs will resist shear or compression. To our knowledge this issue was first investigated by Feng and Sen [91] who considered a disordered lattice in which the particles on nearest neighbor sites interacted through a central force potential. They found that, for the two dimensional triangular lattice, the elastic constants decreased to zero at a bond concentration $p_r \approx 0.58$[1], substantially above the percolation point $p_c = 2\sin(\pi/18) \approx 0.3473$. For a face-centered cubic lattice, the *rigidity percolation* point was found to be $p_r \approx 0.42$ as compared to $p_c \approx 0.198$. It turns out that there are two reasons, one trivial and one more subtle, for this large range in concentrations over which a percolating network is floppy rather than rigid. The simpler reason that these networks are floppy above the percolation point is the central nature of the forces between neighbors. A familiar case in which this occurs is a completely occupied ($p = 1$) square (or simple cubic) lattice. This system has a soft mode — there is no restoring force if all the parallel columns in either the vertical or horizontal direction are tilted at constant separation of the neighbors.

A remarkably simple mean field theory captures the physics of this effect and provides a good estimate of the rigidity percolation concentration. Consider a d-dimensional lattice with sites occupied with probability p and with coordination number z. The number of degrees of freedom of the particles is $f = dpN$, where we have ignored the correction of $O(1)$ due to global translations and rotations. Each particle has, on average, pz nearest neighbors and the total number of constraints on the system is $c = Np(pz/2)$ where the factor of $1/2$ prevents double counting. If $c < f$, we expect that there will be at least one soft mode and, therefore,

$$\frac{Np_r^2 z}{2} = Ndp_r$$

or $p_r = 2d/z$ for site rigidity percolation[2]. This predicts $p_r = 2/3$ for the triangular lattice, $p_r = 1/2$ for the fcc lattice and $p_r = 1$ for both the square

[1] Recent more accurate simulations yield a value closer to $p_r \approx 2/3$.

[2] An analogous calculation for the case of random occupation of bonds with probability p also yields $p_r^B = 2d/z$.

and simple cubic lattices, in respectable agreement with simulations for the square and fcc lattices. We note that, if the potential energy of two bonds with a common vertex depends on the angle between those bonds, then $p_r = p_c$.

The second reason for the loss of rigidity of a network at p_r instead of at p_c is the fact that, in the calculations discussed to this point, we have implicitly taken $T = 0$ — we have focused on the energy of the system rather than on the free energy. As mentioned earlier, the backbone of a percolating cluster has a fractal dimension $D_B < d$ and, very close to p_c, is similar to the tenuous polymer networks found in rubbers — see Figure 13.9. The analog of the crosslinks that stabilize the amorphous phase of rubbers are the lattice points from which more than two bonds emerge. If at nonzero temperature we allow the percolating network to relax, with the boundaries fixed, there is generically an entropic tension that induces a finite shear modulus that in leading order is linear in T and which persists to $p = p_c$. Rigidity percolation is thus seen to be a zero-temperature phase transition.

Several years before the work of Feng and Sen [91], de Gennes [70] had conjectured that the random resistor network and the disordered central force network are in the same universality class. As $p \to p_c$ from above, the conductivity of a random mixture of conductors and insulators vanishes as $\sigma(p) \sim (p - p_c)^t$ where $t \approx 1.31$ ($d = 2$) and $t \approx 2.0$ ($d = 3$). He predicted that the shear modulus of the analogous network of springs $\mu(p) \sim (p - p_c)^t$ with the same exponent t. His argument is very simple. Suppose that sites i and j are connected by a conductor of conductivity σ_{ij} and that the voltages at the two sites are V_i and V_j. Kirchoff's law for the network is

$$\sum_j (V_j - V_i)\sigma_{ij} = 0$$

for all i. If we now replace the conductors with Hookean springs with spring constants k_{ij}, the force-balance conditions for the network are

$$\sum_j k_{ij}\,(\mathbf{u}_i - \mathbf{u}_j) = 0$$

where the $\mathbf{u}$'s are the positions of the respective particles. Each component of these equations is formally the same as the Kirchhoff equations. The reader will note that there is a technical flaw here. Suppose that the potential $\phi(r_{ij})$ from which these forces are derived is of the form

$$\phi(R_{ij}) = \frac{\kappa}{2}\,(r_{ij} - r_0)^2 \,. \qquad (13.59)$$

Then

$$\mathbf{F}_{ij} = -\frac{\partial \phi(r_{ij})}{\partial \mathbf{r}_i} = \kappa \frac{(r_{ij} - r_0)(\mathbf{r}_j - \mathbf{r}_i)}{r_{ij}}$$

i.e., the force constant k_{ij} is a function of the separations r_{ij} and the three components of the force-balance equations in general do not decouple. They do decouple for Gaussian (Hookean) springs $\phi(r_{ij}) = \kappa r_{ij}^2/2$ but not for arbitrary central potentials.

Numerical simulations [246, 247] for disordered networks of particles interacting through potentials of the form (13.59) display (to within numerical accuracy) the critical behavior of the random resistor network in both two and three dimensions. This surprising result can be understood by means of a phenomenological renormalization group calculation. Clearly, only the backbone can transmit forces from one side of the system to the other. As mentioned earlier, the backbone in $d = 2$ has a fractal dimension $D_B \approx 1.64$ which is not very different from that of the Sierpinski gasket $D_{SG} = \ln(3)/\ln(2) \approx 1.585$. For such a gasket, or for any other regular hierarchial fractal, one can carry out an exact renormalization transformation by integrating over the coordinates of successive generations of particles. We refer the reader to [247] for the details of such a calculation. The result is that, for any finite temperature, the equilibrium spacing r_0 of (13.59) iterates to zero and, therefore, at large length scales, the effective interaction between nearest neighbors on the 'backbone' is the Gaussian or Hookean potential. This result is consistent with our intuitive association of this system with a tenuous polymer network.

13.2.4 Conclusion

In this section we have barely scratched the surface of the subject of percolation. We have chosen to discuss the aspects that are formally equivalent to the statistical mechanics of thermally driven phase transitions and a few other topics of interest to us. For a more complete and balanced review, the reader is referred to [287]. Many applications, especially to materials science, are discussed in the book by Sahimi [263].

13.3 Phase Transitions in Disordered Materials

In this section we discuss some aspects of phase transitions in disordered materials. We use as our primary model a crystalline ferromagnet randomly diluted with nonmagnetic atoms. We assume that the magnetic atoms interact through short-range exchange interactions (nearest neighbors) and that the interactions are all ferromagnetic. This is an important simplification, as it at least allows us to determine the ground state. If the system has a mixture of ferromagnetic and antiferromagnetic interactions, even the determination of the ground state may be a difficult if not unsolvable problem. Such systems are discussed in Section 13.4.

To be specific, we consider primarily the Hamiltonian

$$H = -J\sum_{\langle ij\rangle} \epsilon_i \epsilon_j \mathbf{S}_i \cdot \mathbf{S}_j - h\sum_i \epsilon_i S_{iz} \tag{13.60}$$

where $\mathbf{S}_i = (S_{i1}, S_{i2}, \ldots S_{in})$ is an n-component spin and where the variables ϵ_i are either 1, if the site i is occupied by a magnetic atom, or 0, if site i is occupied by a nonmagnetic impurity. We assume that the random variables ϵ_i are uncorrelated, *i.e.*, $\langle\epsilon_i\epsilon_j\rangle = \langle\epsilon_i\rangle\langle\epsilon_j\rangle = p^2$, where p is the concentration of magnetic atoms. We also assume that the disorder is quenched, namely that there is no change in the distribution of magnetic atoms as a function of temperature. The Hamiltonian (13.60) thus encompasses the disordered version of the standard models for magnetism that we have studied in previous chapters ($n = 1$, Ising; $n = 2$, XY; $n = 3$, Heisenberg, etc.).

Physical realizations of the n-vector model can be found in several series of compounds (mostly antiferromagnets, with short-range interactions) such as $\mathrm{Co_pZn_{1-p}Cs_3Cl_5}$, which is a diluted three-dimensional Ising model, $\mathrm{Co_pZn_{1-p}(C_5H_5NO)_6(Cl\,O_4)_2}$ (XY model), and $\mathrm{K\,Mn_pMg_{1-p}F_2}$ (Heisenberg model). A review of experimental work and comparison with theory can be found in [290].

The ground state of (13.60) is clearly the state with all spins aligned with the magnetic field and with a magnetization per site given by

$$m(h = 0^+, T = 0) = \frac{1}{N}\sum_{i=1}^{N} \epsilon_i \langle S_{iz}\rangle|_{h=0^+,T=0} = pS\ . \tag{13.61}$$

We will primarily be interested in the question of whether or not a sharp phase transition, *i.e.*, a unique well-defined critical temperature exists in the presence of disorder; if so, what is the dependence of the critical temperature $T_c(p)$ on concentration, and what, if any, changes are to be expected in the critical exponents. It is clear that with all interactions ferromagnetic, dilution of the system with nonmagnetic impurities can only lower the critical temperature. A crude generalization of mean field theory (Section 3.1) can be constructed immediately. The mean effective field, averaging over both impurity distribution and the Boltzmann distribution, acting on a magnetic atom

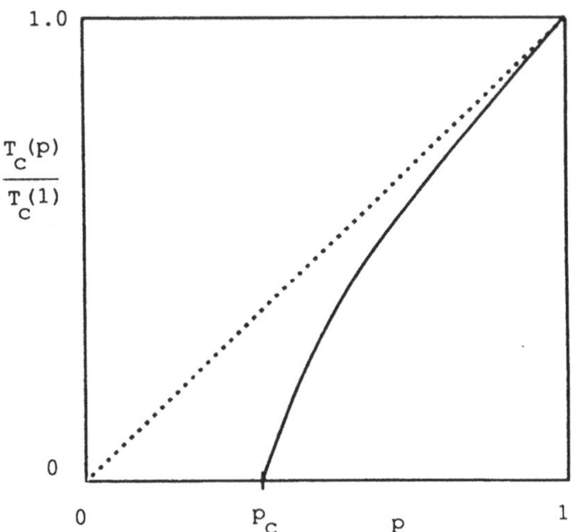

Figure 13.13: Critical temperature of model (13.60) as function of concentration p in mean field theory (dashed curve). The solid curve is qualitative but is consistent with the requirement $T_c = 0$ for $p < p_c$.

in the case of an Ising model is

$$h_{eff} = pqJm + h \tag{13.62}$$

where q is the number of nearest neighbors. Hence

$$\langle S_{iz} \rangle = \tanh\left[\beta(pqJm + h)\right] \tag{13.63}$$

which yields $T_c(p)/T_c(1) = p$ for the critical temperature. This result is obviously wrong in view of the discussion of Section 13.2. There cannot be a finite-temperature phase transition for $p < p_c$, the percolation concentration, and the critical temperature, as a function of p, must be qualitatively like the solid curve in Figure 13.13 rather than the dashed mean field approximation. Before dealing further with these topics, we digress briefly on the statistical formalism necessary for a description of disordered systems.

13.3.1 Statistical formalism and the replica trick

As mentioned above, we will be interested primarily in the case of quenched disorder. Thus the partition function is a function of the variables $\{\epsilon_j\}$,

$$Z = \mathrm{Tr}\exp\left\{-\beta H[\epsilon_i, \mathbf{S}_i]\right\} = Z(h, T, \epsilon_1, \epsilon_2, \ldots, \epsilon_N) \tag{13.64}$$

as is the free energy,

$$G(h,T,\epsilon_1,\epsilon_2,\ldots,\epsilon_N) = -k_BT\ln Z(h,T,\epsilon_1,\epsilon_2,\ldots,\epsilon_N) . \tag{13.65}$$

In practice, it is impossible to calculate the free energy (13.65) for an arbitrary configuration of impurities. To simplify the problem somewhat, we assume that the system can be divided into a large number of subvolumes, each much larger than the correlation length. These subvolumes then constitute an ensemble of statistically independent systems, each with an impurity configuration governed by the same distribution. The result of an experiment will thus yield an average of the observable with respect to the impurity distribution. For example, the measured free energy per site is given by

$$g(h,T,p) = -\frac{k_BT}{N}\langle\ln Z(h,T,\epsilon_1,\epsilon_2,\ldots,\epsilon_N)\rangle \tag{13.66}$$

where the concentration p is a parameter, not a thermodynamic variable. Other thermodynamic functions are derived from (13.66) in the usual way.

On the other hand, if the impurities are annealed rather than quenched, the variables ϵ_i must be treated as dynamical variables on the same footing as the spin variables. The impurity concentration can be fixed through the introduction of a chemical potential and the appropriate partition function is

$$Z(h,T,\mu) = \sum_{\epsilon_i=0,1}\mathrm{Tr}\exp\left\{-\beta H[\epsilon_i,\mathbf{S}_i] + \beta\mu\sum_{i=1}^{N}\epsilon_i\right\} \tag{13.67}$$

with

$$Np = k_BT\frac{\partial}{\partial\mu}\ln Z(h,T,\mu) . \tag{13.68}$$

It is clear that the two cases are fundamentally different.

The calculation of the free energy (13.66) is a very difficult problem. We now describe briefly one of the more intriguing and popular methods used to carry out the average in (13.66) known as the "replica trick" ([121]; see also [290], [180], [16] and [40] for discussions). This method makes use of the formal identity

$$\left\langle\lim_{n\to 0}\frac{x^n-1}{n}\right\rangle = \langle\ln x\rangle . \tag{13.69}$$

Taking x to be the partition function (13.64) with a given set of variables $\epsilon_1,\epsilon_2,\ldots,\epsilon_N$, we see that

$$Z^n = \left[\mathrm{Tr}_{\mathbf{S}_i}e^{-\beta H\{\mathbf{S}_i,\epsilon_i\}}\right]^n = \mathrm{Tr}_{\mathbf{S}_{i,1}}\mathrm{Tr}_{\mathbf{S}_{i,2}}\cdots\mathrm{Tr}_{\mathbf{S}_{i,n}}\exp\left\{-\beta\sum_{\alpha=1}^{n}H[\mathbf{S}_{i,\alpha},\epsilon_i]\right\} . \tag{13.70}$$

The spin variables $S_{i,\alpha}$, for different α, are independent dynamical variables and we essentially have n copies, or replicas, of the system, all with the same configuration of magnetic atoms specified by the set of variables $\{\epsilon_j\}$. One now interchanges the calculation of the average over the impurity distribution and the limit $n \to 0$ in (13.69) and, as well, attempts to analytically continue to $n = 0$ from integer values of n. This is, in fact, the delicate step in the procedure and may, as we shall see in Section 13.4.3, lead to unphysical results at low temperature. Even in this case the replica method proved valuable in leading towards the correct solution (see, for example, [204] and references therein).

If one proceeds, the average over $\{\epsilon_j\}$ can be carried out (in closed form if the distribution is simple enough) and the resulting effective Hamiltonian is translationally invariant but couples spins in different replicas, that is,

$$\int d\epsilon_1 d\epsilon_2 \ldots d\epsilon_N P(\epsilon_1, \epsilon_2, \ldots, \epsilon_N) \mathrm{Tr} \exp\left\{-\beta \sum_{\alpha=1}^{n} H[\mathbf{S}_{i,\alpha}, \epsilon_i]\right\}$$
$$= \mathrm{Tr} \exp\left\{-\beta H'[\mathbf{S}_{i,1}, \mathbf{S}_{i,2}, \ldots, \mathbf{S}_{i,n}]\right\} = Z'(h, T, n) \qquad (13.71)$$

where the functional form of H' depends on the impurity distribution function $P(\epsilon_1, \epsilon_2, \ldots, \epsilon_N)$. The translationally invariant effective Hamiltonian can then be treated in the usual way in mean field theory (Chapter 3) or by renormalization group methods (Chapter 7). In most cases the results obtained by the replica methods are consistent with those arrived at by other means.

13.3.2 Nature of phase transitions

To this point, a few exact results on the statistical mechanics of disordered materials for dimensionality $d > 1$ exist. One of the exceptions is a calculation by McCoy and Wu (reviewed in [197] and [198]) for a two-dimensional Ising model. They considered the case of a constant nearest-neighbor interaction in the horizontal direction on a square lattice and random interactions between spins in different rows. McCoy and Wu found that a well-defined transition temperature exists but that the critical behavior is modified by disorder. For example, the zero-field specific heat is infinitely differentiable but nonanalytic at T_c in contrast to the logarithmic divergence found in the pure model (Section 6.1).

A second interesting analytic result is due to Griffiths [118]. He showed that the magnetization $m(h, T)$ of a disordered Ising model is not an analytic function of h at $h = 0$ for any $T < T_c(p = 1)$. Thus the phase diagram in

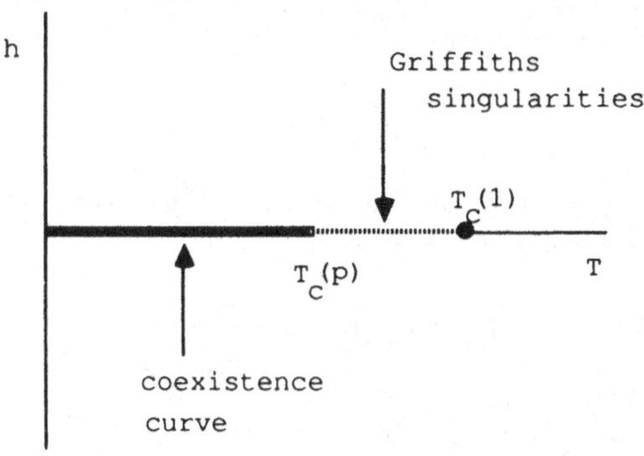

Figure 13.14: Phase diagram of a disordered Ising model in the $h-T$ plane.

the $h-T$ plane of the disordered Ising model is as shown in Figure 13.14. The *Griffiths singularities*, in the region $T_c(p) < T < T_c(1)$, are weak essential singularities which may in fact be unobservable experimentally. They arise, even for $p < p_c$, the percolation concentration, from arbitrarily large clusters of connected spins which occur with finite probability for any nonzero p. This effect of large connected islands is reminiscent of the singular structure (Lifshitz tails) of the density of states near band edges discussed in Section 13.1.

We next mention a heuristic argument due to Harris [126] regarding the nature of phase transitions in disordered materials. We have seen, in Chapters 6 and 7 that the singular part of the free energy obeys a homogeneity or scaling relation of the form

$$g(t,h,u) = |t|^{d/y_1} g\left(t/|t|, h|t|^{-y_h/y_1}, u|t|^{-y_u/y_1}\right) . \tag{13.72}$$

In the case of a disordered system [such as (13.60)], we identify the quantity u with the fluctuation in nearest-neighbor coupling constants, and the fixed point $t = h = u = 0$ with the pure system fixed point. The quantities y_1, y_h, and y_u are the exponents determined from the linearized recursion relations near the pure system fixed point. If the exponent y_u is greater than zero, the perturbation u is relevant and the fixed point is unstable. The renormalization group flow is then usually to another, stable, fixed point and it is conventional to call the exponent $\phi_u = y_u/y_1$ the crossover exponent. Since the fluctuations in the dimensionless coupling constants are temperature-like variables, we expect that $\phi_u = 1$.

It is clear, from (13.72), that in general the variable u becomes important at a characteristic temperature

$$u|t|^{-\phi_u} \approx 1 \,. \tag{13.73}$$

We assume that the system at temperature t has correlation length ξ. A measure of the magnitude of u is therefore

$$u \approx \frac{\sqrt{\int_{\xi^d} d^d r \, \langle (J(\mathbf{r}) - \langle J \rangle)^2 \rangle}}{\xi^d \langle J \rangle} \approx \xi^{-d/2} \tag{13.74}$$

where $J(\mathbf{r})$ is the coupling constant at point $\mathbf{r}$ and where the angular brackets indicate averaging with respect to the impurity distribution function. Using (13.74) in (13.73), we have

$$u|t|^{-\phi_u} \approx u|t|^{-1} \approx |t|^{d\nu/2-1} = |t|^{-\alpha/2}$$

where, in the last step, we have used the hyperscaling relation $d\nu = 2-\alpha$, where α is the specific heat exponent of the pure system. We conclude, therefore, that disorder is relevant if the specific heat exponent α is greater than zero (three-dimensional Ising model), irrelevant if it is less than zero (three-dimensional Heisenberg model and possibly the XY model).

A slightly different way of phrasing the argument above is to think of the system as a collection of domains of linear dimension ξ. The degree of order in each domain is characteristic of a system a distance ΔT from the critical point, and this deviation is determined by t and by the quantity u defined in (13.74). As $t \to 0$, the entire system will approach the critical point simultaneously only if $\Delta T(t, u) \to 0$. It is tempting to conclude that if this does not occur, the transition will not be sharp, but will rather be smeared over some finite temperature interval, and that the sharp power law singularities characteristic of pure systems will be replaced by smooth functions characterized by some width ΔT. However, the evidence (see, *e.g.*, [290]) from renormalization group and Monte Carlo calculations is that a sharp transition exists nevertheless, but with modified critical exponents, namely those of a disorder fixed point.

Another interesting case that we can consider is the model (13.60) with constant exchange interactions but with a random magnetic field with mean value $\langle h_i \rangle = 0$ [138]. In this case one can show that the system may be unstable, at low temperature, against domain formation. The argument goes as follows. The energy cost of introducing a domain of volume L^d into the sample is of order L^{d-1} for an Ising model and of order L^{d-2} for models with continuous symmetry. (The domain wall is diffuse in this case. The spins can rotate continuously over the distance L. Writing $\mathbf{S}_i \cdot \mathbf{S}_j = S^2 \cos(\hat{n}_i \cdot \hat{n}_j) \approx S^2[1 - \frac{1}{2}(\pi/L)^2]$, for i, j nearest neighbors, we immediately arrive at the result that the domain wall energy scales as L^{d-2}.) The sum of the random fields, over a finite region of size L^d, has a finite expectation value which is randomly positive or negative but typically has a magnitude given by $\langle (\sum_i h_i)^2 \rangle^{1/2} \approx L^{d/2}$. Thus the energy gain due to alignment with the local field is of order $L^{d/2}$. Therefore, for large L, domain formation is energetically favorable if

$$\begin{aligned} L^{d/2} &> L^{d-1} \quad \text{(Ising model)} \\ L^{d/2} &> L^{d-2} \quad (n \geq 2) \end{aligned}$$

which implies that the Ising model in a random field will break up into domains for $d < 2$ and models with continuous symmetry for $d < 4$. Note that the entropy, which has not been considered, can only help this process. Domain formation of course implies that the order parameter is zero. The situation of a random magnetic field does not turn out to be of purely academic interest. Imry and Ma [138] discuss a number of possible physical realizations of this case.

We now briefly discuss the dependence of the critical temperature on p, the concentration of magnetic atoms in the sample, assuming that the system has a sharp continuous transition with exponents that may or may not be those of the pure system. Our discussion closely follows that of Lubensky [180]. For p close to 1 it is intuitively clear that T_c will decrease linearly with p. More interesting is the behavior of T_c near the percolation concentration p_c. A scaling argument can be used to find the dependence of T_c on p. We assume that at $T = 0$, $p = p_c$, a crossover occurs between the thermal fixed point (not necessarily the pure system fixed point) and the percolation fixed point. Thus, as in (13.72), we may write the free energy or other functions of interest in the scaled form

$$f(T, p - p_c) = |p - p_c|^{d/y_p} F\left[\frac{g(T)}{|p - p_c|^{\phi}}\right] \tag{13.75}$$

where $g(T)$ and the crossover exponent ϕ remain to be identified, and where y_p is the appropriate percolation exponent equivalent to the thermal exponent y_1. In (13.43) we have seen that for an Ising model at low temperature, $T\chi(0, T)$ is related to the percolation function $S(p)$. Thus we write

$$T\chi(0, T) = |p - p_c|^{-\gamma_p} \Psi\left[\frac{g(T)}{|p - p_c|^{\phi}}\right] \tag{13.76}$$

where γ_p again is the percolation exponent.

To make further progress we use a conceptual picture of percolation clusters due to Skal and Shklovskii [276]. We assume that after all dangling bonds (irrelevant for phase transitions) are removed from the infinite cluster, the resulting *backbone* is schematically of the form shown in Figure 13.15. In this picture relatively dense regions (the black "blobs") are connected by long, essentially one-dimensional strands. We assume that the average distance between the blobs is the percolation correlation length ξ_p and that the number of segments in the connecting strand is[3] L. The spin-spin correlation length $\xi_M(T)$ along one of the one-dimensional segments can be determined analytically. For the one-dimensional Ising model we have (Section 3.6)

$$\xi_M(T) = \frac{-1}{\ln\tanh(J/k_BT)} \approx \frac{1}{2}e^{2J/k_BT} \tag{13.77}$$

at low temperatures. For the classical n-component Heisenberg model Stanley [281] has shown that

$$\xi_M(T) \approx \frac{2n}{n-1}\frac{J}{k_BT}\ . \tag{13.78}$$

This length ξ_M characterizes the magnetic order as long as $\xi_M \ll L$. Once the magnetic correlation length becomes comparable to L, the blobs form a strongly interacting d-dimensional network and the one-dimensional correlations are no longer relevant. We assume, therefore, that

$$T\chi(0,T) = |p-p_c|^{-\gamma_p}\Psi\left[\frac{\xi_M(T)}{L(p)}\right] \tag{13.79}$$

and, further, that $L(p) \sim |p-p_c|^{-\zeta}$. The crossover from percolative to magnetic behavior then occurs at $\xi_M(T) \approx L \approx |p-p_c|^{-\zeta}$, and for the Ising model we therefore find

$$\frac{k_BT_c(p)}{J} \approx -\frac{2}{\zeta\ln|p-p_c|} \tag{13.80}$$

while

$$\frac{k_BT_c(p)}{J} \approx \frac{2n}{n-1}|p-p_c|^{\zeta} \tag{13.81}$$

for the n-vector model. The exponent ζ is thought to be exactly 1 in all dimensions on the basis of renormalization group calculations [180], and if this holds, T_c approaches zero linearly as a function of $p - p_c$ for the n-vector models but, independent of ζ, with an infinite slope for the Ising model. High

[3]One might naively assume that $L \geq \xi_p$ but this leads to a contradiction for the two-dimensional Ising and Potts models. The problem is that the blobs and strings picture is oversimplified. See [62] for a discussion.

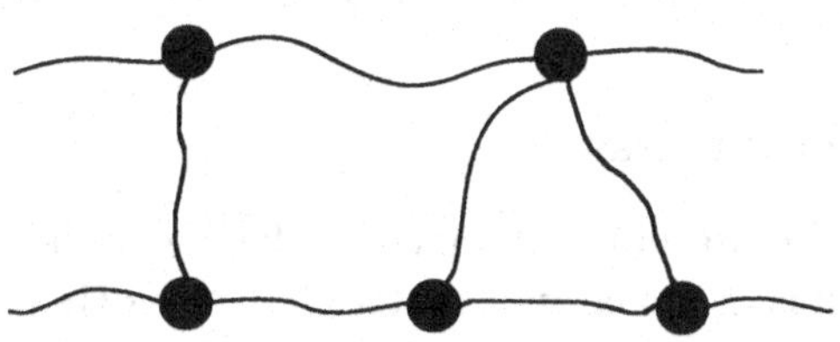

Figure 13.15: Schematic picture of the backbone of the infinite cluster near the percolation threshold.

temperature series and Monte Carlo calculations support these conclusions [290].

We now turn briefly to the calculation of critical exponents by renormalization group methods. The application of the position space renormalization group methods of Chapter 7 to disordered systems is considerably more difficult than in the case of pure materials. Under a rescaling transformation, not only the coupling constants but also the probability distribution of the random interactions is modified and at the fixed points is represented by a *fixed distribution* rather than by a finite set of numbers. We will not discuss the various methods that have been developed to deal with these difficulties but rather, refer the reader to the reviews by Kirkpatrick [153] and Stinchcombe [290].

Field-theoretic and momentum space renormalization group methods have also been applied to the problem of dilute magnets. We again refer to the aforementioned reviews and references therein. The physical picture that has emerged from this work is consistent with the Harris criterion. In pure systems with specific heat exponents greater than zero, the pure system fixed point is unstable with respect to disorder, and a new disorder fixed point with its own set of critical exponents governs the critical behavior of the system. Conversely, the exponents of systems with negative specific heat exponents are unaffected by disorder.

13.4 Strongly Disordered Systems

We now turn to the most difficult and least understood subject, that we have included under the general heading of disordered systems, namely the physics of amorphous or glassy materials and spin glasses. As in the previous sections, we have leaned heavily on recent reviews, in particular those of Anderson [16],

Lubensky [180], Jäckle [140], Binder and Young [40] and Mezard *et al.* [204].

13.4.1 Molecular glasses

A large number of quite different materials form glasses as they are cooled from the melt. Films of monatomic materials such as Ge and Si can be readily prepared in an amorphous form. Metallic glasses, often a mixture of a metal such as Ni or Co with a metalloid such as P can be prepared by rapid cooling techniques at least in certain composition ranges. Intermetallic compounds ($TbFe_2$ and others) form glasses if sputtered onto a cold substrate. The common feature in these preparation techniques is the rapid extraction of thermal energy, and it is possible that even the simplest monatomic liquids would form glasses if heat could be extracted rapidly enough. Phenomenologically, the glass transition is characterized by a rapid freezing out of transport processes at the transition temperature, T_g—the system becomes rigid, diffusion decreases rapidly, the shear viscosity increases dramatically. The transition temperature T_g depends on the cooling rate, as do the low-temperature physical properties such as the specific volume. This is already an indication that in contrast to the phase transitions that we have described previously, the glass transition is not an equilibrium phenomenon. Further evidence for this point of view comes from the observation that below the transition temperature the properties of the glass continue to change as a function of time, although very slowly, and it is believed that given sufficient time the sample will eventually reach the equilibrium low-temperature configuration, namely a crystal.

One of the unusual features of amorphous materials is the low-temperature behavior of the specific heat, which is linear in T, even for insulators, rather than cubic as one expects for the phonon specific heat in three-dimensional crystals. This feature can be understood on the basis of the following conceptual picture, which also qualitatively accounts for the observed kinetic effects alluded to above. A simplified version is shown in Figure 13.16. At high temperature the free energy $A(x)$ of the system, as a function of a number of configurational parameters x, has a unique minimum that is accessible from any point in the configuration space. At lower temperatures it is assumed that this free energy is the more complicated function, with many local minima shown in the lower graph and an absolute minimum (presumably a crystalline state) some distance in configuration space from the high-temperature minimum. It is clear that if the system is rapidly cooled—more rapidly than the characteristic relaxation times—it may become trapped at one of the local minima. Once the free-energy barriers have formed, the probability of a transition

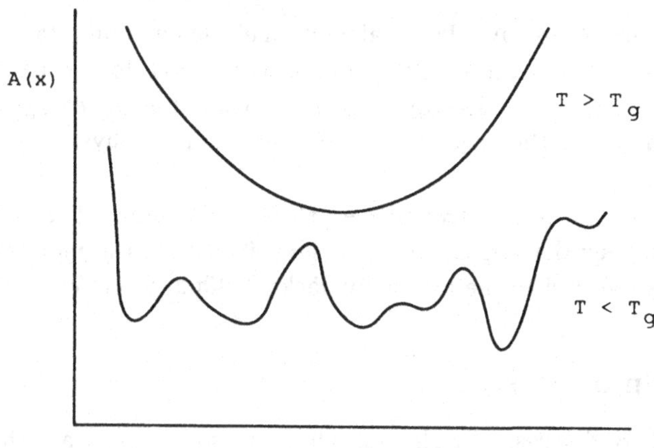

Figure 13.16: Schematic picture of the free energy of a system above and below the glass transition temperature (see the text).

is $\mathcal{O}(\exp\{-\beta\Delta A\})$, where ΔA is the height of the free-energy barrier, and if ΔA is roughly of magnitude 1 eV, the transition rate can become vanishingly small. On the other hand, if the cooling process is slow, the system may be able to follow the absolute free-energy minimum to the ground state.

Returning to the specific heat, we now assume that there are states very close in energy to the local free energy minima and that the density $n(E)$ of these states approaches a constant, $n(0)$, as the energy difference between the excited and local minimum states approaches zero. Note that this dependence is in sharp contrast to the phonon density of states in a crystal which has the functional form $n_{ph}(E) \sim E^2$ for small E. The original picture of these "tunneling states" is due to Anderson *et al.* [18]. Assuming that this picture is correct, we have

$$\langle E(T)\rangle = \int dE' n(E') \frac{E'}{\exp\{\beta E'\} - 1} \approx \frac{\pi^2}{6} n(0)(k_B T)^2 \tag{13.82}$$

and for the specific heat,

$$C(T) \approx \frac{\pi^2}{3} n(0) k_B^2 T \; . \tag{13.83}$$

This formula, without further justification, is purely phenomenological. Anderson *et al.*, however, predicted a number of other effects on the basis of the tunneling model, which seem to be consistent with experiment and which lend support to this picture. In a later calculation of the low-lying states of a model Hamiltonian for a spin glass, Walker and Walstedt [317] found that the density of low-lying excitations indeed approaches a constant as the energy approaches zero. In spin glasses the magnetic specific heat is, as in physical glasses, linear in the temperature.

A convincing microscopically based theory of the glass transition has not to this point been developed. For a review of much of the more recent work, the reader is referred to the article by Jäckle [140] and references therein.

13.4.2 Spin glasses

We now briefly discuss spin glasses, which, in some ways, are the magnetic counterpart of molecular glasses. The first examples of the spin glass transition (see, *e.g.*, [57]) were found in dilute alloy systems such as $Au_{1-x}Fe_x$, with x very small. Experimentally, one sees a rather sharp maximum in the zero-field susceptibility, a broad maximum in the specific heat, and an absence of any long-range order below this spin glass transition temperature, although there is both hysteresis and remanence. Many other materials (including nonmagnetic ones) have since been identified as having a transition of the spin glass type (see [40]).

As in the case of molecular glasses, one interprets the spin glass transition as a freezing out of fluctuations—in the molecular glasses, the freezing out of large-scale structural rearrangements, in the spin glass the freezing out of magnetic transitions. The existence of hysteresis and remanence is already evidence that this occurs. There is, however, one important conceptual difference between the two systems. In the case of molecular glasses, one usually knows that the true equilibrium state at low temperatures is a crystal. In spin glasses there may be no unique ground state. The magnetic atoms in dilute alloys, such as $Au_{1-x}Fe_x$ or $Cu_{1-x}Mn_x$, do not diffuse appreciably at low temperatures. The interaction between magnetic atoms is therefore also frozen or quenched and, in a metallic environment, this interaction is most simply modeled by the Ruderman–Kittel–Kasuya–Yosida (RKKY) interaction, which is of the form [194]

$$H = \sum_{i,j} J(R_{ij})\mathbf{S}_i\cdot\mathbf{S}_j = J_0 \sum_{ij} \frac{\cos(2k_F R_{ij} + \phi)}{R_{ij}^3}\mathbf{S}_i\cdot\mathbf{S}_j \tag{13.84}$$

(*i.e.*, *long range* and *oscillatory*). The RKKY oscillations are caused by the sharpness of the Fermi surface in the same way as the Friedel oscillations

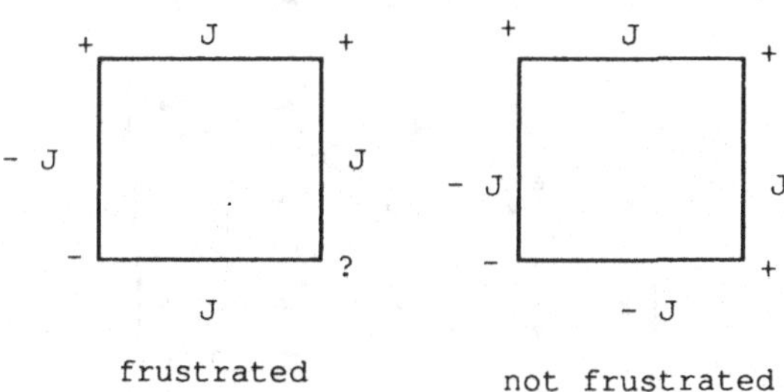

Figure 13.17.

discussed in Chapter 8. A given magnetic atom thus interacts both via ferromagnetic and antiferromagnetic interactions with others, and it is conceivable that no simple ground state analogous to the crystalline state of molecular glasses exists.

This idea can be made more precise and is generally referred to as *frustration*. Consider a simpler model Hamiltonian than (13.84), namely the nearest-neighbor Ising model on a square lattice, but with interactions which are randomly either $+J$ (ferromagnetic) or $-J$ (antiferromagnetic). In Figure 11.14 we show two elementary plaquettes (elementary square units of four spins), one of which is frustrated, one which is not. It is easy to see that no arrangement of spins on the frustrated plaquette will leave all bonds satisfied. A choice of $\sigma = 1$ in the lower right-hand corner leaves the lower horizontal bond in an unfavorable configuration, the opposite choice leaves the right vertical bond in a high-energy state. A plaquette is frustrated if the product of the four coupling constants is negative, not frustrated if it is positive, as can easily be verified explicitly. We note in passing that the nonrandom Ising antiferromagnet on the triangular lattice is frustrated and this model has a finite ground-state entropy per spin and no phase transition.

Consider now the segment of the square lattice shown in Figure 13.17, which contains four frustrated plaquettes labelled by the open circles. The $+$ and $-$ signs between the lattice sites indicate the sign of the interaction between spins. In the absence of a magnetic field we have spin-flip symmetry and can choose the orientation of one of the spins (upper left-hand corner) freely. We next choose the remaining spins, in turn, satisfying all bonds until we reach a frustrated plaquette in which we are forced to make one unfavorable assignment. Since the bond in the unfavorable configuration is shared between two plaquettes, the unfavorable assignment propagates until we arrive at an-

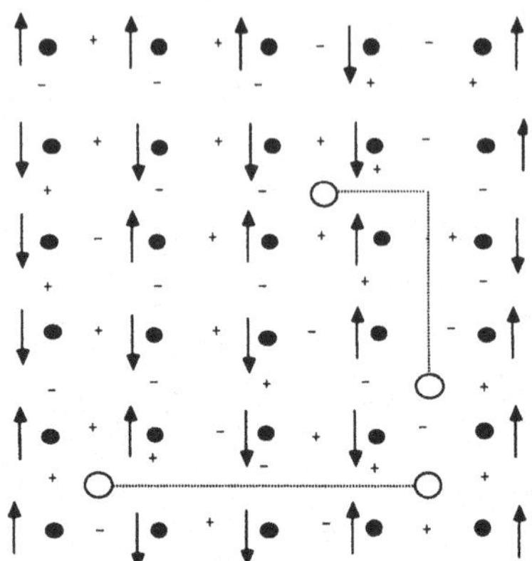

Figure 13.18: Section of square lattice with four frustrated plaquettes.

other frustrated plaquette at which the process terminates. This is indicated in Figure 13.18 by the dotted lines perpendicular to the bonds in unfavorable configurations. The length of these lines or "strings" is a measure of the energy of the spin configuration relative to a state with all bonds satisfied. The problem of determining the ground state is thus equivalent to finding the set of strings, connecting frustrated plaquettes, with shortest total length. This is a nontrivial problem and we can already see from this simple example how a high degree of degeneracy can arise. In Figure 13.19 we show two string configurations with the same energy as the state of Figure 13.18, which, however, imply different orientations of the spins.

Thus a frustrated system is, we believe, characterized by a high degree of ground-state degeneracy. In more realistic models we would also expect to find a high density of low-lying excited states and dynamical effects due to string motion. Further support for the idea that frustration is an important aspect of spin glass ordering comes from the Mattis model [193], which, although random in nature, undergoes an ordinary continuous phase transition. The Hamiltonian of this model is

$$H = -\sum_{i,j} J(R_{ij})\epsilon_i\epsilon_j \mathbf{S}_i\cdot\mathbf{S}_j - h\sum_i S_{iz} \tag{13.85}$$

where $\epsilon_i = +1$ or -1 are quenched random variables. Defining new spins

$$\boldsymbol{\tau}_i = \epsilon_i \mathbf{S}_i$$

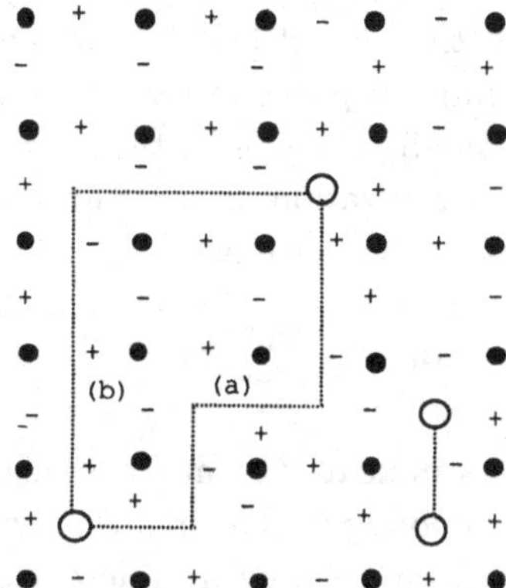

Figure 13.19: String configurations (a) and (b) together with the nearest-neighbor string are each degenerate with that of Figure 13.18. The assignment of coupling constants is, of course, the same.

we obtain

$$H = -\sum_{i,j} J(R_{ij})\boldsymbol{\tau}_i\cdot\boldsymbol{\tau}_j - h\sum_i \epsilon_i \tau_{iz} \tag{13.86}$$

which represents a translationally invariant "pseudo spin" interaction and a Zeeman term with a random magnetic field. For $h = 0$, randomness is completely absent and the partition function for this model with quenched disorder is the same as the partition function of the pure Heisenberg, XY, or Ising model, depending on the number of spin components. The specific heat, internal energy, and other zero-field thermodynamic functions thus have simply the singularities of the pure model.

It is clear that the ground state of (13.86) in zero magnetic field is the state with the τ spins fully aligned; hence there is no frustration. The original spins $\{S\}$ are randomly oriented, but as far as the critical behavior of the system is concerned, the model is equivalent to a ferromagnet and it is the absence of frustration that makes it unsuitable as a model for spin glasses. A model that does seem to have the essential features necessary to describe the spin glass transition is the Edwards–Anderson model [83], which has the Hamiltonian

$$H = -\sum_{i,j} J(R_{ij})\mathbf{S}_i\cdot\mathbf{S}_j - h\sum_i S_{iz} \tag{13.87}$$

where the interaction $J(R_{ij})$ is a random variable, and where the range of the interaction can be varied. We shall not review the various methods that have been devised to deal with this model [40] but do wish to discuss briefly the characterization of the spin glass phase in terms of an order parameter.

It is believed that for small values of $\bar{J}$, the average exchange interaction, there is no spontaneous magnetization or other form of long-range order. It is clear then that expectation values such as

$$\mathbf{m_q} = \frac{1}{N}\sum_i \langle \mathbf{S}_i \rangle e^{i\mathbf{q}\cdot\mathbf{R}_i}$$

where the angular brackets indicate thermal averages, must all be zero. On the other hand, if the spin configuration freezes, the local expectation value $\langle \mathbf{S}_i \rangle$ is well defined but will be oriented in different directions, depending on the site. One choice of order parameter which distinguishes between a high-temperature paramagnetic phase and a low-temperature frozen configuration is the Edwards–Anderson order parameter which arises naturally from the replica approach. This quantity is

$$q \equiv \overline{\langle \mathbf{S}_i \rangle^2} = \frac{1}{N}\sum_i \langle \mathbf{S}_i \rangle^2 \tag{13.88}$$

and it is zero at high temperature, since each individual spin has zero expectation value in the high-temperature phase, and positive in the spin glass state. In the next subsection we discuss a somewhat simpler model, the Sherrington-Kirkpatrick model and attempt to calculate the partition function using the replica method discussed in Section 13.3.1.

13.4.3 Sherrington–Kirkpatrick model

The Sherrington–Kirkpatrick model [266], [152] differs from the Edwards–Anderson model in that the interactions are of infinite range. We also specialize to the case of an Ising model although a generalization to continuous-spin models is possible [152]. The Hamiltonian is

$$H = -\sum_{\{i,j\}} J_{ij} S_i S_j - h\sum_i S_i \tag{13.89}$$

where $S_i = \pm 1$ and where the sum in the first term extends over *all* pairs of particles in an N-site lattice without double-counting. The coupling constants J_{ij} are distributed according to a Gaussian probability distribution:

$$P(J_{ij}) = \frac{1}{\sqrt{2\pi}\,J} \exp\left\{-\frac{J_{ij}^2}{2J^2}\right\} \tag{13.90}$$

where the width J is independent of the separation of sites i and j. The incorporation of a nonzero average interaction $\bar{J}$ presents no additional problems of principle, but complicates the formulas, and we have omitted it for this reason. Mean field theory is exact for infinite range translationally invariant interactions, and one hopes that a theory of the mean-field type will work as well for this model.

Since the disorder is quenched, we must calculate the free energy and average this expression over the distribution (13.90):

$$f = -\frac{k_B T}{N} \prod_{\{ij\}} \int dJ_{ij}\, P(J_{ij}) \ln Z\left[\{J_{ij}\}, h, T\right] \tag{13.91}$$

with

$$Z = \operatorname{Tr} \exp\left\{-\beta H\left[\{J_{ij}\}, \{S_i\}\right]\right\} . \tag{13.92}$$

In order to interchange the order of the integration over coupling constants and the trace over spin configurations, we use the replica trick (Section 13.3.1):

$$\ln Z = \lim_{n \to 0} \frac{1}{n}\left[Z^n - 1\right] \tag{13.93}$$

where n will be an integer larger than 2 in the initial stages of the calculation but, in the end, will be treated as a continuous variable and set to 0. The n-th power of the partition function is then

$$Z^n = \operatorname*{Tr}_{\{S_i^\alpha\}} \exp\left\{\beta \sum_{\alpha=1}^{n}\left[\sum_{\{ij\}} J_{ij} S_i^\alpha S_j^\alpha + h \sum_i S_i^\alpha\right]\right\}$$

where the spins in each of the n replicas are interacting with the same set of coupling constants J_{ij}. Before proceeding with the calculation we show that the Edwards–Anderson order parameter can be expressed as a correlation function between spins on the same site in different replicas. Consider two replicas, labeled α and γ and add a dummy field h' that couples S_i^α and S_i^γ. Then

$$\begin{aligned}\langle S_i \rangle^2 = \frac{\partial}{\partial h'} \ln \operatorname{Tr} \exp\{-\beta[H[\{J_{ij}\}, \{S^\alpha\}] \\ + H[\{J_{ij}\}, \{S^\gamma\}]] + h' S_i^\alpha S_i^\gamma\}|_{h'=0} \end{aligned} \tag{13.94}$$

Using the replica trick as above and averaging over the distribution P we have

$$q \equiv \langle \langle S_i^2 \rangle \rangle_J$$

$$
\begin{aligned}
&= \frac{\partial}{\partial h'} \lim_{n\to 0} \frac{1}{n} \prod_{ij} \int dJ_{ij}\, P(J_{ij})\,[Z^n(h') - 1]\Bigg|_{h'=0} \\
&= \lim_{n\to 0} \frac{1}{n} \prod_{ij} \int dJ_{ij}\, P(J_{ij}) \\
&\quad \times \left[\frac{n \mathrm{Tr}\, S_i^\alpha S_i^\gamma \exp\{-\beta\,[H[\{J_{ij}\},\{S^\alpha\}] + H[\{J_{ij}\},\{S^\gamma\}]]\}}{\mathrm{Tr}\,\exp\{-\beta\,[H[\{J_{ij}\},\{S^\alpha\}] + H[\{J_{ij}\},\{S^\gamma\}]]\}} \right] Z^n \\
&= \langle\langle S_i^\alpha S_i^\gamma \rangle\rangle_J \Big|_{n=0}
\end{aligned} \tag{13.95}
$$

which demonstrates that the order parameter is the $n = 0$ limit of the spin-spin correlation function between replicas. In (13.95) the notation $\langle\langle A\rangle\rangle_J$ indicates both a trace over the spin states and an average over the distribution function P.

We now proceed to evaluate the free energy f, using (13.93). Since the coupling constants are independent random variables, the integrals over J_{ij} factor. A typical term is

$$
\frac{1}{\sqrt{2\pi}\,J} \int_{-\infty}^{\infty} dJ_{ij}\, e^{-J_{ij}^2/2J^2}\, e^{\beta J_{ij}\sum_\alpha S_i^\alpha S_j^\alpha} = e^{\beta^2 J^2/2 \sum_{\alpha,\gamma} S_i^\alpha S_j^\alpha S_i^\gamma S_j^\gamma}\ .
$$

Using the identity

$$
\sqrt{\frac{a}{2\pi}} \int_{-\infty}^{\infty} dx\, e^{-\frac{1}{2}\left(ax^2 - \lambda\sqrt{2}x\right)} = e^{\lambda^2/4N} \tag{13.96}
$$

to integrate over $\{J\}$ thus results in an effective interaction between spins in different replicas

$$
\begin{aligned}
-\beta f &= \lim_{\substack{N\to\infty \\ n\to 0}} \frac{1}{nN} \\
&\quad \times \left[\operatorname*{Tr}_{\{S^\alpha\}} \exp\left\{ \frac{\beta^2 J^2}{2} \sum_{\{ij\}} \sum_{\alpha,\gamma} S_i^\alpha S_j^\alpha S_i^\gamma S_j^\gamma + \beta h \sum_{i,\alpha} S_i^\alpha \right\} - 1 \right] \\
&= \lim_{\substack{N\to\infty \\ n\to 0}} \frac{1}{nN} \left[\operatorname*{Tr}_{\{S^\alpha\}} e^{\mathcal{H}} - 1 \right]\ .
\end{aligned} \tag{13.97}
$$

The first expression in $\mathcal{H}$ consists of $n^2 N(N-1)/2$ terms and we see that in order to obtain an extensive free energy we must let the width of the distribution become smaller as the number of interacting spins increases, *i.e.*, $J = \tilde{J}/\sqrt{N}$.

We also note that the first term in $\mathcal{H}$ can be written in the form

$$\begin{aligned}\sum_{\{i,j\}}\sum_{\alpha,\gamma} S_i^\alpha S_j^\alpha S_i^\gamma S_j^\gamma &= \frac{1}{2}\sum_{\alpha\neq\gamma}\sum_{\{i,j\}} S_i^\alpha S_j^\alpha S_i^\gamma S_j^\gamma + \frac{nN(N-1)}{2} \\ &= \frac{1}{2}\sum_{\alpha\neq\gamma}\left(\sum_i S_i^\alpha S_i^\gamma\right)^2 + \frac{N^2 n - n^2 N}{2} .\end{aligned}$$

Therefore,

$$\mathcal{H} = \frac{\beta^2 \tilde{J}^2 nN}{4} + \frac{\beta^2 \tilde{J}^2}{2N}\sum_{\{\alpha,\gamma\}}\left(\sum_i S_i^\alpha S_i^\gamma\right)^2 + \beta h \sum_{i,\alpha} S_i^\alpha \qquad (13.98)$$

where each replica-pair now appears only once in the second term and where we have dropped a constant term that vanishes in the thermodynamic limit. The evaluation of the trace in (13.97) is complicated by the fact that in $\mathcal{H}$ spins on different sites are coupled. We can remove this coupling at the expense of introducing another set of dummy variables. The identity (13.96), now in reverse with $a = N$ and $\lambda = \sqrt{2}\beta\tilde{J}\sum_i S_i^\alpha S_i^\gamma$ for all pairs α, γ then results in the expression

$$\begin{aligned} e^{\mathcal{H}} &= \exp\left\{\frac{\beta^2\tilde{J}^2 nN}{4} + \beta h\sum_{i,\alpha} S_i^\alpha\right\}\prod_{\{\alpha,\gamma\}}\sqrt{\frac{N}{2\pi}} \\ &\quad\times\int_{-\infty}^{\infty} dx_{\alpha\gamma}\,\exp\left\{-\frac{1}{2}Nx_{\alpha\gamma}^2 + \beta\tilde{J}x_{\alpha\gamma}\sum_i S_i^\alpha S_i^\gamma\right\} . \qquad (13.99)\end{aligned}$$

In (13.99) the spins on different sites are no longer coupled and we can evaluate the trace of (13.99) as the trace over the spins on a single site (*e.g.*, i) raised to the power N. We may therefore drop the site-label i and concentrate on the coupling between spins in different replicas.

Before proceeding further, we schematically outline the rest of the calculation. After the trace over spin degrees of freedom is done we will still have to do the integrals over the $n(n-1)/2$ variables $x_{\alpha\gamma}$. This will be an integral of the form

$$I = \left(\prod_{\{\alpha,\gamma\}}\sqrt{\frac{N}{2\pi}}\int_{-\infty}^{\infty} dx_{\alpha\gamma}\right)\exp\{-\frac{N}{2}\sum_{\{\alpha\gamma\}} x_{\alpha\gamma}^2 + N\Phi(\{x_{\alpha\gamma}\})\} \qquad (13.100)$$

where the function Φ is in principle a function of all the variables $x_{\alpha\gamma}$ because of the trace. In the limit $N\to\infty$ the integrand is sharply peaked at $x_{\alpha\gamma} = x_m$, independent of $\alpha\gamma$ because of *replica symmetry.* Integrals of this kind can be done by the method of steepest descent which yields an expression of the form

$$\begin{aligned} I &= \exp\left[\frac{n(n-1)}{2}\left\{-\frac{N}{2}x_m^2 + N\Phi(\{x_m\})\right\}\right]\left(\prod_i \sqrt{\frac{N}{2\pi}}\int_{-\infty}^{\infty} dx_i\right) \\ &\quad\times \exp\left[-\frac{N}{2}\sum_{j,l} g_{jl}(x_j - x_m)(x_l - x_m)\right] \end{aligned}$$

with

$$g_{jl} = \delta_{jl} - \left.\frac{\partial^2\Phi}{\partial x_j \partial x_l}\right|_{\{x_m\}}$$

The remaining integrations merely produce a constant, independent of N, and this can be ignored as it is insignificant compared to the prefactor, in the thermodynamic limit $N \to \infty$.

In taking the limits $n \to 0$ and $N \to \infty$ we should, in principle, let $n \to 0$ first. We will, however, take the second limit first so that we may replace the $x_{\alpha\gamma}$ by x_m *before* carrying out the trace over the spins. We note that x_m is given by

$$-Nx_{\alpha\gamma} + \beta\tilde{J}\sum_i S_i^\alpha S_i^\gamma\bigg|_{x_{\alpha\gamma}=x_m} = 0 \tag{13.101}$$

or

$$x_m = \beta\tilde{J}q \tag{13.102}$$

where q is the spin-glass order parameter (13.95). Substituting x_m for $x_{\alpha\gamma}$ in (13.99) and, as discussed above, dropping the remaining integrations we obtain

$$\begin{aligned} \mathrm{Tr}\, e^{\mathcal{H}} &= \exp\left\{\frac{N\beta^2\tilde{J}^2}{4}[n - n(n-1)q^2]\right\} \\ &\quad\times \left[\mathrm{Tr}\exp\{\beta H\sum_\alpha S^\alpha + \beta^2\tilde{J}^2 q\sum_{\{\alpha\gamma\}} S^\alpha S^\gamma\}\right]^N . \end{aligned} \tag{13.103}$$

We again use the property (13.96) of Gaussian integrals to decouple the spins in different replicas in the second factor in (13.103). Writing

$$\begin{aligned} \exp\left\{\beta^2\tilde{J}^2 q\sum_{\{\alpha\gamma\}} S^\alpha S^\gamma\right\} &= \exp\left\{-\frac{n\beta^2\tilde{J}^2 q}{2}\right\}\int_{-\infty}^{\infty}\frac{dz}{\sqrt{2\pi}}e^{-z^2/2} \\ &\quad\times \exp\left\{z\beta\tilde{J}\sqrt{q}\sum_\alpha S^\alpha\right\} \end{aligned} \tag{13.104}$$

we obtain

$$\begin{aligned}
\mathrm{Tr}\, e^{\mathcal{H}} &= e^P \left[\mathrm{Tr} \int_{-\infty}^{\infty} \frac{dz}{\sqrt{2\pi}} e^{-z^2/2} \exp\left\{ \left(z\beta\tilde{J}\sqrt{q} + \beta h \right) \sum_\alpha S^\alpha \right\} \right]^N \\
&= e^P \left[\int_{-\infty}^{\infty} \frac{dz}{\sqrt{2\pi}} e^{-z^2/2} (2\cosh \mathcal{Z})^n \right]^N \qquad (13.105)
\end{aligned}$$

where $P = N\beta^2\tilde{J}^2/4(n - n(n-1)q^2) - nN\beta^2\tilde{J}^2 q/2$ and $\mathcal{Z} = z\beta\tilde{J}\sqrt{q} + \beta h$. Finally, we wish to expand the right-hand side in powers of n and therefore write

$$\begin{aligned}
&\left[\int_{-\infty}^{\infty} \frac{dz}{\sqrt{2\pi}} e^{-z^2/2} (2\cosh \mathcal{Z})^n \right]^N \\
&\quad = \exp\left\{ N \ln \int_{-\infty}^{\infty} \frac{dz}{\sqrt{2\pi}} e^{-z^2/2} (2\cosh \mathcal{Z})^n \right\} \\
&\quad \approx \exp\left\{ N \ln \int_{-\infty}^{\infty} \frac{dz}{\sqrt{2\pi}} e^{-z^2/2} (1 + n\ln[2\cosh \mathcal{Z}]) \right\} \\
&\quad \approx \exp\left\{ nN \int_{-\infty}^{\infty} \frac{dz}{\sqrt{2\pi}} e^{-z^2/2} \ln(2\cosh \mathcal{Z}) \right\} .
\end{aligned}$$

We are now in a position to combine these results. Substituting into (13.97) and taking the limit $n \to 0$ we have

$$\beta f = -\frac{\beta^2\tilde{J}^2}{4}(1-q)^2 - \int_{-\infty}^{\infty} \frac{dz}{\sqrt{2\pi}} e^{-z^2/2} \ln(2\cosh \mathcal{Z}) . \qquad (13.106)$$

The order parameter q is given by

$$\begin{aligned}
q = \langle\langle S^\alpha S^\gamma \rangle\rangle &= \left. \frac{\mathrm{Tr}\, S^\alpha S^\gamma e^{\mathcal{H}}}{\mathrm{Tr}\, e^{\mathcal{H}}} \right|_{n\to 0} \\
&= \left. \frac{\int \frac{dz}{\sqrt{2\pi}} e^{-z^2/2} (2\sinh \mathcal{Z})^2 (2\cosh \mathcal{Z})^{n-2}}{\int \frac{dz}{\sqrt{2\pi}} e^{-z^2/2} (2\cosh \mathcal{Z})^n} \right|_{n\to 0} \\
&= \int \frac{dz}{\sqrt{2\pi}} e^{-z^2/2} \tanh^2 \mathcal{Z} . \qquad (13.107)
\end{aligned}$$

We can easily show that for T small enough, the spin-glass order parameter is non-zero. Setting $h = 0$ and expanding the right-hand side of (13.107) in powers of q we have

$$\begin{aligned} q &= \int_{-\infty}^{\infty} \frac{dz}{\sqrt{2\pi}} e^{-z^2/2} \left[z^2 \beta^2 \tilde{J}^2 q - \frac{2}{3} z^4 \beta^4 \tilde{J}^4 q^2 + \ldots \right] \\ &= q \left(\beta \tilde{J}\right)^2 - 2q^2 \left(\beta \tilde{J}\right)^4 + \ldots . \end{aligned} \tag{13.108}$$

Therefore, for small q we have a nonzero solution for $T < T_f = \tilde{J}/k_B$ where T_f is the "spin-freezing" temperature. To show that the spin-glass phase is of lower free energy than the paramagnetic phase we still have to compare free energies for $q = 0$ and $q \neq 0$. We leave this as an exercise.

At this point it seems that we have taken at least a first step toward the development of a theory of spin-glasses. However, the situation is more complicated than it looks at first sight: the low-temperature properties of the solution given above are unphysical. To see this we examine the properties of the free energy near $T = 0$. We first calculate the temperature dependence of the order parameter for small T. Taking $h = 0$ and noting that

$$\begin{aligned} \tanh^2 z\beta \tilde{J}\sqrt{q} &\approx 1 - 4\exp\{-2z\beta \tilde{J}\sqrt{q}\} \qquad z > 0 \\ &\approx 1 - 4\exp\{2z\beta \tilde{J}\sqrt{q}\} \qquad z < 0 \end{aligned}$$

we have

$$q(T) \approx 1 - 8\int_0^{\infty} \frac{dz}{\sqrt{2\pi}} e^{-z^2/2} \exp\{-2\beta \tilde{J} z\sqrt{q}\} \tag{13.109}$$

and, using the asymptotic expansion of the error function, we find

$$q(T) \approx 1 - \sqrt{\frac{2}{\pi}} \frac{k_B T}{\tilde{J}} . \tag{13.110}$$

Substituting this into the expression (13.106) for the free energy we obtain

$$\begin{aligned} f &\approx -\frac{\tilde{J}^2}{4k_B T}(1-q)^2 - k_B T \int_{-\infty}^{\infty} \frac{dz}{\sqrt{2\pi}} e^{-z^2/2} (\beta \tilde{J}\sqrt{q})|z| \\ &\approx -\sqrt{\frac{2}{\pi}}\tilde{J} + \frac{k_B T}{2\pi} . \end{aligned} \tag{13.111}$$

Since $S/N = -\partial f/\partial T$ we see that the entropy per particle is *negative* at $T = 0$ — an obvious indication that something has gone seriously wrong.

The resolution of this difficulty turns out to be rather subtle. The problem can be traced to our assumption of replica-symmetry which we used in the evaluation of the integrals over $x_{\alpha\gamma}$ by steepest descent. We assumed that

$x_{\alpha\gamma} = x_m = \beta \tilde{J} q$ for all $\alpha\gamma$ minimizes the exponent in (13.100). This assumption turns out to be correct only if $q = 0$. In the spin-glass phase it is necessary to look for stationary points of the exponent that break the replica symmetry. The discussion of this theory is beyond the scope of this book and we refer the reader to [204] or [95] for a discussion of replica-symmetry breaking, associated concepts such as ergodicity-breaking, as well as the connection of the spin-glass problem to other areas of research such as optimization and neural networks.

The study of glasses and spin glasses is a fascinating but difficult subject which promises to be an active area of research for many years to come. We hope that we have given the reader at least an introduction to the subject.

13.5 Problems

13.1. *Single particle density of states.* Prove (13.9.

13.2. *Electron States in One Dimension.*

Consider the transfer matrix $\mathbf{T}$ (13.13) for the case of a pure material: $\epsilon_j = \epsilon_A$ for all j.

(a) Show that equation (13.15) can only be satisfied if the eigenvalues of $\mathbf{T}$ are complex or ± 1.

(b) Formulate the one-electron problem with fixed end boundary conditions $A_1 = A_N = 0$ and derive the eigenvalues and eigenfunctions using the transfer matrix approach for a pure type A material.

13.3. *One-Dimensional Liquid: Disordered Kronig–Penney Model.*

Consider the problem of an electron of momentum $\hbar k$ incident from the left on a set of $N + 1$ scatterers with potential

$$V(x) = \sum_{i=0}^{N} \frac{\hbar^2}{2m} u \delta(x - x_i)$$

where $u > 0$ and where the location of the δ-function scatterers is for the moment unspecified except for $x_{i+1} > x_i$. The Schrödinger equation is therefore

$$-\frac{d^2\psi(x)}{dx^2} + \sum_{i=0}^{N} u \delta(x - x_i)\psi(x) = k^2 \psi(x) \ .$$

Let $\Delta_j = x_j - x_{j-1}$, $j = 1, 2, \ldots, N$, and write

$$\begin{aligned}
\psi(x) &= A_0 e^{ik(x-x_0)} + B_0 e^{-ik(x-x_0)} && x < x_0 \\
&= A'_j e^{ik(x-x_{j-1})} + B'_j e^{-ik(x-x_{j-1})} && x_{j-1} < x < x_j \\
&= A_j e^{ik(x-x_j)} + B_j e^{-ik(x-x_j)} && x_{j-1} < x < x_j \\
&= A'_{N+1} e^{ik(x-x_N)} && x \geq x_N
\end{aligned}$$

where $A'_j = A_j \exp -ik\Delta_j$ and $B'_j = B_j \exp ik\Delta_j$.

(a) Apply the usual continuity conditions to the wave function and show that

$$\begin{bmatrix} A_j \\ B_j \end{bmatrix} = \begin{bmatrix} \frac{1}{t} & \frac{r^*}{t^*} \\ \frac{r}{t} & \frac{1}{t^*} \end{bmatrix} \begin{bmatrix} A'_{j+1} \\ B'_{j+1} \end{bmatrix} = \mathbf{Q} \begin{bmatrix} A'_{j+1} \\ B'_{j+1} \end{bmatrix}$$

where $t = |t|e^{i\delta} = [1 - u/2ik]^{-1}$ and $r = -i|r|e^{i\delta} = u/2ik[1 - u/2ik]^{-1}$ are the complex transmission and reflection coefficients of each scatterer. Using the relation between A', B' and A, B given above, we have

$$\begin{bmatrix} A'_j \\ B'_j \end{bmatrix} = \mathbf{M}_j \mathbf{Q} \begin{bmatrix} A'_{j+1} \\ B'_{j+1} \end{bmatrix} \qquad \text{where } \mathbf{M}_j = \begin{bmatrix} e^{-ik\Delta_j} & 0 \\ 0 & e^{ik\Delta_j} \end{bmatrix}$$

contains the spacing between neighboring scatterers.

(b) Define $T_N = |A'_{N+1}|/|A_0|$ and $R_N = |B_0|/|A_0|$ to be the transmission and reflection amplitudes of the array of $N+1$ potentials and show that

$$\frac{1 + R_N^2}{T_N^2} = \frac{|A_0|^2 + |B_0|^2}{|A'_{N+1}|^2} = a_N$$

where

$$\begin{bmatrix} a_N & b_N \\ b_N^* & a_N \end{bmatrix} = \mathbf{Q}^\dagger \mathbf{M}_N^\dagger \mathbf{Q}^\dagger \mathbf{M}_{N-1}^\dagger \cdots \mathbf{Q}^\dagger \mathbf{M}_1^\dagger \mathbf{Q}^\dagger \mathbf{Q} \mathbf{M}_1 \mathbf{Q} \cdots \mathbf{M}_N \mathbf{Q} \,.$$

(c) Show that the matrix elements a_N, b_N obey the recursion relation

$$\begin{aligned}
a_N &= a_{N-1} \frac{1 + |r|^2}{|t|^2} + 2\mathrm{Re} \frac{b_{N-1} r e^{2ik\Delta_N}}{|t|^2} \\
b_N &= 2a_{N-1} \frac{r^*}{(t^*)^2} + \frac{b_{N-1}^* (r^*)^2 e^{-2ik\Delta_N} + b_{N-1} e^{2ik\Delta_N}}{(t^*)^2} \,.
\end{aligned}$$

(d) Show that if the spacings Δ_j are independent random variables governed by a distribution $P(\Delta)$ with the property

$$\int_0^\infty d\Delta P(\Delta) e^{2ik\Delta} = 0$$

the transmission coefficient is zero in the limit $N \to \infty$.

(e) More generally, assume that $P(\Delta) = 1/W$ for $0 < \Delta < W$. Show numerically, by iteration of the recursion relations for a_N and b_N for $u = 1$, $W = 1$ and a range of k, that $T_N(k) \to 0$ as N becomes large.

13.4. *Real Space Renormalization Group for Percolation on the Square Lattice.*

Consider the following nine-site block on the square lattice:

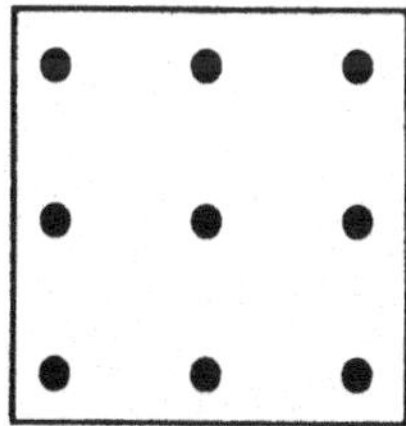

Figure 13.20: Nine-site block on square lattice.

(a) Adapt the real space renormalization group treatment of Section 13.2 to this cluster. Define the block to be occupied if the majority of the sites are occupied, empty otherwise. Find the fixed point and the correlation length exponent ν.

(b) As an alternative, define the cluster to be occupied if a path of connected sites from the left side of the block to the right side exists. Find p_c and ν.

Appendix A

Occupation Number Representation

An important aspect of quantum many-particle systems is the symmetry requirement that identical particles obey either Bose or Fermi statistics. This requirement can be conveniently incorporated into the formalism by making use of what is known as the occupation number or Fock representation, sometimes also referred to as second quantization. The underlying concept is that when one is dealing with a system of identical particles, one cannot specify which particle is in which state, only which states are occupied, and how many particles there are in each.

Let $|i\rangle$ be a complete set of one-particle states with wave functions

$$\phi_i(\mathbf{r}) = \langle \mathbf{r}|i\rangle \ . \tag{A.1}$$

We assume that these states are orthonormal, that is,

$$\langle i|j\rangle = \int d^3r \phi_i^*(\mathbf{r})\phi_j(\mathbf{r}) = \delta_{ij} \ . \tag{A.2}$$

A state in which particle 1 is in state 1, particle 2 is in state 2, and so on, is represented by the product wave function

$$(\mathbf{r}_1, \mathbf{r}_2, \ldots, \mathbf{r}_N|1, 2, 3, \ldots, N) = \phi_1(\mathbf{r}_1)\phi_2(\mathbf{r}_2)\cdots\phi_N(\mathbf{r}_N) \ . \tag{A.3}$$

If the particles are *identical*, this state is indistinguishable from one in which the labels are permuted, for example

$$\phi_{P1}(\mathbf{r}_1)\phi_{P2}(\mathbf{r}_2)\cdots\phi_{PN}(\mathbf{r}_N) \tag{A.4}$$

where P is a permutation operator, or a mapping, of the numbers $1, 2, 3, \ldots, N$ on themselves, for example,

$$P = \begin{pmatrix} 1 & 2 & 3 & 4 & 5 & 6 & 7 & 8 & 9 \\ 9 & 2 & 4 & 6 & 1 & 7 & 3 & 5 & 8 \end{pmatrix} . \tag{A.5}$$

In this example $P1 = 9$, $P2 = 2$, $P3 = 4$, and so on. Each such permutation can be written as a product of *transpositions*. A transposition is a permutation in which only two labels are interchanged, for example,

$$T_{36} = \begin{pmatrix} 1 & 2 & 3 & 4 & 5 & 6 & 7 & 8 & 9 \\ 1 & 2 & 6 & 4 & 5 & 3 & 7 & 8 & 9 \end{pmatrix}$$

and the permutation (A.5) can be written

$$P = T_{89}T_{67}T_{59}T_{46}T_{34}T_{19} .$$

The way a permutation is factorized into transpositions is clearly not unique, but it is possible to classify all permutations as being either even or odd, depending on whether an even or odd number of transpositions are required to achieve it. Altogether, the numbers $1, 2, 3, \ldots, N$ can be permuted in $N!$ different ways. The quantum-mechanical symmetry requirement for a system of identical particles is then that a state must be a linear combination of all possible ways the single-particle state labels can be attached to the particles. Let

$$S_P = \begin{cases} 1 & \text{for } P \text{ even} \\ -1 & \text{odd} \end{cases} . \tag{A.6}$$

For a system of *fermions* the *Pauli exclusion principle* states that the wave function must be odd under transpositions. Consider the situation which one of N fermions is in single-particle state 1, one in state 2, and so on. The corresponding many-particle state can be expressed as

$$|1, 2, \ldots, N\rangle = \nu \sum_P S_P |P1, P2, \ldots, PN) \tag{A.7}$$

where ν is a normalization factor, and where we use the notation that kets $|1, 2, \ldots, N\rangle$ have been properly symmetrized, while $|1, 2, \ldots, N)$ is a many-particle state in which particle 1 is in state 1, particle 2 in state 2, and so on. As an example, consider a state with three particles,

$$\begin{aligned}
(\mathbf{r}_1, \mathbf{r}_2, \mathbf{r}_3|1, 2, 3\rangle &= \nu \sum_P (\mathbf{r}_1, \mathbf{r}_2, \mathbf{r}_3|P1, P2, P3) \\
&= \nu[\phi_1(\mathbf{r}_1)\phi_2(\mathbf{r}_2)\phi_3(\mathbf{r}_3) - \phi_1(\mathbf{r}_1)\phi_3(\mathbf{r}_2)\phi_2(\mathbf{r}_3) \\
&\quad + \phi_2(\mathbf{r}_1)\phi_3(\mathbf{r}_2)\phi_1(\mathbf{r}_3) - \phi_2(\mathbf{r}_1)\phi_1(\mathbf{r}_2)\phi_3(\mathbf{r}_3) \qquad (A.8) \\
&\quad + \phi_3(\mathbf{r}_1)\phi_1(\mathbf{r}_2)\phi_2(\mathbf{r}_3) - \phi_3(\mathbf{r}_1)\phi_2(\mathbf{r}_2)\phi_1(\mathbf{r}_3)] \\
&= \langle \mathbf{r}_1, \mathbf{r}_2, \mathbf{r}_3|1, 2, 3) = \langle \mathbf{r}_1, \mathbf{r}_2, \mathbf{r}_3|1, 2, 3\rangle
\end{aligned}$$

if we require that the states are normalized. In the case of the three-particle system we thus have $\nu = (3!)^{-1/2}$ and in general,

$$\nu = \frac{1}{\sqrt{N!}} \, . \tag{A.9}$$

Equations (A.7) and (A.8) can be expressed as determinants known as *Slater determinants.*

$$\begin{aligned}
&\langle \mathbf{r}_1, \mathbf{r}_2, \ldots, \mathbf{r}_N|1, 2, \ldots, N\rangle \\
&\quad = \frac{1}{\sqrt{N!}} \begin{vmatrix} \langle \mathbf{r}_1|1\rangle & \langle \mathbf{r}_2|1\rangle & \cdots & \langle \mathbf{r}_N|1\rangle \\ \langle \mathbf{r}_1|2\rangle & \langle \mathbf{r}_2|2\rangle & \cdots & \langle \mathbf{r}_N|2\rangle \\ \vdots & \vdots & & \vdots \\ \langle \mathbf{r}_1|N\rangle & \langle \mathbf{r}_2|N\rangle & \cdots & \langle \mathbf{r}_N|N\rangle \end{vmatrix} \, .
\end{aligned} \tag{A.10}$$

Since a determinant is zero when two rows or two columns are equal, it follows that no two fermions can occupy the same state.

In the case of Bose particles the symmetry requirement is that the wave-function should be even under the transposition of two particle or state labels. Consider a state in which n_1 bosons are in state 1, n_2 are in state 2, and so on. Let N be the total number of particles

$$\sum_i n_i = N \, .$$

We obtain properly normalized wave functions if we write

$$|n_1, n_2, \ldots\rangle = \sqrt{\frac{\prod_i n_i!}{N!}} \sum\nolimits'_{distinct\ P} |1, \ldots, 1, 2, \ldots) \, . \tag{A.11}$$

In (A.11) the notation $\sum'$ indicates that only those permutations in which the particles do not remain in the same state are included.

We wish to consider a system of particles that interact through two-body interactions

$$H = \sum_{i=1}^{N} \frac{p_i^2}{2m} + \sum_{i=1}^{N} U(\mathbf{r}_i) + \frac{1}{2}\sum_{i \neq j} v(\mathbf{r}_i - \mathbf{r}_j) . \tag{A.12}$$

The eigenstates of the Hamiltonian can be represented by the simple product states that we used above only when the interparticle interaction is zero. In general, a many-particle eigenstate cannot be expressed in terms of a single symmetrized product wave function. Instead, linear combinations of such wave functions are required. It is then no longer practical to make use of wave functions that make explicit reference as to which particles are in which states. We must also rewrite the Hamiltonian without reference to particle labels. In this process the emphasis is shifted from the wave functions to the operators. The fundamental operators in the occupation number representation are the *creation* and *annihilation* operators. Consider first the case of fermions and let $|0\rangle$ be the vacuum (no particle) state. We define the creation operator $a_1^\dagger$ by the effect it has on a state in which the single-particle states $2, 3, \ldots, N$ are occupied:

$$\begin{aligned} |1,2,3,\ldots,N\rangle &= a_1^\dagger |2,3,\ldots,N\rangle \\ &= a_1^\dagger a_2^\dagger a_3^\dagger \ldots a_N^\dagger |0\rangle . \end{aligned} \tag{A.13}$$

Remembering that this state must be odd under transposition of labels, we must have, for example,

$$|1,2,3,4,\ldots,N\rangle = -|1,2,4,3,\ldots,N\rangle . \tag{A.14}$$

We conclude that the creation operators formally satisfy the algebra

$$a_i^\dagger a_j^\dagger = -a_j^\dagger a_i^\dagger \tag{A.15}$$

when operating on any properly symmetrized state. From (A.15) we see that

$$\left(a_i^\dagger\right)^2 = 0 \tag{A.16}$$

implying that no two fermions can occupy the same state.

We write for the Hermitian conjugate of (A.13)

$$\langle 0| a_N a_{N-1} \cdots a_2 a_1 = \left(a_1^\dagger a_2^\dagger \cdots a_{N-1}^\dagger a_N^\dagger |0\rangle\right)^\dagger . \tag{A.17}$$

Since

$$|i\rangle = a_i^\dagger|0\rangle \tag{A.18}$$

we have

$$\langle 0|a_i|i\rangle = 1 \ . \tag{A.19}$$

If we consider a_i as an operator acting to the right, we see that it has the effect of removing a particle from state i. For this reason we refer to a_i as the annihilation (or destruction) operator, that is,

$$a_i|i\rangle = |0\rangle \tag{A.20}$$

$$a_i a_j = -a_j a_i \ . \tag{A.21}$$

Using all this information we find for $i \neq j$

$$\begin{aligned} a_i^\dagger a_j|j,1,2,\ldots\rangle &= |i,1,2,\ldots\rangle = a_j|j,i,1,2,\ldots\rangle \\ &= -a_j|i,j,1,2\ldots\rangle = -a_j a_i^\dagger|j,1,2,\ldots\rangle \ . \end{aligned} \tag{A.22}$$

We see that when acting on a symmetrized state and for $i \neq j$

$$a_i^\dagger a_j = -a_j a_i^\dagger \ . \tag{A.23}$$

Next consider a many-particle configuration in which the single particle state i is occupied. In this case

$$\begin{aligned} a_i^\dagger a_i|\cdots i\cdots\rangle &= |\cdots i\cdots\rangle \\ a_i a_i^\dagger|\cdots i\cdots\rangle &= 0 \ . \end{aligned} \tag{A.24}$$

Similarly, if i is empty,

$$\begin{aligned} a_i^\dagger a_i|\cdots\rangle &= 0 \\ a_i a_i^\dagger|\cdots\rangle &= |\cdots\rangle \ . \end{aligned} \tag{A.25}$$

Since the single-particle state is either empty or occupied, we must have

$$a_i^\dagger a_i + a_i a_i^\dagger = 1 \ . \tag{A.26}$$

Introducing the *anticommutator*

$$[A,B]_+ \equiv AB + BA \tag{A.27}$$

we can summarize the commutation relations for the fermion creation and annihilation operators as

$$\begin{aligned} \left[a_i^\dagger, a_j^\dagger\right]_+ &= [a_i, a_j]_+ = 0 \\ \left[a_i, a_j^\dagger\right]_+ &= \delta_{ij} \ . \end{aligned} \tag{A.28}$$

We next define the corresponding operators for a system of particles obeying Bose statistics. Consider a many-particle state in which there are n_i particles in the ith single-particle state. We define the annihilation operator b_i through

$$b_i|n_1 \cdots n_i \cdots\rangle = \sqrt{n_i}|n_1 \cdots (n_i - 1) \cdots\rangle \ . \tag{A.29}$$

In the special case where a single-particle state is unoccupied, the annihilation operator returns zero. The Hermitian conjugate operator $b_i^\dagger$ must satisfy

$$b_i^\dagger|n_1 \cdots (n_i - 1) \cdots\rangle = \sqrt{n_i}|n_1 \cdots n_i \cdots\rangle \tag{A.30}$$

or

$$b_i^\dagger|n_1 \cdots n_i \cdots\rangle = \sqrt{n_i + 1}|n_1 \cdots (n_i + 1) \cdots\rangle \tag{A.31}$$

and we see that $b_i^\dagger$ has the interpretation of creation operator. From (A.29) and (A.30) we see that

$$b_i^\dagger b_i|\cdots n_i \cdots\rangle = n_i|\cdots n_i \cdots\rangle \tag{A.32}$$

that is, $b_i^\dagger b_i$ can be interpreted as the *number operator*, while from (A.31) and (A.29) we see that

$$b_i b_i^\dagger|\cdots n_i \cdots\rangle = (n_i + 1)|\cdots n_i \cdots\rangle \ . \tag{A.33}$$

We now define the commutator

$$[A, B]_- \equiv AB - BA \ . \tag{A.34}$$

Since commutators occur much more frequently than anticommutators we will henceforth drop the subscript for the commutator. With this notation we find that the boson creation and annihilation operators satisfy

$$[b_i, b_j] = \left[b_i^\dagger, b_j^\dagger\right] = 0 \tag{A.35}$$

since the many-particle states are symmetric under transposition of particle labels. Similarly, from (A.32) and (A.33) we find

$$\left[b_i, b_j^\dagger\right] = \delta_{ij} \ . \tag{A.36}$$

We have implicitly assumed that the representation (A.1) for the single-particle states is complete, that is,

$$\sum_i |i\rangle\langle i| = 1 \ . \tag{A.37}$$

In addition to the spatial variable $\mathbf{r}$, there is often an internal degree of freedom such as the spin. We must then consider the wave functions to be spinors; for example, for spin-1/2 particles,

$$\langle \mathbf{r}|i\rangle = \phi_i(\mathbf{r},\uparrow)\begin{pmatrix}1\\0\end{pmatrix} + \phi_i(\mathbf{r},\downarrow)\begin{pmatrix}0\\1\end{pmatrix} \tag{A.38}$$

and we write

$$\langle \mathbf{r}\sigma|i\rangle = \phi_i(\mathbf{r},\sigma) \ .$$

It is often convenient to work with creation and annihilation operators that are not tied to any particular representation $|i\rangle$ for the single-particle states. This leads us to define *field operators.* In the case of fermions we define

$$\begin{aligned}\psi(\mathbf{r},\sigma) &= \sum_i \langle \mathbf{r},\sigma|i\rangle a_i = \sum_i \phi_i(\mathbf{r},\sigma)a_i\\ \psi^\dagger(\mathbf{r},\sigma) &= \sum_i \langle i|\mathbf{r},\sigma\rangle a_i^\dagger = \sum_i \phi_i^*(\mathbf{r},\sigma)a_i^\dagger \ .\end{aligned} \tag{A.39}$$

Using the orthonormality of the wave functions, we invert (A.39) to obtain

$$\begin{aligned}a_i &= \int d^3r \sum_\sigma \phi_i^*(\mathbf{r},\sigma)\psi(\mathbf{r},\sigma)\\ a_i^\dagger &= \int d^3r \sum_\sigma \phi_i(\mathbf{r},\sigma)\psi^\dagger(\mathbf{r},\sigma) \ .\end{aligned} \tag{A.40}$$

The completeness requirement (A.37) can be re-expressed as

$$\sum_i \phi_i^*(\mathbf{r}',\sigma')\phi_i(\mathbf{r},\sigma) = \delta(\mathbf{r}-\mathbf{r}')\delta_{\sigma\sigma'} \tag{A.41}$$

and we see that the field operators satisfy the anticommutation relations

$$\begin{aligned}[\psi(\mathbf{r},\sigma),\psi(\mathbf{r}',\sigma')]_+ &= \left[\psi^\dagger(\mathbf{r},\sigma),\psi^\dagger(\mathbf{r}',\sigma')\right]_+ = 0\\ \left[\psi(\mathbf{r},\sigma),\psi^\dagger(\mathbf{r}',\sigma')\right]_+ &= \delta(\mathbf{r}-\mathbf{r}')\delta_{\sigma\sigma'} \ .\end{aligned} \tag{A.42}$$

The analogous operators for (spinless) Bose particles can be written

$$\chi(\mathbf{r}) = \sum_i \phi_i(\mathbf{r})b_i \qquad \chi^\dagger(\mathbf{r}) = \sum_i \phi_i^*(\mathbf{r})b_i^\dagger \tag{A.43}$$

and the commutation relations become

$$\begin{aligned}[\chi(\mathbf{r}),\chi(\mathbf{r}')] &= \left[\chi^\dagger(\mathbf{r}),\chi^\dagger(\mathbf{r}')\right] = 0\\ \left[\chi(\mathbf{r}),\chi^\dagger(\mathbf{r}')\right] &= \delta(\mathbf{r}-\mathbf{r}') \ .\end{aligned} \tag{A.44}$$

We may also express the operator associated with total particle number in terms of field operators

$$N = \sum_i b_i^\dagger b_i = \int d^3r \chi^\dagger(\mathbf{r})\chi(\mathbf{r}) \tag{A.45}$$

for spinless bosons, and

$$N = \sum_i a_i^\dagger a_i = \int d^3r \sum_\sigma \psi^\dagger(\mathbf{r},\sigma)\psi(\mathbf{r},\sigma) \tag{A.46}$$

for fermions. By considering a set of states $|j\rangle$ that are localized within a subvolume Ω and taking the limit $\Omega \to 0$, we identify the particle density operator as

$$n(\mathbf{r}) \equiv \sum_{\text{particles } i} \delta(\mathbf{r}-\mathbf{r}_i) = \chi^\dagger(\mathbf{r})\chi(\mathbf{r}) \tag{A.47}$$

for spinless bosons, and

$$n(\mathbf{r}) = \sum_\sigma \psi^\dagger(\mathbf{r},\sigma)\psi(\mathbf{r},\sigma) \tag{A.48}$$

for fermions. A particularly useful single-particle representation is the momentum representation. In this case the operators are related to the field operators through

$$\begin{aligned} \psi(\mathbf{r},\sigma) &= \frac{1}{\sqrt{V}}\sum_{\mathbf{k}} e^{i\mathbf{k}\cdot\mathbf{r}} a_{\mathbf{k},\sigma} \\ \psi^\dagger(\mathbf{r},\sigma) &= \frac{1}{\sqrt{V}}\sum_{\mathbf{k}} e^{-i\mathbf{k}\cdot\mathbf{r}} a_{\mathbf{k},\sigma}^\dagger \end{aligned} \tag{A.49}$$

and

$$\begin{aligned} a_{\mathbf{k},\sigma} &= \frac{1}{\sqrt{V}}\int d^3r e^{-i\mathbf{k}\cdot\mathbf{r}}\psi(\mathbf{r},\sigma) \\ a_{\mathbf{k},\sigma}^\dagger &= \frac{1}{\sqrt{V}}\int d^3r e^{i\mathbf{k}\cdot\mathbf{r}}\psi^\dagger(\mathbf{r},\sigma) \end{aligned} \tag{A.50}$$

in the case of fermion operators. The expression for Bose operators are analogous.

The Fourier transform of the density operator is

$$\rho(\mathbf{q}) = \int d^3r e^{-i\mathbf{q}\cdot\mathbf{r}} n(\mathbf{r}) = \sum_{\text{particles } i} e^{-i\mathbf{q}\cdot\mathbf{r}_i} \,. \tag{A.51}$$

In the occupation number representation we see that

$$\rho(\mathbf{q}) = \sum_{\mathbf{k},\sigma} a^\dagger_{\mathbf{k}-\mathbf{q},\sigma} a_{\mathbf{k},\sigma} \ . \tag{A.52}$$

We next wish to express the Hamiltonian (A.12) in terms of the field operators. We first note that (A.12) contains two types of terms. We refer to expressions such as

$$\sum_i \frac{p_i^2}{2m} \qquad\qquad \sum_i U(\mathbf{r}_i)$$

as *one-body operators*, since they can be evaluated by adding up single-particle contributions. The last term in (A.12),

$$\frac{1}{2}\sum_{i\neq j} v(\mathbf{r}_{ij})$$

is an example of a *two-body operator*. As before, let $|i\rangle$ be a complete set of single-particle states, characterized by wave functions $\phi_i(\mathbf{r}) = \langle \mathbf{r}|i\rangle$. A one-body operator T can be completely specified through its matrix elements

$$\langle i|T|j\rangle = \int d^3r \phi_i^*(\mathbf{r}) T(\mathbf{r}) \phi_j(\mathbf{r}) \ . \tag{A.53}$$

Similarly, a two-body operator V is fully determined by the matrix elements

$$(ij|V|kl) = \int d^3r_1 d^3r_2 \phi_i^*(\mathbf{r}_1)\phi_j^*(\mathbf{r}_2) V(\mathbf{r}_1,\mathbf{r}_2)\phi_k(\mathbf{r}_2)\phi_l(\mathbf{r}_1) \ . \tag{A.54}$$

Let

$$|n_1,\ldots,n_i,\ldots,n_j,\ldots\rangle \tag{A.55}$$

be a properly symmetrized many-particle state (A.13) or (A.7) in which there are n_1 particles in state 1, n_i particles in state i, and so on (for fermions n_i can, of course, only take on the values 0 and 1). Also, let $c_i^\dagger$, c_i be the appropriate creation and annihilation operators ($a_i^\dagger$, a_i, or $b_i^\dagger$, b_i).

$$\begin{aligned} c_i|n_1,\ldots,n_i,\ldots,n_j,\ldots\rangle &= \sqrt{n_i}|n_1,\ldots,(n_i-1),\ldots,n_j,\ldots\rangle \\ c_i^\dagger|n_1,\ldots,(n_i-1),\ldots,n_j,\ldots\rangle &= \sqrt{n_i}|n_1,\ldots,n_i,\ldots,n_j,\ldots\rangle \ . \end{aligned} \tag{A.56}$$

A little reflection should convince the reader that

$$\sum_{\text{particles } p} T(\mathbf{r}_p) \tag{A.57}$$

has nonzero matrix elements only between states in which at most one particle is in a different state than in (A.55) and that the matrix elements of (A.57) are identical to those of

$$\sum_{\text{states } i,j} \langle i|T|j\rangle c_i^\dagger c_j \ . \tag{A.58}$$

This allows us to write formally

$$\sum_{\text{particles } p} T(\mathbf{r}_p) = \sum_{\text{states } i,j} \langle i|T|j\rangle c_i^\dagger c_j \ . \tag{A.59}$$

As a check, consider an operator V which is diagonal in the $\mathbf{r}$-representation. We have, using (A.47) (similar results obtain with the analogous expression for fermions)

$$\sum_p V(\mathbf{r}_p) = \int d^3r V(\mathbf{r}) n(\mathbf{r}) = \int d^3r V(\mathbf{r}) \chi^\dagger(\mathbf{r}) \chi(\mathbf{r}) \ . \tag{A.60}$$

Substitution of (A.43) then yields (A.59). Let T be an arbitrary Hermitian one-body operator. One can always express T in terms of a generalized density, in the representation in which T is diagonal, multiplied by the appropriate eigenvalue. By use of completeness and orthogonality relations, one recovers (A.59).

In the case of two-body operators we note that

$$\sum_{p \neq p'} V(\mathbf{r}_p, \mathbf{r}_{p'}) \tag{A.61}$$

when acting on any state can only produce states in which at most two particles have changed state. The operator (A.61) is therefore indistinguishable from

$$\sum_{i,j,k,l} (ij|V|kl) c_i^\dagger c_j^\dagger c_k c_l \tag{A.62}$$

and we can formally write

$$\sum_{p \neq p'} V(\mathbf{r}_p, \mathbf{r}_{p'}) = \sum_{i,j,k,l} (ij|V|kl) c_i^\dagger c_j^\dagger c_k c_l \ . \tag{A.63}$$

It is instructive to verify (A.63) by re-expressing the two-body operator in terms of density operators. We have

$$\sum_{p\neq p'} V(\mathbf{r}_p,\mathbf{r}_{p'}) = \sum_{p,p'} V(\mathbf{r}_p,\mathbf{r}_{p'}) - \sum_{p} V(\mathbf{r}_p,\mathbf{r}_p)$$
$$= \int d^3r \left(\int d^3\mathbf{r}' V(\mathbf{r}.\mathbf{r}')n(\mathbf{r}') - V(\mathbf{r},\mathbf{r})\right) n(\mathbf{r})$$
$$= \int d^3r \int d^3\mathbf{r}' V(\mathbf{r},\mathbf{r}')[\chi^\dagger(\mathbf{r}')\chi(\mathbf{r}')\chi^\dagger(\mathbf{r})\chi(\mathbf{r})$$
$$- \delta(\mathbf{r}-\mathbf{r}')\chi^\dagger(\mathbf{r})\chi(\mathbf{r})]$$
$$= \int d^3r \int d^3\mathbf{r}' V(\mathbf{r},\mathbf{r}')\chi^\dagger(\mathbf{r}')\chi^\dagger(\mathbf{r})\chi(\mathbf{r})\chi(\mathbf{r}') \tag{A.64}$$

where we have made use of the commutation relations (A.44) of the field operators. Note that since an even number of transpositions are involved, the argument leading to (A.64) holds equally well for fermions. Substituting the definitions of the field operators we recover (A.63).

We can now re-express the Hamiltonian (A.12), assuming that the potentials have Fourier transforms given by

$$U(\mathbf{q}) = \int d^3r e^{-i\mathbf{q}\cdot\mathbf{r}} U(\mathbf{r})$$
$$v(\mathbf{q}) = \int d^3r_{ij} e^{-i\mathbf{q}\cdot(\mathbf{r}_i-\mathbf{r}_j)} v(\mathbf{r}_i - \mathbf{r}_j) \ . \tag{A.65}$$

Then

$$H = \sum_{\mathbf{k},\sigma} \frac{\hbar^2 k^2}{2m} c^\dagger_{\mathbf{k},\sigma} c_{\mathbf{k},\sigma} + \frac{1}{V}\sum_{\mathbf{k},\mathbf{q},\sigma} U(\mathbf{q}) c^\dagger_{\mathbf{k}+\mathbf{q},\sigma} c_{\mathbf{k},\sigma}$$
$$+ \frac{1}{V}\sum_{\mathbf{k},\mathbf{k}',\mathbf{q},\sigma,\sigma'} v(\mathbf{q}) c^\dagger_{\mathbf{k}+\mathbf{q},\sigma} c^\dagger_{\mathbf{k}'-\mathbf{q},\sigma'} c_{\mathbf{k}',\sigma'} c_{\mathbf{k},\sigma} \ . \tag{A.66}$$

As a final example, we express the pair distribution function (5.25) and structure factor (5.28) as expectation values of products of annihilation and creation operators, and evaluate the expectation values for the special case of a noninteracting Fermi gas at $T = 0$. We have from (5.20) and (5.25)

$$g(\mathbf{r}) = \frac{V}{\langle N\rangle^2}\left\langle \sum_{i\neq j} \int d^3x \delta(\mathbf{r}_i - \mathbf{x})\delta(\mathbf{r}_j - \mathbf{x} - \mathbf{r})\right\rangle \tag{A.67}$$

or

$$g(\mathbf{r}) = \frac{V}{\langle N\rangle^2}\left\langle \int d^3x \left(\sum_i \delta(\mathbf{r}_i - \mathbf{x}) \sum_j \delta(\mathbf{r}_j - \mathbf{x} - \mathbf{r}) \right.\right.$$
$$\left.\left. - \sum_i \delta(\mathbf{r}_i - \mathbf{x})\delta(\mathbf{r}_i - \mathbf{x} - \mathbf{r}) \right) \right\rangle .$$

From (A.47) and (A.48) we obtain

$$\begin{aligned} g(\mathbf{r}) &= \frac{V}{\langle N\rangle^2} \left\langle \int d^3x n(\mathbf{x}) n(\mathbf{x}+\mathbf{r}) - \frac{V}{\langle N\rangle}\delta(\mathbf{r}) \right\rangle \\ &= \frac{V}{\langle N\rangle^2} \left\langle \int d^3x \sum_{\sigma,\sigma'} \psi^\dagger(\mathbf{x},\sigma')\psi(\mathbf{x},\sigma')\psi^\dagger(\mathbf{x}+\mathbf{r},\sigma)\psi(\mathbf{x}+\mathbf{r},\sigma) \right\rangle \\ &\quad - \frac{V}{\langle N\rangle}\delta(\mathbf{r}) . \end{aligned}$$

Using the commutation relations (A.42), this expression reduces to

$$g(\mathbf{r}) = \frac{V}{\langle N\rangle^2} \left\langle \int d^3x \sum_{\sigma,\sigma'} \psi^\dagger(\mathbf{x}+\mathbf{r},\sigma)\psi^\dagger(\mathbf{x},\sigma')\psi(\mathbf{x},\sigma')\psi(\mathbf{x}+\mathbf{r},\sigma) \right\rangle . \tag{A.68}$$

In the momentum representation (A.49) and (A.50) the corresponding equation becomes

$$g(\mathbf{x}) = \frac{1}{\langle N\rangle^2} \sum_{\mathbf{k},\mathbf{p},\mathbf{q},\sigma,\sigma'} e^{i\mathbf{q}\cdot\mathbf{x}} \langle a^\dagger_{\mathbf{p}+\mathbf{q},\sigma} a^\dagger_{\mathbf{k}-\mathbf{q},\sigma'} a_{\mathbf{k},\sigma'} a_{\mathbf{p},\sigma}\rangle . \tag{A.69}$$

An expression for the structure factor can be obtained from (5.28) and (A.51):

$$S(\mathbf{q}) = \frac{1}{\langle N\rangle} \langle \rho(\mathbf{q})\rho(-\mathbf{q})\rangle - \langle N\rangle \delta_{\mathbf{q},0} .$$

Substitution of (A.52) yields

$$S(\mathbf{q}) = \frac{1}{\langle N\rangle} \left\langle \sum_{\mathbf{k},\mathbf{p},\sigma,\sigma'} a^\dagger_{\mathbf{k}-\mathbf{q},\sigma'} a_{\mathbf{k},\sigma} a^\dagger_{\mathbf{p}+\mathbf{q},\sigma} a_{\mathbf{p},\sigma} \right\rangle - \langle N\rangle \delta_{\mathbf{q},0} .$$

By making use of the commutation relations this expression can be rewritten as

$$S(\mathbf{q}) - 1 = \frac{1}{\langle N\rangle} \left\langle \sum_{\mathbf{k},\mathbf{p},\sigma,\sigma'} a^\dagger_{\mathbf{p}+\mathbf{q},\sigma} a^\dagger_{\mathbf{k}-\mathbf{q},\sigma'} a_{\mathbf{k},\sigma'} a_{\mathbf{p},\sigma} \right\rangle - \langle N\rangle \delta_{\mathbf{q},0} \tag{A.70}$$

which in terms of the field operators becomes

$$S(\mathbf{q}) - 1 = \frac{1}{\langle N\rangle}\Big\langle \sum_{\sigma,\sigma'} \int d^3r \int d^3x e^{i\mathbf{q}\cdot\mathbf{r}} \psi^\dagger(\mathbf{x}+\mathbf{r},\sigma)$$
$$\psi^\dagger(\mathbf{x},\sigma')\psi(\mathbf{x},\sigma')\psi(\mathbf{x}+\mathbf{r},\sigma)\Big\rangle - \langle N\rangle \delta_{\mathbf{q},0} \,. \qquad (A.71)$$

We now turn to the problem of evaluating the expressions for $g(\mathbf{r})$ and $S(\mathbf{q})$ in the special case of a noninteracting gas of spin-$\frac{1}{2}$ fermions. The ground state has $N/2$ particles in each spin state, filling up the lowest momentum states up to a Fermi wave vector k_F. We write

$$g(\mathbf{x}) = \sum_{\sigma,\sigma'} g_{\sigma\sigma'}(\mathbf{x}) = g_{\uparrow\uparrow}(\mathbf{x}) + g_{\downarrow\uparrow}(\mathbf{x}) + g_{\uparrow\downarrow}(\mathbf{x}) + g_{\downarrow\downarrow}(\mathbf{x}) \qquad (A.72)$$

where

$$g_{\uparrow\uparrow}(\mathbf{x}) = \frac{1}{\langle N\rangle^2} \sum_{\mathbf{k},\mathbf{p},\mathbf{q}} e^{i\mathbf{q}\cdot\mathbf{x}} \langle 0| a^\dagger_{\mathbf{p}+\mathbf{q},\uparrow} a^\dagger_{-\mathbf{k}-\mathbf{q},\uparrow} a_{\mathbf{k},\uparrow} a_{\mathbf{p},\uparrow} |0\rangle$$

and $|0\rangle$ represents the ground state. To obtain nonzero matrix elements we must have $k,\, p > k_F$ and either $q = 0$ or $\mathbf{k} - \mathbf{q} = \mathbf{p}$. In the former case the matrix element is 1 in the latter case -1. We obtain, assuming that the ground state is an eigenstate of the number operator with eigenvalue N,

$$g_{\uparrow\uparrow}(\mathbf{x}) = \frac{1}{N^2}\left\{\frac{N}{2}\left[\frac{N}{2} - 1\right]\right\} - \frac{1}{N^2} \sum_{k<k_F,\, p<k_F,\, \mathbf{p}\neq\mathbf{k}} e^{i(\mathbf{k}-\mathbf{p})\cdot\mathbf{x}} \,.$$

This expression can be cast into the form

$$g_{\uparrow\uparrow}(\mathbf{x}) = \frac{1}{4} - \left| \frac{1}{N} \sum_{p<k_F} e^{i\mathbf{p}\cdot\mathbf{x}} \right|^2 \,. \qquad (A.73)$$

Note that $g_{\uparrow\uparrow}(0) = 0$ in agreement with the Pauli exclusion principle. Clearly, $g_{\uparrow\uparrow}(\mathbf{x}) = g_{\downarrow\downarrow}(\mathbf{x})$. Next

$$g_{\uparrow\downarrow}(\mathbf{x}) = \frac{1}{N^2} \sum_{\mathbf{k},\mathbf{p},\mathbf{q}} e^{i\mathbf{q}\cdot\mathbf{x}} \langle 0| a^\dagger_{\mathbf{p}+\mathbf{q},\uparrow} a^\dagger_{\mathbf{k}-\mathbf{q},\downarrow} a_{\mathbf{k},\downarrow} a_{\mathbf{p},\uparrow} |0\rangle \,.$$

Here the only possibility is $q = 0$ and

$$g_{\uparrow\downarrow}(\mathbf{x}) = g_{\downarrow\uparrow}(\mathbf{x}) = \frac{1}{N^2}\left(\frac{N}{2}\right)^2 = \frac{1}{4} \,. \qquad (A.74)$$

Collecting terms, we find

$$g(\mathbf{x}) = 1 - 2\left|\frac{1}{N}\sum_{p<k_F} e^{i\mathbf{p}\cdot\mathbf{x}}\right|^2 . \tag{A.75}$$

The sum in (A.75) can be evaluated analytically to give

$$\begin{aligned}\frac{1}{N}\sum_{p<k_F} e^{i\mathbf{p}\cdot\mathbf{x}} &= \frac{V}{N}\int_{p<p_F} e^{i\mathbf{p}\cdot\mathbf{x}}\frac{d^3p}{(2\pi)^3} \\ &= \frac{V}{2\pi^2 N}\left(\frac{1}{x^3}\sin k_F x - \frac{k_F}{x^2}\cos k_F x\right) .\end{aligned} \tag{A.76}$$

Similar arguments yield for the structure factor

$$S(q) = 1 - \frac{2}{N}\sum_{p<k_F,|\mathbf{p}+\mathbf{q}|<k_F} 1 = 1 - \frac{2V}{N(2\pi)^3}\int_{p<k_F,|\mathbf{p}+\mathbf{q}|<k_F} d^3p .$$

The reader may wish to verify that this expression holds both for $q = 0$ and $q \neq 0$. Evaluating the integral, we find that

$$\begin{aligned} S(q) &= \tfrac{3q}{4k_F} - \tfrac{q^3}{16k_F^3} \quad &&\text{for } q < 2k_F \\ S(q) &= 1 &&\text{for } q \geq 2k_F . \end{aligned} \tag{A.77}$$

Note that the result $S(0) = 0$ for $T = 0$ is expected from (5.33) and (5.27).

Bibliography

[1] ABRAHAM, F. F. [1986]. Adv. Phys. **35**:1.

[2] ABRAHAM, F. F., RUDGE, W. E., AUERBACH, D. J. and KOCH, S. W. [1984]. Phys. Lett. **52**:445.

[3] ABRAHAM, F. F., RUDGE, W. E. and PLISCHKE, M. [1989]. Phys. Rev. Lett. **62**:1757.

[4] ABRAHAM, F. F. and NELSON, D. R. [1990]. J. Phys. France **51**:2653.

[5] ABRIKOSOV, A. A., GORKOV, L. P. and DZYALOSHINSKY, I. E. [1965]. *Quantum Field Theoretical Methods in Statistical Physics.* London: Pergamon Press.

[6] ADLER, S. L. [1962]. Phys. Rev. **126**:413.

[7] AHARONY, A. [1976]. In *Phase Transitions and Critical Phenomena,* Vol. 6, eds. C. Domb and M. S. Green. New York:Academic Press.

[8] AHARONY, A. [1986]. In *Directions in Condensed Matter Physics,* Vol. 1, eds. G. Grinstein and G. F. Mazenko. Singapore:World Scientific.

[9] ALBEN, R. [1972]. Amer. J. of Phys. **40**:3.

[10] ALDER, B. J. and HECHT, C. E. [1969]. J. Chem. Phys. **50**:2032

[11] ALLEN, M. P. and TILDESLEY, D. J. [1987]. *Computer Simulation of Liquids.* Oxford:Oxford University Press.

[12] ALS-NIELSEN, J. [1985]. Z. Phys. **B61**:411.

[13] AMIT, D.J, GUTFREUND, H. and SOMPOLINSKY, H. [1985]. Phys. Rev. Lett. **55**:1530.

[14] ANDERSON, M.H., ENSHER, J.H., MATTHEWS, J.R., WIEMAN, C.E. and CORNELL, E.A. [1995]. Science **298**:198.

[15] ANDERSON, P. W. [1958]. Phys. Rev. **109**:1492.

[16] ANDERSON, P. W. [1979]. In *Ill Condensed Matter*, eds. R. Balian, R. Maynard, and G Toulouse. Amsterdam:North-Holland.

[17] ANDERSON, P. W. [1984]. *Basic Notions in Condensed Matter Physics.* Menlo Park, Calif.:Benjamin-Cummings.

[18] ANDERSON, P. W., HALPERIN, B. E. and VARMA, C. M. [1971]. Philos. Mag. **25**:1.

[19] ANDRONIKASHVILI, E. L. and MAMALADZE, YU. G. [1966]. Rev. Mod. Phys. **3**:567.

[20] ANIFRANI, J.-C, LE FLOC'H, C., SORNETTE, D. and SUILLARD B. [1995]. J. Phys. I (France) **5**:631-8.

[21] ASHCROFT, N. and MERMIN, N. D. [1976]. *Solid State Physics.* New York:Holt, Rinehart and Winston.

[22] BAKER, G. A., NICKEL, B. G. and MEIRON, D. I. [1976]. Phys. Rev. Lett. **36**:1351.

[23] BALESCU, R. [1975]. *Equilibrium and Nonequilibrium Statistical Mechanics.* New York:Wiley.

[24] BARBER, M. N. [1983]. In *Phase Transitions and Critical Phenomena*, Volume 8. eds. C. Domb and J.L. Lebowitz. New York:Academic Press.

[25] BARDEEN, J., COOPER, L. N. and SCHRIEFFER, J. R. [1957]. Phys. Rev. **108**:1175.

[26] BARKER, J. A. and HENDERSON, D. [1967]. J. Chem. Phys. **43**:4714.

[27] BARKER, J. A. and HENDERSON, D. [1971]. Mol. Phys. **21**:187.

[28] BARKER, J. A. and HENDERSON, D. [1976]. Rev. Mod. Phys. **48**:587.

[29] BAXTER, R. J. [1973]. J. Phys. **C6**:L445.

[30] BAXTER, R. J. [1982]. *Exactly Solved Models in Statistical Mechanics.* New York:Academic Press.

[31] BEAGLEHOLE, D. [1979]. Phys. Rev. Lett. **43**:2016.

[32] BEDNORZ, J.G. and MÜLLER, K.A. [1986]. Z. Phys. B Condensed Matter, **64**:189.

[33] BERETTA, E., Y. TAKEUCHI, Y. [1995]. J. Math. Biol. **33**:250.

[34] BERRY, M. V. [1981]. Ann. Phys. **131**:163.

[35] BETHE, H. [1935]. Proc. R. Soc. London **A150**:552.

[36] BETTS, D. D. [1974]. In *Phase Transitions and Critical Phenomena*, Vol. 3, eds. C. Domb and M. S. Green. New York:Academic Press.

[37] BIJL, A. [1940]. Physica **7**:860.

[38] BINDER, K. [1986]. *Monte Carlo Methods in Statistical Mechanics*, 2nd ed. Berlin:Springer-Verlag.

[39] BINDER, K. and STAUFFER, D. [1984]. In *Applications of the Monte Carlo Method in Statistical Physics*, ed. K. Binder. Berlin:Springer-Verlag.

[40] BINDER, K. and YOUNG, A. P. [1986]. Rev. Mod. Phys. **58**:801.

[41] BINDER, K. and HEERMANN, D. W. [1988]. *Monte Carlo Simulation in Statistical Physics.* Berlin:Springer Verlag.

[42] BINDER, K. [1984]. in *Finite Size Scaling and Numerical Simulation of Statistical Systems*, ed. V. Privman. Singapore:World Scientific.

[43] BISHOP, D. J. and REPPY, J. D. [1978]. Phys. Rev. Lett. **40**:1727.

[44] BLÖTE, H. W. J. and SWENDSEN, R. H. [1979]. Phys. Rev. **B20**:2077.

[45] BLUME, H., EMERY, V. J. and GRIFFITHS, R. B. [1971]. Phys. Rev. **A4**:1071.

[46] BOAL, D. and RAO, M. [1992]. Phys. Rev. A **45**:R6947.

[47] BOAL, D. [2002]. *Mechanics of the Cell.* Cambridge: Cambridge University Press.

[48] BOGOLIUBOV, N. N. [1947]. J. Phys. USSR **11**:23.

[49] BORTZ, A.B., KALOS M.H. and LEBOWITZ, J.L. [1975]. J. Comput. Phys., **17**, 10.

[50] BOUCHAUD, J.-P. and GEORGES, A. [1990]. Physics Reports **195**:127-293.

[51] BRADLEY, C.C., SACKETT, C.C. and HULET, R.G. [1997]. Phys. Rev. Lett. **78**:985.

[52] BREZIN, E., LE GUILLOU, J. C. and ZINN JUSTIN, J. [1976]. In *Phase Transitions and Critical Phenomena*, Vol. 6, eds. C. Domb and M. S. Green. New York:Academic Press.

[53] CAHN, J. W. and HILLIARD, J. E. [1958]. J. Chem. Phys. **28**:258.

[54] CALLAWAY, J. [1991]. *Quantum Theory of the Solid State*, 2nd ed. New York:Academic Press.

[55] CALLEN, H. B. [1985]. *Thermodynamics and an Introduction to Thermostatistics*, 2nd ed. New York:Wiley.

[56] CAMP, W. J. and VAN DYKE, J. P. [1976]. J. Phys. **A9**:L73.

[57] CANELLA, V. and MYDOSH, J. A. [1972]. Phys. Rev. **B6**:4220.

[58] CARDY, J. [1996]. *Scaling and Renormalization in Statistical Physics*, Cambridge: Cambridge University Press.

[59] CHANDRASEKHAR, S. [1992]. *Liquid Crystals.* Cambridge:Cambridge University Press.

[60] CHAPELA, G., SAVILLE, C., THOMPSON, S. M. and ROWLINSON, J. S. [1977]. J. Chem. Soc. Faraday Trans. 2 **73**:1133.

[61] CHOU, T. and LOHSE, D. [1999]. Phys. Rev. Lett. **82** 3552.

[62] CONIGLIO, A. [1981]. Phys. Rev. Lett. **46**: 250.

[63] COOPER, L. N. [1956]. Phys. Rev. **104**:1189.

[64] COWLEY, R. A. and WOODS, A. D. B. [1971]. Can. J. Phys. **49**:177.

[65] CREUTZ, M. [1983]. Phys. Rev. Lett. **50**:1411.

[66] CRISTOPH, M. and HOFFMANN, K. H. [1993]. J. Phys. A: Math. Gen. **26**:3267.

[67] DAMASCELLI, A., HUSSAIN, Z. and SHEN, Z.-X. [2003] Rev. Mod. Phys. **75**:473.

[68] DAVIS, K.B., MEWES, M.-O., ANDREWS, M.R., van DRUTEN, N.J., DURFEE, D.S., KURN, D.M. and KETTERLE, W. [1995]. Phys. Rev. Lett. **75**:3969.

[69] DE GENNES, P. G. [1963]. Solid State Commun. **1**:132.

[70] DE GENNES, P. G. [1976]. J. Phys. Paris Lett. **37**:L1.

[71] DE GENNES, P. G. [1993] and PROST, J. *The Physics of Liquid Crystals* 2nd ed. Oxford:Clarendon.

[72] DE GENNES, P. G. [1989]. *Superconductivity of Metals and Alloys*, Redwood City:Addison Wesley.

[73] DE GENNES, P. G. [1979]. *Scaling Concepts in Polymer Physics.* Ithaca:Cornell University Press.

[74] DE GENNES, P. G. and TAUPIN, C. [1982]. J. Phys. Chem. **86**:2294.

[75] DES CLOISEAUX, G. and JANNINK, J. F. [1990]. *Polymers in Solution: Their Modelling and Structure*, Oxford:Clarendon Press.

[76] DOI, M. and EDWARDS, S. F. [1986]. *The Theory of Polymer Dynamics.* Oxford:Clarendon.

[77] DOLGOV, O. V, KIRZHNITZ, D. A. and MAKSIMOV, E. G. [1981]. Rev. Mod. Phys. **53**:81.

[78] DOMB, C. [1974]. In *Phase Transitions and Critical Phenomena.* Vol. 3, eds. C. Domb and M. S. Green. New York:Academic Press.

[79] DOMB, C. and HUNTER, D. L. [1965]. Proc. Phys. Soc. **86**:1147.

[80] DUNN, A. G., ESSAM, J. W. and RITCHIE, D. S. [1975]. J. Phys. **C8**:4219.

[81] DURETT, R. [1999]. *Essentials of Stochastic Processes.* Berlin: Springer.

[82] DYSON, F. J. [1958]. Phys. Rev. **102**:1217, 1230.

[83] EDWARDS, S. F. and ANDERSON, P. W. [1975]. J. Phys. **F5**:965.

[84] ELLIOTT, R. J., KRUMHANSL, J. A. and LEATH, P. L. [1974]. Rev. Mod. Phys. **46**:465.

[85] ESSAM, J. W. [1972]. In *Phase Transitions and Critical Phenomena*, Vol. 2, eds. C. Domb and M.S. Green. New York:Academic Press.

[86] EVANS, M.R., FOSTER, D.P. , GODRÉCHE, C. and MUKAMEL D. [1995]. J. Stat. Phys. **80**:69-103.

[87] EVANS, R. [1979]. Adv. Physics **28**:143.

[88] EVANS, R. [1992]. In *Fundamentals of Inhomogeneous Fluids*, ed. D. Henderson. New York:Marcel Dekker.

[89] FARMER, J. D. [1990]. Physica D **42**:153.

[90] FELLER, W. [1968]. *An Introduction to Probability Theory and its Applications*, 3rd ed. New York:Wiley.

[91] FENG, S. and SEN, P.N. [1984], Phys. Rev. Lett. 52: 216.

[92] FERRENBERG, A. M. and SWENDSEN, R. H. [1989]. Computers in Physics, Sept/Oct p. 101.

[93] FETTER, A. L. and WALECKA, J. D. [1971]. *Quantum Theory of Many Particle Systems.* New York:McGraw-Hill.

[94] FEYNMAN, R. P. [1954]. Phys. Rev. **94**:262.

[95] FISCHER, K. H. and HERTZ, J. A. [1991]. *Spin Glasses.* Cambridge University Press.

[96] FISHER, M. E. [1974]. Rev. Mod. Phys. **4**:597.

[97] FISK, S. and WIDOM, B. [1968]. J. Chem. Phys. **50**:3219.

[98] FLAPPER, D. P. and VERTOGEN, G. [1981]. Phys. Rev. **A24**:2089.

[99] FLAPPER, D. P. and VERTOGEN, G. [1981]. J. Chem. Phys. **75**:3599.

[100] FLORY, P. [1969]. *Statistical Mechanics of Chain Molecules.* New York: Wiley.

[101] FOWLER, R. H. and GUGGENHEIM, E. A. [1940]. Proc. R. Soc. London **A174**:189.

[102] FREED, K. F. [1987]. *Renormalization Group Theory of Macromolecules.* New York:Wiley.

[103] FREEMAN, J. A. [1994]. *Simulating Neural Networks with Mathematica.* Reading:Addison Wesley.

[104] FRISKEN, B. J., BERGERSEN, B. and PALFFY-MUHORAY, P. [1987]. Mol. Cryst. Liq. Cryst. **148**:45.

[105] FRÖHLICH, H. [1950]. Phys. Rev. **79**:845.

[106] GAUNT, D. S. and GUTTMANN, A. J. [1974]. In *Phase Transitions and Critical Phenomena*, Vol. 3, eds. C. Domb and M. S. Green. New York:Academic Press.

[107] GELL-MANN, M. and BRUECKNER, K. [1957]. Phys. Rev. **139**:407.

[108] GEMAN, S. and GEMAN, D. [1984]. IEEE PAMI **6**:721.

[109] GILLESPIE, D. T.[2001]. J. Chem. Phys. **115** 1716.

[110] GILLESPIE, D. T.[1976]. J. Comput. Phys. **22** 403.

[111] GLENDENNING, P. and PERRY, L.P. [1997]. J. Math. Biol. **35**:359.

[112] GLOSLI, J. and PLISCHKE, M. [1983]. Can. J. Phys. **61**:1515.

[113] GOLDSTEIN, H. [1980]. *Classical Mechanics*, 2nd ed. Reading, Mass.:Addison-Wesley.

[114] GORISHNY, S. G., LARIN, S. A. and TKACHOV, F. V. [1984]. Phys. Lett. **A101**:120.

[115] GOULD, H. and TOBOCHNIK, J. [1989]. Computers in Physics Jul/Aug p. 82.

[116] GREYTAK, T. J. and KLEPPNER, D. [1984]. In *New Trends in Atomic Physics*, Les Houches Summer School, 1982, eds. G. Greenberg and R. Stora. Amsterdam:North-Holland.

[117] GRIFFITHS, R. B. [1967]. Phys. Rev. **158**:176.

[118] GRIFFITHS, R. B. [1969]. Phys. Rev. Lett. **23**:17.

[119] GRIFFITHS, R. B. [1970]. Phys. Rev. Lett. **24**:1479.

[120] GRIFFITHS, R. B. and PEARCE, P. A. [1978]. Phys. Rev. Lett. **41**:917.

[121] GRINSTEIN, G. [1974]. AIP Conf. Proc. **24**:313.

[122] GUBERNATIS, J. E. and TAYLOR, P. L. [1973]. J. Phys. C **6**:1889.

[123] GUGGENHEIM, E. A. [1965]. Mol. Phys. **9**:199.

[124] GUGGENHEIM, E. A. [1967]. *Thermodynamics. An Advanced Treatment for Chemists and Physicists*, 5th ed. Amsterdam:North-Holland.

[125] HANSEN, J. P. and MCDONALD, I. R. [1986]. *Theory of Simple Liquids*, 2nd ed. London:Academic Press.

[126] HARRIS, A. B. [1974]. J. Phys. **C7**:L167.

[127] HARRIS, A. B., LUBENSKY, T. C., HOLCOMB, W. K. and DASGUPTA, C. [1975]. Phys. Rev. Lett. **35**:327.

[128] HAYMET, A. D. J. [1992]. In *Fundamentals of Inhomogeneous Fluids*, ed. D. Henderson. New York:Marcel Dekker.

[129] HERTZ, J., KROGH A., PALMER, R.G. [1991]. *Introduction to the theory of neural computation*, Redwood City: Addison Wesley, Lecture notes Volume I Sanna Fe Institute.

[130] HICKS, C. P. and YOUNG, C. L. [1977]. J. Chem. Soc. Faraday Trans. 2 **73**:597.

[131] HIRSCHFELDER, J. O., CURTISS, C. F. and BIRD, R. B. [1954]. *Molecular Theory of Gases and Liquids*, New York:Wiley.

[132] HO, J. T. and LITSTER, J. D. [1969]. Phys. Rev. Lett. **22**:603.

[133] HOHENERG, P. C. [1967]. Phys. Rev. **158**:383.

[134] HOPFIELD, J. J. [1982]. Proc. Natl. Acad. Sci (USA) **79**:2554.

[135] HOUTAPPEL, R. M. F. [1950]. Physica **1**:425.

[136] HUANG, K. [1987]. *Statistical Mechanics*, 2nd ed. New York: Wiley.

[137] HUSIMI, K. and SYOZI, I. [1950]. Prog. Theor. Phys. **5**:177, 341.

[138] IMRY, Y. and MA, S. K. [1975]. Phys. Rev. Lett. **35**:1399.

[139] ISHII, K. [1973]. Prog. Theor. Phys. Suppl. **53**:77.

[140] JACKLE, J. [1986]. Rep. Prog. Phys. **49**:171.

[141] JASNOW, D. and WORTIS, M. [1968]. Phys. Rev. **176**:739.

[142] JAYNES, E. T. [1957]. Phys. Rev. **106**:620.

[143] JOHANSEN, A., SORNETTE, D., WAKITA, H., TSUNOGAI, U., NEUMAN, W.I. and SALEUR, H. [1996]. J. de Physique I **6**:1391.

[144] JOSE, J. V., KADANOFF, L. P., KIRKPATRICK, S. and NELSON, D. R. [1977]. Phys. Rev. **B16**:1217.

[145] KADANOFF, L. P. [1971]. In *Proceedings of 1970 Varenna Summer School on Critical Phenomena*, ed. M. S. Green. New York:Academic Press.

[146] KADANOFF, L. P., GOTZE, W., HAMBLEN, D., HECHT, R., LEWIS, E. A. S., PALCIAUKAS, V. V., RAYL, M., SWIFT, J., ASPNES, D. and KANE, J. [1967]. Rev. Mod. Phys. **39**:395.

[147] KAMAL, S., BONN, D.A., GOLDENFELD, N., HITSCHFELD, P.I., LIANG, R. and HARDY, W.N. [1993]. Phys. Rev. Lett. **73**:1845.

[148] KANTOR, Y., KARDAR, M. and NELSON, D. R. [1986], [1987]. Phys. Rev. Lett. **57**:791; Phys. Rev A **35**:3056.

[149] KANTOR, Y. and NELSON, D. R., [1987]. Phys. Rev. Lett. **58**:2774; Phys. Rev. A **36**:4020.

[150] KASTELEYN, P. W. and FORTUIN, C. M. [1969]. J. Phys. Soc. Japan Suppl. **16**:11.

[151] KIMURA, M. [1955]. Proc. National Acad. Sc. **41:**145.

[152] KIRKPATRICK, S. and SHERRINGTON, D. [1978]. Phys. Rev. B **17**:4384.

[153] KIRKPATRICK, S. [1979]. In *Ill Condensed Matter*, eds. R. Balian, R. Maynard and G. Toulouse. Amsterdam:North Holland.

[154] KIRKPATRICK, S., GELATT, C. D. and VECCHI, M. P. [1983]. Science **220**:671.

[155] KIRZHNITZ, D. A. [1976]. Usp. Fiz. Nauk **119**:357 (Engl. transl. Sov. Phys. Usp.) **19**:530.

[156] KITTEL, C. [1986]. *Introduction to Solid State Physics*, 6th Ed. New York:Wiley.

[157] KLEIN, M. J. [1990], Physics Today, September issue p. 40.

[158] KOSTERLITZ, J. M. and THOULESS, D. J. [1973]. J. Phys. **C6**:1181.

[159] KROLL, D. M. and GOMPPER, G. [1992]. Science **255**:968.

[160] KUBO, R. [1957]. J. Phys. Soc. Japan **12**:570.

[161] KUBO, R., ICHIMURA, H., USUL, T. and HASHIZUME, N. [1965]. *Statistical Mechanics, An Advanced Course.* Amsterdam:North Holland.

[162] KUZNETETSOV, Yu. A., and PICCARDI, C. [1994], J. Math. Biol. **32** 109-121 (1994).

[163] LANDAU, L. D. [1941]. J. Phys. USSR **5**:71.

[164] LANDAU, L. D. [1947]. J. Phys. USSR **11**:91.

[165] LANDAU, L. D. and LIFSHITZ, E. M. [1980]. *Statistical Physics*, parts I and 2. Oxford:Pergamon Press.

[166] LAST, B. J. and THOULESS, D. J. [1973]. J. Phys. **C7**:715.

[167] LAWRIE, I. D. and SARBACH, S. [1984]. In *Phase Transitions and Critical Phenomena*, Vol. 9, eds. C. Domb and J. L. Lebowitz. New York:Academic Press.

[168] LEBOWITZ, J. L. and LIEB, E. H. [1969]. Phys. Rev. Lett. **22**:631.

[169] LEBOWITZ, J. L., PERCUS, J. K. and VERLET, L. [1967]. Phys. Rev. **153**:250.

[170] LEE, J. and KOSTERLITZ, J. M. [1991]. Phys. Rev. **B43**:3265.

[171] LE DOUSSAL, P. and RADZIHOVSKY, L. [1992]. Phys. Rev. Lett. **69**:1209.

[172] LIEB, E. H. [1976]. Rev. Mod. Phys. **48**:553.

[173] LE GUILLOU, J. C. and ZINN-JUSTIN, J. [1987]. J. Phys. Paris. **48**:19.

[174] LEE, P. A. and RAMAKRISHNAN, T. V. [1985]. Rev. Mod. Phys. **57**:287.

[175] LEIBLER, S. [1989]. In *Statistical Mechanics of Membranes and Surfaces*, eds. D. Nelson, T. Piran and S. Weinberg. Singapore:World Scientific, p. 46.

[176] LICCIARDELLO, D. C. and THOULESS, D. J. [1975]. J. Phys. **C8**:4157.

[177] LIFSHITZ, I. M. [1964]. Adv. Phys. **13**:483.

[178] LIPOWSKY, R. and GIRARDET, M. [1990]. Phys. Rev. Lett. **65**:2893.

[179] LONGUET-HIGGINS, H. C. and WIDOM, B. [1965]. Molecular Physics. **8**:549.

[180] LUBENSKY, T. [1979]. In *Ill Condensed Matter*, eds. R. Balian, R. Maynard and G. Toulouse. Amsterdam:North-Holland.

[181] LUDWIG, D., JONES, D.D. and HOLLING, C.S.[1978], J. Animal Ecology **47**:315.

[182] MA, S.-K. [1976a]. *Modern Theory of Critical Phenomena.* New York:Benjamin.

[183] MA, S.-K. [1976b]. Phys. Rev. Lett. **37**:461.

[184] MA, S.-K. [1985]. *Statistical Mechanics.* Singapore:World Scientific.

[185] MAHAN, G. D. [1990]. *Many Particle Physics*, 2nd ed. New York:Plenum.

[186] MAIER, W. and SAUPE, A. [1959]. Z. Naturforsch. **A14**:882.

[187] MAIER, W. and SAUPE, A. [1960]. Z. Naturforsch. **A15**:287.

[188] MAITLAND, G. C., RIGBY, M., SMITH, E. B. and WAKEHAM, W. A. [1981]. *Intermolecular Forces, Their Origin and Determination.* Oxford: Clarendon Press.

[189] MANDELBROT, B.B. [1982]. *The Fractal Geometry of Nature.* San Francisco:W. H. Freeman.

[190] MANDELBROT, B.B. [1997]. *Fractals and Scaling in Finance; Discontinuity, Concentration, Risk*, Berlin:Springer.

[191] MATSUBARA, T. and YONEZAWA, F. [1967]. Prog. Theor. Phys. **37**:1346.

[192] MATSUDA, H. and ISHII, K. [1970]. Prog. Theor. Phys. Suppl. **45**:56.

[193] MATTIS, D. C. [1976]. Phys. Lett. **A56**:421.

[194] MATTIS, D. C. [1981]. *The Theory of Magnetism 1.* Berlin:Springer-Verlag.

[195] MAYER, J. E. and MAYER, M. G. [1940]. *Statistical Mechanics.* New York:Wiley.

[196] MAZO, R.M. [1975] The J. Chem. Phys. **62**:4244.

[197] McCOY, B. [1972]. In *Phase Transitions and Critical Phenomena*, Vol. 2, eds. C. Domb and M. S. Green. New York:Academic Press.

[198] McCOY, B. and WU, T. T. [1973]. *The Two-Dimensional Ising Model.* Cambridge, Mass:Harvard University Press.

[199] McCULLOCH, W. S. and PITTS, W. [1943]. Bull. Math. Biophys. **5**:115.

[200] MCKENZIE, S. [1975]. J. Phys. **A8**:L102.

[201] MCNELLY, T. B., ROGERS, S. B, CHANNIN, D. J., ROLLEFSON, R. J., GOUBAU, W. M., SCHMIDT, G. E., KRUMHANSL, J. A. and POHL, R. O. [1970]. Phys. Rev. Lett. **24**:100.

[202] MERMIN, N. D. [1968]. Phys. Rev. **176**:250.

[203] MERMIN, N. D. and WAGNER, H. [1966]. Phys. Rev. Lett. **17**:1133.

[204] MEZARD, M., PARISI, G. and VIRASORO, M. A., [1987]. *Spin Glass Theory and Beyond.* Singapore:World Scientific.

[205] MILNER, P. M. [1993]. Scientific Amer. January p. 124.

[206] MIROWSKI, P.[1989]. *More heat than ligh: Economics as Social Physics, Physics as Nature's Economics*, Cambridge: Cambridge University Press.

[207] MONTROLL, E.W. and WEST, B.J.[1987]. *On an Enriched Collection of Stochastic Processes*, Chapter 2 in E.W. Montroll and J.L. Lebowitz eds. , *Fluctuation Phenomena*, Amsterdam: North Holland.

[208] MOTT, N. E. [1967]. Adv. Phys. **16**:49.

[209] MÜLLER, B. and REINHARDT, J. [1990]. *Neural Networks, An Introduction.* Springer Verlag:Berlin.

[210] MURRAY, J.D.[1993]. *Mathematical Biology*, Springer Biomathematics texts **19** 2nd ed (1993).

[211] MURRAY, C. A. and VAN WINKLE, D. H. [1987]. Phys. Rev. Lett. **58**:1200.

[212] NAUENBERG, M. and NIENHUIS, B. [1974]. Phys. Rev. Lett. **33**:944.

[213] NELSON, D. R. [1983]. In *Phase Transitions and Critical Phenomena*, Vol. 7, eds. C Domb and J. L. Lebowitz New York:Academic Press.

[214] NELSON, D. R. and HALPERIN, B. I. [1979]. Phys. Rev. **B19**:2457.

[215] NEWMAN, M.E.J.[2002]. Phys. Rev. **E 64**:016128.

[216] NICKEL, B. G. [1982]. In *Phase Transitions*, Cargese 1980, eds. M. Levy, J.-C. Le Gouillou and J. Zinn-Justin. New York:Plenum, p. 291.

[217] NIEMEIJER, T. H. and VAN LEEUWEN, J. M. J. [1976]. In *Phase Transitions and Critical Phenomena*, Vol. 6, eds. C. Domb and M. S. Green. New York:Academic Press.

[218] NIGHTINGALE, M. P. [1976]. Physica **83A**, 561.

[219] NODA, I., KATO, N., KITANO, T. and NAGASAWA, M. [1981]. Macromolecules **14**:668.

[220] ONSAGER, L. [1931]. Phys. Rev. **37**:405; **38**:2265.

[221] ONSAGER, L. [1936]. J. Am. Chem. Soc. **58**:1486.

[222] ONSAGER, L. [1944]. Phys. Rev. **65**:117.

[223] ONSAGER, L. [1949]. Nuovo Cimento Suppl. **2**:249.

[224] OONO, Y. [1985]. Adv. Chem. Phys. **61**:301.

[225] OPECHOWSKI, W. [1937]. Physica **4**:181.

[226] OSBORNE, D. V. [1950]. Proc. Phys. Soc. London **A63**:909.

[227] PALFFY-MUHORAY, P, BARRIE, R, BERGERSEN, B, CARVALHO I. and FREEMAN M.. [1984]. J.Stat.Phys. **35:**119.

[228] PALFFY-MUHORAY, P. and BERGERSEN, B. [1987]. Phys. Rev. **A35**:2704.

[229] PALFFY-MUHORAY, P. and DUNMUR, D. A. [1983]. Mol. Cryst. Liq. Cryst. **97**:337.

[230] PARISI, G. [1986]. J. Phys. A: Math. Gen. **19**:L675.

[231] PARR, R. G. and YANG, W. [1989]. *Density-Functional Theory of Atoms and Molecules.* Oxford:Clarendon Press.

[232] PASTOR-SATORRAS, R. and VESPIGNANI, A. [2001]. Phys. Rev. **E 63**:066117.

[233] PASTOR-SATORRAS, R. and VESPIGNANI, A. [2004]. *Evolution and Structure of the Internet; a Statistical Physics Approach.* Cambridge: Cambridge University Press.

[234] PATHRIA, R. K. [1983]. Can. J. Phys. **61**:228.

[235] PATHRIA, R. K. [1972]. *Statistical Mechanics* New York:Pergamon Press.

[236] PAWLEY, G. S., SWENDSEN, R. H., WALLACE, D. J. and WILSON, K. G. [1984]. Phys. Rev. **B29**:4030.

[237] PEIERLS, R. E. [1936]. Proc. Cambridge Philos. Soc. **32**:477.

[238] PELITI, L. and LEIBLER, S. [1985]. Phys. Rev. Lett. **54**:1690.

[239] PENROSE, O. [1951]. Philos. Mag. **42**:1373.

[240] PENROSE, O. and ONSAGER, L. [1956]. Phys. Rev. **104**:576.

[241] PETHIC, C. and SMITH H. [2002]. *Bose Condensation in Dilute Gases.* Cambridge: Cambridge University Press.

[242] PFEUTY, P. and TOULOUSE, G. [1977]. *Introduction to the Renormalization Group and to Critical Phenomena.* London:Wiley.

[243] PINES, D. and NOZIÈRES, P. [1966]. *The Theory of Quantum Liquids.* New York:Benjamin.

[244] PLISCHKE, M. and BOAL, D. [1989]. Phys. Rev. A**38**:4943.

[245] PLISCHKE, M., HENDERSON, D. and SHARMA, S. R. [1985]. In *Physics and Chemistry of Disordered Systems*, eds. D. Adler, H. Fritsche, and S. R. Ovshinsky. New York:Plenum.

[246] PLISCHKE, M. and JOÓS, B., Phys. Rev. Lett.**80**:4907.

[247] PLISCHKE, M., VERNON, D.C., JOÓS, B. and ZHOU, Z. [1999] Phys. Rev. **E 60**:3129.

[248] POTTS, R. B. [1952]. Proc. Cambridge Philos. Soc. **48**:106.

[249] PRESS, W. H., TEUKOLSKY, S. A., VETTERLING, W. T. and FLANNERY, B. P. [1992]. *Numerical Recipes*, 2nd ed. Cambridge:Cambridge University Press.

[250] PRIEST, R. G. and LUBENSKY, T. C. [1976]. Phys. Rev. **B13**:4159.

[251] PRIESTLEY, E. B., WOITOWICZ, P. J. and SHENG, P. [1974]. *Introduction to Liquid Crystals.* New York:Plenum.

[252] RASETTI, M. [1986]. *Modern Methods in Statistical Mechanics.* Singapore:World Scientific.

[253] REDNER, S. [2001]. *A Guide to First-Passage Processes*, Cambridge: Cambridge University Press.

[254] REICHL, L. [1998]. *A Modern Course in Statistical Physics.* Second Ed. New York: Wiley.

[255] RILEY, K.F., HOBSON, M.P. and BENCE, S.J. [2202]. *Mathematical methods for physics and engineering*, second edition Cambridge: Cambridge University Press.

[256] RISKEN, H. [1996] *The Fokker–Planck Equation*, Berlin:Springer, 2nd e. 3rd printing.

[257] RØNNOW, H.M., PARTHASARATHY, R., JENSEN, J., AEPPLI, G., ROSENBAUM, T.F. and McMORROW, D.F. [2005]. Science **308**:389.

[258] ROWLINSON, J. S. and WIDOM, B. [1982]. *Molecular Theory of Capillarity.* Oxford:Clarendon Press.

[259] RUBINSTEIN, R. and COLBY, R.R. [2003]. *Polymer Physics*, Oxford:Oxford University Press.

[260] RUDNICK, I. [1978]. Phys. Rev. Lett. **40**:1454.

[261] RUSHBROOKE, G. S. [1963]. J. Chem. Phys. **39**:842.

[262] SACHDEV, S. [1999]. *Quantum Phase Transitions.* Cambridge: Cambridge University Press.

[263] SAHIMI, M. [1994]. *Applications of Percolation Theory*, London: Taylor and Francis.

[264] SAITO, Y. and MULLER-KRUMBHAAR, H. [1984]. *In Applications of the Monte Carlo Method in Statistical Physics*, ed. K. Binder. Berlin:Springer-Verlag.

[265] SALEUR, H., SAMMIS, C.G., SORNETTE, D. [1996]. J. Geophys. Res., **101**:17661-77.

[266] SHERRINGTON, D. and KIRKPATRICK, S. [1975]. Phys. Rev. Lett. **35**:104.

[267] SCHICK, M., WALKER, J. S. and WORTIS, M. [1977]. Phys. Rev. **B16**:2205.

[268] SCHMIDT, C. F., SVOBODA, K., LEI, N., PETSCHE, I. B., BERMAN, L. E., SAFINYA, C. R. and GREST, G. S. [1993]. Science **259**:952.

[269] SCHRIEFFER, J. R. [1964]. *Superconductivity.* New York:Benjamin.

[270] SCHULTZ, T., MATTIS, D. and LIEB, E. [1964]. Rev. Mod. Phys. **36**:856.

[271] SHANNON, C. E. [1948]. Bell Syst. Tech. J. **27**:379, 623. (Reprinted in *The Mathematical Theory of Communication.* Urbana:University of Illinois Press.)

[272] SHARIPOV, F. [1996]. J. Vac. Sci. Technol. **A20**:814.

[273] SILVERA, I. F. and WALRAVEN, J. T. M. [1986]. In *Progress in Low Temperature Physics*, Vol. 10, ed. D. Brewer. Amsterdam:North-Holland.

[274] SINAI, IA. [1966]. In *Statistical Mechanics, Foundations and Applications*, ed. T. Bak. IUPAP meeting, Copenhagen, 1966.

[275] SINAI, IA. [1970]. Russ. Math. Surv. **25**:137.

[276] SKAL, A. S. and SHKLOVSKII, B. I. [1975]. Sov. Phys. Semicond. **8**:1029.

[277] SMITH, E. and FOLEY, D.K.. [2002]. *Is utility theory so different from thermodynamics?*, Santa Fe Working Paper SFI no. 02-04-016.

[278] SORNETTE, D. [1998]. Phys. Rep. **297**:239.

[279] SORNETTE, D. [2003]. *Why Stock Markets Crash: critical Events in Complex Financial Systems.* Princeton: Princeton University Press.

[280] SOVEN, P. [1967]. Phys. Rev. **156**:809.

[281] STANLEY, H. E. [1969]. Phys. Rev. **179**:501.

[282] STANLEY, H. E. [1971]. *Introduction to Phase Transitions and Critical Phenomena.* Oxford:Oxford University Press.

[283] STANLEY, H. E. [1974]. In *Phase Transitions and Critical Phenomena,* Vol. 3, eds. C Domb and M. S. Green. New York:Academic Press.

[284] STANLEY, H. E. and OSTROWSKY, N., eds. [1985]. *On Growth and Form.* Hingham, Mass.:Martinus Nijhoff.

[285] STANLEY, H. E., REYNOLDS, P. J., REDNER, S. and FAMILY, F. [1982]. In *Real Space Renormalization*, eds. T. W. Burkhardt and J. M. J. van Leeuwen. Berlin:Springer-Verlag.

[286] STAUFFER, D. [1979]. Phys. Rep. **54**:1.

[287] STAUFFER, D. and AHARONY, A. [1992]. *Introduction to Percolation Theory.* 2nd ed. London:Taylor and Francis.

[288] STEPHENS, M. J. and STRALEY, J. P. [1974]. Rev. Mod. Phys. **46**:617.

[289] STIEFVATER, T., MÜLLER, K-R. and KÜHN, R. [1996]. Physica **A** **232**:61.

[290] STINCHCOMBE, R. B. [1983]. *In Phase Transitions and Critical Phenomena,* Vol. 7, eds. C. Domb and J. L. Lebowitz. New York:Academic Press.

[291] STRALEY, J. P. [1974]. Phys. Rev. **A10**:1881.

[292] SWENDSEN, R. H. [1979]. Phys. Rev. Lett. **42**:461.

[293] SWENDSEN, R. H. [1984]. Phys. Rev. Lett. **52**:1165.

[294] SWENDSEN, R. H. [1984]. J. Stat. Phys. **34**:963.

[295] SWENDSEN, R. H. and KRINSKY, S. [1979]. Phys. Rev. Lett. **43**:177.

[296] SUZUKI, M. [1976]. Prog. Theor. Phys. **56**:1454.

[297] SUZUKI, M. [1985]. Phys. Rev. **B31**:2957.

[298] SZU, H. and HARTLEY, R. [1987]. Phys. Lett. **122A**:157.

[299] TEMPERLEY, H. N. V., ROWLINSON, J. S. and RUSHBROOKE, G. S. [1968]. *Physics of Simple Liquids.* Amsterdam:North-Holland.

[300] THIELE, E. T. [1963]. J. Chem. Phys. **39**:474.

[301] TIMUSK, T. and STATT, B. [1999]. Rep. Prog. Phys. **62**61.

[302] THOULESS, D. J. [1979]. In *Ill Condensed Matter*, eds. R Balian, R. Maynard and G. Toulouse. Amsterdam:North-Holland.

[303] TOBOCHNIK, J. and CHESTER, G. V. [1979]. Phys. Rev. **B20**:3761.

[304] TODA, M., KUBO, R. and SAITO, N. [1983]. *Statistical Physics 1, Equilibrium Statistical Mechanics*, Springer Series in Solid-State Sciences, Vol. 30. Berlin:Springer-Verlag.

[305] TSALLIS, C. [1988]. J. Stat Phys. **52**:479.

[306] TSUEI, C.C. and KIRTLEY, J.R. [2000]. Rev. Mod. Phys. **72**:969.

[307] UHLENBECK, G. E. and FORD, G. W. [1963]. *Lectures in Statistical Mechanics*, Lectures in Applied Mathematics, Vol 1. Providence, R. I.:American Mathematical Society.

[308] VAN DER WAALS, J. D. [1893]. Verh. K. Wet. Amsterdam. (English translation in J. Stat. Phys. **20**:197, 1979.)

[309] van KAMPEN, N.G. [1981]. *Stochastic processes in physics and chemistry*, North Holland.

[310] van KAMPEN, N.G. [1981]. J. Stat. Phys. **24:**175.

[311] VASHISHTA, P. and SINGWI, K. S. [1972]. Phys. Rev. **B6**:875.

[312] VELICKY, B., KIRKPATRICK, S. and EHRENREICH, H. [1968]. Phys. Rev. **175**:747.

[313] VERLET, L. [1968]. Phys. Rev. **163**:201

[314] VINCENTINI-MISSONI, M. S. [1972]. *Phase Transitions and Critical Phenomena*, Vol. 2, eds. C. Domb and M. S. Green. New York:Academic Press.

[315] WAISMAN, E. and LEBOWITZ, J. [1970]. J. Chem. Phys. **52**:4707.

[316] WAISMAN, E. and LEBOWITZ, J. [1972]. J. Chem. Phys. **56**:3086, 3093.

[317] WALKER, L. R. and WALSTEDT, R. E. [1977]. Phys. Rev. Lett. **38**:514.

[318] WANNIER, G. H. [1966]. *Statistical Physics.* New York:Wiley.

[319] WEEKS, J. D. and GILMER, G. H. [1979]. Adv. Chem. Phys. **40**:157.

[320] WEGNER, F. J. [1973]. Phys. Rev. **B5**:4529.

[321] WEN, X., GARLAND, C. W., HWA, T., KARDAR, M., KOKUFOTA, E., LI, Y., ORKISZ, M. and TANAKA, T. [1992]. Nature **355**:426.

[322] WERTHEIM, M. S. [1963]. Phys. Rev. Lett. **10**:321.

[323] WERTHEIM, M. S. [1971]. J. Chem. Phys. **55**:4291.

[324] WIDOM, B. [1965]. J. Chem. Phys. **43**:3898.

[325] WILSON, K. G. [1971]. Phys. Rev. **B4**:3174, 3184.

[326] WILSON, K. G. and FISHER, M. E. [1972]. Phys. Rev. Lett. **28**:240.

[327] WILSON, K. G. and KOGUT, J. [1974]. Phys. Rep. **12**:75.

[328] WISER, N. [1963]. Phys. Rev. **129**:62.

[329] WOJTOWICZ, P. J. and SHENG, P. [1974]. Phys. Lett. **A48**:235.

[330] WU,F.Y. [1982]. Rev Mod. Phys. **54**:236.

[331] YANG, C. N. [1952]. Phys. Rev. **85**:809.

[332] YANG, C. N. [1962]. Rev. Mod. Phys. **34**:694.

[333] ZASPEL, Z. E. [1990]. Amer. J. Phys. **58**:992.

[334] ZIMAN, J. M. [1964]. *Principles of the Theory of Solids.* Oxford:Clarendon Press.

[335] ZIMAN, J. M. [1979]. *Models of Disorder.* Cambridge:Cambridge University Press.

Index

Absolute temperature scale, 7
Absorbing boundary, 328
Absorbing state, 129, 309
Absorption, 467
Action-angle variables, 30
Adiabatic
  compressibility, 14
  compression, 26
  process, 4, 5, 9, 10, 25, 28
  susceptibility, 19
Adsorbed monolayers, 268–272
Alben model, 104
Allele, 305, 330
Alloys, 113
Amorphous materials, 551–565
Ampère's law, 19, 230
Anderson localization, 519–525
Anisotropic Heisenberg model, 300
Anisotropic relaxation time, 501
Annealed disorder, 515, 545
Annihilation operators, 465, 572–574
Anomalous diffusion, 333
Anticommutator, 573
Argon gas, 351
Artificial boundaries, 328, 331
Assignment problem, 376
Assocative memory, 371
Autocatalytic systems, 371
Axial vectors, 494

Bachelier-Wiener process, 324, 336
Backbone, 533, 542, 550
Bardeen Cooper Schrieffer (BCS) theory, 226, 442–452
  BCS ground state, 445–449
  finite temperature BCS theory, 449–452
Basin of attraction, 292
BBGKY hierarchy, 157, 158
BEG model, 121, 123
Bending rigidity, 408, 409, 413–415
Beta–brass, 110, 113, 515
Bethe approximation, 71–76, 80, 101
Bethe lattice, 533
BGY-equation, 158
Biaxial phase, 120
Binary alloys, 110, 516, 530
Binary tree, 362
Birth and death process, 305–309, 313, 345

Birth rate, 307
Black–body radiation, 58
Blobs and strings picture, 550
Bloch function, 498, 527
Bloch's theorem, 517
Block diagonal matrix, 358
Block spin, 278
Blume–Emery–Griffiths model, 121, 123
Bogoliubov theory of superfluidity, 439–442
Bogoliubov transformation, 441, 450, 477
Bohm–Staver velocity, 489
Bohr–Sommerfeld quantization, 33
Boltzmann constant, 50
Boltzmann equation, 498–507
  dc conductivity, 500–502
  thermal conductivity, 503–507
  thermoelectric effects, 503–507
Boltzmann statistics, 45, 46
Boltzmann–Shannon entropy, 50, 51
Bond percolation, 530, 531, 534
Born–Green–Yvon equation, 158
Bose condensation, 422–429
  in arbitrary dimension, 429
Bose gas, 439–442, 475–477, 508
Bose–Einstein distribution, 46
Bose–Einstein integral, 422
Bose–Einstein statistics, 43
Bragg–Williams approximation, 67–71, 80, 82, 87, 127, 137
Branched polymers, 418
Branching process, 309–313
Brayton cycle, 26
Brillouin zone, 280–282, 526
Broken symmetry, 66, 83
  in BCS theory, 448
Brownian dynamics, 350, 353
Brownian motion, 323–325, 345
Brownian particle, 354
Burger's vector, 232

Canonical
  distribution, 37, 38, 52, 354
  ensemble, 29, 35–39, 41, 43, 44, 47, 55, 57, 357
  partition function, 38, 43, 44, 56, 57, 249
  transformation, 33, 516
  variable, 30
Capillary length, 170
Capillary waves, 163, 169, 415
Carnot cycle, 5–8, 24
Carrying capacity, 129
Catastrophic events, 257, 258
Cauchy distribution, 378
Causality, 471, 474
Cellular automaton, 372
Chapman-Kolmogorov equation, 311
Characterisitic length, 399
Chemical potential, 2, 3, 10, 24, 25, 34, 36, 40, 41, 45, 121, 126, 129, 367, 402, 422–424, 441
Chemical reactions, 362
Classical vectors, 226
Classical–quantum correspodnence, 186
Classifier systems, 371
Clausius
  formulation of second law, 5, 6

Clausius–Clapeyron equation, 23, 426
Cluster approximation, 101
Cluster integral, 146, 147, 150, 159, 181
Cluster methods, 362
Coarse graining, 498
Coexistence of phases, 20–23, 80, 92, 105, 125–127, 129, 137
Coherence length, 168, 454
Coherent potential approximation, 525–530
Colloidal suspensions, 233
Combinatorical optimization, 376
Complete set of states, 46, 47
Compressibility, 1, 14, 15, 29, 42, 43, 97, 125, 154–156, 356, 367
  equation of state, 155
  isothermal, 27, 156, 366
Computer simulations, 349–382
  molecular dynamics, 350–357
  Monte Carlo methods, 357–370
Comtinuous phase transition, 68
Concave functions, 17, 27, 28
Concentrated solutions, 403, 405
Conditional probability, 311, 324, 327
Conductivity, 490–492, 500–502, 533, 541
  temperature dependence, 514
Configuration integral, 144, 145, 147, 148, 161
Connected cluster, 514
Conservation of energy, 3
Conservative molecular dynamics, 351
Constant energy simulation, 355
Constant pressure ensemble, 60
Constitutive relation, 321
Content addressible memory, 371
Continuous symmetry, 226, 229
Continuum time evolution, 306
Convex function, 28, 114
Cooper problem, 443–444
Correlation function, 96, 97, 162, 174, 183, 215, 217, 220, 227–229, 235, 274, 283, 284, 286
  current-current, 492
  direct, 158, 161, 177, 179, 180
  equilibrium, 468, 490
  Gaussian model, 283–284
  Heisenberg model, 477–480
  noninteracting electrons, 579–582
  pair correlation function in liquids, 155, 157, 159, 161, 178, 179
  retarded, 464
  spin-spin, 75, 76, 80, 82, 220
Correlation length, 82, 96, 98, 215–218, 220, 221, 223, 225, 240, 243, 295
Correlation time, 367
Corresponding states, 126–128
Cost function, 376, 377
Coulomb gas, 161
  two dimensions, 231
Coulomb repulsion, 60
$CrBr_3$, 217, 219
Creation operators, 465, 572–574
Critical dimensionality, 93, 97, 98
Critical exponent, 246
Critical exponents, 197, 200, 209–214, 220, 222–225, 229, 232, 235, 237, 245, 275, 289, 369
  complex, 257

correlation length, 96
critical isotherm, 257, 301
crossover exponent, 300, 547–548
disordered systems, 546–551
Flory, 393
Gaussian, 410
Gaussian model, 282–284
Heisenberg model, 210
Ising model, 209, 247
Landau theory, 98
linear polymers, 399
mean field theory, 101, 197, 282
n-vector model to order $\epsilon = 4 - d$, 290
order parameter, 67, 93, 197, 257, 301
percolation, 535, 536, 549
roughness, 411, 414
specific heat, 76, 197, 233, 246, 255, 266, 301, 548
susceptibility, 197, 255, 290, 301, 392
tricritical point, 93, 98, 104
XY-model, 210, 247
Critical isotherm, 76, 235
Critical line, 244
Critical opalescense, 156
Critical point, 20, 21, 24, 74, 75, 77, 82, 83, 91, 93–95, 98, 119, 123, 183, 197, 199, 200, 212, 216–221, 231, 243, 365, 370, 399
liquid vapor, 168
van der Waals, 125
Critical slowing down, 362, 367
Critical temperature, 64, 120, 194, 207, 212, 221, 222, 229, 233, 234, 243, 255, 278
two dimensional Ising model, 73
Critical velocity of superfluids, 432
Crosslinking, 541
Crossover behavior, 220
Crumpled membranes, 406–409, 412 414
Cubic symmetry, 100
Cubic term in Landau expansion, 86
Cumulant expansion, 258, 285, 288
first order approximation, 262, 264, 295
second order approximation, 26 292, 296
Curie constant, 3
Curie law, 3, 25
Curvature energy, 417

Damping coefficient, 353, 354
Dangling bonds, 532, 550
Death rate, 307
Debye screening, 507
Debye temperature, 506
Degeneracy, 43, 59, 98
Density functional theory, 168
Density matrix, 29, 46–48
Density of states, 517–530
Density operator, 47, 576
Density-functional methods, 171–18
Detailed balance, 306, 359–362, 467, 499
Diagrammatic expansion, 530
Diamond fractal, 248–258
Diatomic gas, 56
Diatomic molecule, 58, 59
Dielectric function, 56
diatomic gas, 57
electron gas, 473–475

local field correction, 475
metals, 487–490
metals, 490
Diffusion, 321, 328
Dilute solutions, 403
Diluted ferrmomagnet, 542
Dimerization, 105
Dipolar hard spheres, 161
Dipolar interactions, 291
Dipole moment, 56
Direct correlation function, 158, 161, 177, 179, 180
Director field, 117
Discrete time Markov processes, 358–359
Dislocations, 232
Disordered systems, 513–567
single particle states, 515–530
Dissipation, 467
Distinguishable particles, 32, 33
Distribution function, 116, 144, 152–154, 157–160, 162, 163, 181
single particle, 498
Domain wall, 70, 101
Domains, 548
Double tangent construction, 114, 115, 165, 167
Drift term, 321
Drift velocity, 502
Drude model, 485
Dynamic structure factor, 465–472
electron gas, 480–487
Heisenberg ferromagnet, 477–480
ideal Bose gas, 508
interacting bosons, 475–477

Earthquakes, 258
Edwards model, 394, 395, 413
Edwards–Anderson order parameter, 558
Effective bond length, 418
Effective mass, 502
Efficiency, 5–7, 25–27
Einstein oscillator, 59
Einstein realtion, 325
Elastic scattering, 153–155, 499
Elasticity theory, 413
Electrolytes, 161
Electromotive field, 504
Electron gas, 480–487
pairing, 442–449
screening, 480–485, 507–508
Thomas–Fermi approximation, 483, 507–508
Electron states
disordered alloys, 515–530, 565–567
one dimension, 565
Ellipsometry, 169
Emission, 467
Energy fluctuations
for an ideal gas, 58
Energy gap in superconductors, 445–452
Ensemble average, 32
Entanglement, 403
Enthalpy, 65, 182, 361
Entropic elasticity, 383, 390, 406
Entropy, 1, 2, 5, 7–10, 12, 15–17, 21, 28, 32–36, 39, 40, 50–53, 65, 70, 111, 124, 392, 497
Bragg–Williams approximation, 67
concavity, 17, 27
ideal gas, 34, 55, 124
information theoretic, 48, 52

maximum principle, 9, 59
mixing, 27, 33, 55, 128, 400
orientational, 116, 138
production, 492–494
statistical definition, 32
vortices, 229, 231
Epidemiology, 305, 316
Epsilon expansion, 279–295, 297, 409, 414
Equal-area construction, 126, 127
Equation of state, 1, 2, 23, 28, 56
compressibility, 155
fluctuations, 156
hard sphere system, 150
ideal gas, 25, 125, 145, 156
mixtures, 128
Tonks gas, 181
van der Waals, 124, 125, 127, 143, 166
virial, 2, 182
Equilibrium, 1–3, 5, 9, 16–18, 22, 27, 29, 30, 40, 52, 306
chemical, 3
concentration of defects, 53
fixed S, V and N, 28
fixed T, P and N, 17, 53
fixed T, V and N, 11, 53
gas liquid, 27, 105
mechanical, 3
thermal, 3
Equipartition, 55, 170, 469
Ergodic hypothesis, 31, 48, 54, 349
Euler summation formula, 59
Exact differential, 4, 130
Exchange and correlation energy, 474, 486–487
Excitation spectrum, 430–432, 439–442, 477, 486, 509
Extended states, 514, 517
Extensive variables, 2, 5, 10, 12, 13, 28, 34, 35, 39, 49, 51
Extinction, 330

f-sum rule, 471–472, 508
Factorial moments, 310
Faraday's law, 19
Fermi energy, 45
Fermi golden rule, 305
Fermi–Dirac statistics, 43
Ferromagnet, 21, 22
Feynman diagram techniques, 290
Fick's law, 130, 493
Field operators, 575
Finite lattice method, 267
Finite-size scaling, 218–223, 277, 364 365, 368–370
First integrals, 30
First law of thermodynamics, 3–4, 6, 8, 9
First order transition, 85, 87, 91, 104, 118, 126, 364
First passage time distribution, 332, 336
Fixed point, 237, 240–248, 251, 252, 263, 278, 290
XY-model, 300
disorder, 548
Gaussian, 287, 290
n-vector, 300
Floating monolayer, 231
Flory theory, 391, 395, 403, 409, 410, 418
Flory–Huggins parameter, 401
Flory–Huggins theory, 400, 403
Fluctuation dissipation theorem, 39, 467–469
Fluctuation equation of state, 155, 156

Fluctuations, 17, 25, 29, 39, 42, 43, 58, 64, 94, 99, 155, 365
  Landau–Ginzburg theory, 64, 94, 98
  particle number, 367
  volume, 361, 366
Fluid membranes, 383
Fokker–Planck equation, 313–347
Forbidden energy region, 519
Forecasting, 258
Form factor, 154
Fountain effect, 437
Fourier transform, 333
Fractal dimension, 532, 533
Fractal lattice, 301
Free energy functional, 174
Freely jointed chain, 389, 390
Friedel oscillations, 484
Frustration, 270, 555–558
Fuel cell, 11
Functional
  differentiation, 171–173, 176, 177
  free energy, 174–179

Gauche configuration, 384
Gaussian chain, 383, 389–393, 395, 405, 419
Gaussian distribution, 369, 370, 379, 389
Gaussian fixed point, 395
Gaussian model, 281–284, 286
Gaussian noise, 353
Gaussian Process, 324
Gegenbauer polynomial, 330
Generalized force, 492, 497
Generalized susceptibility, 464–469
Generating function, 310, 333, 334
Geometric phase transition, 531
Gibbs dividing surface, 165
Gibbs J. W., 5
Gibbs paradox, 33, 43, 55
Gibbs phase rule, 1, 23
Gibbs potential, 11, 13, 17, 18, 22, 28, 53, 54, 83, 94, 126
  for an ideal gas, 60
  for magnetic systems, 19
Gibbs, J.W, 130
Gibbs–Duhem equation, 1, 12, 22, 35, 41, 42, 58, 164, 436
Ginzburg criterion, 64, 97, 98, 456
Glasses, 551–558
Golden rule, 499
Goldstone boson, 229
Grand canonical
  ensemble, 29
  ensemble, 40–43, 47, 52, 57–59, 145, 269
  partition function, 41, 44–46, 58
Grand potential, 12, 41, 42
Gravitational collapse, 60
Green's function, 468, 527, 528
Green-Kubo formalism, 354, 357
Griffiths singularities, 546
Ground state, 480
  energy, 470, 487

Hamiltonian dynamics, 30
Hard core interaction, 124, 386, 391, 406, 411
Hard sphere fluids, 150, 158, 161
Harmonic oscillator, 44, 54, 55, 59
Harris criterion, 547–548
Heap, 362
Heat, 3, 5
Heat current, 503
Hebb rule, 372
Hebb, D.O., 372

Heisenberg
  equation of motion, 491
Heisenberg model, 64, 65, 99, 100, 202, 206, 209, 210, 223, 234, 247, 290, 297–300, 369, 477–480, 509, 550, 557
  anisotropic, 477
  high temperature expansion, 210
  spin waves (magnons), 477–480, 509
  three-dimensional, 210
Heisenberg model, 104
Heisenberg operators, 463
Heisenberg picture, 463
Helium monolayer, 269
Helmholtz free energy, 10, 38, 52, 94, 137, 164, 402, 470
  ideal gas, 57, 104
  scaling properties, 212
  van der Waals, 124
Heterogeneous diffusion, 340–343
Hierarchical organization, 257, 258
High T_c superconductors, 98
High temperature expansion, 200, 202, 210
  Heisenberg model, 209, 234
  Ising model, 218
  susceptibility, 205
Histogram methods, 363, 364, 379
Homogeneous functions, 56
Homogeneous mixing, 317
Hooke's law, 383
Hookean springs, 541, 542
Hopfield model, 371–375
Hopping, 524
Human nervous system, 371
Hypergeometric equation, 330
Hypernetted chain approximation, 160, 161
Hyperscaling, 217, 218, 548

Ideal Bose gas, 226, 422–429, 456–457
Ideal gas
  entropy, 34
  entropy of mixing, 33
  free energy functional for, 176
  Gibbs potential, 60
  Helmholtz free energy, 57, 104
  law, 2, 28, 125, 145, 380, 403
  particle number fluctuations, 58
Ideal gas law, 405
Identical particles, 62
Impurity band, 521
Information theory, 29, 48, 52
Inhibition, 371
Inhomogeneous liquid, 144, 163, 171, 178–180
Integrable systems, 30, 31, 33
Intensive variables, 2, 3, 12, 35, 42
Interacting fermion problem, 188
Interaction range, 224, 225
Interface
  liquid-vapor, 144, 163–165, 167–170, 172, 181
Internal energy, 2, 4, 9, 10, 12, 14, 28, 58
  convexity of, 28
Inversion temperature, 182
Irrelevant scaling field, 247, 287, 290, 394, 408
Irreversible process, 4, 5, 7, 8
Ising antiferromagnet, 113, 271
Ising chain, 381
Ising ferromagnet, 112, 113
Ising model, 63, 65, 67, 71, 74, 75, 98, 101, 113, 358, 359, 369

absence of phase transition one-dimensional, 70
Bragg Williams approximation, 68
critical exponents, 218
dimensionality of the lattice, 69
finite-size scaling, 220
high temperature expansion, 207, 209, 210
Kadanoff block spins, 215
low temperature expansion, 206
mean field theory, 64, 83, 84
one-dimensional, 71, 73, 74, 77, 80–82, 101–103, 105, 184, 238–242, 295, 381
spin 1, 102
three-dimensional, 199, 209, 291
two-dimensional, 70, 75, 101, 183–199, 233, 235, 258–266
two-dimensional Ising antiferromagnet, 268–272
Isobaric processes, 4
Isochoric processes, 4
Isolated system, 8, 30, 31, 35
Isotherm
van der Waals, 125
Isothermal compressibility, 27, 42, 156
Isothermal process, 4, 5, 10, 14
Isothermal susceptibility, 211

Jordan–Wigner transformation, 188, 199
Joule cycle, 26
Joule–Thompson process, 182
Jump moments, 321, 323

Kadanoff block spins, 215, 218, 241
Kelvin
formulation of second law, 5, 6
Kimura-Weiss model, 329–330
Kinetic coefficients, 493–498
Kirchhoff equations, 541
Kirkwood superposition approximation, 158
Kohn-Hohenberg theorem, 174
Kolmogorov backwards equation, 311, 312
Kondo problem, 292
Kosterlitz–Thouless transition, 183, 210, 226–233
Kramers equation, 326–328, 345
Kramers escape rate, 337–339
Kramers–Kronig relations, 471, 473
Kramers-Moyal expansion, 323
Kronig–Penney model, 565
Kubo formula, 490–492
Kuhn length, 385

Lagrange multiplier, 51
Lamé coefficients, 413
Lamellar phases, 415
Landau approximation
time dependent, 130
Landau expansion
cubic term, 86
Landau theory of phase transitions, 63, 64, 77, 83–87, 95, 97, 99, 100, 104, 114, 118, 183, 197, 271
Maier–Saupe model, 86, 137
tricritical point, 90, 98, 122
Landau–Ginzburg theory, 64, 94, 144, 168, 235, 283
of superconductivity, 453–456
of the liquid vapor interface, 166

Landau–Ginzburg–Wilson Hamiltonian, 285, 291, 297
Langevin equation, 353
Latent heat, 20, 104
Lattice constant, 203
Lattice gas, 269, 400
Laws of thermodynamics, 3–9
Le Châtelier's principle, 18
Leapfrog algorithm, 352
Legendre transformation, 10, 41
Lennard–Jones potential, 145, 160, 269, 350, 351
Lever rule, 114
Levy-Smirnov distribution, 337
Lifshitz tail, 521, 528
Light scattering, 153, 414
Limbo state, 331
Lindhard function, 482–485
Linear response theory, 461–511
Linearization, 245, 246, 253, 266, 275, 290, 300, 502, 547
Lipid bilayers, 415
Liquid crystals, 91, 100, 114, 137
Liquid membranes, 406, 415, 417
Liquid-vapor interface, 144, 156, 163, 164, 167–169, 172, 181
Liquids, 143–182
Local field correction, 56, 475
Localization, 514, 519, 521–525, 530
Log–periodic oscillations, 257, 258
London penetration depth, 455
Long-range interactions, 291, 386, 393
Lorentz distribution, 378
Low temperature expansions, 206, 209, 211
Lower critical dimension, 410
Lyotropic liquid crystals, 415
Magnetic field, 2, 65
  Onsager relations, 494
Magnetic induction, 230
Magnetic work, 18, 20
Magnetization, 2, 18, 21, 65
Magnons, 477–480, 509
Maier–Saupe model, 114–120, 123, 137
Majority rule, 260, 266–268, 273
Marginal scaling fields, 247
Markov process, 303–306, 350, 368
Markov processes, 381
Master equation, 304–306, 313, 314, 322, 323, 329, 331
Maximum current phase, 135
Maximum entropy principle, 9, 16, 17, 27, 36, 48, 52, 59
Maxwell construction, 126, 165
Maxwell relations, 1, 12–15
Maxwell-Boltzmann velocity distribution, 327, 381
Mayer function, 145–147, 158
Mayer J.E., 145
Mean field theory, 63–107, 183, 184
  based on Ornstein–Zernike equation, 168
  critical behavior, 74
  critical exponents, 74–76, 217
  fluids, 123
  for interfaces, 164
  misleading aspects, 69, 71, 87
  neglect of long range correlations, 74, 75, 80, 94, 183
  polymer solutions, 400–403
  response functions, 472–490
  spinodals, 127
  superconductivity, 449–452
  van der Waals theory of liquids, 123, 143, 166

weakly interacting Bose gas, 475–477, 508
Mean spherical approximation, 160
Meissner effect, 455
Melting of two-dimensional solids, 231–233
Membranes, 405–420
and branched polymers, 418
Edwards model, 409
Flory theory, 410
fluid, 415–418
persistence length, 416
phantom, 406–409
self-avoiding, 409–414
tethered, 405–414
Memory, 372
Mermin–Wagner theorem, 210, 509
Mesoscopic system, 303
Microcanonical ensemble, 29–35, 38, 49, 51, 52
partition function, 40
Microemulsions, 415
Microscopic reversibility, 494–498
Migdal–Kadanoff transformation, 296–297
Mixing
entropy, 33, 55
hypothesis, 31, 33
van der Waals theory, 127
Mobility, 324
Mobility edge, 523–525
Molecular dynamics, 350–357, 379, 381
Molecular motors, 134
Momentum space renormalization group, 551
Monte Carlo method, 306, 350, 399, 409
Monte Carlo methods, 357–370
data analysis, 365–370
finite-size scaling, 368–370
histogram methods, 363–364
Markov processes, 358–359
Metropolis algorithm, 359–362
neural networks, 371–375
simulated annealing, 376–379
traveling salesman, 376–379
Monte Carlo renormalization, 272–275, 539
Monte Carlo simulations, 368, 535, 536, 548, 551
Multicomponent order parameter, 64, 100

n-vector model, 99, 292, 297, 383, 543–551
application to linear polymers, 395
Narural boundary, 328
Nematic liquid crystals, 100, 114, 117, 129
Net reproduction rate, 315
Neural networks, 371–375
Neutral evolution, 329
Neutron scattering, 153, 478
Noise function, 354
Noise strength, 353
Non-equilibrium phenomena, 492
Noninteracting bosons, 45
Noninteracting fermions, 44
Normal coordinates, 55
Normal modes, 419
Normal systems, 35
Nosé-Hoover dynamics, 350
NTP ensemble, 361, 366
Number operator, 574

Occupation number, 44, 467

Occupation number representation, 569–582
Off-diagonal long range order, 428, 449
Ohm's law, 493, 503
One step process, 306, 307, 313
One-body operators, 577
Onsager relations, 494–498, 510
Onsager solution
  two-dimensional Ising model, 73, 184–200, 207
Onsager, L., 268
Open systems, 12, 25
Optimized Monte Carlo method, 364
Order of phase transitions, 64
Order parameter, 63–65, 67, 68, 71, 73, 74, 80, 83–86, 90, 92, 94, 95, 97, 99, 100, 104, 105, 111–113, 116, 118–123, 126, 137, 138, 184, 194, 210, 217, 364, 368
  Edwards–Anderson, 558
  percolation, 534
  superfluids, 428, 433
Order–disorder transition, 110
Ordering in alloys, 109, 110, 112, 113
Ornstein–Zernike equation, 144, 158, 160, 161, 168, 177, 178
Ornstein-Uhlenbeck process, 327
Osmotic pressure, 402–405

Padé approximant, 207, 234
Pair connectedness, 534, 536
Pair correlation function, 151, 155, 157, 469
Pair distribution function, 80, 114, 115, 153, 154, 157–160, 162, 181, 472, 508, 579
Pair potential, 114, 124, 144, 161, 350
Paramagnet, 3, 25
Particle current, 321
Pauli principle, 44
Pauli spin operators, 185
Pawula's theorem, 323
Peierls, R.E., 71
Peltier effects, 493
Percolation, 524, 530–542, 549–551, 567
  backbone, 533
  critical exponents, 536
  scaling theory, 534–536
Percus–Yevick equation, 159, 160
Periodic boundary conditions, 238, 357, 516, 518, 519
Permeability, 19, 20
Permittivity, 230
Persistence length, 385, 386, 405, 416, 417
Perturbation theory of liquids, 144, 160–163, 180
Phantom membranes, 406, 408, 409, 411, 413
Phase diagram, 20, 21, 23, 92, 120, 129
Phase separation, 113, 115, 120, 127, 137
Phase space, 30, 32, 33, 40, 51, 54
Phase transitions, 20, 65, 85, 105, 106, 183, 219, 220, 229, 542–551
  continuous, 222
  disordered materials, 542–551
Phenomenological renormalization g 275, 542
Phonons, 480
  metals, 487–490

Planar magnets, 232, 233
Plasma frequency, 485
Plasmons, 480–485, 508
Polarization, 56, 474
Polyethylene, 384, 385
Polymers, 383–420
    Θ point, 402
    connection to n-vector model, 395–399
    critical exponents, 399
    dense solutions, 400–405
        mean field theory, 400–403
    Edwards model, 394–395
    entropic elasticity, 390
    excluded volume effects, 391–395
    Flory theory, 391–395
    Flory–Huggins theory, 400–403
    freely jointed chain, 386–389
    good and poor solvents, 393
    linear polymers, 384–405
    osmotic pressure, 402
    self-avoiding walks, 391
Position space renormalization group, 282, 551
Position-space renormalization group, 258–275
Potts model, 87, 100, 106, 107, 271, 534
Predator-prey interaction, 305
Pressure
    statistical definition, 34
Pressure equation of state, 356
Probability current, 321, 322, 325, 326
Probability distribution, 48–52, 357, 361, 369, 389
    end-to-end distance, 388
Projection operator, 259, 275
Propagator, 468
Pseudopotential, 115, 119
PVT system, 11–14, 20–24

q-state Potts model, 107
Quantum fluids, 421–459
Quantum phase transitions, 225
Quantum states, 33, 43, 44, 46, 59
Quantum statistics, 40, 43, 44
Quantum systems, 469
Quantum-classical correspondence, 186
Quasi-elastic scattering, 506
Quasistatic process, 4
Quenched disorder, 515, 543–546, 557, 559

Radius of gyration, 387, 392, 394, 399, 407, 411
Raising and lowering operators, 313
Random fields, 548
Random number generator, 379
Random resistor network, 541, 542
Random walk, 102, 103, 359, 388, 390, 391
    biased, 385
Rayleigh particle, 326–328
Reciprocal lattice, 475
Recursion relation, 242–244, 246, 250, 253, 273, 282, 288, 290, 292
Recursion relations, 246, 248, 263, 266, 275, 296, 299
Recursion series, 257
Red blood cells, 405, 414
Reduced distribution function, 151, 157, 178
Reflecting boundary condition, 325
Regular matrix, 358, 360

Relativistic ideal gas, 61
Relaxation time, 368
anisotropic, 501
Relaxation time approximation, 499, 504, 510
Relevant scaling field, 247, 266, 395, 410
Remanence, 554
Renormalization flow, 240, 241, 244, 246, 251, 252, 264, 290
Renormalization group, 237–300, 395, 417, 548, 550
ϵ-expansion, 279–292, 297–300
anisotropic n-vector model, 297–300
cumulant approximation, 258–266, 295
finite lattice methods, 267–268
fixed points, 240, 242–248
Migdal–Kadanoff method, 296–297
momentum space, 551
Monte Carlo renormalization, 272–275
one-dimensional Ising model, 238–242, 295
percolation, 538–539, 567
phenomenological, 275–279
position space, 551
real space, 538
scaling and universality, 246–248
Replica symmetry, 375
Replica trick, 544–546, 558–565
replica symmetry breaking, 565
Reservoir, 5
Response function, 1, 14, 15, 39
frequency dependent, 461–490
Return probability, 335
Reversibility
microscopic, 494, 498
Reversible process, 4–11, 25
Rigid band approximation, 525
Rigidity percolation, 540
RKKY interaction, 554
Rotating bucket experiment, 434
Rotational quantum number, 59
Rotational symmetry, 99
Rubber, 390, 541
Rupture, 258
Rushbrooke inequality, 211, 212

S^4 model, 284–290
Sackur–Tetrode equation, 34, 62
Scaling, 246–248, 284
fields, 245, 247, 289
theory of percolation, 534–536
Scaling fields, 132
Scaling laws, 183, 209–211, 214–223, 235
Scaling relation, 369
Schottky defect, 53
Schrödinger equation, 330, 343
Screening, 480–485, 507–509
Second law of thermodynamics, 3–9, 26
Second neighbor interaction, 243, 247
Second order phase transition, 68, 87, 95, 104, 120, 364
Second quantization, 188, 569
Second sound, 438–439
Seebeck effect, 493
Self-adjoint operators, 344
Self-avoiding membranes, 406, 409, 411, 413, 415, 417
Self-avoiding random walk, 391–393, 395, 396, 398, 420

Semiclassical approximation, 499
Semiclassical quantization, 32, 33, 51, 499
Semidilute solutions, 403–405
Semipermeable membrane, 402
Sensitivity to initial conditions, 31, 353
Separation of variables, 330, 332
Series expansion, 199, 200, 202, 206, 207, 209–211, 218, 224, 235, 399, 536–538
  analysis of series, 200, 206, 207, 215
  high temperature, 397
  specific heat, 210
  susceptibility, 202, 225, 234, 290, 399
Series exspansion
  percolation, 538
Shannon, C.E., 50
Shear force, 540
Shear modulus, 415
Sherrington–Kirkpatrick model, 558–565
Simulated annealing, 376–379, 382
Simulations, 349–382
SIR model, 316–321, 345
Site percolation, 530
Slab geometry, 222, 223
Slater determinant, 44, 571
Smectic phases, 117
Solid-solid solutions, 137
Solvent, 402
Solvents
  good and poor, 393, 394, 402, 403, 405
Spanning cluster, 530
Specific heat, 1, 14, 19, 25, 27, 28, 59, 75, 84, 200, 209, 211, 220, 356, 357, 365
  Bethe approximation, 101
  critical exponent, 76, 216
  one-dimensional Ising model, 80
  scaling laws, 221
  series expansion, 210
  tricritical point, 92
  two-dimensional Ising model, 184, 196, 197, 233, 234
Spectrin network, 414
Spin glass, 371
Spin glasses, 554–565
  Sherrington–Kirkpatrick model, 558–565
Spin waves, 229, 232, 235, 477–480, 509
Spinodal, 127, 131
Spontaneous magnetization, 66, 79, 257
Spontaneous process, 4, 8, 9, 11, 18
Spruce budworm model, 129–132, 307, 345
Stability, 1, 15, 17, 18, 25, 28, 290
  local, 27
Stable fixed point, 275, 287
Star graph, 150
State variable, 1–4, 8, 10, 11, 20
Statistical ensembles, 29–62
Steady state, 2, 306, 315, 322, 324, 325
Steady state probability, 358, 359
Steepest descent, 376, 377
Stimulation, 371
Stirling's formula, 34, 400
Stochastic processes, 303–347, 357
Stochastic variable, 304
Stock market crashes, 258
Stokes law, 326
Stored pattern, 372, 373

Strong fields, 499
Structure factor
 dynamic, 462–490, 508
 static, 153, 156, 161, 469, 470, 579–582
Subdiffusion, 333
Sublattices, 113, 270
Substitutional disorder, 516
Sum rules, 470–472
Superconductivity, 100, 168, 226, 228, 442–456, 490
 critical fluctuations, 98
 BCS theory, 445–452
 Cooper problem, 443–444
 Landau–Ginzburg theory, 453–456
Superdiffusion, 333
Superfluid films, 228
Superfluidity, 100, 430–442, 458–459
 elementary excitations, 430–432, 439–442
Superionic conductors, 134
Superleak, 436
Superposition approximation, 158
Surface roughening, 171
Surface tension, 163–165, 168, 169
Survival probability, 309, 312
Susceptibility, 1, 74, 97, 367, 369, 397–399
 anisotropic, 119
 critical exponent, 75, 210, 211, 213, 217, 220, 221, 235
 electric, 465
 frequency dependent, 462–490
 generalized, 462
 magnetic, 19, 20, 465
 one-dimensional Ising model, 80
 scaling properties, 220, 223
 series expansion, 204, 205, 224, 235
Swelling, 391, 392, 403, 404
Symmetry breaking, 104, 290, 291
Synapse, 371
Synchrotron radiation, 169
System size parameter, 313

T-matrix, 527, 528
Temperature
 fluctuations, 355
 scale, 7, 24
 statistical definition, 34, 36
Tethered membranes, 383, 406, 407, 413, 415, 417, 418
Thermal conductivity, 503–507, 510
Thermal expansion coefficient, 14, 27
Thermal wavelength, 35, 45, 46, 57, 144
Thermodynamic
 equilibrium, 5
 potentials, 9–12
 process, 4
 stability, 15
 variables, 2, 13
Thermodynamic limit, 2, 34, 35, 78, 79, 257, 276, 365, 368, 384, 388, 389, 399, 419, 531
Thermodynamics, 1–28
Thermoelectric effects, 493, 503–507 510
Thomas–Fermi approximation, 483, 484, 507
Three-body interactions, 143
Tight binding method, 515
Time reversal symmetry, 494
Tolerance, 36, 52
Tonks gas, 181
Trans configuration, 384

Transfer matrix, 107, 276, 278, 565
  one-dimensional Ising model, 78, 102, 103, 106
  one-dimensional Schrödinger equation, 517–523
  two-dimensional Ising model, 184–186, 191, 192, 194, 199, 218, 233
Transition probability, 360
Transition rate, 305, 467, 499
Transport coefficients
  symmetry, 509
Transport properties, 490–507
Transposition, 377
Traveling salesman problem, 376–379, 382
Triangular lattice, 259, 267, 270, 271
Tricritical point, 90–93, 98, 104–106, 120, 121, 123
Triple point, 20, 24
Tsallis, C., 51
Tunneling, 499, 524
Two-body forces, 144
Two-body operator, 577
Type I and type II superconductors, 455

Umklapp, 475
Universality, 183, 209, 223, 225, 246–248, 279, 291, 536, 541
Update rule, 372
Upper critical dimension, 98, 386, 409
Utility, 130

Vacancy, 53, 54, 59
van der Waals equation, 2, 123–127, 143
van der Waals forces, 393
van der Waals theory, 131, 393
van der Waals theory of interfaces, 163, 165, 172
Vapor pressure, 27
Variational principles, 29
Velocity relaxation time, 327
Velocity-velocity correlation, 354, 357
Velocity-velocity correlation function, 327
Velocity-Verlet algorithm, 352
Vibrational degree of freedom, 59
Virial, 355
  coefficients, 3, 145, 158
  equation of state, 2
  expansion, 144–150, 159, 161, 181, 182
Virial equation of state, 356
Virial expansion, 403
Virtual crystal approximation, 525
Viscosity, 353
Vortex unbinding, 231
Vortices
  in superfluid Helium, 435
  two-dimensional XY model, 229–232

Wannier states, 515, 527
Weak-coupling approximation in superconductivity, 452
Weiss molecular field theory, 64, 66, 115, 117
White dwarf, 60
Wiedemann–Franz law, 506
Wilson, K.G., 241
Work, 3–6, 9–11, 18, 26

X-ray scattering, 414
XY fixed point, 247

XY model, 206, 210, 211, 223, 224, 226, 300

Zeeman energy, 65
Zero temperature Monte Carlo simulation, 372
Zero temperature quench, 377, 378
Zeroth law of thermodynamics, 3, 36